Quasiconformal Mappings and Sobolev Spaces

Mathematics and Its Applications (*Soviet Series*)

Volume 54

Quasiconformal Mappings and Sobolev Spaces

by

V. M. Gol'dshtein and Yu. G. Reshetnyak

Institute of Mathematics,
Siberian Branch of the U.S.S.R. Academy of Sciences,
Novosibirsk, U.S.S.R.

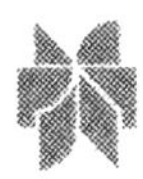

KLUWER ACADEMIC PUBLISHERS
DORDRECHT / BOSTON / LONDON

Library of Congress Cataloging in Publication Data

Gol'dshtein, V. M. (Vladimir Mikhaĭlovich)
[Vvedenie v teoriiu funktsiĭ s obobshchennymi proizvodnymi i kvazikonformnye otobrazheniia. English]
Quasiconformal mappings and Sobolev spaces / by V.M. Gol'dshtein and Yu.G. Reshetnyak.
p. cm. -- (Mathematics and its applications. Soviet series ; 54)
Revised translation of: Vvedenie v teoriiu funktsiĭ s obobshchennymi proizvodnymi i kvazikonformnye otobrazheniia.
Includes bibliographical references.
ISBN 0-7923-0543-4 (alk. paper)
1. Quasiconformal mappings. 2. Functions. I. Reshetniak, IUriĭ Grigor'evich. II. Title. III. Series: Mathematics and its applications (Kluwer Academic Publishers). Soviet series ; 54.
QA360.G6213 1990
515--dc20 89-71684

ISBN 0-7923-0543-4

Published by Kluwer Academic Publishers,
P.O. Box 17, 3300 AA Dordrecht, The Netherlands.

Kluwer Academic Publishers incorporates
the publishing programmes of
D. Reidel, Martinus Nijhoff, Dr W. Junk and MTP Press.

Sold and distributed in the U.S.A. and Canada
by Kluwer Academic Publishers,
101 Philip Drive, Norwell, MA 02061, U.S.A.

In all other countries, sold and distributed
by Kluwer Academic Publishers Group,
P.O. Box 322, 3300 AH Dordrecht, The Netherlands.

Printed on acid-free paper

This is the revised translation of the original work
ВВЕДЕНИЕ В ТЕОРИЮ ФУНКЦИЙ С ОБОБЩЕННЫМИ ПРОИЗВОДНЫМИ И КВАЗИКОНФОРМНЫЕ ОТОБРАЖЕНИЯ
Published by Nauka Publishers, Moscow, © 1983

Translated from the Russian by O. Korneeva
Typeset by Rosenlaui Publishing Services, Inc.

Printed in The Netherlands

SERIES EDITOR'S PREFACE

'Et moi, ..., si j'avait su comment en revenir, je n'y serais point allé.'

Jules Verne

The series is divergent; therefore we may be able to do something with it.

O. Heaviside

One service mathematics has rendered the human race. It has put common sense back where it belongs, on the topmost shelf next to the dusty canister labelled 'discarded non-sense'.

Eric T. Bell

Mathematics is a tool for thought. A highly necessary tool in a world where both feedback and non-linearities abound. Similarly, all kinds of parts of mathematics serve as tools for other parts and for other sciences.

Applying a simple rewriting rule to the quote on the right above one finds such statements as: 'One service topology has rendered mathematical physics ...'; 'One service logic has rendered computer science ...'; 'One service category theory has rendered mathematics ...'. All arguably true. And all statements obtainable this way form part of the raison d'être of this series.

This series, *Mathematics and Its Applications*, started in 1977. Now that over one hundred volumes have appeared it seems opportune to reexamine its scope. At the time I wrote

> "Growing specialization and diversification have brought a host of monographs and textbooks on increasingly specialized topics. However, the 'tree' of knowledge of mathematics and related fields does not grow only by putting forth new branches. It also happens, quite often in fact, that branches which were thought to be completely disparate are suddenly seen to be related. Further, the kind and level of sophistication of mathematics applied in various sciences has changed drastically in recent years: measure theory is used (non-trivially) in regional and theoretical economics; algebraic geometry interacts with physics; the Minkowsky lemma, coding theory and the structure of water meet one another in packing and covering theory; quantum fields, crystal defects and mathematical programming profit from homotopy theory; Lie algebras are relevant to filtering; and prediction and electrical engineering can use Stein spaces. And in addition to this there are such new emerging subdisciplines as 'experimental mathematics', 'CFD', 'completely integrable systems', 'chaos, synergetics and large-scale order', which are almost impossible to fit into the existing classification schemes. They draw upon widely different sections of mathematics."

By and large, all this still applies today. It is still true that at first sight mathematics seems rather fragmented and that to find, see, and exploit the deeper underlying interrelations more effort is needed and so are books that can help mathematicians and scientists do so. Accordingly MIA will continue to try to make such books available.

If anything, the description I gave in 1977 is now an understatement. To the examples of interaction areas one should add string theory where Riemann surfaces, algebraic geometry, modular functions, knots, quantum field theory, Kac-Moody algebras, monstrous moonshine (and more) all come together. And to the examples of things which can be usefully applied let me add the topic 'finite geometry'; a combination of words which sounds like it might not even exist, let alone be applicable. And yet it is being applied: to statistics via designs, to radar/sonar detection arrays (via finite projective planes), and to bus connections of VLSI chips (via difference sets). There seems to be no part of (so-called pure) mathematics that is not in immediate danger of being applied. And, accordingly, the applied mathematician needs to be aware of much more. Besides analysis and numerics, the traditional workhorses, he may need all kinds of combinatorics, algebra, probability, and so on.

In addition, the applied scientist needs to cope increasingly with the nonlinear world and the extra mathematical sophistication that this requires. For that is where the rewards are. Linear models are honest and a bit sad and depressing: proportional efforts and results. It is in the non-linear world that infinitesimal inputs may result in macroscopic outputs (or vice versa). To appreciate what I am hinting at: if electronics were linear we would have no fun with transistors and computers; we would have no TV; in fact you would not be reading these lines.

There is also no safety in ignoring such outlandish things as nonstandard analysis, superspace and anticommuting integration, p-adic and ultrametric space. All three have applications in both electrical engineering and physics. Once, complex numbers were equally outlandish, but they frequently proved the shortest path between 'real' results. Similarly, the first two topics named have already provided a number of 'wormhole' paths. There is no telling where all this is leading - fortunately.

Thus the original scope of the series, which for various (sound) reasons now comprises five subseries: white (Japan), yellow (China), red (USSR), blue (Eastern Europe), and green (everything else), still applies. It has been enlarged a bit to include books treating of the tools from one subdiscipline which are used in others. Thus the series still aims at books dealing with:

- a central concept which plays an important role in several different mathematical and/or scientific specialization areas;
- new applications of the results and ideas from one area of scientific endeavour into another;
- influences which the results, problems and concepts of one field of enquiry have, and have had, on the development of another.

Ever since their introduction by Sobolev in the 1930's, Sobolev function spaces have grown in importance and by now their theory is a substantial part of mathematical analysis with many applications. The first three chapters of this book deal with the spaces of functions with generalized derivatives, with, perhaps, particular emphasis on the theory of nonlinear capacity.

Quasiconformal mappings in their turn also have a well deserved reputation in terms of their applicability to a wide range of problems. This book is about the interrelations between the two; more precisely, it reflects the authors' interests and research in the uses of Sobolev spaces to study various questions of geometry in general and in the theory of quasiconformal mappings such as the description of sets of removable singularities of such mappings.

The shortest path between two truths in the real domain passes through the complex domain.

J. Hadamard

La physique ne nous donne pas seulement l'occasion de résoudre des problèmes ... elle nous fait pressentir la solution.

H. Poincaré

Never lend books, for no one ever returns them; the only books I have in my library are books that other folk have lent me.

Anatole France

The function of an expert is not to be more right than other people, but to be wrong for more sophisticated reasons.

David Butler

Bussum, January 1990

Michiel Hazewinkel

Table of Contents

PREFACE

The concept of a generalized derivative of a function in several variables was introduced by S. L. Sobolev as early as the 1930s. Let U be a domain in an n-dimensional Euclidean space $\mathbf{R}^n$ and let $f(x)$ and $g(x)$ be functions defined and summable in the sense of Lebesgue on the set U. The function g is said to be a generalized derivative $\frac{\partial^{k_1+k_2+\cdots+k_n} f}{\partial x_1^{k_1} \partial x_2^{k_2} \cdots \partial x_n^{k_n}}$ of the function f if for every finitely differentiable function $\varphi(x)$ vanishing outside some compact subset of U, the equality

$$\int_U f(x) \frac{\partial^{k_1+k_2+\cdots+k_n} \varphi}{\partial x_1^{k_1} \partial x_2^{k_2} \cdots \partial x_n^{k_n}}(x) dx = (-1)^{k_1+\cdots+k_n} \int_U f(x) \varphi(x) dx$$

is valid.

The function f is said to belong to a class $W_p^l(U)$ if it has in U all generalized derivatives of order l, with these derivatives being integrable in U of degree p. The classes $W_p^l(U)$ were introduced by Sobolev, who established some of their basic properties. Among them, the central place is occupied by imbedding theorems that connect the classes $W_p^l(U)$ with other known classes of functions. At present, the theory of spaces of the functions of W_p^l is an extensive part of mathematical analysis and has numerous applications in various problem of mathematical analysis and in the theory of partial differential equations.

This book reflects the authors' interests and presents their original investigations. At the same time, the presentation is arranged so that the book (at least the first three chapters) may be used to understand the space theory of functions with generalized derivatives. The reader should be familiar with the traditional university course of mathematical analysis and the elements of functional analysis (including the theory of L_p spaces).

The authors' interest in Sobolev spaces is due to some investigations in geometry "in the large" and in the theory of quasiconformal mappings. Some problems concerning the relation between the theory of spatial quasiconformal mappings and the theory of W_p^l classes are also considered in this book.

The book consists of six chapters. Chapter 1 presents terminology, notations, and auxiliary results used throughout the book. In particular, theorems of differentiation of measures in an n-dimensional space and some preliminary information about generalized functions (distributions, in the sense of L. Schwartz) are given.

The second chapter deals with the foundations of the theory of classes of functions with generalized derivatives. The main peculiarities of our presentation are as follows. The spaces W_p^l are studied for a wider class of domains than in other investigations concerning the subject. In the theory of W_p^l spaces, some special integral representations of functions are used (such representations were first constructed by Sobolev). This chapter discusses in detail the methods of constructing such integral representations. In particular, integral representations of a function are given by means of some special differential operators. Among other results in this chapter, let us point out a general theorem concerning differentiability almost everywhere of functions with generalized derivatives. Here, differentiability is understood in a stronger sense than in the classical investigations in the theory of functions, and therefore, the theorem proved in this book may be used to obtain numerous special results.

The third chapter concerns the theory of nonlinear capacity. This notion allows us to give an exact description, in terms of Lebesgue integration theory, of the convergence character in the "pointwise" sense. This convergence arises as the corollary of norm convergence of functions in the W_p^l spaces. In this case, nonlinear capacity plays a role similar to that of measure in the Lebesgue integral. Capacity here is essentially a finer characteristic than measure. In a rougher variant, it allows us to characterize the "discontinuity" set of functions of the W_p^l class in terms of the Hausdorff measures. For the W_p^l spaces, analogies of the classical theorems of Egorov and Luzin have been obtained in which nonlinear capacity plays the role of measure. Presentation of the abstract variant is given. The obtained results remain valid for a number of functional spaces, such as for the Besov–Nickolsky classes. On the other hand, the methods of capacity are used below to study relations between differential and geometric characteristics of quasiconformal mappings. At the end of the chapter, capacity estimates for a number of concrete sets are given. These estimates are adjusted to study the Sobolev spaces and the Besov–Nickolsky spaces.

The fourth chapter studies the problem of density of extremal functions (for capacity of pairs of connected compact sets) in the spaces of functions with first generalized derivatives. It is proved possible to approximate functions from W_p^1 by linear combinations of extremal functions for capacity whose gradients have nonintersecting supports. This approximation is further used to obtain the description of sets of removable singularities of quasiconformal and quasiisometric mappings.

The fifth chapter investigates the problem of change of the independent variable for functions of the W_p^1 classes. We are interested in the following problem: What minimal requirements should be imposed on a mapping so that it transfers the functions of the W_p^1 class into the functions of the same class? At the beginning of the chapter, general information is given about the multiplicity function, degree of mapping, and change of variables in a multiple integral. The final part of the chapter presents a full description of the change of independent variables preserving the class W_p^1. For $p = n$, they are quasiconformal homeomorphisms; for $p \neq n$, quasiisometric homeomorphisms.

The sixth chapter deals with the problem of extension to the entire space $\mathbf{R}^n$ (the class being preserved) of functions from the space $W_p^1(G)$ assigned in a domain $G \subset \mathbf{R}^n$. The presentation is mainly given in the abstract variant which allows us, in addition to the W_p^1 spaces, to consider the Besov–Nickolsky spaces as well. Principal attention is paid to the necessary conditions of extension that, for planar finitely-connected domains, coincide with the sufficient ones (for the first generalized derivatives).

The authors are grateful to A. Yu. Kopeleva, T. G. Latfullin, A. S. Romanov, and V. K. Sitnikov, whose assistance promoted the publication of this book.

NOTATION

$\mathbf{R}$	the set of real numbers
$\mathbf{N}$	the set of natural numbers
$\mathbf{R}^n$	the n-dimensional vector space
$\mathbf{C}$	the set of complex numbers
$\mathbf{C}^n$	a complex vector space of dimension n
$\langle x, y \rangle$	the scalar product of the vectors x and y
$\lvert x \rvert$	the length of x
$\rho(x, A)$	the distance from the point x to the set A
$\rho(A, B)$	the distance between the sets A and B
$\bar{A}$	the closure of A
A°	the interior of A
∂A	the boundary of A
$\overline{B}(a, r)$	a closed ball with the centre a and radius r
$B(a, r)$	an open ball with the centre a and radius r
$A_1 \times A_2 \times \cdots \times A_n$	the totality of points $(x_1, x_2, \ldots, x_n)$, where $x_1 \subset A_1, x_2 \subset A_2, x_n \subset A_n$
$S(a, r)$	a sphere with the centre a and radius r
$Q(a, r)$	an open cube with the centre a and the edge length r
$\overline{Q}(a, r)$	a closed cube with the centre a and the edge length r
e_i	the vector of $\mathbf{R}^n$
$\lvert A \rvert$	the Lebesgue exterior measure of A
$m(A), \mu(A)$	the Lebesgue measure of A
$L_p(A)$	the totality of all measurable real functions $f(x)$ defined almost everywhere on A and integrable in the power p (Lebesgue space)
$\lVert f \rVert_{L_p(A)}$	the norm of the function on L_p
$L_{p,\mathrm{loc}}(U)$	the totality of all functions which are locally integrable in U in the power p

W_p^l	Sobolev spaces
$B_{p,p}^l$	Besov spaces
$S'(f)$	a set of all $x \subset X$ for which $f(x) \neq 0$
$S(f)$	the closure of the set $S'(f)$, the support of the function f
$C^K(U)$	the totality of all functions $f : U - \mathbf{R}^n$, with derivatives of the order $\leqslant k$ and the standard norm
$C_0^k(U)$	the totality of all test functions in U of the class $C^k(U)$
$C^{k,\alpha}(U)$	the totality of all functions $f \subset C^k(U)$ whose derivatives of order k locally satisfy in U the Hölder condition with the exponent α
$C_0^{k,\alpha}(U)$	the totality of all functions $f \subset C_0^k(U)$ whose derivatives of order k locally satisfy in U the Hölder condition with the exponent α
$C^\infty(U)$	the totality of all functions in U with derivatives of all orders and standard topology
$C_0^\infty(U)$	the totality of all test functions on (U) of the class $C^\infty(U)$
$C(U)$	the totality of all continuous functions on U
$C_0(U)$	the totality of all continuous test functions on U
$C^{0,\alpha}(U)$	the totality of all continuous functions on U, each of them locally satisfying in U the Hölder conditions with the exponent α
$C_0^{o,\alpha}(U)$	the totality of all functions from $C_0(U)$, each of them locally satisfying in U the Hölder condition with the exponent α
U_h	the totality of all $x \subset U$ for which $\rho(x, \partial U) > h$
$K_h * f$	the average function f_h with kernel K for the function f
A^+	a μ-positive set
A^-	a μ-negative set
$\frac{d\lambda}{d\mu}$	the function ϕ (Theorem 3.2), the derivative of the measure λ with respect to the measure μ
$D_R\mu(a)$	the R-density of the measure μ at the point a
$D_L\mu(a)$	the L-density of the measure μ at the point a
$\langle f, \phi \rangle$	the magnitude $f(\phi)$ if f is a generalized function in U for $\phi \subset C_0(U)$
J	a class of domains in $\mathbf{R}^n$

$J(r,R)$	the class of domains with the inner radius r and the outer radius R
$\Omega(x,\xi)$	the solution $Y(x)$ of the matrix equation (2.46)
$Q_1 v(x)$	the deformation tensor corresponding to the vector field $v(x)$
$Q_1 v(x)$	the tensor of conformal deformation corresponding to the vector field $v(x)$
$M_t(K)$	the least of the numbers M
$W_p^L(U)$	the totality of all functions with generalized derivatives of order $\leqslant 1$ that are integrable in the power p
$\tilde{C}_P^l(F_0, F_1, \mathbf{R}^n)$	weak variational $(1,p)$-capacity
$\overline{C}_p^1(F_0, F_1, \mathbf{R}^n)$	strong variational $(1,p)$-capacity
$\mathrm{Cap}_{T,p}(E,V)$	variational capacity induced by a linear positive operator T
$\mathrm{Cap}_{T,p}(E,V)$	capacity induced by a linear positive operator T
$\mathrm{Cap}_{l,p}(E)$	$(1,p)$ capacity
$C_p^l(F_0, F_1, \mathbf{R}^n)$	variational $(1,p)$-capacity
$\mathrm{Cap}_{T,p}(E,V)$	dual capacity
I_1	the Bessel potential
$M_{T,p}$	the set of admissible functions for the (T,p)-capacity
$M_F(F_0, F_1, G)$	the set of admissible functions for the variational capacity
$\mathrm{app}(df_x)$	the approximate differential
$N_f(y,E)$	the multiplicity function
$\mu(y,\phi,U)$	the degree of mapping
$\chi(y, f(\partial G))$	the linking index

CHAPTER 1

PRELIMINARY INFORMATION ABOUT INTEGRATION THEORY

§1 Notation and Terminology

1.1. Sets in $\mathbf{R}^n$

We use the following standard notations for some important sets: **R** is the set of all real numbers, **N** is the set of natural numbers, **C** is the set of all complex numbers, $\emptyset$ is the empty set. The symbol $\mathbf{R}^n$ denotes the n-dimensional vector space of points $x = (x_1, x_2, \ldots, x_n)$, where $x_i \in \mathbf{R}$, $i = 1, 2, \ldots, n$. For arbitrary vectors $x = (x_1, x_2, \ldots, x_n)$, $y = (y_1, y_2, \ldots, y_n)$, we set

$$\langle x, y\rangle = x_1y_1 + x_2y_2 + \ldots + x_ny_n.$$

The magnitude $\langle x, y\rangle$ is called the scalar product of the vectors x and y. For $x \in \mathbf{R}$, $|x|$ denotes the length of x, i. e., $|x| = (\langle x, y\rangle)^{1/2}$.

Let X be an arbitrary set, $A \subset X$. The symbol χ_A denotes the indicator of the set A in X, i.e., χ_A is a function in X such that $\chi_A(x) = 1$ for $x \in A$, and $\chi_A(x) = 0$ for $x \notin A$.

If P is a property or a condition, A is a set, then the expression $\{x \in A | P(x)\}$ means the totality of all elements of the set A having the property $P(x)$ (correspondingly, satisfying the condition P(x)).

A sequence whose common term is x_m, and the index m runs the set **N** is denoted by (x_m), $m = 1, 2, \ldots$, or by (x_m), $m \in \mathbf{N}$.

Let X be a metric space, and let ρ be its metric. For an arbitrary set $A \subset X$ and $x \in \chi$, $\rho(x, A)$ is the distance from the point x to the set A, that is,

$$\rho(x, A) = \inf_{y \in A} \rho(x, y).$$

For arbitrary subsets A and B of X, we set

$$\rho(A, B) = \inf_{\substack{x \in A, \\ y \in B}} \rho(x, y).$$

The number $\rho(A, B)$ is said to be the distance between the sets A and B. For $A \subset X$, let $d(A)$ denote the diameter of A, that is,

$$d(A) = \sup_{\substack{x \in A, \\ y \in A}} \rho(x, y).$$

In the special case $X = \mathbf{R}^n$, we have $\rho(x, A) = \inf_{y \in A} |x - y|$, $\rho(A, B) = \inf_{x \in A, y \in B} |x - y|$, and, finally,

$$d(A) = \sup_{\substack{x \in A, \\ y \in A}} |x - y|.$$

For the set A in the metric space X, the totality of all points $x \in X$ such that $\rho(x, A) < h$, where $h > 0$, is called an h-neighbourhood of the set A and is denoted by $U_h(A)$. The set of all points $x \in X$ for which $\rho(x, A) \leqslant h$ is said to be a closed h-neighbourhood of A and is denoted below as $\bar{U}_h(A)$. Obviously, $U_h(A)$ is an open set, $\bar{U}_h(A)$ is a closed one, $A \subset U_h(A) \subset \bar{U}_h(A)$.

Let X be a topological space. For an arbitrary set $A \subset X$, the symbol $\bar{A}$ denotes the closure of A, A° is the interior of A, ∂A is the boundary of A, $\partial A = \bar{A}/A^\circ$. A connected open set $U \subset X$ is called a domain. The set $A \subset X$ is called a closed domain in X if its interior A° is a domain, and the closure of A° coincides with A. If the closed domain A is a compact set, then A is said to be a compact domain in X. Let U be an open set in X. Then A is said to lie strictly inside U if $\bar{A}$ is compact and $U \supset \bar{A}$.

For $a \in \mathbf{R}^n$, $r > 0$, we put

$$B(a, r) = \{x \in R^n \mid |x - a| < r\},$$
$$\bar{B}(a, r) = \{x \in R^n \mid |x - a| \leqslant r\},$$
$$S(a, r) = \{x \in R^n \mid |x - a| = r\}.$$

The sets $B(a,r)$, $\bar{B}(a,r)$, and $S(a,r)$ are called, respectively, an open ball, a closed ball, and a sphere with the centre a and radius r.

Let the sets $A_1, A_2, \ldots, A_n \subset \mathbf{R}$ be given. The totality of all points $x = (x_1, x_2, \ldots, x_n) \in \mathbf{R}^n$ for which $x_1 \in A_1$, $x_2 \in A_2, \ldots,$ $x_n \in A_n$ is denoted by the symbol $A_1 \times A_2 \times \cdots \times A_n$. In the case where each of the sets $A_j, j = 1, 2, \ldots, n$ is a segment, the product $A_1 \times A_2 \times \cdots \times A_n$ is called an n-dimensional rectangle or simply a rectangle in $\mathbf{R}^n$. Here, if the lengths of the segments A_j are the same, the rectangle $A_1 \times A_2 \times \cdots \times A_n$ is said to be a cube.

The symbols $Q(a,r)$ and $\bar{Q}(a,r)$, respectively, denote an open and a closed cube with the centre a and with the edge length r, i.e.,

$$Q(a, r) = (a_1 - r/2, a_1 + r/2) \times \times (a_2 - r/2, a_2 + r/2) \times \ldots \times (a_n - r/2, a_n + r/2),$$
$$\bar{Q}(a, r) = [a_1 - r/2, a_1 + r/2] \times \times [a_2 - r/2, a_2 + r/2] \times \ldots \times [a_n - r/2, a_n + r/2].$$

Let S be an arbitrary set. For $i, j \in S$ we put $\delta_j^i = 1$ if $i = j$, $\delta_j^i = 0$ if $i \neq j$. The function $\delta : S \times S \to \mathbf{R}$ thus defined is called the Kronecker symbol on S. In general, the introduced function δ is applied in the case where S is a segment $\{1, 2, \ldots, n\}$ of the set $\mathbf{N}$.

The symbol e_i, $i = 1, 2, \ldots, n$, denotes the vector $(\delta_i^1, \delta_i^2, \ldots, \delta_i^n)$ of $\mathbf{R}^n$. The vectors $e_1, e_2, \ldots, e_n$ form the basis of $\mathbf{R}^n$, which is said to be canonical.

In some problems below, it will be necessary to consider complex-valued functions. Let us introduce the necessary notations. By $\mathbf{C}^n$ we denote a complex vector space of points $z = (z_1, z_2, \ldots, z_n)$ where, $z_1, z_2, \ldots, z_n$ are complex numbers. For $z = (z_1, z_2, \ldots, z_n)$, we have

$$j(z) = (\operatorname{Re} z_1, \operatorname{Re} z_2, \ldots, \operatorname{Re} z_n, \operatorname{Im} z_1, \operatorname{Im} z_2, \ldots, \operatorname{Im} z_n).$$

The mapping $z \to j(z)$ establishes the one-to-one correspondence between $\mathbf{C}^n$ and $\mathbf{R}^{2n}$. The correspondence is called canonical. It is sometimes convenient to consider the space $\mathbf{C}^n$ as the space $\mathbf{R}^{2n}$ if we identify the point $z \in \mathbf{C}^n$ with the point $j(z) \in \mathbf{R}^{2n}$. For $z \in \mathbf{C}^n$, we put

$$|z| = \left(\sum_{i=1}^{n} |z_i|^2 \right)^{1/2} = |j(z)|.$$

1.2. Classes of Functions in $\mathbf{R}^n$

The presentation below assumes that the reader knows the main facts of measure theory and of the theory of the Lebesgue integral in $\mathbf{R}^n$. Notations used here coincide with the traditional ones.

For an arbitrary set $A \in \mathbf{R}^n$, the symbol $|A|$ denotes its Lebesgue exterior measure.

Let $p \geqslant 1$, $p \in \mathbf{R}$, A be a measurable set in $\mathbf{R}^n$. The symbol $L_p(A)$ denotes the totality of all measurable real functions $f(x)$ defined almost everywhere on A, and such that

$$\int_A |f(x)|^p \, dx < \infty.$$

For $f \in L_p(A)$ we set:

$$\|f\|_{L_p(A)} = \left(\int_A |f(x)|^p dx\right)^{1/p}.$$

Let U be an arbitrary open set in $\mathbf{R}^n$, and $f : U \to \mathbf{R}$ is a measurable function defined in U almost everywhere. The function f is said to be locally integrable in U in the power $p \geqslant 1$ if f is integrable in the power p on every compact set $A \subset U$. This is obviously equivalent to the following. For every point $x \in U$, there exists $\delta > 0$ such that the function f is integrable in the power p on the ball $B(x, \delta)$. The totality of all functions which are locally integrable in U in the power p is denoted by $L_{p,\mathrm{loc}}(U)$ or by $L_{p,\mathrm{loc}}$ if no misunderstanding is possible.

The classical inequalities of Hölder and Minkowski for integrals are supposed to be known.

Let X be a topological space, let Y be a vector space. Suppose that a function $f : X \to Y$ is given. Let $S'(f)$ be a set of all $x \in X$ for which $f(x) \neq 0$. The closure of the set $S'(f)$ is denoted by $S(f)$ and is called the support of the function f. The function $f : X \to Y$ is said to be a test function if its support is compact.

Let a set $A \subset \mathbf{R}^n$ and a function $f : A \to \mathbf{R}^k, k \geqslant 1$ be given. The function f is said to be a test function in A if f is a test function as a mapping of the topological space A which possesses topology induced from $\mathbf{R}^n$ in the vector space $\mathbf{R}^k$.

Suppose that a function $f : A \to \mathbf{R}^k$ is given, where $A \subset \mathbf{R}^n$, $A \neq \emptyset$, $k \geqslant 1$. The function $w : [0, \infty) \to \mathbf{R}$ is called the continuity modulus of the function f if w is a nondecreasing function, $w(t) \to 0$ by $t \to 0$, and for any $x_1, x_2 \in A$ the inequality

$$|f(x_1) - f(x_2)| \leqslant w(|x_1 - x_2|).$$

is valid. If the function $f : A \to \mathbf{R}^k$ has the continuity modulus $w(t) = Ct^\alpha$, where $C = \text{const}$, $0 < \alpha \leqslant 1$, then f is said to satisfy the Hölder condition with the exponent α, and the constant C (in the case $\alpha = 1$, it satisfied the Lipschitz condition with the constant C).

If every point $x \in A$ has a neighbourhood U in $\mathbf{R}^n$ such that the restriction of f on the set $U \cap A$ satisfies the Hölder condition with the exponent α, then f is said to locally satisfy on A the Hölder condition with the exponent α (to satisfy the Lipschitz condition in the case $\alpha = 1$).

Let $A \subset \mathbf{R}^n$. Suppose that $f : A \to \mathbf{R}^k$ is bounded and satisfies the Hölder condition with the exponent α, $0 < \alpha \leqslant 1$. Then we set

$$\|f\|_{C^\alpha} = \|f\|_{0,\alpha} = \max\left\{\sup_{x \in A} |f(x)|, \sup_{x_1, x_2 \in A} \frac{|f(x_1) - f(x_2)|}{|x_1 - x_2|^\alpha}\right\}.$$

If the set $A \subset \mathbf{R}^n$ is compact and f locally satisfies on A the Hölder condition with the exponent α, then f satisfies on the set A the Hölder condition on the whole with the exponent α. Indeed, let

$$C = \sup_{x\in A, y\in A} \frac{|f(x) - f(y)|}{|x - y|^\alpha}.$$

It is necessary to prove that C is finite. Let (x_m, y_m), $m = 1, 2, \dots$ be a sequence of pairs of points of the set A such that for $m \to \infty |f(x_m) - f(y_m)| / |x_m - y_m|^\alpha \to C$. Due to compactness of A one may, without loss of generality, assume that the sequences (x_m) $m \in \mathbf{N}$, and (y_m) $m \in \mathbf{N}$ are converging ones, $x_m \to x_0 \in A$, $y_m \to y_0 \in A$ for $m \to \infty$. If $x_0 \neq y_0$, then we obtain that

$$C = \lim_{m\to\infty} (|f(x_m) - f(y_m)| / |x_m - y_m|^\alpha) = |f(x_0) - f(y_0)| / |x_0 - y_0|^\alpha.$$

Let $x_0 = y_0$. Let us find a neighbourhood U of the point x_0 such that for any $x', x'' \in U \cap A$, $|f(x') - f(x'')| \leqslant C'|x' - x''|$ where $C' < \infty$. For sufficiently large m, the points x_m and y_m lie in the neighbourhood U. Hence it follows that $C \leqslant C'$ so that in this case $C \leqslant \infty$ as well. Q. E. D.

Let $U \subset \mathbf{R}^n$ be an open set and let $k \geqslant 0$ be an integer. The symbol $C^k(U)$ denotes the totality of all functions $f : U \to \mathbf{R}^m$, $m \geqslant 1$; each of these functions has in U all the partial derivatives of order k, with these derivatives in U being continuous and bounded. The totality of all test functions in U of the class $C^k(U)$ is denoted by the symbol $C_0^k(U)$. The set of all functions $f \in C^k(U)(f \in C_0^k(U))$, whose derivatives of order k all locally satisfy in U the Hölder condition with the exponent α, is denoted by the symbol $C^{k,\alpha}(U)$ (respectively, by $C_0^{k,\alpha}(U)$).

We shall say that $f : U \to \mathbf{R}^m, m \geqslant 1$, belongs to the class $C^\infty(U)$ if the function f everywhere in U has the continuous kth derivatives for each integer $k \geqslant 1$. The function f is said to belong to the class $C_0^\infty(U)$ if f is a test function and belongs to the class $C^\infty(U)$. If $f \in C_0^\infty(U)$, then $f \in C^k(U)$ for each $k \geqslant 1$.

The totality of all continuous (continuous test) functions $f : U \to \mathbf{R}^m$ is denoted by the symbol $C(U)$ (respectively, by $C_0(U)$). By $C^{0,\alpha}(U)(C_0^{0,\infty}(U))$, we denote the totality of all functions from $C(U)$ (respectively, from $C_0(U)$), each of them locally satisfying in U the Hölder condition with the exponent α.

The following proposition is true.

Lemma 1.1. *A set of real functions $C_0(\mathbf{R}^n)$ is everywhere dense in $L_p(\mathbf{R}^n)$ for every $p \geqslant 1$, i.e., for every function $f \in L_p(\mathbf{R}^n)$ for any $\varepsilon > 0$ one can show the function $\varphi \in C_0(\mathbf{R}^n)$ such that*

$$\|f - \varphi\|_{L_p(\mathbf{R}^n)} \leqslant \varepsilon.$$

Then a vector in $\mathbf{R}^n$ whose coordinates are nonnegative integers is called an n-dimensional multiindex. Strictly speaking, the term "multiindex" for such vectors is only applied in a certain context—below we shall explain what kind of context it is. For the multiindex $\alpha = (\alpha_1, \alpha_2, \dots, \alpha_n)$, we set

$$|\alpha| = \alpha_1 + \alpha_2 + \ldots + \alpha_n, \qquad \alpha! = \alpha_1!\alpha_2! \ldots \alpha_n!.$$

If $x = (x_1, x_2, \ldots, x_n)$ is an arbitrary vector in $\mathbf{R}^n$, then we let $x^\alpha = x_1^\alpha x_2^\alpha \ldots x_n^\alpha$. (In this notation, the expression 0°, if any, is supposed to be equal to 1.) Finally, the symbol $D^\alpha f$ denotes the partial derivative $\frac{\partial^\alpha f}{\partial x_1^{\alpha_1} \ldots \partial x_n^{\alpha_n}}$.

Let the multiindices $\alpha = (\alpha_1, \alpha_2, \ldots, \alpha_n)$, $\beta = (\beta_1, \beta_2, \ldots, \beta_n)$ be given. We write $\alpha \leqslant \beta$, if $\alpha_i \leqslant \beta_i$ for every $i = 1, 2, \ldots, n$.

Let $U \subset \mathbf{R}^n$ be an open set, and $\varphi \in C_0^k(U)$. Let us extend the function φ to all $\mathbf{R}^n$ setting $\varphi(x) = 0$ for $x \notin U$. It is easy to show that the constructed extension is the function of the class $C_0^k(\mathbf{R}^n)$. Therefore, in the following we always suppose the functions of the class $C_0^k(U)$ to be defined in all $\mathbf{R}^n$ and to be equal to zero outside of U.

On the set of functions $C^k(U)$, we define a norm where for $f : U \to \mathbf{R}^m$, f belongs to the class $C^k(U)$,

$$\|f\|_{C^k(U)} \equiv \|f\|_k = \sup_{x \in U, |\alpha| \leqslant k} |D^\alpha f(x)|.$$

The lowest upper bound here is taken by the set of all $x \in U$ and all α such that $|\alpha| \leqslant k$. Let $f \in C^{k,\lambda}(U)$. Then we set

$$\|f\|_{C^{k,\lambda}(U)} = \|f\|_{k,\lambda} = \max\left\{ \|f\|_k, \max_{|\alpha|=k} \|D^\alpha f\|_{0,\lambda} \right\}.$$

Let A be an arbitrary set in $\mathbf{R}^n$. Then we say that the function $f : A \to \mathbf{R}^m$ belongs to the class $C^k(C^{k,\alpha}$ if $0 < \alpha \leqslant 1)$, $k > 0$ is an integer, if there exist an open set $U \supset A$ and a function $f^* : U \to \mathbf{R}^m$ belonging to the class $C^k(U)$ (respectively, to the class $C^{k,\alpha}(U)$) such that $f^*(x) = f(x)$ at every point x of the set A.

For an arbitrary function $f : A \to \mathbf{R}^m$ where $A \subset \mathbf{R}^n$, we put

$$\|f\|_{L_\infty(A)} = \operatorname{ess\,sup}_{x \in A} |f(x)|.$$

If the function f is continuous, then instead of $\|f\|_{L_\infty(A)}$, we also write $\|f\|_{C(A)}$.

Let $U \in \mathbf{R}^n$ be an open set in $\mathbf{R}^n$. The mapping $f : U \to \mathbf{R}^m$ is said to be differentiable at the point $x_0 \in U$ if there exists a linear mapping $L : \mathbf{R}^n \to \mathbf{R}^m$ such that $f(x) = f(x_0) + L(x - x_0) + \alpha(x)|x - x_0|$ where $\alpha(x) \to 0$ for $x \to x_0$. The mapping L in this case is called the differential of the mapping f at the point x and is denoted by the symbol $f'(x_0)$. If f is differentiable at the point x_0, then the partial derivatives $\frac{\partial f}{\partial x_i}(x_0)$ are defined. For every vector $h = (h_1, h_2, \ldots, h_n) \in \mathbf{R}^n$, we have:

$$f^1(x_0)(h) = \sum_{i=1}^n \frac{\partial f}{\partial x_i}(x_0) h_i.$$

Every mapping $f : U \to \mathbf{R}^m$ of the class $C^1(U)$ (U is an open set in $\mathbf{R}^n$) is locally Lipschitz. Indeed, let us arbitrarily take $x_0 \in U$. Let us find $\delta > 0$ such that the closed cube $\bar{Q}(x_0, \delta) \subset U$. For arbitrary $x, y \in \bar{Q}(x_0, \delta)$, we have

$$|f(y) - f(x)| = |\int_0^1 \frac{d}{dt} f[x + t(y - x)]dt|$$
$$= |\sum_{i=1}^n \frac{\partial f}{\partial x_i}[x + t(y - x)(y_i - x_i)\, dt|. \qquad (*)$$

Since the functions $\frac{\partial f}{\partial x_i}$ are continuous, they are bounded on the cube $\bar{Q}(x_0, \delta)$. Let $|\frac{\partial f}{\partial x_i}(x)| \leqslant M < \infty$, $i = 1, 2, \dots, n$ for all $x \in \bar{Q}(x_0, \delta)$. If $x \in \bar{Q}(x_0, \delta), y \in \bar{Q}(x_0, \delta)$, then for every $t \in [0, 1]$ the point $x + t(y - x) \in \bar{Q}(x_0, \delta)$, and from the equality (*), we obviously obtain:

$$|f(y) - f(x)| \leqslant \sum_{i=1}^n \int_0^1 M |y_i - x_i|\, dt = M \sum_{i=1}^n |y_i - x_i| \leqslant M\sqrt{n}|y - x|,$$

and we have that f in the cube $Q(x_0, \delta)$ satisfies the Lipschitz condition with the constant $C = M\sqrt{n}$. Q. E. D.

§2 Some Auxiliary Information about Sets and Functions in $\mathbf{R}^n$

2.1. Averaging of Functions

Let us now describe the procedure of smoothing the functions introduced by Sobolev. We call it the averaging operation. This operation allows us to associate to every function f defined and locally integrable in an open set $U \subset \mathbf{R}^n$ some functions of the class C^∞ defined for the case $U = \mathbf{R}^n$ on the entire space $\mathbf{R}^n$, for $U \neq \mathbf{R}^n$—on some subset of U.

Let U be an open set in $\mathbf{R}^n$, and $h > 0$, $h \in \mathbf{R}$. Denote by $\hat{U}_h$ the totality of all $x \in U$ for which $\rho(x, \partial U) > h$. The set $\hat{U}_h$ is open. For $h_1 < h_2$, $\hat{U}_{h_1} \supset \bar{U}_{h_2}$, and the union of all sets $\hat{U}_h$ coincides with U. For every compact set $A \subset U$, one can show such a number $\delta > 0$ that for $h < \delta$, $A \subset \hat{U}_h$. Generally speaking, for some values of h, the set $\hat{U}_h$ may turn out to be empty. The lowest upper bound of the values of h for which $\hat{U}_h$ is nonempty is called the inner radius of U and is denoted by $\rho(U)$.

To define the averaging operation, let us assign a function $K : \mathbf{R}^n \to \mathbf{R}$ such that the following conditions are satisfied. The function K belongs to the class $C_0^\infty(\mathbf{R}^n)$, its support is contained in the ball $B(0, 1)$ and

$$\int_{R^n} K(x)\, dx = 1. \qquad (2.1)$$

The function $K(x)$ is called an averaging kernel.

Let us give an example. Let the function $\varphi : \mathbf{R} \to \mathbf{R}$ be defined as follows:

$$\varphi(t) = \begin{cases} e^{1/t} & \text{for } t < 0, \\ 0 & \text{for } t \geqslant 0. \end{cases}$$

Then $\varphi \in C^\infty(\mathbf{R})$, and the function

$$K(x) = \gamma\varphi(|x|^2 - 1/4), \tag{2.2}$$

where γ is chosen so that the equality (2.1) should be valid, is an averaging kernel.

Let U be an open set in $\mathbf{R}^n$, and let $f : U \to \mathbf{R}^m$ be a function of the class $L_{1,\mathrm{loc}}(U)$. Let us put for $x \in \hat{U}_h, h < \rho(U)$

$$(K_h * f)(x) = \int_{|t|<1} f(x + ht) K(t)\, dt \tag{2.3}$$

$$= \frac{1}{h^n} \int_{|z-x|\leqslant h} f(z) K\left(\frac{z-x}{h}\right) dz = \frac{1}{h^n} \int_U f(z) K\left(\frac{z-x}{h}\right) dz.$$

The integrals here make sense due to the local integrability of f. In the case where no misunderstanding is possible, we just write f_h instead of $K_h * f$. The function f_h is called the average function with kernel K for the function f.

If $f \in L_{1,\mathrm{loc}}(\mathbf{R}^n)$, then the function $K_h * f$ is defined in $\mathbf{R}^n$ everywhere.

The function $K_h * f$ belongs to the class C in the domain $\hat{U}_h$ which directly follows from the second expression for $K_h * f$ and from the known theorems for functions represented by integrals depending on a parameter. Here, for every differentiation operator D^α, we have

$$D^\alpha (K_h * f)(x) = \frac{1}{h^n} \int_U f(z) D_x^\alpha K\left(\frac{z-x}{h}\right) dz. \tag{2.4}$$

The index x in the expression D_x^α means that the differentiation is performed by the variable x.

If the function f is a test function in U and $h < \rho(\partial U, S(f))$, then the function f_h is a test function in $\hat{U}_h$, and $S(f_h) \subset U_h[S(f)]$.

Theorem 2.1. *Let U be an open set in $\mathbf{R}^n$, and let $f : U \to \mathbf{R}$ be a continuous function. Then the function $f_h = K_h * f$ for $h \to 0$ converges to f uniformly on every compact set $A \in U$. If f is a test function in U, then $f_h \to f$ uniformly in U for $h \to 0$.*

Proof. Let $A \subset U$ be a compact set, $\delta = \rho(A, \partial U) > 0$. We assume that $0 < h < \delta$. Let us denote by V the closed h-neighbourhood of the set A. The set V is compact, therefore f is uniformly continuous on V. Let $w(t)$ be the continuity modulus of f on V, so that $|f(x_1) - f(x_2)| \leqslant w(|x_1 - x_2|)$ for any $x_1, x_2 \in V$.

Let $x \in A$. Then, if $h < \delta$ and $|t| < 1$, the point $x + ht \in V$, hence it follows that for such x,

$$|(K_h * f)(x) - f(x)| \leqslant \int_{|t|<1} |f(x+ht) - f(x)| K(t)dt \leqslant Mw(h),$$

where

$$M = \int_{|t|\leqslant 1} |K(t)|\, dt.$$

Thus,

$$|(K_h * f)(x) - f(x)| \leqslant Mw(h)$$

for all $x \in A$ for $0 < h < \delta$. Since $w(h) \to 0$ for $h \to 0$, this proves that $K_h * f \to f$ uniformly on A.

If f is a test function, let 2η be the distance from the set $S(f)$ to the boundary of U. For A let us take the closed η-neighbourhood of the set $S(f)$. According to what was proved above, $K_h * f \to f$ uniformly on A. For $0 < h < \eta$, $(K_h * f)(x) = 0$ outside of A, whence it is clear that $K_h * f \to f$ uniformly on the entire $\mathbf{R}^n$.

This completes the proof of the theorem.

Theorem 2.2. *Let U be an open set in $\mathbf{R}^n$, and let f be a function of the class $L_{p,\mathrm{loc}}(U)$; $A \subset U$ is compact, $\delta = \rho(A, \partial U)$, $A_h = \hat{U}_h(A)$ is the h-neighbourhood of the set A. Then, if $0 < h < \delta$,*

$$\|f_h\|_{L_p(A)} \leqslant M \|f\|_{L_p(A_h)}, \tag{2.5}$$

where

$$M = \int_{\mathbf{R}^n} |K(t)|\, dt,$$

and for $h \to 0$, $\|f_h - f\|_{L_p(A)} \to 0$. If $U = \mathbf{R}^n$, and $f \in L_p(\mathbf{R}^n)$, then $f_h \in L_p(\mathbf{R}^n)$,

$$\|f_h\|_{L_p(\mathbf{R}^n)} \leqslant M \|f\|_{L_p(\mathbf{R}^n)}, \tag{2.6}$$

and for $h \to 0$,

$$\|f_h - f\|_{L_p(\mathbf{R}^n)} \to 0.$$

Proof. First let us consider the case where $U = \mathbf{R}^n$ and $f \in L_p(\mathbf{R}^n)$. We have:

$$f_h(x) = \int_{\mathbf{R}^n} f(x+ht) K(t)\, dt,$$

and so for all x,

$$|f_h(x)| \leqslant \int_{\mathbf{R}^n} |f(x+ht)|\, |K(t)|\, dt. \tag{2.7}$$

Applying the Hölder inequality we hence obtain that for every $p \geqslant 1$,

$$|f_h(x)|^p \leqslant \int_{\mathbf{R}^n} |f(x+ht)|^p |K(t)|\, dt \left(\int_{\mathbf{R}^n} |K(t)|\, dt\right)^{p-1}. \tag{2.8}$$

By integrating inequality (2.8) term by term, we obtain (2.6).

Let us prove that if $f \in L_p(\mathbf{R}^n)$, then $\|f_h - f\|_{L_p(\mathbf{R}^n)} \to 0$ for $h \to 0$. Let us arbitrarily assign $\varepsilon > 0$. Due to Lemma 1.1, there exists a function $g \in C_0(\mathbf{R}^n)$ such that

$$\|f - g\|_{L_p(\mathbf{R}^n)} < \frac{\varepsilon}{2+M}.$$

Let $B = S(g)$. According to Theorem 2.1, the functions $g_h = K_h * g$ converge to g in $\mathbf{R}^n$ uniformly. For $0 < h < h_0$, $S(g_h) \subset U_{h_0}(B)$. Hence it follows that

$$\|g_h - g\|_{L_p(\mathbf{R}^n)} \to 0$$

by $h \to 0$; thus there exists $h_1 > 0$ such that for $0 < h < h_1$, $\|g_h - g\|_{L_p(\mathbf{R}^n)} < \varepsilon/(2+M)$. For $0 < h < h_1$, we have

$$\|f_h - f\|_{L_p(\mathbf{R}^n)} \leqslant \|f_h - g_h\|_{L_p(\mathbf{R}^n)} + \|g_h - g\|_{L_p(\mathbf{R}^n)} + \|g - f\|_{L_p(\mathbf{R}^n)}.$$

Since $\|f_h - g_h\|_{L_p(\mathbf{R}^n)} \leqslant M\|f - g\|_{L_p(\mathbf{R}^n)}$, hence it follows that for $0 < h < h_1$,

$$\|f_h - f\|_{L_p(\mathbf{R}^n)} < \frac{(M+2)\varepsilon}{M+2} \leqslant \varepsilon.$$

This proves that $\|f_h - f\|_{L_p(\mathbf{R}^n)} \to 0$ for $h \to 0$.

Now let us consider the general case. Let $U \subset \mathbf{R}^n$ be an open set, $f \in L_{p,\mathrm{loc}}(U)$. Let $A \subset U$ be compact, $\delta = \rho(A, \partial U)$. Let us assign h arbitrarily such that $0 < h < \delta$. The set $A_h = \overline{U}_h(A)$ is compact and is contained in U. Let us introduce an auxiliary function $\tilde{f} : \mathbf{R}^n \to \mathbf{R}$ by setting $\tilde{f}(x) = 0$ for $x \notin A_h$ and $\tilde{f}(x) = f(x)$ for $x \in A_h$. The function $\tilde{f}$ belongs to the class $L_p(\mathbf{R}^n)$. The function $\tilde{f}_h = K_h * \tilde{f}$ on the set A coincides with f_h. Hence it follows that

$$\|f_h\|_{L_p(A)} = \|\tilde{f}_h\|_{L_p(A)} \leqslant \|\tilde{f}_h\|_{L_p(\mathbf{R}^n)} \leqslant M\|\tilde{f}\|_{L_p(\mathbf{R}^n)} = M\|f\|_{L_p(A_h)},$$

and estimate (2.5) is proved.

According to the results proved above, $\|\tilde{f}_h - \tilde{f}\|_{L_p(\mathbf{R}^n)} \to 0$ for $h \to 0$. We have: $\|\tilde{f}_h - \tilde{f}\|_{L_p(\mathbf{R}^n)} \geqslant \|\tilde{f}_h - \tilde{f}\|_{L_p(A)} = \|f_h - f\|_{L_p(A)}$, thus $\|f_h - f\|_{L_p(A)} \to 0$ for $h \to 0$. This proves the theorem.

Corollary. *The set $C_0^\infty(\mathbf{R}^n)$ is dense everywhere in $L_p(\mathbf{R}^n)$ for any $p \geqslant 1$.*

Proof. Let $f \in L_p(\mathbf{R}^n)$. Let us arbitrarily assign $\varepsilon > 0$. By it, there exists a function $g \in C_0(\mathbf{R}^n)$ such that $\|f - g\|_{L_p(\mathbf{R}^n)} < \varepsilon/2$. Due to the theorem, there exists $h > 0$ such that $\|g - g_h\|_{L_p(\mathbf{R}^n)} < \varepsilon/2$. The function $\varphi = g_h \in C_0^\infty(\mathbf{R}^n)$, and $\|f - \varphi\|_{L_p(\mathbf{R}^n)} < \varepsilon$. The corollary is proved.

2.2. The Whitney Partition Theorem

Below, every rectangle of the form $H = [a_1, a_1 + l) \times [a_2, a_2 + l) \times \cdots \times [a_n, a_n + l)$ is called an n-dimensional cube in $\mathbf{R}^n$. The symbol l denotes the edge length of the cube H. The cube H is said to be a binary cube of rank r, where $r \geqslant 1$ is an integer if $l = 2^{-r}$, and $a_i = k_i 2^{-r}$, where $k_1, k_2, \ldots, k_n$ are integers.

Let us give some simple properties of binary cubes.

1) For every point $x \in \mathbf{R}^n$, there exists a binary cube of rank r containing x.

2) Let H_1 and H_2 be two binary cubes of ranks r_1 and r_2, respectively, besides, $r_1 \leqslant r_2$. Then, if H_1 and H_2 have common points, $H_1 \supset H_2$.

From Property 2 it follows that the binary cube of rank r containing an arbitrary point $x \in \mathbf{R}^n$ is unique so that for every r, the binary cubes of rank r form the partition of $\mathbf{R}^n$ into pairwise nonintersecting sets.

3) The diameter of every binary cube of rank r equals $\sqrt{n}2^{-r}$.

4) The cube $Q(0, 2m)$, where m is an integer, intersects $(2^{r+1}m)^n$ binary cubes of rank r.

5) The set of all binary cubes is countable.

From Property 4 it follows, in particular, that every bounded set in $\mathbf{R}^n$ intersects a finite number of binary cubes of the same rank r.

Theorem 2.3 (The Whitney partition theorem). *Let U be a bounded open set in $\mathbf{R}^n$. Then there exists a sequence of pairwise nonintersecting cubes (Q_m), $m = 1, 2, \ldots$, such that*

$$\bigcup_{m=1}^{\infty} Q_m = U$$

and for every m, the inequalities

$$d(Q_m) \leqslant \rho(Q_m, \partial U) < 3d(Q_m)$$

are valid.

Proof. First, suppose that U is contained in the open cube $Q_0 = (0,1) \times (0,1) \times \cdots \times (0,1)$. The general case is obviously reduced to this by a dilatation and a parallel translation of the set U.

Let H be an arbitrary binary cube contained in U. The cube H is said to be maximal if $d(H) \leqslant \rho(H, \partial U)$, and for every binary cube $H' \supset H$, this inequality does not hold. Let us show that every point $x \in U$ belongs to at least one maximal cube. Let us arbitrarily take $x \in U$. Since the set U is open, then there exists $\delta > 0$ such that the ball $B(x, \delta) \subset U$. Let us denote a binary cube of rank r by H_r such that $x \in H$. Then, if $\sqrt{n}/2^{-r} \leqslant \delta/2$, the cube $H_r \subset U$ and $d(H_r) \leqslant \rho(H_r, \partial U)$. Let r_0 be the lowest of the numbers r for which $H_r \subset U$ and $d(H_r) \leqslant \rho(H_r, \partial U)$. It is obvious that $r_0 > 0$, since $H_{r_0} \subset Q(0,1)$, and the cube H_{r_0} is the maximal one.

If the cube H is maximal, then no binary cube $H' \subset H$, distinct from H, is maximal. Hence it follows that different maximal cubes do not intersect.

We see that the union of maximal cubes coincides with U; consequently, they form the partition of U. Let us number the cubes of the constructed partition, and let Q_m be a cube with the number m. For every r the rank r_m of the cube Q_m is positive.

Let H be a binary cube of rank $r_m - 1$, containing Q_m. Then, due to the definition of the maximal cube, the inequality $\rho(H, \partial U) < d(H) = 2d(Q_m)$ is true. Let us take an arbitrary point $x \in H$. We have $\rho(Q_m, \partial U) \leqslant \rho(x, Q_m) + \rho(x, \partial U)$.

Further, $\rho(x, Q_m) \leqslant d(Q_m)$, whence

$$\rho(Q_m, \partial U) \leqslant d(Q_m) + \rho(x, \partial U).$$

Since x is an arbitrary point of the cube H, it follows that

$$\rho(Q_m, \partial U) \leqslant d(Q_m) + \rho(H, \partial U) < d(Q_m) + d(H) = 3d(Q_m).$$

Thus, for every m,

$$d(Q_m) \leqslant \rho(Q_m, \partial U) < 3d(Q_m).$$

The theorem is proved.

2.3. Partition of Unity

Let U be an open set in $\mathbf{R}^n$, and let $(\varphi_m)_{m \in \mathbf{N}}$ be a sequence of functions defined in U. We say that (φ_m) is the partition of unity in the set U if:

1) each of the functions φ_m is nonnegative and belongs to the class $C_0^\infty(U)$;

2) every point $x \in U$ has a neighbourhood G in which only a finite number of the functions φ_m is distinct from zero;

3) for all $x \in U$, the equality

$$\sum_{m=1}^{\infty} \varphi_m(x) = 1$$

is valid.

The partition of unity $(\varphi_m)_{m \in \mathbf{N}}$ in an open set $U \in \mathbf{R}^n$ is said to be subjected to an open covering $(U_\tau)_{\tau \in T}$ of the set U if for every m there exists τ such that $S(\varphi_m) \subset U_\tau$.

Theorem 2.4. *For every open covering of an open set $U \subset \mathbf{R}^n$, there exists the partition of unity in U that is subjected to this covering.*

Here, for every compact set $A \subset U$, the set of the functions of the partition of unity whose supports contain the points of A is finite.

Proof. Let $(U_\tau)_{\tau\in T}$ be a family of open sets that form a covering of the set $U \subset \mathbf{R}^n$, i.e., a family such that $U \subset \cup_{\tau\in T} U_\tau$.

First let us construct some sequence of compact sets (K_m). If $U = \mathbf{R}^n$, then we set $K_m = \bar{B}(0, m)$. If $U \neq \mathbf{R}^n$, then we choose an arbitrary point $a \in U$. Let $h = \rho(a, \partial U)$. Let us denote the totality of all points $x \in U$ by K_m such that $\rho(x, \partial U) \geqslant \frac{h}{m}$ and $|x| \leqslant m + |a|$. It is obvious that K_m is compact, $K_m \subset K_{m+1}$ for every m. For every compact set $A \subset U$, there exists $m_0 \in \mathbf{N}$ such that for all $x \in A, |x| \leqslant m_0$, and (for the case $U \neq \mathbf{R}^n$) $\rho(x, \partial U) \geqslant \frac{h}{m_0}$. Obviously, $A \subset K_{m_0}$. Since the compact set $A \subset U$ was chosen arbitrarily, it follows that $\cup_{m=1}^{\infty} K_m = U$.

A ball $B(x, r) \subset U$ is said to be admissible if there exists $\tau \in T$ such that the closed ball $\bar{B}(x, r) \subset U \cap U_\tau$.

Every point $x \in K_1$ is the centre of some admissible ball. Since K_1 is compact, according to the Borel theorem, there exists a finite system $B_1 = B(a_1, r_1)$, $B_2 = B(a_2, r_2), \ldots, B_{\nu_1} = B(a_{\nu_1}, r_{\nu_1})$ of the admissible balls which covers K_1. We set $H_1 = \cup_{\nu=1}^{\nu_1} B_\nu$. Suppose that for some $s \in \mathbf{N}$, a number ν_s and a finite sequence of balls $B_1, B_2, \ldots, B_{\nu_s}$ are defined, each of them being admissible; if $H_s = \cup_{\nu=1}^{\nu_s} B_\nu$, then the inclusion $K_s \subset H_s$ is valid. The set

$$\bar{H}_s = \bigcup_{\nu=1}^{\nu_s} \bar{B}_\nu$$

is compact and is contained in U, since each of the balls $\bar{B}_\nu$ is contained in U. Hence it follows that m exists such that $\bar{H}_s \subset K_m$. Let us arbitrarily choose $m > s$ for which this inclusion takes place. The set $K_m \backslash H_s$ is compact and does not intersect the set $K_s \subset H_s$. For every point $x \in K_m \backslash H_s$, there exists an admissible ball with centre x that does not intersect the set K_s. According to the Borel theorem, there exists a finite system of admissible balls $B_{\nu_s+1}, B_{\nu_s+2}, \ldots, B_{\nu_{s+1}}$, none of which has common points with the set K_s. This system covers the set $K_m \backslash H_s$. Thus, ν_{s+1} was defined, and besides, for every $\nu \leqslant \nu_{s+1}$, $\nu \in \mathbf{N}$, some admissible ball $\bar{B}$ was defined. From the construction it is clear that

$$H_{s+1} = \bigcup_{\nu=1}^{\nu_{s+1}} B_\nu \supset K_m \supset K_{s+1}.$$

By induction, we define an increasing sequence of numbers $\nu_1 < \nu_2 < \cdots < \nu_s < \ldots$ and a sequence of admissible balls $(B_\nu)_{\nu\in\mathbf{N}}$. For every s,

$$U \supset \bigcup_{\nu=1}^{\nu_s} B_\nu \supset K_{\nu_s},$$

whence it obviously follows that

$$\bigcup_{\nu=1}^{\infty} B_\nu = U.$$

Let $A \subset U$ be compact. Then there exists s such that $A \subset K_s$. From the construction it is clear that for $\nu > \nu_s$, the ball B_ν does not intersect the set K_s, i.e., we obtain that every compact set $A \subset U$ only intersects a finite number of balls B_ν.

Now one can easily complete the proof. Let φ be a function defined in Subsection (2.1), i.e., $\varphi(t) = e^{1/t}$ for $t < 0$, $\varphi(t) = 0$ for $t \geqslant 0$. Let a_ν be the centre of the ball B_ν, and $r_\nu > 0$ be its radius. Let us set $\theta_\nu(x) = \varphi\left(|x - a_\nu|^2 - r_\nu^2\right)$. The function θ_ν belongs to the class $C^\infty(\mathbf{R}^n)$, for all $x \in B_\nu$, and $\theta_\nu(x) = 0$ for $x \in B_\nu$, whence it follows that the support θ_ν is the closed ball $\bar{B}_\nu$. Let us set

$$\theta(x) = \sum_{\nu=1}^{\infty} \theta_\nu(x). \tag{*}$$

Let $x_0 \in U$ and $\delta > 0$ be such that the closed ball $\bar{B}(x_0, \delta) \subset U$. According to what was proved above, there exists $\bar{\nu}$ such that for $\nu > \bar{\nu}$, $B_\nu \cap B(x_0, \delta) = \emptyset$. Hence it follows that for all $x \in B(x_0, \delta)$:

$$\theta(x) = \sum_{\nu=1}^{\bar{\nu}} \theta_\nu(x).$$

Due to the arbitrariness of $x_0 \in U$, this allows us to conclude that the series (*) converges for all $x \in U$, and the function $\theta \in C^\infty(U)$. For every $x \in U$, there exists ν such that $x \in B_\nu$, so $\theta_\nu(x) > 0$. Hence it follows that $\theta(x) > 0$ for all $x \in U$. Let us set

$$\varphi_\nu(x) = \frac{\theta_\nu(x)}{\theta(x)}.$$

Then the function φ_ν belongs to the class $C^\infty(U)$, its support is the ball $\bar{B}_\nu$, and for all $x \in U$,

$$\sum_{\nu=1}^{\infty} \varphi_\nu(x) = 1,$$

and $\varphi_\nu(x) \geqslant 0$ for every $x \in U$ for all $\nu \in \mathbf{N}$. As follows from the properties of the sequence of balls (B_ν), the sequence of functions $(\varphi_\nu)_{\nu \in \mathbf{N}}$ also satisfies other conditions in the definition of partition of unity, which is subjected to the open covering $(U_\tau)_{\tau \in T}$.

This completes the proof of the theorem.

§3 General Information about Measures and Integrals

3.1. Notion of a Measure

Let us give the basic definitions and results of the general theory of measure.

Let X be an arbitrary set and let S be a nonempty family of subsets of X. The family S is said to be a ring if for any two sets $A, B \in S$, we have: $A \cup B \in S$ and $A \backslash B \in S$. Due to the identity $A \cap B = A \backslash (A \backslash B)$, we obtain that if S is a ring, then $A \cap B \in S$ as well. By induction one can easily verify that the union and the intersection of any finite number of sets from S belong to S. We have $A \backslash A = \emptyset$, hence it follows that the empty set is an element of every ring S.

The ring S is called a σ-ring if the union of any of the utmost countable family of sets from S belongs to S. If the set X, whose subsets are the elements of the σ-ring S, itself belongs to S, then the σ-ring S is called a σ-algebra.

If the intersection of any countable family of sets of the ring S belongs to S, then S is said to be a δ-ring.

Every σ-ring is a δ-ring. In general, the inverse is wrong. For instance, the totality of all bounded subsets of $\mathbf{R}$ is a δ-ring, but it is not a σ-ring.

A pair (X, S), where X is a set and S is a δ-ring of its subsets, is called a measurable space.

Let a measurable space (X, S) be given. A function of the set $\mu : S \to \mathbf{R}$ is called a measure if for every sequence $(A_m), m = 1, 2, \dots$ of pairwise nonintersecting sets from S, such that $\cup_{m=1}^{\infty} A_m$ belongs to S,

$$\mu\left(\bigcup_{m=1}^{\infty} A_m\right) = \sum_{m=1}^{\infty} \mu(A_m) \tag{3.1}$$

holds.

If $\mu : S \to \mathbf{R}$ is a measure in the measurable space (X, S), then $\mu(\emptyset) = 0$, and for any finite number of pairwise nonintersecting sets $A_1, A_2, \dots, A_m$, the equality

$$\mu\left(\bigcup_{k=1}^{m} A_k\right) = \sum_{k=1}^{m} \mu(A_k)$$

holds. The measure $\mu : S \to \mathbf{R}$ is said to be nonnegative if $\mu(A) \geqslant 0$ for any $A \in S$.

Let there be a measurable space (X, S). In the case where S is a σ-algebra, every measure $\mu : S \to \mathbf{R}$ is bounded. (This fact follows from the results of Subsection 3.2). Therefore, if we need the theory of measure in which unbounded measures are possible (such as, for instance, the Lebesgue measure in $\mathbf{R}^n$), then it is necessary to give up the requirement that S should be a δ-algebra. The presentation of the theory of measure is often based on the notion of a σ-ring. In this case, every pair (X, S), where S is a σ-ring, is called a measurable space. A measure is then every function $\mu : S \to \mathbf{R}$ such that either for all $A \in S$, $-\infty < \mu(A) \leqslant \infty$, or for all $A \in S$, $-\infty \leqslant \mu(A) < \infty$ (thus, infinite values of $\mu(A)$ are admitted for one sign only), and for every sequence (A_m) of pairwise nonintersecting sets from S, equality (3.1) is valid. (It is also necessary that μ is not identically equal to ∞.) In such a version, measure may be presented as an unbounded function. Here, however, it occurs to be bounded from one side (from above, if for all $A \in S$, $\mu(A) < \infty$, and from below if for all $A \in S$, $\mu(A) > -\infty$). Some important special cases in such a presentation of the measure theory turn out to be missing. For instance, the indeterminate integral of a function $f : \mathbf{R}^n \to \mathbf{R}$ which is integrable by the Lebesgue measure on every compact subset $\mathbf{R}^n$ is not, in general, a measure in the theory where the basic class of the sets S is considered to be a σ-ring. However, if we require that S be a δ-ring, then no such difficulties will arise.

Let (X, S) be a measurable space (S is a δ-ring of subsets of X), $A \in S$. The totality of all sets $E \in S$ contained in A is a σ-ring which we denote by $S(A)$. The function $f : A \to \bar{\mathbf{R}}$ is said to be measurable with respect to S or, in short, is called S-measurable if for every Borel set E, the set $f^{-1}(E) \in S$.

The notion of the function integrable in the sense of Lebesgue relative to the nonnegative measure assigned on a σ-algebra, and the notion of the Lebesgue integral of the function relative to measure for this case, and all their basic properties (in particular, theorems about passage to the limit for the Lebesgue integrals, etc.) are supposed to be known.

The totality of all functions f defined on the set $A \in S$ and such that the function $|f|^p$, $p \geqslant 1$, is integrable by A with respect to the measure $\mu \geqslant 0$, is denoted by the symbol $L_p(A, \mu)$. In this case, we also set

$$||f||_{L_p(A,\mu)} = \left(\int_A |f(x)|^p d\mu(x)\right)^{1/p}.$$

Let us note the following useful fact. Let $f \in L_1(A, \mu)$. Suppose that the function f is nonnegative. For $t > 0$, we let

$$E_f(t) = \{x \in A \mid f(x) > t\}; \quad \bar{E}_f(t) = \{x \in A \mid f(x) \geqslant t\}.$$

Then we have

$$\int_A f(x)\, \mu(dx) = \int_0^\infty \mu[E_f(t)]\, dt = \int_0^\infty \mu[\bar{E}_f(t)]\, dt \tag{3.2}$$

(the Cavalieri–Lebesgue formula).

3.2. Decompositions in the Sense of Hahn and Jordan

The proofs of the main results of this and of the following subsection are based on some statement about ordered sets.

An ordered set is a pair $(X, \prec)$ where X is an arbitrary set, $\prec$ is some transitive relation in X. This implies that for some pairs x, y of elements of X, it is indicated that the relation $x \prec y$ is valid; besides, if $x, y, z \in X$ are such that $x \prec y$ and $y \prec z$, then $x \prec z$. An element a of the ordered set X is said to be maximal if X contains no elements x such that $a \prec x$.

Lemma 3.1. (Lemma about the maximal element). *Let $(X, \prec)$ be an ordered set. Suppose that X is nonempty and that the following conditions are satisfied:*

1) There exists a function $f : X \to \mathbf{R}$ such that if $x \prec y$, then $f(x) < f(y)$.

2) For every sequence (x_m), $m = 1, 2, \ldots$ of the elements of the set X, such that for every m, $x_m \prec x_{m+1}$, there exists $x \in X$ for which $x_m \prec x$ for all $m = 1, 2, \ldots$.

Then X has a maximal element.

Proof. Let all the conditions of the lemma be satisfied. Let us construct, by induction, some finite or infinite sequence $x_1, x_2, \ldots, x_n, \ldots$ such that for every $n > 1$, $x_{n-1} \prec x_n$. x_1 is chosen arbitrarily. Suppose that for some n, $x_1, x_2, \ldots, x_n$ are defined; besides, $x_1 \prec x_2 \prec \cdots \prec x_n$. Let $H_n = \{y \in X | x_n < y\}$. If H_m is empty, then x_n is the maximal element of X, and the lemma is proved. Suppose that $H_n \neq \emptyset$. Let us put $L_n = \sup_{y \in H_n} f(y)$ and, as x_{n+1}, choose an element y of the set H_n such that if $L_n = \infty$, then $f(y) > n$, but if L_n is finite, then $f(y) \geqslant L_n - \frac{1}{n}$. If the sequence $x_1, x_2, \ldots, x_n, \ldots$ defined by the above construction proves to be finite, its last term is the maximal element of X. Let us consider the case where the sequence (x_n), $n = 1, 2, \ldots$, is infinite. For every n, $x_n \prec x_{n+1}$ by construction; thus, due to condition 2, there exists $y \in X$ such that $x_n \prec y$ for all n. Let us prove that y is the maximal element of X. Suppose that it is not so. Then there exists $y' \in X$ such that $y \prec y'$. Due to condition 1 we have: $f(y) < f(y')$. For every n, $x_n \prec y'$, so $y' \in H_n$; therefore, $f(y') \leqslant L_n$ for all n. Let n be such that $f(y) \leqslant n$ and $f(y') - f(y) > \frac{1}{n}$. For this n, $L_n < \infty$ because otherwise the inequalities $f(y) > f(x_{n+1}) > n$ hold. Thus, $f(x_{n+1}) > L_n - \frac{1}{n}$, whence $L_n \geqslant f(y') > f(y) > L_n - \frac{1}{n}$. This contradicts the fact that $f(y') - f(y) > \frac{1}{n}$. Therefore, the assumption that y is not the maximal element of X leads to contradiction. The lemma is proved.

Corollary. *Suppose that the ordered set $(X, \prec)$ satisfies all conditions of the lemma. Then, if $x \in X$ is not the maximal element of X, there exists the maximal element y of the set X such that $x \prec y$.*

Proof. Let $x \in X$ and let x not be the maximal element of X. We set $X' = \{y \in X | x \prec y\}$. The ordered set $(X', \prec)$ satisfies all conditions of the lemma, so it has the maximal element y. This y is obviously the maximal element of the initial set X as well. Here $x \prec y$. Corollary is proved.

Let μ be a measure in a measurable space (X, S). The set $A \in S$ is called μ-positive if for $E \in S(A)$, $\mu(E) \geqslant 0$; it is said to be μ-negative if for any $E \in S(A)$, $\mu(E) \leqslant 0$.

Theorem 3.3. *Let μ be a measure in a measurable space (X, S). For every set $A \in S$, there exist the sets $A^+ \in S(A)$ and $A^- \in S(A)$ such that A^+ is μ-positive and A^- is μ-negative, $A^+ \cup A^- = A$, $A^+ \cup A^- = \phi$.*

Remark. The sets A^+ and A^- mentioned in the theorem are said to form the decomposition in the sense of Hahn of the set A corresponding to the measure μ.

Proof. First, let us prove that every set $A \in S$ contains a μ-positive set $A' \in S$ such that $\mu(A) \leqslant \mu(A')$. If A is μ-positive, we may take $A' = A$. To prove the given statement in the case where A is not μ-positive, let us use the corollary of Lemma 3.1. In the set S, we introduce the relation of order $\prec$ by letting $E_1 \prec E_2$ where $E_1, E_2 \in S'$, if $E_1 \supset E_2$ and $\mu(E_1) < \mu(E_2)$. It is easy to verify that the relation is actually the relation of order and that the ordered space $(S, \prec)$ satisfies the conditions of Lemma 3.1. Every μ-positive set $A \in S$ is the maximal element of the ordered set $(S, \prec)$. Indeed, if A is μ-positive, then for every $B \subset A$, $B \in S$, we have: $\mu(B) = \mu(A) - \mu(A \backslash B) \leqslant \mu(A)$, since $\mu(A \backslash B) \geqslant 0$; hence it is clear that S does not contain elements B such that $A \prec B$. Conversely, let A be the maximal element of S. Let us arbitrarily take $E \subset A$, $E \in S$. We have: $\mu(E) = \mu(A) - \mu(A \backslash E)$. Since A is the maximal element of S and $A \backslash E \subset A$, the inequality $\mu(A) < \mu(A \backslash E)$ does not hold, because otherwise the relation $A \prec (A \backslash E)$ is valid in spite of the fact that A is maximal. Thus, $\mu(A \backslash E) \leqslant \mu(A)$, and consequently, A is μ-positive.

If the set $A \in S$ is not μ-positive, then, applying the corollary of Lemma 3.1, we obtain that there exists $E' \in S(A)$, which is the maximal element of S. According to what was proved above, the set E' is μ-positive and $\mu(A) < \mu(E')$.

Let us take arbitrarily $A \in S$. According to what was proved above, A contains μ-positive subsets. The totality of all μ-positive sets of $E \in S(A)$ is denoted by $S^+(A)$. In the set $S^+(A)$, we introduce the relation of order as follows. Let $E_1, E_2 \in S^+(A)$. Then we assume that $E_1 \prec E_2$ if $E_1 \subset E_2$ and $\mu(E_1) < \mu(E_2)$. It is easy to see that the relation $\prec$ is the relation of order in this case as well, and the ordered set $(S^+(A), \prec)$ satisfies the condition of the lemma about the maximal element. Let P be a maximal element of the space $(S^+(A), \prec)$, $Q = A \backslash P$. Let us show that the set Q is μ-negative. Suppose that this is not the case. Then there exists $E \subset Q$ such that $\mu(E) > 0$. Let $H \subset E$ be a μ-positive set such that $\mu(E) \leqslant \mu(H)$. The set $P' = P \cup H$ is μ-positive, $A \supset P'$ and $\mu(P') = \mu(P) + \mu(H) > \mu(P)$. We have: $P' \supset P$, thus $P \prec P'$. This contradicts the fact that, by condition, Q is the maximal element of the set $S^+(A)$. Therefore, the admission that Q is not a μ-negative set leads to contradiction. The sets $A^+ = P, A^- = Q$ obviously form the desired decomposition of the set A. The theorem is proved.

A measurable space (X, S) (S is a δ-ring of subsets of X) is called σ-finite if X is the union of a countable number of sets belonging to S. If (X, S) is σ-finite, then X may be represented as a union of some increasing sequence of sets belonging to S. Indeed, let (Y_m), $m = 1, 2, \ldots$ be an arbitrary sequence of elements of S such that $X = \bigcup_{m=1}^{\infty} Y_m$. Let us set $X_m = \bigcup_{k=1}^{\infty} Y_k$. The sequence (X_m), $m = 1, 2, \ldots$ is increasing, for every m, $X_m \in S$ and $\bigcup_{m=1}^{\infty} X_m = \bigcup_{m=1}^{\infty} Y_m = X$. By setting $A_1 = X_1, A_m = X_m \setminus X_{m-1}$ for $m > 1$, we obtain a sequence (A_m), $m = 1, 2, \ldots$ of pairwise nonintersecting sets from S such that $\bigcup_{m=1}^{\infty} A_m = S$.

If the measurable space (X, S) is σ-finite, then Theorem 3.3 admits some refinement. Namely, we have the following:

Corollary. *If a measurable space (X, S) is σ-finite, then for every measure $\mu : S \to \mathbf{R}$, there exist sets Y and Z such that $Y \cap Z = \emptyset, Y \cup Z = X$, and for every $E \in S$, $E \cap Y \in S, E \cap Z \in S$, and besides, $\mu(E \cap Y) \geqslant 0, \mu(E \cap Z) \leqslant 0$.*

Proof. Let $(A_m), m = 1, 2 \ldots$ be a sequence of pairwise nonintersecting sets of S such that $\bigcup_{m=1}^{\infty} A_m = X$. For each m, let us construct the partition in the sense of Hahn, $A_m = A_m^+ \cup A_m^-$ of the set A_m. The sets $Y = \bigcup_{m=1}^{\infty} A_m^+$ and $Z = \bigcup_{m=1}^{-}$ are obviously the desired ones.

Remark. If Y and Z are the sets mentioned in the formulation of the corollary, then it is obvious that for every $E \in S$, the set $E \cap Y$ is μ-positive, the set $E \cap Z$ is μ-negative.

Let μ be a measure in a measurable space (X, S). Let us arbitrarily take $A \in S$. Let us set

$$\mu^+(A) = \sup_{E \in S(A)} \mu(E).$$

Let $A = A^+ \cup A^-$ be the partition of the set A in the sense of Hahn. For arbitrary $E \in S(A)$, we have:

$$\mu(E) = \mu(E \cap A^+) + \mu(E \cap A^-) \leqslant \mu(E \cap A^+) \leqslant \mu(A^+).$$

For $E = A^+$, $\mu(E) = \mu(A^+)$, and consequently we obtain

$$\mu^+(A) = \mu(A^+).$$

In particular, it follows from this that $\mu^+(A)$ is a nonnegative and finite magnitude. Thus, on the totality of the sets S, some function μ^+ is defined. This function is called the upper variation of the measure μ. Let us prove that μ^+ is also a measure. Let $(A_m), m = 1, 2, \ldots$ be an arbitrary sequence of pairwise nonintersecting sets from S, such that $A = \bigcup_{m=1}^{\infty} A_m \in S$. For every m, we have the decomposition in the sense of Hahn of the set $A_m, A_m = A_m^+ \cup A_m^-$. Let us set $P = \bigcup_{m=1}^{\infty} A_m^+$, $Q = \bigcup_{m=1}^{\infty} A_m^-$. Then obviously, the set P is μ-positive, Q is μ-negative, $P \cap Q = \emptyset$, $P \cup Q = A$, so that P and Q form the decomposition in the sense of Hahn of the set A. We have

$$\mu^+(A) = \mu(P) = \sum_{m=1}^{\infty} \mu(A_m^+) = \sum_{m=1}^{\infty} \mu^+(A_m),$$

and thus we have proved that μ^+ is a measure.

From the definition of the magnitude $\mu^+(A)$, it follows that for every $A \in S$, $\mu(A) \leqslant \mu^+(A)$, and we let $\mu^-(A) = \mu^+(A) - \mu^-(A)$. The function of the sets μ^- is obviously a measure, too. Here $\mu^-(A) \geqslant 0$ for all $A \in S$. The measure μ^- is called the lower variation of the measure μ. Obviously, $\mu = \mu^+ - \mu^-$, and thus we obtain that every measure on S may be represented as the difference of two nonnegative measures. The constructed representation

$$\mu = \mu^+ - \mu^-$$

is called the decomposition in the sense of Jordan of the measure μ.

If μ is a measure on S, then for every $A \in S$, we have $\mu(A) = \mu(A^+) - \mu(A^-)$, where $A^+ \cup A^- = A$ is the decomposition in the sense of Hahn of the set A. We have: $\mu(A^+) = \mu^+(A)$. Hence we obtain

$$\mu^-(A) = \mu^+(A) - \mu(A) = -\mu(A^-).$$

We put

$$|\mu|(A) = \mu^+(A) + \mu^-(A). \tag{3.3}$$

The measure $|\mu|$ thus defined is called the complete variation of the measure μ. According to what was proved above, we have:

$$|\mu|(A) = |\mu(A^+)| + |\mu(A^-)|, \tag{3.4}$$

where $A = A^+ \cap A^-$ is the decomposition in the sense of Hahn of the set A.

Let $f : A \to \mathbf{R}$, where $A \in S$ is an S-measurable function. Then f is said to be integrable with respect to the measure μ if f is integrable relative to each of the measures μ^+ and μ^-. In this case, we set

$$\int_A f(x) d\mu = \int_A f(x) d\mu^+ - \int_A f(x) d\mu^-.$$

From the above definition it follows, in particular, that if f is integrable by the measure μ, then it is also integrable by the measure $|\mu| = \mu^+ + \mu^-$. We have:

$$\begin{aligned} \left|\int_A f(x)\, d\mu\right| &\leqslant \left|\int_A f(x)\, d\mu^+\right| + \left|\int_A f(x)\, d\mu^-\right| \\ &\leqslant \int_A |f(x)|\, d\mu^+ + \int_A |f(x)|\, d\mu^- = \int_A |f(x)|\, d|\mu|. \end{aligned} \tag{3.5}$$

It should also be noted that if the measure $\mu : S \to \mathbf{R}$ is nonnegative, then $\mu^+ = \mu$, $\mu^- = 0$, and $|\mu| = \mu^+$.

3.3. The Radon–Nikodym Theorem and the Lebesgue Decomposition of Measure

Let us arbitrarily assign a measurable space (X, S).

Let $\mu : S \to \mathbf{R}$ be a measure on a σ-ring S. A set $K \subset X$ is called μ-negligible if for every $E \in S$, the set $K \cap E \in S$, and $|\mu|(K \cap E) = 0$.

Let the measure $\lambda : S \to \mathbf{R}$ be given. The measure λ is said to be absolutely continuous with respect to μ if for every $E \in S$ such that $|\mu|(E) = 0$, $|\lambda|(E) = 0$ as well. The measure λ is called singular with respect to μ if there exists a μ-negligible set K such that for every $E \in S$, $\lambda(E) = \lambda(E \cap K)$. The measure λ may be said to be singular with respect to the measure μ if it is concentrated on some μ-negligible set.

Suppose that the function $f : X \to \bar{\mathbf{R}}$ is S-measurable and integrable on every set $E \in S$. By setting for $E \in S$,

$$\mu f(E) = \int_E f(x) d\mu,$$

we obtain a function μf defined on S. Due to the known properties of the Lebesgue integral, the function μf is a measure. We call it the indefinite integral of the function f with respect to the measure μ.

Below, the following simple supposition will come in handy.

Lemma 3.1. *Let $\mu : S \to \mathbf{R}$ be a measure in a measurable space (X, S), $A \in S$, $\varphi : A \to \mathbf{R}$, $f : A \to \mathbf{R}$ are S-measurable functions. Suppose that f is integrable by A with respect to the measure μ, and for $E \in S(A)$, we put*

$$\mu f(E) = \int_E f(x)\, d\mu.$$

The function μf is a measure on $S(A)$. Then the function φ will be integrable by A with respect to the measure μf if and only if the function φf is integrable with respect to the measure μ. Here

$$\int_A \varphi(x)\, d\mu f = \int_A \varphi(x) f(x)\, d\mu.$$

Proof. First let us suppose that the functions φ and f and the measure μ are nonnegative. Suppose that the function φ is a stepfunction; that is, there exist pairwise nonintersecting sets $E_i \in s, i = 1, 2, \dots, m$ such that $\cup_{k=1}^{m} E_k = A$ and φ is constant on every one of the sets $E_i, \varphi(x) = \lambda_i \in \mathbf{R}$ for $x \in E_i$. Then we have:

$$\int_A \varphi(x)\, d\mu f = \sum_{i=1}^{m} \lambda_i \int_{E_i} f(x)\, d\mu = \int_A \varphi(x) f(x)\, d\mu.$$

Consequently, the lemma is true for the given case. If $\varphi \geqslant 0$ is an arbitrary S-measurable function, then there exists an increasing sequence $\varphi_k : A \to \mathbf{R}$, $k = 1, 2, \ldots$ of stepfunctions such that $\varphi_k(x) \to \varphi(x)$ for $k \to \infty$ for all $x \in A$. For every k according to what was proved above, we have

$$\int_A \varphi_k(x) d\mu f = \int_A \varphi_k(x) f(x) d\mu. \tag{3.6}$$

Suppose that the function φ is integrable by the measure μf. For $k \to \infty$,

$$\int_A \varphi_k(x)\, d\mu f \to \int_A \varphi(x)\, d\mu f. \tag{3.7}$$

The sequence of functions $(\varphi_k f)$, $k = 1, 2, \ldots$ is increasing, and as $k \to \infty$, $\varphi_k(x) f(x) \to \varphi(x) f(x)$ for all $x \in A$. Due to (3.6) and (3.7), the sequence of integrals

$$\int_A \varphi_k(x) f(x)\, d\mu$$

is bounded. So the function φf is integrable, and

$$\begin{aligned}\int_A \varphi(x) f(x)\, d\mu &= \lim_{k \to \infty} \int_A \varphi_k(x) f(x)\, d\mu \\ &= \int_A \varphi_k(x) f(x)\, d\mu = \int_A \varphi(x) f(x)\, d\mu.\end{aligned}$$

Conversely, if φf is integrable by A with respect to μ, then

$$\int_A \varphi_k(x) f(x)\, d\mu \to \int_A \varphi(x) f(x)\, d\mu.$$

Using the arguments of the previous case, we obtain that φ is integrable by the measure μf.

Let the functions φ, f and the measure μ be alternating. Let $A^+ \cup A^-$ be the decomposition in the sense of Hahn of the set A, corresponding to the measure μ. Let $A^+ = A_1$, $A^- = A_2$, for $i = 1, 2$, let $A_{i1} = \{x \in A_i | f(x) > 0\}$, $A_{i2} = \{x \in A_i | f(x) \leqslant 0\}$, and for $i, j = 1, 2$, let $A_{ij1} = \{x \in A_{ij} | \varphi(x) > 0\}$, $A_{ij2} = \{x \in A_{ij} | \varphi(x) \leqslant 0\}$. The sets A_{ijk}, $i, j, k = 1, 2$ are pairwise nonintersecting, each of them is an element of S, and their union coincides with A. Therefore, an S-measurable function $F : A \to \mathbf{R}$ is integrable by A with respect to some measure if and only if it is integrable with respect to this measure by each of the sets A_{ijk}. On the set A_{ijk}, the functions φ and f and the measure μ have constant signs. Due to what was proved above, it follows that φ is integrable by A_{ijk} with respect to the measure μf if and only if φf is integrable by A_{ijk} with respect to the measure μ. Here

$$\int_{A_{ijk}} \varphi(x)\, d\mu = \int_{A_{ijk}} \varphi(x) f(x)\, d\mu.$$

Hence the statement of the lemma obviously follows.

Theorem 3.2. (The Radon–Nikodym theorem). *Suppose that a measurable space (X, S) is σ-finite and $\mu : S \to \mathbf{R}$ is an arbitrary measure on S. Then, if the measure $\lambda : S \to \mathbf{R}$ is absolutely continuous with respect to μ, there exists an S-measurable function $\varphi : X \to \bar{\mathbf{R}}$ such that for every $E \in S$, the equality*

$$\lambda(E) = \int_E \varphi(x)\, d\mu$$

is valid. Here, if the measures λ and μ are nonnegative, the function φ may also be considered to be nonnegative.

Proof. First let us consider the case when each of the measures μ and λ is non-negative, and the set X is an element of S. Let us denote by $\sum$ the totality of all non-negative S-measurable functions $\varphi : X \to \bar{\mathbf{R}}$ such that for every $E \in S$ there holds the inequality

$$\int_E \varphi(x)\, d\mu \leqslant \lambda(E).$$

The set of functions $\sum$ is nonempty since, for instance, the function $\varphi \equiv 0$ belongs to $\sum$. Let us introduce the relation $\prec$ on the set $\sum$, setting $\varphi_1 \prec \varphi_2$ for $\varphi_1, \varphi_2 \in \sum$ if and only if the following two conditions hold:

1) For every $x \in X \quad \varphi_1(x) \leqslant \varphi_2(x)$
2) $\int_X \varphi_1(x) d\mu < \int_X \varphi_2(x) d\mu$.

The relation $\prec$ is the order relation. On the ordered set $(\sum, \prec)$, the real function $I : \varphi \in \sum \to \int_X \varphi(x) d\mu$ is defined. If $\varphi_1 < \varphi_2$, then $I(\varphi_1) < I(\varphi_2)$. Let $(\varphi_m), m = 1, 2, \ldots$, be a sequence of functions belonging to $\sum$, such that for every m, $\varphi_m \prec \varphi_{m+1}$. This means that for any $x \in X$, $\varphi_m(x) \leqslant \varphi_{m+1}(x)$, that is, the numerical sequence $(\varphi_m(x))$, $m = 1, 2, \ldots$, is the increasing one; thus, for every $x \in X$, there exists the limit $\operatorname{Lim} \varphi_m(x) = \varphi(x) \in \mathbf{R}$. The function $\varphi : X \to \mathbf{R}$ thus obtained is S-measurable. For every $E \in S$, we have

$$\int_E \varphi_m(x) d\mu \leqslant \lambda(E). \tag{3.8}$$

In particular,

$$\int_X \varphi_m(x) d\mu \leqslant \lambda(X) < \infty$$

for all m. Hence it follows that the limit function φ is integrable with respect to the measure μ. By taking the limit in inequality (3.8) for $m \to \infty$, we obtain that for any $E \in S$, the inequality

$$\int_E \varphi(x) d\mu \leqslant \lambda(E)$$

holds. Thus the function φ belongs to the class $\sum$. For every m, we have $\varphi_m(x) \leqslant \varphi(x)$ for all x, and

$$\int_X \varphi_m(x)d\mu < \int_X \varphi_{m+1}(x)d\mu \leqslant \int_X \varphi(x)d\mu.$$

This means that $\varphi_m \prec \varphi$ for every m. Thus we obtain that for the ordered set $(\sum, \prec)$ the second condition of the lemma about the maximal element is satisfied as well. Hence it follows that $(\sum, \prec)$ has a maximal element. Let $\varphi_0 : X \to \bar{\mathbf{R}}$ be this element.

Let us prove that for every $E \in S$,

$$\int_E \varphi_0(x)d\mu = \lambda(E).$$

Let us set

$$\lambda(E) - \int_E \varphi_0(x)d\mu = \theta(E).$$

According to the definition of the class $\sum$, $\theta(E) \geqslant 0$ for all $E \in S$. It is also obvious that $\theta(E) \leqslant \lambda(E)$ for any $E \in S$. If the set $E \in S$ is such that $\mu(E) = 0$, then, since the measure λ is absolutely continuous with respect to μ, for this E, $\lambda(E) = 0$ as well; so, $\theta(E) = 0$. It is necessary to prove that $\theta(E) \equiv 0$. Suppose that this is not so. Then there exists a set $A \in S$ such that $\theta(A) > 0$. For this A, $\mu(A) > 0$ as well. Let $\varepsilon > 0$ be such that $\theta(A) - \varepsilon\mu(A) > 0$. Consider the measure $\theta_\varepsilon = \theta - \varepsilon\mu$. Let $A = A^+ \cup A^-$ be the decomposition of the set A in the sense of Hahn, which corresponds to the measure θ_ε. We have $\theta(A^+) \geqslant \theta_\varepsilon(A^+) \geqslant \theta_\varepsilon(A) > 0$. Hence it follows that $\mu(A^+) > 0$. Let us put $\varphi = \varphi_0 + \varepsilon x_{A^+}$ (x_{A^+} is the indicator of the set A^+). For all $x \in X$, $\varphi_0(x) \leqslant \varphi(x)$, and

$$\int_X \varphi(x)\, d\mu = \int_X \varphi_0(x)\, d\mu + \varepsilon\mu(A^+) > \int_X \varphi_0(x)\, d\mu.$$

For arbitrary $E \in S$, we have

$$\int_E \varphi_0(x)d\mu + \theta(E) = \lambda(E),$$

and then

$$\begin{aligned}\int_E \varphi(x)\, d\mu &= \int_{E\setminus A^+} \varphi(x)\, d\mu + \int_{E\cap A^+} \varphi(x)\, d\mu \\ &= \int_{E\setminus A^+} \varphi_0(x)\, d\mu + \int_{E\cap A^+} (\varphi_0(x) + \varepsilon)\, d\mu \leqslant \lambda(E\setminus A^+) \\ &+ \int_{E\cap A^+} \varphi_0(x)\, d\mu + \varepsilon\mu(E \cap A^+).\end{aligned}$$

Since the set A^+ is θ_ε-positive, then $\theta_\varepsilon(E \cap A^+) \geqslant 0$, that is, $\varepsilon \mu(E \cap A^+) \leqslant \theta(E \cap A^+)$, and we obtain

$$\begin{aligned}\int_E \varphi(x)\, d\mu &\leqslant \lambda(E \backslash A^+) + \int_{E\cap A^+} \varphi_0(x)\, d\mu \\ &\quad + \theta(E \cap A^+) \\ &= \lambda(E\backslash A^+) + \lambda(E \cap A^+) = \lambda(E).\end{aligned}$$

Since $E \in S$ is taken arbitrarily, then we obtain that $\varphi \in \sum$. From what was proved above it follows that $\varphi_0 \prec \varphi$. This contradicts the fact that φ_0 is a maximal element of $\sum$. So, the supposition that θ is not identically equal to zero leads to contradiction, therefore

$$\int_E \varphi_0(x)\, d\mu = \lambda(E)$$

for every $E \in S$. Since φ_0 is integrable with respect to μ, then for the set E_0 of such $x \in X$ for which $\varphi_0(x) = \infty$ we have: $\mu(E_0) = 0$. By changing the values of the function for E_0, we obtain an everywhere finite function φ such that $\int_E \varphi(x)\, d\mu = \lambda(E)$ for all $E \in S$. For the given case the theorem is proved. The constructed function φ is nonnegative.

Now, consider the case where the measures λ and μ are nonnegative, but X is not an element of S. Since the measurable space (X, S) is σ-finite, X admits the representation $X = \cup_{m=1}^{\infty} X_m$, where $X_m \in S$ for every m and the sets X_m do not pairwise intersect. On the σ-ring $S(X_m)$, the measure λ is absolutely continuous with respect to the measure μ, therefore, for every m there exists a function $\varphi_m : x_m \to \mathbf{R}$ such that for any $E \in S(X_m)$,

$$\lambda(E) = \int_E \varphi_m(x)\, d\mu.$$

Let the function $\varphi : X \to \mathbf{R}$ be such that for every $m = 1, 2, \ldots,$ $\varphi(x) = \varphi_m(x)$ for all $x \in X_m$. Let us show that the function φ is the desired one. Indeed, for every $E \in S$, we have

$$\begin{aligned}\lambda(E) &= \sum_{m=1}^{\infty} \lambda(E \cap X_m) = \sum_{m=1}^{\infty} \int_{E\cap X_m} \varphi_m(x)\, d\mu \\ &= \sum_{m=1}^{\infty} \int_{E\cap X_m} \varphi(x)\, d\mu = \int_E \varphi(x)\, d\mu\end{aligned}$$

Q. E. D.

Here the function φ is nonnegative.

Now let us consider the case where the measure λ has an arbitrary sign, and the measure μ is nonnegative. We have the decomposition of the measure

$\lambda = \lambda^+ - \lambda^-$ in the sense of Jordan. Let $A \in S$, and $A = A^+ \cup A^-$ is the decomposition of the set A in the sense of Hahn, which corresponds to the measure λ. If $\mu(A) = 0$, then $\mu(A^+) = 0$ and $\mu(A^-) = 0$ as well; hence it follows that $\lambda^+(A) = \lambda(A^+) = 0$ and $\lambda^-(A) = \lambda(A^-) = 0$, too; that is, the measures λ^+ and λ^- are also absolutely continuous with respect to the measure μ. Let the functions $\psi : X \to \mathbf{R}$ and $\theta : X \to \mathbf{R}$ be such that for every $E \in S$,

$$\lambda^+(E) = \int_E \psi(x)\, d\mu, \quad \lambda^-(E) = \int_E \theta(x)\, d\mu.$$

Then

$$\lambda(E) = \int_E (\psi(x) - \theta(x))\, d\mu$$

for any $E \in S$, and the function $\varphi = \psi - \theta$ is the desired one.

It only remains to consider the case where λ and μ are the measures of an arbitrary sign. If the measure λ is absolutely continuous with respect to μ, then it is absolutely continuous with respect to $|\mu|$ as well; therefore, there exists a function φ_0 such that for every $E \in S$,

$$\lambda(E) = \int_E \varphi_0(x)\, d\mu.$$

Let $X = Y \cup Z$ be the decomposition in the sense of Hahn of the space X which corresponds to the measure μ. Then for every $E \in X$, the set $E \cap Y$ is μ-positive, and $E \cap Z$ is μ-negative, $Y \cap Z = \emptyset$. Let us put $j(x) = 1$ for $x \in Y$, and $j(x) = -1$ for $x \in Z$. For every $E \in S$, we have:

$$|\mu|(E) = \int_E j(x)\, d|\mu|.$$

Let us apply Lemma 3.2, setting $\varphi = \varphi_0, f = j$. We obtain

$$\lambda(E) = \int_E \varphi_0(x) j(x)\, d\mu,$$

and the function $\varphi = \varphi_0 \circ j$ is the desired one.

This completes the proof of the theorem.

Remark 1. The function φ in the Radon–Nikodym theorem is determined by the measures λ and μ being assigned uniquely to within the values on a μ-negligible set.

We leave the proof to the reader.

Remark 2. The function φ mentioned in Theorem 3.2 is also called the derivative of the measure λ with respect to the measure μ. In connection with this, we use for it the notation

$$\varphi = \frac{d\lambda}{d\mu}.$$

Theorem 3.3. *Let λ and μ be measures in a σ-finite measurable space (X, S). Then there exist the function $\varphi : X \to \mathbf{R}$ and the measure $\delta : S \to \mathbf{R}$ such that the measure δ is singular with respect to μ, and for every $E \in S$, the equality:*

$$\lambda(E) = \int_E \varphi(x)\, d\mu + \delta(E) \tag{3.8'}$$

holds (the Lebesgue decomposition of the measure λ with respect to the measure μ).

Proof. Let $w = |\lambda| + |\mu|$. Each of the measures λ and μ is absolutely continuous with respect to the measure w, therefore, there exist the functions $u : X \to \mathbf{R}$ and $v : X \to \mathbf{R}$ such that for any $E \in S$

$$\lambda(E) = \int_E u(x)\, dw, \qquad \mu(E) = \int_E v(x)\, dw.$$

Let K be the set of all $x \in X$ for which $v(x) = 0$. For every $E \in S$, the set $K \cap E \subset S$ and $\mu(K \cap E) = 0$, so that K is a μ-negligible set. Let $\varphi(x) = \frac{u(x)}{v(x)}$ for $x \notin K$, $\varphi(x) = 0$ for $x \in K$. The function φ^v is integrable with respect to the measure w by any set $E \in S$. Due to Lemma 3.1 it follows that φ is integrable with respect to the measure $\mu = w$ by any $E \in S$. Here, if $E \cap K = \emptyset$, then

$$\int_E \varphi(x)\, d\mu = \int_E v(x)\varphi(x)\, dw = \int_E u(x)\, dw = \lambda(E). \tag{3.9}$$

Let us set

$$\lambda(E) - \int_E \varphi(x)\, d\mu = \delta(E).$$

For every $E \in S$, we have

$$\delta(E) = \delta(E \backslash K) + \delta(E \cap K).$$

Since $(E \backslash K) \cap K = \emptyset$, then due to (3.9), $\delta(E \backslash K) = 0$, therefore, $\delta(E) = \delta(E \cap K)$. Since the set K is μ-negligible, it means that the measure δ is singular and the theorem is proved.

Remark. The measure δ in equality (3.8) is uniquely determined by the assignment of the measures λ and μ. The function φ is uniquely restored by the measures λ and μ to within the values on a μ-negligible set.

Indeed, suppose that the functions φ_1 and φ_2 and the measures δ_1 and δ_2 are such that for every E

$$\lambda(E) = \int_E \varphi_1(x)\, d\mu + \delta_1(E) = \int_E \varphi_2(x)\, d\mu + \delta_2(E),$$

the measures δ_1 and δ_2 being singular with respect to μ. Hence

$$\delta_1(E) - \delta_2(E) = \int_E (\varphi_1(x) - \varphi_2(x))\, d\mu.$$

The measure $\delta_1 - \delta_2$ is singular with respect to μ, and from the latter equality it follows that it is also absolutely continuous with respect to μ. Hence it follows that $\delta_1 - \delta_2 = 0$, therefore, $\delta_1 = \delta_2$ and

$$\int_E (\varphi_1(x) - \varphi_2(x))\, d\mu = 0$$

for every $E \in S$. Hence it follows that $\varphi_1(x) - \varphi_2(x) = 0$ almost everywhere in the sense of the measure μ. Q. E. D.

§4 Differentiation Theorems for Measures in $\mathbf{R}^n$

4.1. Definitions

The goal of the present subsection is to prove the classical theorems about the differentiation of measures in $\mathbf{R}^n$. As a rule, these theorems are proved by means of the so-called Vitali–Lebesgue covering theorem. The presentation below uses only the partial case of the general Vitali theorem in which the covering of a set by balls is considered.

Let us give some definitions. Let us assign an arbitrary open set $U \in \mathbf{R}^n$. Let us denote by $\mathcal{L}(U)$ the totality of Borel sets contained in U, and $\mathcal{L}_0(U)$ stands for the totality of the sets from $\mathcal{L}(U)$ which lie strictly inside U. It is obvious that $\mathcal{L}(U)$ is a σ-ring, and $\mathcal{L}_0(U)$ is a δ-ring.

The measurable space $(U, \mathcal{L}_0(U))$ is σ-finite. Indeed, for $m \in \mathbf{N}$, let

$$A_m = \{x \in U \mid \rho(x, \partial U) \geqslant 1/m \ \&\ |x| \leqslant m\}.$$

The sets A_m are compact and $\cup_{m=1}^{\infty} A_m = U$. Every measure defined on the δ-ring $\mathcal{L}_0(U)$ is said to be a measure in the open set U.

The symbol m herein denotes the Lebesgue measure in $\mathbf{R}^n$. The words "almost everywhere," if not noted otherwise, imply: "almost everywhere in the sense of the Lebesgue measure in $\mathbf{R}^n$."

Let f be a measurable function defined almost everywhere in the open set $U \subset \mathbf{R}^n$. Then for every $h \in \mathbf{R}$, the sets $\{x \in U | f(x) < h\}, \{x \in U | f(x) \leqslant h\}$ are measurable in the sense of Lebesgue. We say that the function f is measurable in the sense of Borel if these sets belong to the class $\mathcal{L}(U)$ for any $h \in \mathbf{R}^n$. As is established in the theory of the Lebesgue integral, for every measurable function f there exists a measurable (in the sense of Borel) function $\tilde{f}$ such that $\tilde{f}(x) = f(x)$ almost everywhere in U. All measurable functions below are supposed to be measurable in the sense of Borel.

Let $f \in L_{1,\mathrm{loc}}(U)$. The function of the set

$$\mu_f : E \in \mathfrak{B}_0(U) \longmapsto \int_E f(x)\,dx$$

is a measure in U which is called the indefinite integral of the function f. It is easy to verify that the complete variation of the measure μ_f is the measure

$$\mu_{|f|} : E \longmapsto \int_E |f(x)|\,dx.$$

Let μ be an arbitrary measure in the open set $U \subset \mathbf{R}^n$ and let $x \in U$. Let us set

$$\bar{D}\mu(x) = \overline{\lim_{r\to 0}} \frac{\mu[\bar{B}(x,r)]}{|\bar{B}(x,r)|},$$

$$\underline{D}\mu(x) = \underline{\lim_{r\to 0}} \frac{\mu[\bar{B}(x,r)]}{|\bar{B}(x,r)|}.$$

If $-\infty < \underline{D}\mu(x) = \bar{D}\mu(x) < \infty$, then the measure μ is said to be differentiable at the point x, and the number $D\mu(x) = \underline{D}\mu(x)$ is called the density of the measure at the point x.

Let us introduce two more conditions of differentiability of measure. They are stronger than the above one and successively strengthen each other.

Let $a \subset U, F \in \mathcal{L}_0(U)$. Let $r(a,F) = \sup_{x\in F}|a - x|$. The ball $B(a,r)$, where $r = r(a,F)$, contains F. Let us set

$$\alpha(F) = \frac{m(F)}{m[\bar{B}(a,r)]}.$$

The sequence of sets (F_ν), $\nu = 1,2,\ldots$, where $F_\nu \in \mathcal{L}_0(U)$ for all ν, is said to regularly contract to the point $a \in U$ if $r(a,F_\nu) \to 0$ for $\nu \to \infty$ and if there exists $\alpha_0 > 0$ such that $\alpha(F_\nu) \geqslant \alpha_0$ for all ν. Suppose that μ is an arbitrary measure in U. The measure μ is said to be R-differentiable at the point $a \in U$ if there exists a number $l \in \mathbf{R}$ such that for every sequence (F_ν) of the sets from $\mathcal{L}_0(U)$, the relation $\mu(F_\nu)/|F_\nu|$ tends to l for $\nu \to \infty$. This sequence (F_ν) is regularly contracting to the point a. The number l is called the R-density of the measure μ at the point a and is denoted by the symbol $D_R\mu(a)$.

Let μ again be a measure in U. Let us take an arbitrary point $a \in U$ and the number $h > 0$, and let P_h be the mapping of $X \to a + hX$. We have $P_h(0) = a$ and $P_h[\bar{B}(0,1)] = \bar{B}(\bar{a},h)$. For sufficiently small $h > 0$, namely for $h < \rho(a,\partial U)$, the set $P_h[\bar{B}(0,1)]$ is contained in U. Let us construct some measure μ_h by putting

$$\mu_h(E) = 1/h^n \mu[P_h(E)].$$

The measure μ_h is defined in the set $P_h^{-1}(U)$. If μ is the definite integral of the function $f \in L_{1,\mathrm{loc}}(U)$, then

$$\mu_h(E) = 1/h^n \int_{P_h(E)} f(x)\,dx.$$

By performing the change of the integration variable according to the formula $x = a + hX$, we obtain in this case

$$\mu_h(E) = \int_E f(a + hX)\,dX.$$

In particular, if $\mu = m$ is the Lebesgue measure in $\mathbf{R}^n$, then μ_h coincides with m for any $h > 0$.

The measure μ is said to be L-differentiable at the point a if there exists a number $l \in \mathbf{R}$ such that the value of the complete variation of the measure $\mu_h - lm$ on the ball $B(0,1)$ tends to zero for $h \to 0$. This number l is called the L-density of the measure μ at the point a and is denoted by the symbol $D_L\,\mu(a)$.

Lemma 4.1. *If a measure μ is L-differentiable at a point $a \in U$, then it is also R-differentiable at this point, besides*

$$D_R\mu(a) = D_L\mu(a).$$

Proof. Suppose that the measure μ is L-differentiable at the point a. Let (F_ν), $\nu = 1, 2, \ldots$, be an arbitrary sequence of sets from $\mathcal{L}_0(U)$ which is regularly contracting to the point a. Let $r_\nu = r(a, F)$, $B_\nu = B(a, r_\nu)$ and let $\alpha_0 > 0$ be such that $|F_\nu|/|B_\nu| \geqslant \alpha_0$ for all ν. Let $l = D_L\mu(a)$. Let $H_\nu = P_{r_\nu}^{-1}(F_\nu)$. For every ν, $H_\nu \in \mathcal{L}_0(\mathbf{R}^n)$ and $\bar{B}(0,1) \supset H_\nu$. We have

$$m(H_\nu)/m[\bar{B}(0,1)] = m(F_\nu)/m(B_\nu) \geqslant \alpha_0 > 0,$$

therefore,

$$m(H_\nu) \geqslant \beta = \alpha_0 m(\bar{B}(0,1)) = \text{const}.$$

Let $l = D_L\mu(a)$. Then

$$\left|\frac{\mu(F_\nu)}{|F_\nu|} - l\right| = \left|\frac{\mu_{r_\nu}(H_\nu)}{m(H_\nu)} - l\right| \leqslant \frac{1}{\beta}|\mu_{r_\nu}(H_\nu) - lm(H_\nu)|$$

$$\leqslant \frac{1}{\beta}|\mu_{r_\nu} - lm|(H_\nu) \leqslant \frac{1}{\beta}|\mu_{r_\nu} - lm|[\bar{B}(0,1)].$$

Since l is the L-density of the measure μ at the point a, then $|\mu_{r_\nu} - lm|(\bar{B})$ for $\nu \to \infty$. Hence we obtain that $\mu(F_\nu)/|F_\nu| \to l$ for $\nu \to \infty$. Q. E. D.

Let us also introduce some characteristics of behavior at a point of an arbitrary real function defined almost everywhere in the open set $U \subset \mathbf{R}^n$. Let us take an arbitrary point $a \in U$ and let $0 < h < \rho(a, \partial U)$. Then for almost all vectors $X \in B(0,1)$, the number $f_h(X) = f(a + hX)$ is defined. Thus we obtain a family of real functions (f_h) defined in the closed ball $B(0,1)$. The structure of the function f near the point x may be characterized if we study the behavior of the family (f_h) for $h > 0$. For instance, the condition that the number l is the limit of the function f at the point a is, as can easily be seen, equivalent to the following one: for $h \to 0$ the functions f_h uniformly converge to l on the set $B(0,1)\setminus\{0\}$. Let us give the general definition. Let $\mathcal{R}$ be a topological space whose elements are real functions defined almost everywhere in U, and such that for any $l \in \mathbf{R}$ the function identically equal to l belongs to $\mathcal{R}$. The number $l \in \mathbf{R}$ is said to be the limit of the function f at the point a in the sense of convergence in $\mathcal{R}$, and we write $l = \lim(\mathcal{R})f(x)$ if there exists $h_0 > 0$ such that for $0 < h < h_0$ the function $f_h \in \mathcal{R}$ and for $h \to 0$, f_h converges in the sense of the topology of the space $\mathcal{R}$ to the function identically equal to l. In the case $f(a) = \lim_{x\to a}(\mathcal{R})f(x)$, f is said to be continuous at the point a in the sense of convergence in $\mathcal{R}$.

Let us give some examples. Let M be a vector space of bounded real functions $F : \overline{B}(0,1) \to \mathbf{R}$ and let the topology of M be defined by the norm

$$\|F\| = \sup_{0<|x|\leq 1} |F(x)|.$$

The convergence in the topology of the space M is the usual uniform convergence in the ball $\bar{B}(0,1)$ with the punctured centre. Therefore, due to the above remark, the limit in the sense of convergence in M is the same as the usual limit of the function f at the point x.

Let $\mathcal{M}$ be the totality of all measurable functions in the ball $\bar{B}(0,1)$. For $F, G \in \mathcal{M}$, we put

$$\rho(F, G) = \int_{\bar{B}(0,1)} \frac{|F(x) - G(x)|}{1 + |F(x) - G(x)|}\, dx.$$

Thus, some metric is defined in $\mathcal{M}$. Convergence in the sense of this metric is nothing but convergence by measure. The notion of limit in the sense of convergence in $\mathcal{M}$ coincides with the well-known notion of approximative limit. (For the definition of approximative limit see [70]. We give the readers the proof that the definition of the approximative limit given in [70] is equivalent to that given here).

Suppose that f is a function of the class $L_{p,\mathrm{loc}}(U)$. Then for sufficiently small h, the function $f_h : X \to f(x + hX)$ belongs to the class $L_p(B(0,1))$. The limit of the function f at the point x in the sense of convergence in $L_p(B(0,1))$ is called the L_p-limit of f at the point x. If f is continuous at the point x in the sense of convergence in $L_p(B(0,1))$, then x is also said to be the Lebesgue L_p-point of the function f (in the case $p = 1$, it is said to be the Lebesgue point of f). The condition $l \in \mathbf{R}$ is the L_p-limit of the function f at the point x, due to the above definition, implies that

$$\lim_{h\to 0} \int_{B(0,1)} |f(x + hX) - l|^p\, dX = 0.$$

The condition: $l = \lim(L_1)f(x)$, where $f \in L_{1,loc}(U)$, as can easily be shown, is equivalent to the following one: l is the L-density of the measure μ_f of the indefinite integral of the function f.

4.2. The Vitali Covering Lemma

Let a set $A \in \mathbf{R}^n$ and a family $\mathcal{F}$ of closed balls in $\mathbf{R}^n$ be given. Then the family of balls $\mathcal{F}$ is said to cover the set A in the sense of Vitali if for every point $x \in A$ there exists a sequence $(B(x, r_\nu))$, $\nu = 1, 2\ldots,$ of closed balls with the centre x such that each of them belongs to $\mathcal{F}$, and $r_\nu \to 0$ for $\nu \to \infty$.

The following lemma is a special case of the classical Vitali–Lebesgue covering theorem.

Lemma 4.2. *Let A be a bounded set in $\mathbf{R}^n$ and let $\mathcal{F}$ be a family of closed balls covering A in the sense of Vitali. Then there exists a finite or infinite sequence (B_m), $m = 1, 2, \ldots,$ of pairwise nonintersecting balls such that for every m, $B_m \in \mathcal{F}$, and*

$$|A \setminus \bigcup_m B_m| = 0.$$

Proof. Let the set A and the family of balls $\mathcal{F}$ satisfy the conditions of the lemma. The set A is bounded, so there exists $R > 0$ such that $A \subset B(0, R)$. Without loss of generality, we may assume that the balls making up the family $\mathcal{F}$ are all contained in the ball $B(0, R)$. The desired sequence of balls is constructed by induction. A ball B_1 is chosen arbitrarily. Suppose that for some m, the balls $B_1, B_2, \ldots, B_m$ are given, here $B_i \cap B_j = \emptyset$ for $i \neq j$, and $B_j \in \mathcal{F}$ for every $j = 1, 2, \ldots, m$. Let $P_m = B_1 \cup B_2 \cup \cdots \cup B_m$. If $A \subset P_m$, then $A \setminus P_m = \emptyset$, therefore $|A \setminus P_m| = 0$; so, in this case, the required sequence of balls has already been constructed. Suppose that A is not contained in P_m. Let us put $A_m = A \setminus P_m$. The set P_m is closed, therefore, $B(0, R) \setminus P_m$ is an open set. Let $\mathcal{F}_m$ be the totality of all balls from $\mathcal{F}$ which are contained in $B(0, R) \setminus P_m$. It is easy to see that the family of balls $\mathcal{F}_m$ covers the set A_m in the sense of Vitali. Let us denote by λ_m the lowest upper bound of radii of the balls belonging to $\mathcal{F}_m$. It is obvious that $\lambda_m > 0$ and $\lambda_m \leqslant R$. For B_{m+1} we choose from the balls belonging to $\mathcal{F}_m$ the ball whose radius exceeds $\lambda_m/2$. It is clear that $B_{m+1} \cap B_j = \emptyset$ for all $j = 1, 2, \ldots, m$.

The above construction is either interrupted at some step or leads to an infinite sequence of balls (B_m), each of them being an element of $\mathcal{F}$. In the former case, the sequence of balls is obviously the desired one. Let us consider the latter case. Let r_m be the radius of the ball B_m, let B'_m be the ball concentric to B_m with the radius $3r_m$. The balls B_m do not pairwise intersect, each of them is contained in the ball B(0, R). Hence we conclude that

$$\sum_{m=1}^{\infty} |B_m| = \left| \bigcup_{m=1}^{\infty} B_m \right| < \infty.$$

For every m, $|B'_m| = 3^n |B_m|$, and so the series $\sum_{m=1}^{\infty} |B'_m|$ converges. Let us put

$$Q = A \setminus \bigcup_{m=1}^{\infty} B_m.$$

It is necessary to prove that $|Q| = 0$. Suppose, on the contrary, that $|Q| > 0$. Let us take $m_0 \in \mathbf{N}$ such that

$$\sum_{k=m_0+1}^{\infty} |B_k'| < |Q|.$$

This inequality allows us to conclude that there exists a point $a \in Q$ such that $a \notin B_k'$ for all $k \geqslant m_0 + 1$. Obviously, $a \notin P_{m_0}$; therefore, there exists $r > 0$ such that the ball $\bar{B}(a,r)$ belongs to $\mathcal{F}_{m_0}$. We have $\bar{B}(a,r) \cap B_j = \emptyset$ for all $j = 1, 2, \ldots, m_0$. Let us show that there exist the values of m such that $\bar{B}(a,r) \cap B_m \neq \emptyset$. Indeed, suppose, on the contrary, that $\bar{B}(a,r) \cap B_m$ is empty for all m. Then $\bar{B}(a,r) \in \mathcal{F}_m$ for all m; therefore, $\lambda_m > r > 0$ for all m. Since $\sum_{m=1}^{\infty} |B_m| < \infty$, then $|B_m| \to 0$ for $m \to \infty$; consequently, $r_m \to 0$ for $m \to \infty$ as well. For every m, we have $r_{m+1} > \lambda_m/2 > r/2 > 0$. Thus we obtain the contradiction. Consequently, there exist numbers $m > m_0$ for which the ball $\bar{B}(a,r)$ contains the points of the ball B_m. Let m_1 be the smallest of such m. We have $m_1 > m_0$ and $\bar{B}(a,r) \cap B_m = \emptyset$ for $m < m_1$. Hence it follows that $\bar{B}(a,r) \in \mathcal{F}_{m_1-1}$, therefore $r \leqslant \lambda_{m_1} - 1$. Let b be a point of the ball $B(a,r)$ which belong to the ball B_{m_1}. We have $|a-b| < r < \lambda_{m_1} - 1$. The point a lies outside the ball B'_{m_1} whose radius equals $3r_{m_1}$, this ball is concentric to the ball B_{m_1}. Hence it follows that $|a-b| \geqslant 3r_{m_1} - r_{m_1} = 2r_{m_1}$. For every $m > 1$, $r_m > \lambda_{m-1}/2$, so $|a-b| > \lambda_{m_1-1}$. Since, on the other hand, $|a-b| \leqslant \lambda_{m_1-1}$, we obtain the contradiction.

Thus, the assumption that $|Q| > 0$ leads to contradiction. Therefore $|Q| = 0$. Q. E. D.

4.3. The L_p-Continuity Theorem for Functions of the Class $L_{p,\mathrm{loc}}$

Lemma 4.3. *Let μ be a nonnegative measure in $\mathbf{R}^n$. Suppose that at every point x of a bounded set $E \subset \mathbf{R}^n$, the inequality $\bar{D}\mu(x) > K$ holds, where $K > 0, K \in \mathbf{R}$. Then for any set $A \supset E, A \subset \mathcal{L}$, we have $\mu(A) \geqslant K|E|$.*

Proof. Let all conditions of the lemma be satisfied, and let $A \supset E, A \in \mathcal{L}_0(\mathbf{R}^n)$. Let us assign arbitrarily $\varepsilon > 0$. Due to the regularity property of the measure for this ε, there exists an open set $G \supset A$ such that $\mu(G) < \mu(A) + \varepsilon$. Let $\mathcal{F}$ be the totality of all closed balls B contained in G and such that $\mu(B) > K|B|$. Let us take an arbitrary point $x \in E$. Then there exists a sequence of balls $(B(x, r_m))$, $m = 1, 2, \ldots$, such that $r_m \to 0$ for $m \to \infty$ and

$$\lim_{m\to\infty} \frac{\mu[\bar{B}(x, r_m)]}{|\bar{B}(x, r_m)|} = \bar{D}\mu(x) > K.$$

Without loss of generality one may assume that $\bar{B}(x, r_m) \subset G$ and $\mu[B(x, r_m)]/|B(x, r_m)| > K$ for all m. This implies that the ball $B(x, r_m)$ for every m is the element of $\mathcal{F}$. Consequently, we obtain that the family of balls $\mathcal{F}$ covers the set E in the sense of Vitali. Due to Lemma 4.2, there exists a sequence (B_m) of pairwise nonintersecting balls from $\mathcal{F}$ such that $|E \setminus \cup B_m| = 0$. We have $|E| \leqslant \sum_m |B_m|$. For every m, $\mu(B_m) > K|B_m|$. Hence, we conclude that $\mu(A) + \varepsilon \geqslant \mu(G) \geqslant \sum_m \mu(B_m) > K \sum_m |B_m| \geqslant K|E|$, therefore $\mu(A) + \varepsilon > K|E|$. Since $\varepsilon > 0$ is arbitrary, we hence obtain that $\mu(A) \geqslant K|E|$. The lemma is proved.

Let $U \subset \mathbf{R}^n$ be an open set, $f \in L_{1,\mathrm{loc}}(U)$. Let $x \in U$ and $0 < r < \rho(x, \partial U)$. Then f is integrable on the closed ball $B(x, r)$. Let us put

$$m_r f(x) = \frac{1}{|\bar{B}(x, r)|} \int_{\bar{B}(x,r)} f(y)\, dy.$$

Lemma 4.4. *Let U be an open set in $\mathbf{R}^n$, $f \in L_{1,\mathrm{loc}}(U)$. Then for almost all $x \in U$, the limit $\lim_{r \to 0} m_r f(x)$ exists and equals $f(x)$.*

Proof. First let us consider the case where the set U is bounded and f is integrable on U. Let us put

$$\overline{m}f(x) = \overline{\lim_{r \to 0}}\, m_r f(x), \qquad \underline{m}f(x) = \underline{\lim_{r \to 0}}\, m_r f(x).$$

We have to prove that for almost all $x \in U$, $\bar{m}f(x) = \underline{m}f(x) = f(x)$.

Let T be the totality of all points $x \in U$ for which $\bar{m}f(x) > f(x)$. We prove that $|T| = 0$. First suppose that f is nonnegative in U. For $k, l \in \mathbf{N}$, let

$$U_{kl} = \left\{ x \in U \,\middle|\, f(x) \leqslant \frac{k}{l} \right\},$$

$$T_{kl} = \left\{ x \in U_{kl} \,\middle|\, \overline{m}f(x) > \frac{k+1}{l} \right\}.$$

Let us arbitrarily assign $\varepsilon > 0$ and construct an open set $G \supset T_{kl}$ such that $|G| < |T_{kl}| + \varepsilon$. Let $H = G \cap U_{kl}$. By applying Lemma 4.3 to the measure μ_f (the indefinite integral of the function f), due to the fact that $\bar{D}\mu_f(x) = \bar{m}f(x)$, we have

$$\int_H f(x)\, dx \geqslant \frac{k+1}{l} |T_{kl}|. \tag{4.1}$$

For all $x \in H$, $f(x) \leqslant k/l$, whence we obtain that

$$\int_H f(x)\, dx \leqslant \frac{k}{l} |H| \leqslant \frac{k}{l} |G| \leqslant \frac{k}{l} (|T_{kl}| + \varepsilon). \tag{4.2}$$

From (4.1) and (4.2), it follows that $|T_{kl}| < \varepsilon$, and since ε is arbitrary, $|T_{kl}| = 0$. It is obvious that $T = \bigcup_{k,l \in \mathbf{N}} T_{kl}$, and, consequently, $|T| = 0$.

Let f be an arbitrary function from $L_1(U)$. For $k \in \mathbf{N}$, we let $f_k(x) = (f(x)+k)^+ = \max\{f(x)+k, 0\}$ and $T_k = \{x \in U | \overline{m}f_k(x) > f_k(x)\}$. According to what was proved above, $|T_{kl}| = 0$ for every $k \in \mathbf{N}$. Let $x_0 \in T$, i.e., $\bar{m}f(x_0) > f(x_0)$. Let us find $k \in \mathbf{N}$ such that $f(x_0) + k > 0$. We obviously have $m_r f_k(x_0) \geqslant m_r f_k(x_0) + k$, and consequently $m f_k(x_0) \geqslant \bar{m} f(x_0) + k \geqslant f(x_0) + k$, i.e., $x_0 \in T_k$. Since $X_0 \in T$ is taken arbitrarily, we obtain that $T \subset \bigcup_{k=1}^{\infty} T_k$, therefore $|T| = 0$.

Thus, we have proved that $\bar{m}f(x) \leqslant f(x)$ almost everywhere in U. By replacing f by $-f$, we obtain that $\bar{m}(-f)(x) \leqslant -f(x)$ almost everywhere in U. We have $\bar{m}(-f)(x) = -\underline{m}f(x)$; consequently, $\underline{m}f(x) \geqslant f(x)$ almost everywhere in U. Since $\underline{m}f(x) \leqslant \bar{m}f(x)$ for all x, therefore, $\underline{m}f(x) = \bar{m}f(x) = f(x)$ almost everywhere in U; and so for the case $f \in L_1(U)$, the lemma is proved.

Let $f \in L_{1,\text{loc}}(U)$. Let us put for $m \in \mathbf{N}$,

$$U_m = \left\{ x \in U \mid \rho(x, \partial U) > \frac{1}{m} \,\&\, |x| < m \right\}.$$

For every m, the set U_m lies strictly inside U, therefore $f \in L_1(U_m)$. Let S_m be a set of all $x \in U_m$ such that $f(x)$ is not the limit of $m_r f(x)$ for $r \to 0$, $S = \bigcup S_m$. According to what was proved above, $|S_m| = 0$ for all m, therefore $|S| = 0$. For every $x \in U \setminus S$, it is obvious that $f(x) = \lim_{r \to 0} m_r f(x)$.

The lemma is proved.

Theorem 4.1. *Let U be an open set in $\mathbf{R}^n$ and let f be a function of the class $L_{p,\text{loc}}(U)$. Then for almost all $x \in U$, the function f is continuous at the point x in the sense of convergence in L_p.*

Proof. Let α be a rational number. The function $|f(x) - \alpha|^p$ is locally integrable in U; therefore, according to Lemma 4.4, for $h \to 0$,

$$\frac{1}{\sigma_n h^n} \int_{|t|<h} |f(x+t) - \alpha|^p \, dt \to |f(x) - \alpha|^p \tag{4.3}$$

for almost all $x \in U$. Here σ_n is the volume of the ball $B(0, 1)$. Let us perform the change of variable on the right-hand side of the integral, setting $t = hX$, where $|X| < 1$. As a result, we obtain

$$\frac{1}{\sigma_n} \int_{|X|<1} |f(x + hX) - \alpha|^p \, dX \to |f(x) - \alpha|^p \tag{4.4}$$

for $h \to 0$ for almost all $x \in U$. We denote by E_α the set of $x \in U$ for which relation (4.4) does not hold. It is obvious that $E_\alpha = 0$. Let E be the union of all sets E_α. Then $|E| = 0$.

Let $x \notin E$. Let us arbitrarily assign $\varepsilon > 0$ and denote by β a rational number such that $|f(x_0) - \beta| < \varepsilon/2$. In the inequality

$$\left\{\frac{1}{\sigma_n}\int\limits_{|X|<1}|f(x_0+hX)-f(x_0)|^p\,dX\right\}^{\frac{1}{p}}$$

$$\leqslant\left\{\frac{1}{\sigma_n}\int\limits_{|X|<1}|f(x_0+hX)-\beta|^p\,dX\right\}^{\frac{1}{p}}$$

$$+\left\{\frac{1}{\sigma_n}\int\limits_{|X|<1}|\beta-f(x_0)|^p\,dX\right\}^{\frac{1}{p}}$$

the first summand on the right-hand side tends to $|f(x_0)-\beta|<\varepsilon/2$ for $h\to 0$ due to the fact that $x_0\notin E$. The second summand is equal to $|f(x_0)-\beta|$. Thus, the right-hand side is smaller than ε for sufficiently small h. Therefore, we obtained that for $x_0\notin E$, the expression

$$\left\{\frac{1}{\sigma_n}\int\limits_{|X|<1}|f(x_0+hX)-f(x_0)|^p\,dX\right\}^{\frac{1}{p}}\to 0$$

for $h\to 0$.

This completes the proof of the theorem.

Corollary 1. *Let f be a function of the class $L_{p,\mathrm{loc}}(U)$ where U is an open set in $\mathbf{R}^n$. Then for almost all $X\in U$, the function f has at the point x the limit in the sense of convergence in L_p.*

Due to what was said in Subsection 4.1, for the case $p=1$, the statement of the theorem implies that for almost all $x\in U$, the number $f(x)$ is the L-density at the point x of the indefinite integral of the function f. By applying Lemma 4.1 we obtain the following proposition:

Corollary 2. *Let f be a function of the class $L_{p,\mathrm{loc}}(U)$ where U is an open set in $\mathbf{R}^n$. For almost all points $x\in U$, the following statement is valid. For any sequence of Borel sets (F_ν) which is regularly contracting to the point x, the magnitude $f(x)$ is equal to the limit*

$$\lim_{\nu\to\infty}\frac{1}{|F_\nu|}\int\limits_{F_\nu}f(t)\,dt.$$

Let A be an arbitrary set in $\mathbf{R}^n$. The symbol $|A|$ denotes the outer measure of the A, i.e., the greatest lower boundary of measure of open sets containing A. It is obvious that if $A\subset B$, then $|A|\leqslant|B|$. Suppose that $|A|$ is finite. According to the definition, there exists a sequence of open sets (U_m), $m=1,2,\ldots$, such that for every m, $A\subset U_m$ and $|U_m|\leqslant|A|+\frac{1}{m}$. The set $H=\bigcap_{m=1}^{\infty}U_m$ is measurable, $|H|=|A|$ and $A\subset H$, and we obtain that if $|A|<\infty$, then there exists a measurable set $H\supset A$ such that $|H|=|A|$.

The point x of the set $A \subset \mathbf{R}^n$ is called the density point of the set A if

$$\lim_{r \to 0} \frac{|A \cap B(x,r)|}{|B(x,r)|} = 1.$$

Corollary 3. *For every set $A \subset \mathbf{R}^n$, the totality of all points $x \in A$ which are not density points of A is a set of zero measure.*

Proof. First let us suppose that the set A is bounded. Then $|A| < \infty$. Let us construct a measurable set $H \supset A$ such that $|H| = |A|$. For every point $x \in A$, $|H \cap B(x,r)| = |A \cap B(x,r)|$. Indeed, suppose that for some $x \in A$ and $r > 0$, this is not so. Then $|A \cap B(x,r)| < |H \cap B(x,r)|$. Let E' be a measurable set such that $E' \supset A \cap B(x,r)$ and $|E'| = |A \cap B(x,r)|$. Let us put $E = E' \cap B(x,r)$. E is also measurable, $A \cap B(x,r) \subset E \subset E'$, whence it follows that $|E| = |E'| = |A \cap B(x,r)|$. The measurable set $H' = E \cup (H \cap B(x,r))$ contains A, and its measure equals $|E| + |H \backslash B(x,r) < |H \cap B(x,r)| + |H \backslash B(x,r)| = |H|$. This contradicts the definition of H, and consequently, $|A \cap B(x,r)| = |H \cap B(x,r)|$.

The indicator-function χ_H of the set H is an integrable function; therefore, due to Corollary 2,

$$\frac{1}{|B(x,r)|} \int_{B(x,r)} \chi_H(y)\, dy = \frac{|H \cap B(x,r)|}{|B(x,r)|}$$

for $r \longrightarrow 0$ tends to the limit equal to $\chi_H(x)$ for almost all x. Let E_0 be a set of $x \in \mathbf{R}^n$ such that this is not so. We have: $|E_0| = 0$. For every point $x \in A \backslash E_0$, we have:

$$\frac{|A \cap B(x,r)|}{|B(x,r)|} = \frac{|H \cap B(x,r)|}{|B(x,r)|} \longrightarrow 1$$

for $r \longrightarrow 0$; that is, x is the density point of the set A. Thus, we have constructed the set E_0 such that $|E_0| = 0$ and every point $x \in A \backslash E_0$ is the density point of the set. This completes the proof of the corollary for the case where $|A| < \infty$.

In the above arguments, A was supposed to be a bounded set. Suppose that this is not the case and $A_m = A \cap B(0,m)$ where $m \in \mathbf{N}$. For every point $x \in A_m$, we have $|A_m \cap B(x,r)| \leqslant |A \cap B(x,r)|$ whence it follows that if x is the density point of the set A_m, then x is the density point of the set A as well. According to what was proved above, for each m there exists a set $E_m \subset A_m$ such that $|E_m| = 0$, and every point $x \in A_m \backslash E_m$ is the density point of A_m. Let $E = \cup_{m=1}^{\infty} E_m$. Then $|E| = 0$. Let $x \in A \backslash E$. Let us find natural m such that $|x| < m$. Then $x \in A_m$ and $x \notin E_m$. Thus, x is the density point of the set A_m, and therefore, also of the set A.

The corollary is proved.

4.4. The Differentiability Theorem for the Measure in $\mathbf{R}^n$

Lemma 4.5. *Let U be an open set in $\mathbf{R}^n$, and let λ be a real measure in U, which is singular with respect to the Lebesgue measure m. Then the measure $|\lambda|$ (the variation of λ) is also singular with respect to m, and for almost all $x \in U$, the L-density of the measure λ does exist and equals zero.*

Proof. Let the measure λ in the set U be singular with respect to the Lebesgue measure m. Let us arbitrarily take $E \in \mathcal{L}_0(U)$. Then there exists $K \subset E$ such that $|K| = 0$, and for every $E' \subset E$ for which $E' \cap K = \emptyset$, we have $\lambda(E') = 0$. Let $E' \subset E$ and $E' \cap K = \emptyset$. Then for any set $H \subset E'$, $\lambda(H) = 0$, whence it follows that $\lambda_+(E) = \sup_{H \subset E'} \lambda(H) = 0$, and therefore, $\lambda_-(E') = \lambda_+(E') - \lambda(E') = 0$. Consequently, $|\lambda|(E') = 0$ as well. Thus, the singularity of the measure $|\lambda|$ is established.

Let λ be a measure singular with respect to the Lebesgue measure. The measure $|\lambda|$ is singular according to what was proved above. The measurable space $(U, \mathcal{L}_0(U))$ is σ-finite; therefore there exists $K \subset U$ such that $|K| = 0$, and for every $E \subset U$ the intersection $K \cap E \in \mathcal{L}_0(U)$, and $|\lambda|(E \backslash K) = 0$. We put

$$E = \{x \in U \,|\, \overline{D}|\lambda|(x) > 0\}; \qquad E_m = \{x \in U \,|\, \overline{D}|\lambda|(x) > 1/m\}.$$

Suppose that for some $m \in \mathbf{N}$, $|E_m| > 0$. Then there exists an open V lying strictly inside U, such that $|V \cap E_m| > 0$. Let $V \cap E_m = A$, $B = A \backslash K$, $G = V \backslash K$. Then $|B| = |A|$. According to Lemma 4.3, we have $|\lambda|(G) \geqslant \frac{1}{m}|B| > 0$. On the other hand, $|\lambda|(G) = 0$. Thus, we obtained the contradiction, and so $|E_m| = 0$. Since $E = \cup_{m=1}^{\infty} E_m$, $|E| = 0$ as well.

For almost all $x \in U$, we obtained $\overline{D}|\lambda|(x) = 0$. Let the point $x \in U$ be such that $\overline{D}|\lambda|(x) = 0$. Let us consider the measure $\lambda_h(E) = (1/h^n)\,\lambda(P_h(E))$, where P_h is the mapping $X \to x + hX$. It is easy to see that the variation of the measure λ_h is the measure $|\lambda|_h : E \to \frac{1}{h^n}|\lambda|[P_h(E)]$. We have

$$|\lambda|_h(\overline{B}(0, 1)) = \frac{1}{h^n}|\lambda|(\overline{B}(x, h)) = \frac{\sigma_n |\lambda|[\overline{B}(x, h)]}{m[\overline{B}(x, h)]}, \tag{4.5}$$

where σ_n is the volume of the ball $B(0,1)$. Since $\overline{D}\,|\lambda|\,(x) = 0$, then the right-hand side of equality (4.5) tends to zero for $h \to 0$; therefore, $|\lambda|_h(B(0,1)) \to 0$ for $h \to 0$. This implies that $D_L\lambda(x) = 0$.

This completes the proof of the lemma.

Theorem 4.2. *Every measure μ in an open set U is L-differentiable almost everywhere in U. In this case, if the Lebesgue decomposition of the measure has the form*

$$\mu(E) = \int_E f(x)\,dx + \delta(E),$$

where $f \in L_{1,\mathrm{loc}}(U), \delta$ is the singular measure with respect to the Lebesgue measure, then

$$D_L\mu(x) = f(x)$$

for almost all $x \in U$.

Proof. According to Theorem 4.2 and Lemma 4.5, the measures $\mu_f : E \to \int_E f(x)\,dx$ and δ are differentiable almost everywhere in U, and $D_L\mu_f(x) = f(x)$, $D_L\delta(x) = 0$ almost everywhere in U. Hence it obviously follows that the measure μ is L-differentiable almost everywhere in U, and

$$D\mu(x) = D_L\mu_f(x) + D_L\delta(x) = f(x)$$

almost everywhere in U.

This proves the theorem.

Corollary. *Every measure μ in an open set $U \in \mathbf{R}^n$ is R-differentiable almost everywhere in U.*

To prove the corollary it suffices to refer to Lemma 4.1.

§5 Generalized Functions

5.1. Definition and Examples of Generalized Functions

The notion of a generalized function was introduced by Sobolev. Further elaboration of the theory of generalized functions was given by L. Schwartz, I. M. Gel'fand and others.

Let us assign an open set $U \subset \mathbf{R}^n$. Let $C_0^\infty(U)$ be a set of infinitely differentiable test functions whose supports are contained in U. A sequence φ_ν, $\nu = 1, 2, \ldots$, converges to 0 in $C_0^\infty(U)$ if the functions and all their derivatives uniformly tend to zero for $\nu \to \infty$, and if there exists a compact set K such that for every ν, $S(\varphi_\nu) \subset K$.

A sequence of functions φ_ν, $\nu = 1, 2, \ldots$, is said to converge in $C_0^\infty(U)$ to the function $\varphi \in C_0^\infty(U)$ if the difference $\varphi_\nu - \varphi$ converges to zero in $C_0^\infty(U)$. If φ_ν converges to φ in $C_0^\infty(U)$, then we write $\varphi_\nu \to \varphi$ in $C_0^\infty(U)$.

The mapping $f : C_0^\infty(U) \to \mathbf{R}^m$, where $m \geqslant 1$ is called a generalized function (or distribution) in the set U with values in $\mathbf{R}^m$ if f is linear, i. e., for any $\varphi_1, \varphi_2 \in C_0^\infty(U)$ and $\alpha_1, \alpha_2 \in \mathbf{R}$, the equality $f(\alpha_1\varphi_1 + \alpha_2\varphi_2) = \alpha_1 f(\varphi_1) + \alpha_2 f(\varphi_2)$ holds and f is continuous in the following sense: for every sequence (φ_ν), $\nu = 1, 2, \ldots$, of functions from $C_0^\infty(U)$ which converges to zero in the space $C_0^\infty(U)$, the magnitude $f(\varphi_\nu)$ tends to zero for $\nu \to \infty$.

If f is a generalized function in U and $\varphi \in C_0^\infty(U)$, then the magnitude $f(\varphi)$ is also denoted by the symbol $< f, \varphi >$. Following [27], we use the expression

$$\int_U f(x)\varphi(x)dx$$

to denote the magnitude $f(\varphi)$.

Generalized functions with values in $\mathbf{R}$ are called real. The totality of all generalized functions in the open set U is denoted by the symbol $D(U, \mathbf{R}^m)$. In the case $m = 1$, we simply write $D(U)$ instead of $D(U, \mathbf{R}^m)$.

Let us give some examples.

1. Let f be a function of the class $L_{1,\mathrm{loc}}(U)$. For every function $\varphi \in C_0^\infty(U)$, the number

$$\tilde{f}(\varphi) = \int_U f(x)\varphi(x)\,dx \tag{5.1}$$

is defined. The functional $\tilde{f} : C_0^\infty(U) \to \mathbf{R}$ thus defined is obviously linear and if $\varphi_\nu \to 0$ in $C_0^\infty(U)$, then $\tilde{f}(\varphi) \to 0$. Thus, $\tilde{f}$ is a generalized function in U. If for the function $f, g \in L_{1,\mathrm{loc}}(U)$, the generalized functions $\tilde{f}$ and $\tilde{g}$ coincide, i.e.,

$$\int_U \varphi(x) f(x)\,dx = \int_U \varphi(x) g(x)\,dx$$

for any function $\varphi \in C_0^\infty(U)$, then $f(x) = g(x)$ almost everywhere. Thus, the linear functional $\tilde{f}$ determines the function f uniquely to within the values on a set of zero measure. Below, every function $f \in L_{1,\mathrm{loc}}(U)$ is identified with the generalized function $\tilde{f}$, which is determined by f according to (5.1).

2. Let U be an open set in $\mathbf{R}^n$, and let $\mu : \mathcal{L}_0(U) \to \mathbf{R}$ be an arbitrary measure in U. For every function $\varphi \in C_0^\infty(U)$, the number

$$\langle \mu, \varphi \rangle = \int_U \varphi(x)\,\mu(dx) \tag{5.2}$$

is defined. The correspondence $\varphi \to \langle \mu, \varphi \rangle$, as can easily be seen, is some generalized function with values in $\mathbf{R}$. Let us note that if the measures μ_1 and μ_2 are such that $\langle \mu_1, \varphi \rangle = \langle \mu_2, \varphi \rangle$ for any function $\varphi \in C_0^\infty(U)$, then $\langle \mu_1, \varphi \rangle = \langle \mu_2, \varphi \rangle$ for any function $\varphi \in C_0(U)$. This fact can easily be verified if we approximate an arbitrary function $\varphi \in C_0(U)$ by the functions of the class $C_0^\infty(U)$. Hence, it follows that if $\langle \mu_1, \varphi \rangle = \langle \mu_2, \varphi \rangle$ for any function $\varphi \in C_0(U)$, then the measures μ_1 and μ_2 coincide. Below, the generalized function $\varphi \to \langle \mu, \varphi \rangle$ is identified with the measure μ. The following criterion of the generalized function being a measure is true.

Lemma 5.1. *The function $f \in D(U, \mathbf{R}^k)$ coincides with some measure iff for every open G lying strictly inside U, there exists a constant $M(G) < \infty$ such that for any test function whose support is contained in G, the inequality*

$$|\langle f, \varphi \rangle| \leqslant M(G)\,\|\varphi\|_{C(U)}$$

is valid.

Proof. The necessity of the lemma's condition is obvious. Let us prove its sufficiency. Let $f \in D(U, \mathbf{R}^k)$ satisfy the condition of the lemma. Let us show that in this case there exists a continuous linear functional $f^* : C_0(U) \to \mathbf{R}$ such that $\langle f^*, \varphi \rangle = \langle f, \varphi \rangle$ for $\varphi \in C_0(U)$. Indeed, let us take an arbitrary function $\varphi \in C_0(U)$. Let G be a h_0-neighbourhood of the support of the function φ where $h_0 < \rho(S(\varphi), \partial U)$. Then G lies strictly inside U. Let us assign an arbitrary averaging kernel K, and let $\varphi_h = K_h * \varphi$ where $0 < h < h_0$. Then $S(\varphi_h) \subset G$ and for $h > 0$, $\varphi_h \to \varphi$ uniformly in G. For any h_1, h_2, where $0 < h_1 < h_0$, $0 < h_2 < h_0$, we have

$$|\langle f, \varphi_{h_1} \rangle - \langle f, \varphi_{h_2} \rangle| \leqslant M(G) \|\varphi_{h_1} - \varphi_{h_2}\|_{C(U)}.$$

Hence the existence of the finite limit $\lim_{h \to 0} \langle f, \varphi_h \rangle$ follows. We assume that $\langle f^*, \varphi \rangle = lim_{h \to 0} \langle f, \varphi \rangle$. It is easy to see that the functional f^* defined by the above method is linear. If $\varphi \in C_0^\infty(U)$, then for $h \to 0$, the functions φ_h converge to the function $\varphi \in C_0(U)$. Hence it follows that for $\varphi \in C_0(U)$, $\langle f^*, \varphi \rangle = \lim_{h \to 0} \langle f, \varphi_h \rangle = \langle f, \varphi \rangle$. For every h, we have $|\langle f, \varphi_h \rangle \leqslant M(G) \|\varphi\|_{C(U)}$, whence it follows that $\langle f^*, \varphi \rangle| \leqslant M(G) \|\varphi\|_{C(U)}$ for all $\varphi \in C_0(U)$, and therefore f^* is a continuous linear functional on $C_0(U)$. From this, it follows that there exists a measure $\mu : \mathcal{L}_0(U) \to \mathbf{R}$ such that for any $\varphi \in C_0(U)$, $\langle f^*, \varphi \rangle = \int_U \varphi(x) d\mu(x)$. In particular, we obtain that for any $\varphi \in C_0^\infty(U)$,

$$\langle f, \varphi \rangle = \langle f^*, \varphi \rangle = \int_U \varphi(x) \mu(dx).$$

The lemma is proved.

3. Let $U = \mathbf{R}^n$. For an arbitrary function $\varphi \in C_0^\infty(\mathbf{R}^n)$, we put

$$\langle \delta, \varphi \rangle = \varphi(0).$$

This defines the mapping $\delta : C_0^\infty(\mathbf{R}^n) \to \mathbf{R}$, which is obviously a generalized function. Denote by μ_0 a measure in $\mathbf{R}^n$ defined as follows: $\mu_0(E) = 1$ if $0 \in E$, and $\mu_0(E) = 0$ if $0 \notin E$. For every $\varphi \in C_0^\infty$, the equality

$$\langle \delta, \varphi \rangle = \int_{R^n} \varphi(x) \mu_0(dx).$$

is valid. The generalized function δ is called the Dirac δ-function. As we see, it coincides with the measure μ_0.

The generalized function δ first appeared in the studies of theoretical physics. On the basis of the notion of the generalized function, Sobolev managed to justify the seemingly risky manipulations with the δ-function performed by the physicists.

The investigation of generalized functions with values in $\mathbf{R}^k$ where $k > 1$, in many cases may be reduced to the investigation of real generalized functions due to the following simple remark. Let $f \in D(U, \mathbf{R}^k)$. For every $\varphi \in C_0^\infty(U)$, the vector $f(\varphi)$ has the components $f_1(\varphi), f_2(\varphi), \ldots, f_k(\varphi)$, thus in C_0^∞ the functionals f_j, $j = 1, 2, \ldots, k$, are defined, each of them being a generalized function. The functionals $f_1, f_2, \dots, f_k$ are said to be the components of the generalized function f. On the contrary, if the real generalized functions $f_1, f_2, \dots, f_k$ are given, then, by putting for $\varphi \in C_0^\infty(U)$,

$$\langle f, \varphi\rangle = (\langle f_1, \varphi\rangle, \dots, \langle f_k, \varphi\rangle),$$

we obtain a generalized function with the values in $\mathbf{R}^k$.

5.2. Operations with Generalized Functions

The following assumption will come in handy.

Lemma 5.2. *Let $(\mathcal{U}_\tau)_{\tau\in T}$ be an arbitrary family of open sets, $\mathcal{U} = \cup_{\tau\in T}\mathcal{U}_\tau$. Every function $\varphi \in C_0^\infty(U)$ can be represented in the form*

$$\varphi = \varphi_1 + \varphi_2 + \cdots + \varphi_k$$

where each of the functions φ_m, $m = 1, 2, \dots, k$, belongs to the class $C_0^\infty(U)$, with the support of φ_m being contained in the set U_{τ_m} for some $\tau_m \in T$, for any $m = 1, 2, \dots, k$.

Proof. Let (λ_m), $m = 1, 2, \dots$, be the partition of unity in U corresponding to the open covering $(\mathcal{U}_\tau)_{\tau\in T}$ of the set U. The set $S(\varphi)$ is compact; therefore, only a finite number of functions λ_m take values that are distinct from zero on the set $S(\varphi)$. Let $k \in \mathbf{N}$ be such that if $m > k$, then $\lambda_m(x) = 0$ for all $x \in S(\varphi)$. We have $\sum_{m=1}^k \lambda_m(x) = 1$ for all $x \in S(\varphi)$. Let us set $\varphi_m = \lambda_m\varphi$. Then $\varphi = \sum_{m=1}^k \varphi_m$ for every $\varphi_m \in C_0^\infty(U)$ and $S(\varphi_m) \subset S(\lambda_m)$. Due to the definition of the partition of unity from every m, there exists $\tau_m \subset T$ such that $S(\lambda_m) \subset U_{\tau_m}$.The constructed representation $\varphi = \sum_{m=1}^k \varphi_m$ is obviously the desired one. The lemma is proved.

5.2.1. Differentiation of Generalized Functions

First, let us present a proposition concerning the functions of the class $C^r, r \geqslant 1$. For every function $f : U \to \mathbf{R}$ of the class C^r and for any function $\varphi \in C_0^\infty(U)$, the equality

$$\int_U D^\alpha f(x)\,\varphi(x)\,dx = (-1)^{|\alpha|} \int_U f(x)\,D^\alpha\varphi(x)\,dx, \tag{5.3}$$

holds. In the case where U is an open cube, the equality (5.3) may easily be obtained via integration by parts. The general case is reduced to this by means of Lemma 5.2. Namely, as the family (U_τ) let us take the totality of all open cubes contained in U. The union of all these cubes coincides with U, and due to Lemma 5.2, every function $\varphi \in C_0^\infty(U)$ may be represented in the form $\varphi = \varphi_1 + \varphi_2 + \cdots + \varphi_k$, where the functions $\varphi_m \in C_0^\infty(U)$ are such that $S(\varphi_m)$ is contained in some cube $Q_m \subset U$. For every m, we have:

$$\int_{Q_m} D^\alpha f(x)\varphi_m(x)\,dx = (-1)^{|\alpha|} \int_{Q_m} f(x) D^\alpha \varphi_m(x)\,dx.$$

Since outside of Q_m, φ_m turns into zero together with its derivatives, the integrals in the latter equality do not change if we take the entire set U as the integration domain. Summing up by m, we obtain (5.3).

Formula (5.3) is a kind of a model by which the derivative of a generalized function is defined.

Let $f \in D(U)$. If $\varphi \in C_0^\infty(U)$, then $D^\alpha\varphi \in C_0^\infty(U)$, too, for any multiindex α; therefore, the number $(-1)^{|\alpha|}\langle f, D^\alpha\varphi\rangle$ is defined. The mapping $\varphi \to (-1)^{|\alpha|}\langle f, D^\alpha\varphi\rangle$ of the space $C_0^\infty(U)$ in $\mathbf{R}$ is denoted by the symbol $D^\alpha f$. It is easy to verify that the mapping $D^\alpha f$ is linear; also, if $\varphi_\nu \to 0$ in $C_0^\infty(U)$, then $\langle D^\alpha f, \varphi_\nu\rangle = (-1)^{|\alpha|}\langle f, D^\alpha\varphi_\nu\rangle \to 0$ for $\nu \to \infty$. Therefore, $D^\alpha f$ is a generalized function in $\mathbf{R}$.

Let D^α and D^β be two arbitrary differentiation operators.Then for every generalized function f, we have (whatever the function $\varphi \in C_0^\infty(U)$ might be):

$$\begin{aligned}\langle D^\alpha(D^\beta f), \varphi\rangle &= (-1)^{|\alpha|}\langle D^\beta f, D^\alpha\varphi\rangle \\ &= (-1)^{|\alpha|+|\beta|}\langle f, D^{\alpha+\beta}\varphi\rangle = \langle D^{\alpha+\beta} f, \varphi\rangle.\end{aligned}$$

Therefore, $D^\alpha(D^\beta f) = D^{\alpha+\beta} f$. Let us give some examples.

5.2.1.1. Let $n = 1$, $U = \mathbf{R}$, $\theta(x) = 1$ for $x \geqslant 0$, $\theta(x) = 0$ for $x < 0$. For an arbitrary function $\varphi \in C_0^\infty(U)$, we have:

$$\int_{-\infty}^{\infty} \theta(x)\varphi'(x)\,dx = \int_0^\infty \varphi'(x)\,dx = -\varphi(0).$$

So, for every function $\varphi \in C_0^\infty$,

$$\langle \frac{d\theta}{dx}, \varphi\rangle = -\varphi(0) = -\langle \delta, \varphi\rangle,$$

that is, $\frac{d\theta}{dx} = \delta$.

5.2.1.2. This example may be considered as a multidimensional generalization of Example 1.1. Let G be a domain in $\mathbf{R}^n$ whose boundary is a smooth $(n-1)$-dimensional manifold F. Let us find first-order generalized derivatives of the function χ_G, which is the indicator of the set G. For the case $n = 1$ and $G = (0, \infty)$, we have solved this problem already. For $x \in F$, let $\nu(x)$ be a unit vector of the inner (with respect to G) normal of the surface of F at a point x. The symbol $\sigma(E)$ stands for the area (the $(n-1)$-dimensional one) of a Borel set $E \subset F$. On the basis of the Ostrogradsky formula for an arbitrary function $\varphi \in C_0^\infty(\mathbf{R}^n)$ we have:

$$\int_G \chi_G(x)\frac{d\varphi}{dx_i}(x)\,dx = -\int_F \nu_i(x)\varphi(x)\,d\sigma(x).$$

Hence, we conclude that derivative $\frac{d\chi_G}{dx_i}$ is a measure μ_i in $\mathbf{R}^n$ such that for every Borel set $E \subset \mathbf{R}^n$,

$$\mu_i(E) = \int_{E\cap F} \nu_i(x)\, d\sigma(x).$$

Let us introduce the generalized vector-function,

$$\operatorname{grad} \chi_G = \left(\frac{\partial \chi_G}{\partial x_1}, \frac{\partial \chi_G}{\partial x_2}, \dots, \frac{\partial \chi_G}{\partial x_n}\right)$$

(the generalized gradient of the function χ_G). Due to the above, $\operatorname{grad} \chi_G$ is a measure in $\mathbf{R}^n$, such that for every $E \subset \mathbf{R}^n$,

$$\operatorname{grad} \chi_G(E) = \int_{E\cap F} \nu(x)\, d\sigma(x).$$

5.2.1.3. Let there be an interval $U = (a,b) \subset \mathbf{R}$ and a function $f : (a,b) \to \bar{\mathbf{R}}$ integrable by any closed interval $[\alpha,\beta] \subset (a,b)$. The function f defines some generalized function which we also denote by f. For $\varphi \in C_0^\infty(U)$,

$$\langle f, \varphi\rangle = \int_a^b f(x)\varphi(x)\, dx.$$

Let us define some function $F : (a,b) \to \mathbf{R}$. Fix an arbitrary point $x_0 \in (a,b)$ and put $F(x_0) = 0$, and

$$F(x) = \int_{x_0}^{x} f(t)\, dt \quad \text{for} \quad x > x_0,$$

$$F(x) = -\int_{x_0}^{x} f(t)\, dt \quad \text{for} \quad x < x_0.$$

For any $x_1, x_2 \in (a,b), x_1 < x_2$, due to the known properties of an integral, we have:

$$F(x_2) - F(x_1) = \int_{x_1}^{x_2} f(t)\, dt.$$

The function $F : (a,b) \to \mathbf{R}$ is continuous. Let us prove that following the definition of the derivative of a generalized function, the equality $\frac{dF}{dx} = f$ is valid. In other words, it is necessary to prove that for every function $\varphi \in C_0^\infty(U)$, the equality

$$\int_a^b F(x)\varphi'(x)\, dx = -\int_a^b f(x)\varphi(x)\, dx \tag{5.4}$$

holds. Let $\varphi \in C_0^\infty(U)$. The support φ is contained in some closed segment $[\alpha, \beta] \subset U$. We have

$$\int_\alpha^\beta \varphi'(x)\,dx = \varphi(\beta) - \varphi(\alpha). \tag{5.5}$$

Since for $x \notin [\alpha, \beta]$, $\varphi(x) = 0$ and $\varphi'(x) = 0$, then due to (5.5),

$$\begin{aligned}
\int_a^b F(x)\varphi'(x)\,dx &= \int_\alpha^\beta F(x)\varphi'(x)\,dx \\
&= \int_\alpha^\beta [F(x) - F(\alpha)]\varphi'(x)\,dx \\
&= \int_\alpha^\beta \left(\int_\alpha^x f(t)\,dt\right)\varphi'(x)\,dx.
\end{aligned}$$

Let us put $\theta(u) = 1$ for $u \geqslant 0$, $\theta(u) = 0$ for $u < 0$. Then

$$\int_\alpha^\beta \left(\int_\alpha^x f(t)\,dt\right)\varphi'(x)\,dx = \int_\alpha^\beta \left(\int_\alpha^\beta \theta(x-t)f(t)\varphi'(x)\,dt\right)dx.$$

The function of two variables $\theta(x-t)f(t)\varphi'(x)$ in the rectangle $[\alpha, \beta] \times [\alpha, \beta] \subset \mathbf{R}^2$ is measurable. We have $|\theta(x-t)f(t)\varphi'(x)| \leqslant |f(t)||\varphi'(x)|$. Application of the Fubini theorem yields that the product $|f(t)||\varphi'(x)|$ is integrable on this rectangle, therefore, the function $\theta(x-t)f(t)\varphi'(x)$ is also integrable. Due to the Fubini theorem, we have:

$$\begin{aligned}
\int_\alpha^\beta \left(\int_\alpha^\beta \theta(x-t)f(t)\,dt\right)\varphi'(x)\,dx &= \int_\alpha^\beta \left(\int_\alpha^\beta \theta(x-t)\varphi'(x)\,dx\right)f(t)\,dt \\
&= \int_\alpha^\beta \left(\int_t^\beta \varphi'(x)\,dx\right)f(t)\,dt \\
&= -\int_\alpha^\beta \varphi(t)f(t)\,dt
\end{aligned}$$

whence (5.4) obviously follows.

5.2.2. The Operation of Shifting a Generalized Function

Let $a \in \mathbf{R}^n$, τ_a be the mapping $x \to x + a$ (parallel transfer to the vector a). Let us put $V = \tau_a^{-1}(U)$. For every locally integrable in U function f, we have:

$$\int_V f(x+a)\,\varphi(x)\,dx = \int_U f(x)\varphi(x-a)\,dx \tag{5.6}$$

for any function $\varphi \in C_0^\infty(V)$. We set $(\tau_a^* f)(x) = f(x+a)$. Correspondingly, the equality (5.6) may be written as follows: $\langle \tau_a^* f, \varphi \rangle = \langle f, \tau_{-a}^* \varphi \rangle$.

This relation is a model to define $\tau_a f$ for an arbitrary generalized function f. Namely, if f is a generalized function on the set $V = \tau_a^{-1}(U)$, then the symbol $\tau_a^* f = f \circ \tau_a$ denotes the linear functional on $C_0^\infty(U)$ defined by the relation $\langle \tau_a^* f, \varphi\rangle = \langle f, \tau_{-a}^* \varphi\rangle$ for every $\varphi \in C_0^\infty(U)$. Obviously, $\tau_a^* \varphi$ satisfies the continuity condition from the definition of a generalized function.

5.2.3. *Multiplication by the Function of the Class $C^\infty(U)$*

If f is a generalized function in an open set U, and λ is the function of the class $C^\infty(U)$, then the product of f by λ is a generalized function defined by the condition

$$\langle \lambda f, \varphi\rangle = \langle f, \lambda\varphi\rangle$$

for every function $\varphi \in C_0^\infty(U)$.

5.3. Support of a Generalized Function. The Order of Singularity of a Generalized Function

Let us arbitrarily fix an open set $\Omega \subset \mathbf{R}^n$. Let f be a generalized function in Ω. The function f is said to vanish on the open set $U \subset \Omega$ if for every function $\varphi \in C_0^\infty(\Omega)$, such that $S(\varphi) \subset U$ we have: $\langle f, \varphi\rangle = 0$. Let U_0 be the union of all open sets $U \subset \Omega$, on each of which f turns into zero. The set U_0 is open. Let us prove that f vanishes on U_0. Indeed, let φ be an arbitrary function of the class $C_0^\infty(\Omega)$ such that its support $S(f) = A \subset U_0$. The sets $U \subset \Omega$, on which f vanishes, form an open covering of U_0. Due to Lemma 5.2, φ admits the representation $\varphi = \sum_{m=1}^k \varphi_m$ in which the functions $\varphi_m \subset C_0^\infty(U)$ are such that for every m $S(\varphi_m) \subset U_m$, where $U_m \subset \Omega$ is such that f vanishes on U_m. We have $\langle f, \varphi\rangle = \sum_{m=1}^k \langle f, \varphi_m\rangle$, that is $\langle f, \varphi\rangle = 0$. Since $\varphi \in C_0^\infty(U_0)$ was taken arbitrarily, this proves that f vanishes on U_0.

Thus, constructed is an open set U_0 such that f vanishes on it, and any other open subset Ω with the same properties is contained in U_0. We assume $\Omega \backslash U_0 = S(f)$. The set $S(f)$ is closed with respect to Ω. It is called the support of the generalized function f. The generalized function f is said to be compactly supported if the set $S(f)$ is compact.

Let f be a compactly supported generalized function in an open set Ω. By definition, f is a linear functional on the set $C_0^\infty(\Omega)$. This functional admits extension onto the set $C^\infty(\Omega)$. Indeed, let $A = S(f)$. Let us construct a function $\lambda \in C_0^\infty(\Omega)$ which equals 1 in the neighbourhood of A. Such a function λ may be obtained as follows. Let $r > 0$ be such that $\rho(x, \partial\Omega) > r$ for all $x \in A$. Let us put $r = 3h$ and let $H = \bar{U}_h(A)$, $V = U_{2h}(A)$. The set H is closed, V is open, the closure of V is compact and is contained in Ω. The function $\lambda \in C^\infty(\mathbf{R}^n)$ such that $\lambda(x) = 1$ for $x \in H$, $\lambda(x) = 0$ for $x \notin V$, and $0 \leqslant \lambda(x) \leqslant 1$ for all x is the desired one. The existence of the function λ which has all these properties follows from Theorem 2.6. Let $\varphi \in C^\infty(\Omega)$. Then the function $\lambda\varphi \in C_0^\infty(\Omega)$, and therefore the magnitude

$\langle f, \lambda\varphi\rangle$ is defined. It is independent of the choice of λ. Indeed, let λ_1 and λ_2 be two functions of the class $C_0^\infty(\Omega)$, each of which equals 1 in the neighbourhood of the set A. Let $\varphi \in C_0^\infty(\Omega)$. The function $(\lambda_1 - \lambda_2)\varphi$ belongs to $C_0^\infty(\Omega)$. It vanishes in the neighbourhood of the set $A = S(f)$, and therefore its support is contained in $\Omega \backslash A$. Hence it follows that $\langle f, (\lambda_1 - \lambda_2)\varphi\rangle = 0$, and therefore, $\langle f, \lambda_1\varphi\rangle = \langle f, \lambda_2\varphi\rangle$. Q. E. D.

Every compactly supported generalized function f in the set Ω is uniquely extended onto $C^\infty(\mathbf{R}^n)$ if for $\varphi \in C^\infty(\mathbf{R}^n)$, we put $\langle f, \varphi\rangle = \langle f, \varphi\,|\Omega\rangle$ ($\varphi|\Omega$ is the restriction upon φ on the set Ω).

Let φ_m, $m = 1, 2, \ldots$, be an arbitrary sequence of functions from $C^\infty(\Omega)$. We say that φ_m tends to zero in $C^\infty(\Omega)$ if for $m \to \infty$, the functions φ_m and all their derivatives tend to zero in the set Ω, with the convergence being uniform on any compact set $A \subset \Omega$.

Let f be a compactly supported generalized function in Ω. Then for every sequence (φ_m), $m = 1, 2, \ldots$, converging to zero in $C^\infty(\Omega)\,\langle f, \varphi_m\rangle \to 0$ for $m \to \infty$. Indeed, let us assign an arbitrary function $\lambda \in C_0^\infty(\Omega)$ which is equal to 1 in the neighbourhood of the set $S(f)$, and let φ_m be a sequence of functions from $C^\infty(\Omega)$ converging to zero in $C^\infty(\Omega)$. Then for $m \to \infty$, $\lambda\varphi_m \to 0$ in $C^\infty(\Omega)$ as well. The supports of the functions $\lambda\varphi_m$ are contained in the compact set $S(\lambda) \subset \Omega$. Thus, $\lambda\varphi_m \to 0$ in $C_0^\infty(\Omega)$. Hence it follows that $\langle \lambda f, \varphi_m\rangle\langle f, \lambda\varphi_m\rangle$ for $m \to \infty$. Q. E. D.

The statement proved above admits inversion; namely, if f is a linear functional on $C^\infty(U)$ such that for every sequence of functions (φ_m), $m = 1, 2, \ldots$, converging to zero in $C^\infty(U)$ for $m \to \infty\,\langle f, \varphi_m\rangle \to 0$, then f is a compactly supported generalized function.

For an arbitrary function $\varphi \in C_0^\infty(\Omega)$ we put:

$$\|\varphi\|_m = \max_x\{\max_{|\alpha| \leqslant m} |D^\alpha\varphi(x)|\}. \tag{5.7}$$

Let f be a generalized function in an open set $\Omega \subset \mathbf{R}^n$ and let $U \subset \Omega$ be an open set. Then f is said to be a generalized function of finite order on the set U if there exist an integer $m \geqslant 0$ and a constant $K < \infty$ such that for every function $\varphi \in C_0^\infty(\Omega)$ for which $S(\varphi) \subset U$, the inequality

$$|\langle f, \varphi\rangle| \leqslant K\,\|\varphi\|_m \tag{5.8}$$

holds. The smallest value of m for which there exists a number $K < \infty$ such that inequality (5.8) holds for any function $\varphi \in C_0^\infty(\Omega)$ with a support contained in U is called the order of singularity of the generalized function f on the set U.

An arbitrary, generalized function may have no finite order. Let us define, for instance, a generalized function f in $\mathbf{R}^n$, setting for an arbitrary test function

$$\langle f, \varphi\rangle = \sum_{k=0}^{\infty} \frac{\partial^k\varphi}{\partial x_1^k}(kl_1)$$

where $e_1 = (1, 0, \dots, 0)$. It is easy to see that the given generalized function f has no finite order in $\mathbf{R}^n$. Nevertheless, the following theorem is valid.

Theorem 5.1. *Every generalized function f in a set Ω has finite singularity order on any open set U lying strictly inside Ω.*

Proof. Let f be a generalized function in Ω, and let U be an open set lying strictly inside Ω. By definition, this implies that $\overline{U}$ is compact and $\Omega \supset \overline{U}$. Suppose, contrary to fact, we are going to prove that f has no finite order in U. This implies that no matter what $m \in \mathbf{N}$ we take, there exists no constant $K < \infty$ such that inequality (5.8) is valid for the given m for all functions $\varphi \in C_0^\infty(\Omega)$ for which $S(\varphi) \subset U$. Consequently, for every m there exists a function $\varphi_m \in C_0^\infty(\Omega)$ such that $S(\varphi) \subset U$ and

$$|\langle f, \varphi_m \rangle| > m \|\varphi_m\|_m.$$

We put

$$\psi_m = \frac{\varphi_m}{m\|\varphi_m\|_m}.$$

Then $\|\psi_m\|_m = \frac{1}{m}, |\langle f, \psi_m \rangle| > 1$ for all m. For $m \to \infty$ the sequence of functions ψ_m tends to zero in $C_0^\infty(\Omega)$. So, $\langle f, \psi_m \rangle \to 0$ for $m \to \infty$, and consequently, we obtain the contradiction. Thus, the theorem is proved.

Corollary. *Every compactly supported generalized function f on the set Ω has finite order in Ω.*

Proof. Suppose that the generalized function f is compactly supported in Ω, and let $A = S(f)$. The set A is compact and $\Omega \supset A$. Let $\delta > 0$ be such that $U_{3\delta}(A) \subset \Omega$. Let us set $U_{2\delta}(A) = G$, and let $H = \bar{U}_\delta(A)$. It is obvious that $G \supset H$. Let ψ be a function of the class C^∞ such that $\psi(x) = 1$ for $x \in H$, $\psi(x) = 0$ for $x \notin G$. For an arbitrary function $\psi \subset C_0^\infty(\Omega)$, we have:

$$\langle f, \varphi \rangle = \langle f, (1 - \psi)\varphi \rangle + \langle f, \psi\varphi \rangle.$$

Since the function $(1 - \psi)\varphi$ vanishes on the set $U_\delta(A)$, then its support is contained in the set $\Omega \backslash A$, and consequently, $\langle f, (1 - \psi)\varphi \rangle = 0$, that is, we obtain for any function that $\varphi \in C_0^\infty(\Omega) \langle f, \varphi \rangle = \langle f, \psi\varphi \rangle$. The support of the function $\varphi\psi$ is contained in the compact set $\bar{U}_{2,\delta} \subset \Omega$. Due to the theorem, there exist constants $K < \infty$ and $m \in \mathbf{N}$ such that

$$|\langle f, \psi\varphi \rangle| \leqslant K \|\psi\varphi\|_m.$$

It can be easily verified that $\|\psi\varphi\|_m \leqslant C \|\varphi\|_m$, where $C = \text{const} < \infty$, and we see that for any function $\varphi \in C_0(U)$,

$$|\langle f, \varphi \rangle| \leqslant K_1 \|\varphi\|_m,$$

where $K_1 = \text{const} < \infty$. The corollary is proved.

5.4. The Generalized Function as a Derivative of the Usual Function. Averaging Operation

Theorem 5.2. *For every compactly supported generalized function f in $\mathbf{R}^n$, there exists a function F of the class $L_{2,\mathrm{loc}}(\mathbf{R}^n)$ such that $f = D^\nu F$ for some multiindex ν.*

Proof. Let f be a compactly supported generalized function in $\mathbf{R}^n$, and let $m \geqslant 0$ be its singularity order. Let us find $r > 0$ such that $S(f) \subset B(0,r)$, and let λ be a function of the class $C_0^\infty(B(0,r))$ which equals 1 in the neighbourhood of the compact set $S(f)$. For the integer $k \geqslant 0$, we put $B_k = B(0, r+k)$. In particular, $B_0 = B(0,r)$.

Let $l \geqslant 1$ be an integer. Then we put $\delta_l(x) = (-1)^l \frac{x^{l-1}}{(l-1)!}$ for $x > 0$, and $\delta_l(x) = 0$ for $x \leqslant 0$. Let $\varphi \in C_0^\infty(\mathbf{R})$. We set

$$I_l\varphi(x) = \int_{-\infty}^{\infty} \delta_l(t-x)\varphi(t)\,dt = -\int_x^\infty \frac{(x-t)^{l-1}}{(l-1)!}\varphi(t)\,dt. \tag{5.9}$$

It is obvious that $I^l\varphi$ is a function of the class C^∞, with the equality

$$I^l\left(\frac{d^l\varphi}{dx^l}\right) \equiv \varphi \tag{5.10}$$

being valid. Actually, let a point $x_0 \in \mathbf{R}$ be such that the support φ is contained in the interval $(-\infty, x_0)$. Then at the point x_0, the function φ and all its derivatives vanish. Applying the Taylor formula with the residual term in the integral form, we obtain for every x,

$$\varphi(x) = \int_{x_0}^{x} \frac{(x-t)^{l-1}}{(l-1)!}\varphi^{(l)}(t)\,dt = -\int_x^{x_0} \frac{(x-t)^{l-1}}{(l-1)!}\varphi(t)\,dt.$$

The right-hand integral does not change if we take the point ∞ as the upper limit. Hence, due to (5.9), (5.10) follows.

Let $\alpha = (\alpha_1, \alpha_2, \ldots, \alpha_n)$ be an n-dimensional multiindex such that all its components are distinct from zero. For $x = (x_1, x_2, \ldots, x_n) \in \mathbf{R}^n$, we put $\delta_\alpha(x) = \delta_{\alpha_1}(x_1)\delta_{\alpha_2}(x_2) \times \ldots \delta_{\alpha_n}(x_n)$. For an arbitrary function $\varphi \in C_0^\infty(\mathbf{R}^n)$, let

$$I_\alpha\varphi(x) = \int_{\mathbf{R}^n} \delta_\alpha(t-x)\varphi(t)\,dt = \int_{\mathbf{R}^n} \delta_\alpha(t)\varphi(t+x)\,dt.$$

The function $I_\alpha\varphi$ belongs to the class $C^\infty(\mathbf{R}^n)$. In general, $I_\alpha\varphi$ is not necessarily a compactly supported function. If the sequence (φ_ν), $\nu = 1, 2, \ldots$, converges to zero in $C_0^\infty(\mathbf{R}^n)$, them the sequence of functions $(I_\alpha\varphi_\nu)$ converges to zero in $C^\infty(\mathbf{R}^n)$. This can be easily obtained from the second representation for $I_\alpha\varphi$ and from the classical theorems about integrals depending on the parameter. From equality (5.10), by applying the Fubini theorem, we obtain that the equality

$$I_\alpha(D^\alpha \varphi) = \varphi \tag{5.11}$$

holds for every function $\varphi \in C_0^\infty(\mathbf{R}^n)$. Let us also note that for every multiindex $\beta = (\beta_1, \beta_2, \ldots, \beta_n)$ such that $\beta_i < \alpha_i$ for every i, the equality

$$D^\beta(I_\alpha \varphi(x)) = \int_{\mathbf{R}^n} \delta_\alpha(t) D^\beta \varphi(t+x)\,dt = (-1)^\beta \int_{\mathbf{R}^n} D^\beta \delta_\alpha(t)\varphi(t+x)\,dt$$

is valid. Simple calculations show that $D^\beta \delta_\alpha = (-1)^{|\beta|}\delta_{\alpha-\beta}$, whence we conclude that

$$D^\beta(I_\alpha \varphi)(x) = I_{\alpha-\beta}\varphi. \tag{5.12}$$

Let $\alpha = (\alpha_1, \alpha_2, \ldots, \alpha_n)$ be an n-dimensional multiindex such that $\alpha_i > 0$ for all $i = 1, 2, \ldots, n$. Denote by $I_\alpha f$ a linear functional on $C_0^\infty(\mathbf{R}^n)$ defined by the condition

$$\langle I_\alpha f, \varphi\rangle = \langle f, I_\alpha \varphi\rangle$$

for every $\varphi \in C_0^\infty(\mathbf{R}^n)$. The right-hand side of this equality makes sense since the generalized function f is compactly supported. If $\varphi_k \to 0$ in $C_0^\infty(\mathbf{R}^n)$, then $I_\alpha \varphi_k \to 0$ in $C^\infty(\mathbf{R}^n)$. Due to the fact that f is compactly supported, it follows that $\langle f, I_\alpha \varphi_k\rangle \to 0$ for $k \to \infty$. By definition $\langle f, I_\alpha \varphi_k\rangle = \langle I_\alpha f, \varphi_k\rangle$, and therefore, $\langle I_\alpha f, \varphi_k\rangle \to 0$ for $k \to \infty$; that is, the functional $I_\alpha f$ is continuous. Note also its following property. We have

$$\begin{aligned}\langle D^\alpha(I_\alpha f), \varphi\rangle &= (-1)^{|\alpha|}\langle I_\alpha f, D^\alpha \varphi\rangle \\ &= (-1)^{|\alpha|}\langle f, I_\alpha D^\alpha \varphi\rangle = (-1)^{|\alpha|}\langle f, \varphi\rangle.\end{aligned}$$

due to equality (5.11). Hence we obtain

$$D^\alpha(I_\alpha f) = f. \tag{5.13}$$

Let us denote by ν an n-dimensional multiindex all components of which are equal to $m+1$, and let $F = I_\nu f$. We show that the generalized function F is a function of the class $L_{2,\mathrm{loc}}(\mathbf{R}^n)$. The set of functions $C_0^\infty(B_k)$ is everywhere dense in $L_2(B_k)$. Let us show that for every function $\varphi \in C_0^\infty(B_k)$, the estimate

$$|\langle F, \varphi\rangle| \leqslant M_k \|\varphi\|_{L_2(B_k)} \tag{5.14}$$

holds, where $M_k = \text{const} < \infty$. Indeed, we have

$$|\langle F, \varphi\rangle| = |\langle f, \lambda I_\nu \varphi\rangle| \leqslant C\|\lambda I_\nu \varphi\|_m. \tag{5.15}$$

The function λ has compact support and belongs to the class $C^\infty(\mathbf{R}^n)$. Hence it follows that all its derivatives are bounded. They vanish for $x \notin B_0$. Due to this, when defining the quantities $\|\lambda I_\nu \varphi\|_m$, it suffices to consider the values of the function $\lambda I_\nu\varphi$ and of its derivatives at the points $x \in B_0$ only. For every α such that $|\alpha| \leqslant m$, we obviously have

$$|D^\alpha \lambda I_\nu \varphi(x)| \leqslant C \max_{|\beta|\leqslant m} |D^\beta(I_\nu\varphi(x))|. \tag{5.16}$$

For every x, due to (5.12), we have

$$D^\beta I_\nu\varphi(x) = (-1)^{|\beta|} I_{\nu-\beta}\varphi(x) = \int_{\mathbf{R}^n} \delta_{\nu-\beta}(t-x)\varphi(t)\,dt.$$

If $t \in B_k$ and $x \in B_0$, then $|\delta_{\nu-\beta}(t-x)| \leqslant M_k' < \infty$, where $M_k' < \infty$ only depends on k, we obtain that for $x \in B_0$,

$$|D^\beta I_\nu\varphi(x)| \leqslant M_k' \int_{B_0} |\varphi(t)|\,dt \leqslant M_k' |B_0|^{1/2} \left(\int_{B_0} |\varphi|^2\,dt\right)^{1/2},$$

that is, for $x \in B_0$,

$$|D^\beta I_\alpha\varphi(x)| \leqslant M_k'' \|\varphi\|_{L_2(B_k)}, \tag{5.17}$$

where $M_k = \text{const} < \infty$. From (5.16) and (5.17), it follows that for all x,

$$\|D^\alpha(\lambda I_\alpha\varphi)\|_{C(\mathbf{R}^n)} \leqslant f M_k'' \|\varphi\|_{L_2(B_k)}$$

for every α with $|\alpha| \leqslant m$; consequently, $\lambda I_\nu\varphi(x)\|_m < M_k''\|\varphi\|_{L_2(B_k)}$. Due to (5.15), (5.14) follows. Since $C_0^\infty(B_k)$ is everywhere dense in $L_2(B_k)$, then due to (5.14), the linear functional F is uniquely extended onto $L_2(B_k)$. Due to the Riesz theorem for the general form of the linear functional in the space L_2, it follows that there exists a functions $\widetilde{F}_k \in L_2(B_k)$ such that for every function $\varphi \in C_0^\infty(B_k)$,

$$\langle F, \varphi\rangle = \int_{B_k} \widetilde{F}_k(x)\varphi(x)\,dx$$

for any $k = 1, 2, \ldots$. If $\varphi \in C_0^\infty(B_k)$, then φ belongs to the class $C_0^\infty(B_{k+1})$ as well, whence it follows that

$$\langle F, \varphi\rangle = \int_{B_k} \widetilde{F}_k(x)\varphi(x)\,dx = \int_{B_{k+1}} \widetilde{F}_{k+1}(x)\varphi(x)\,dx,$$

that is, for every function $\varphi \in C_0(B_k)$,

$$\int_{B_k} \widetilde{F}_k(x)\varphi(x)\,dx = \int_{B_{k+1}} \widetilde{F}_{k+1}(x)\varphi(x)\,dx = \int_{B_k} \widetilde{F}_{k+1}(x)\varphi(x)\,dx.$$

(We used here the fact that $\varphi(x) = 0$ for $x \notin B_k$.) Since $C_0^\infty(B_k)$ is everywhere dense in $L_2(B_k)$, it follows that $\widetilde{F}_{k+1}(x) = \widetilde{F}_k(x)$ almost everywhere on the ball B_k. Let us suppose that $\widetilde{F}_{k+1}(x) = \widetilde{F}_k(x)$ for all $x \in B_k$. This can obviously be obtained if we change the values of each of the functions F_k on a set of zero measure. Let $\widetilde{F}$ now be a function in $\mathbf{R}^n$ defined by the following condition: if $x \in B_k$, then $\widetilde{F}(x) = \widetilde{F}_k(x)$. This condition uniquely defines the function F. It is obvious that $F \in L_{2,\text{loc}}(\mathbf{R}^n)$, and for every $\varphi \in C_0^\infty(\mathbf{R}^n)$,

$$\langle F, \varphi \rangle = \int_{\mathbf{R}^n} \widetilde{F}(x)\varphi(x)\, dx.$$

This implies that, as a generalized function, $\widetilde{F}$ coincides with F. Since $D^\nu F = f$, this completes the proof of the theorem.

Let us define the averaging operation of a generalized function. Let U be an open set in $\mathbf{R}^n$. For every $h > 0$, the set $\hat{U}_h = \{x \in U | \rho(x, \partial U) > h\}$ is defined (see Subsection 2.1). Below, we assume $0 < h < \rho(U)$, where $\rho(U)$ is the inner radius of U. Let us assign an arbitrary averaging kernel K, and let $\widetilde{K}(x) = K(-x)$. Let f be a generalized function in U. For every $x \in \hat{U}_h$, the function $K_h : t \to \frac{1}{h^n} K(\frac{t-x}{h})$ belongs to the class $C_0^\infty(U)$; therefore the number

$$f_h(x) = \langle f, K_h \rangle = \int_U f(t) \frac{1}{h^n} K\left(\frac{t-x}{h}\right) dt$$

is defined. Thus, in U_h some functions f_h is defined which we also denote by the symbol $K_h * f$, and we call it the averaging of the generalized function f by means of the kernel K.

Theorem 5.3. *For every generalized function f in an open set U of $\mathbf{R}^n$ for $0 < h < \rho(U)$ for every averaging kernel K, the function $f_h = K_h * f$ belongs to the class $C^\infty(\hat{U}_h)$. For any function $\varphi \in C_0^\infty(\hat{U}_h)$, the equality*

$$\langle f_h, \varphi \rangle = \langle f, \widetilde{K}_h * \varphi \rangle \tag{5.18}$$

is valid.

Proof. First let us put $U = \mathbf{R}^n$, and let f be a compactly supported generalized function in $\mathbf{R}^n$. In this case, $\hat{U}_h = \mathbf{R}^n$. Due to Theorem 5.2, there exists a function $F \in L_{2,\text{loc}}(\mathbf{R}^n)$ such that f is the derivative of the function F considered as a generalized function, $f = D^\nu F$ for some ν. We have

$$f_h(x) = \langle f, K_h \rangle = \langle D^\nu F, K_h \rangle = (-1)^\nu \langle F, D^\nu K_h \rangle.$$

For every x, $K_h(t) = \frac{1}{h^n} K(\frac{t-x}{h})$,and therefore, $D^\nu K_h(t) = \frac{1}{h^n} D_t^\nu (K(\frac{t-x}{h}))$ (the index t in the notation D_t^ν shows that differentiation is held by the variable t). Since F is the function of the class $L_{2,\text{loc}}(\mathbf{R}^n)$, the magnitude $\langle F, D^\nu K_h \rangle$ is equal to the usual Lebesgue integral

$$\int_{\mathbf{R}^n} F(t) \frac{1}{h^n} D_t^\nu \left(K\left(\frac{t-x}{h}\right)\right) dt.$$

Thus,

$$f_h(x) = (-1)^{|\nu|} \int_{\mathbf{R}^n} F(t) \frac{1}{h^n} D_t^\nu \left(K \left(\frac{t-x}{h} \right) \right) dt. \tag{5.19}$$

The second factor in the integrand is the function of the class C^∞. Due to the classical theorems for the differentiation of integrals depending on the parameter, it follows that f_h is the function of the class C^∞.

Let $\varphi \in C_0^\infty(\mathbf{R}^n)$. By multiplying both parts of equality (5.19) by $\varphi(x)$ and by using the Fubini theorem, we obtain

$$\langle f_h, \varphi \rangle = (-1)^{|\nu|} \int_{\mathbf{R}^n} F(t) \left(\int_{\mathbf{R}^n} \frac{1}{h^n} D_t^\nu K \left(\frac{t-x}{h} \right) \varphi(x)\, dx \right) dt.$$

Due to the theorems known for integrals depending on the parameter, we have:

$$\int_{\mathbf{R}^n} \frac{1}{h^n} D_t^\nu \left(K \left(\frac{t-x}{h} \right) \right) \varphi(x)\, dx = D^\nu \left(\int_{\mathbf{R}^n} \frac{1}{h^n} K(\frac{t-x}{h}) \varphi(x)\, dx \right)$$

$$= D^\nu (\widetilde{K}_h * \varphi)(t).$$

Thus, $\langle f_h, \varphi \rangle = (-1)^\nu \langle F, D^\nu(\widetilde{K}_h * \varphi) \rangle = \langle D^\nu F, \widetilde{K}_h * \varphi \rangle = \langle f, \widetilde{K}_h * \varphi \rangle$ and thus equality (5.18) has been established.

For the case where a generalized function f is compactly supported and $U = \mathbf{R}^n$, the theorem is proved. Let us consider the general case.

First, let us note that since the support of the function K is compact and is contained in the open ball $B(0,1)$, it is contained in some smaller ball $B(0,r)$, where $0 < r < 1$. Let us put $1 - r = 4\varepsilon$.

Let us take an arbitrary point $x_0 \in \hat{U}_h$. Let us put $V = B(x_0, \varepsilon h)$. Let $x \in V$ and $|t - x_0| \geqslant h(1-3\varepsilon)$. Then $|t-x| \geqslant |t-x_0| - |x-x_0| > h(1-3\varepsilon) - h\varepsilon = hr$, and therefore, $\frac{|t-x|}{h} > r$. According to the definition of r,it follows that for such x and t, the equality $K(\frac{t-x}{h}) = 0$ is valid. This allows us to conclude that for $x \in V$ the support of the function $K_h : t \to \frac{1}{h^n} K(\frac{t-x}{h})$ is contained in the closed ball $B(x_0, h(1-3\varepsilon))$. Let the function λ of the class $C^\infty(\mathbf{R}^n)$ be such that $\lambda(x) = 1$ for $x \in \bar{B}(x_0, h(1-2\varepsilon))$, $\lambda(x) = 0$ for $x \notin B(x_0, h(1-\varepsilon))$, and $0 \leqslant \lambda(x) \leqslant 1$ for all x. The existence of such a function λ follows from Theorem 2.6. Let us put $f_1 = \lambda f$, $f_2 = (1-\lambda) f$. For all $x \in \hat{U}_h$, we have

$$(K_h * f)(x) = (K_h * f_1)(x) + (K_h * f_2)(x).$$

For $x \in V$, the support of the function $K_h : t \to \frac{1}{h^n} K(\frac{t-x}{h})$ is contained in the closed ball $B(x_0, h(1-3\varepsilon))$. The function $(1-\lambda)$ and the generalized function $f_2 = (1-\lambda) f$ vanish in the open ball $B(x_0, h(1-2\varepsilon)) \supset B(x_0, h(1-3\varepsilon))$. Hence it follows that $(K_h * f_2)(x) = 0$ for $x \in V$, that is $(K_h * f)(x) = (K_h * f_1)(x)$ in the ball $V = B(x_0, h\varepsilon)$. The generalized function f_1 is compactly supported, and therefore, it is extended onto $C^\infty(\mathbf{R}^n)$. According to what was proved above, $K_h * f_1 \in C^\infty(\mathbf{R}^n)$. Thus, we obtain that in some neighbourhood V of the point x_0 the function $K_h * f$ belongs to the class $C^\infty(V)$, and since $x_0 \in \hat{U}_h$ was taken arbitrarily, then from what was proved above, it follows that $K_h * f$ is the function of the class C^∞ in the set $\hat{U}_h$. The first statement of the theorem is proved.

In order to prove the second statement, let us again take an arbitrary point $x_0 \in \hat{U}_h$ and its neighbourhood $V = B(x_0, \varepsilon h)$, where ε is defined as above. For $x \in V$, we have $(K_h * f)(x) = (K_h * f_1)(x)$. (Here f_1 and f_2 have the same meaning as before). Let the function $\varphi \in C_0^\infty(\hat{U}_h)$ be such that $S(\varphi) \subset V$. Then $\langle K_h * f, \varphi\rangle = \langle K_h * f_1, \varphi\rangle = \langle f_1, \tilde{K}_h * \varphi\rangle$, because for compactly supported generalized functions equality, (5.18) is proved. If $x \in V$ and $|t - x_0| \geqslant (1-3\varepsilon)h$, then $K(\frac{t-x}{h}) = 0$, whence it follows that the support of the function

$$\langle \tilde{K}_h * \varphi\rangle(t) = \int_{\mathbf{R}^n} \frac{1}{h^n} K(\frac{t-x}{h})\varphi(x)\,dx$$

is contained in the closed ball $B(x_0(1-3\varepsilon)h)$. The generalized function f_2 vanishes in some neighbourhood of this ball; therefore, $\langle f_2, \tilde{K}_h * \varphi\rangle = 0$. At the same time, $\langle K_h * f_2\rangle(x) = 0$ for $x \in V$, whence it follows that $\langle K_h * f_2, \varphi\rangle = 0$. Thus, we obtain that

$$\langle K_h * f_1, \varphi\rangle = \langle f_1, \tilde{K}_h * \varphi\rangle \qquad \langle K_h * f_2, \varphi\rangle = \langle f_2, \tilde{K}_h * \varphi\rangle.$$

If we sum up these equalities term by term, we obtain that $\langle K_h * f, \varphi\rangle = \langle f, K_h * \varphi\rangle$.

Thus, equality (5.18) is proved for the case where the support of φ is sufficiently small, namely, $S(\varphi) \subset V = B(x_0, \varepsilon h)$, where $x_0 \in \hat{U}_h$. The general case is easily reduced to this. Let $\varphi \in C_0^\infty(\hat{U}_h)$. Then, according to Lemma 5.2, φ admits the representation $\varphi = \varphi_1 + \varphi_2 + \cdots + \varphi_k$ where each of the functions $\varphi_1, \varphi_2, \ldots, \varphi_k$ belongs to the class $C_0^\infty(U)$, and the support of φ_m being contained in the ball $B(x_m, \varepsilon h)$. So, according to what was proved above, $\langle K_h * f, \varphi_m\rangle = \langle f, \tilde{K}_h * \varphi_m\rangle$ for every $m = 1, 2, \ldots, k$. Summing up by m, we then obtain (5.18). This completes the proof of the theorem.

Corollary. *Let f be a generalized function in U. Then the equality*

$$D^\alpha(K_h * f) = K_h * D^\alpha f$$

is valid.

Proof. For every function $\varphi \in C_0^\infty(\hat{U}_h)$, we have

$$\langle D^\alpha(K_h * f), \varphi\rangle = (-1)^{|\alpha|}\langle K_h * f, D^\alpha\varphi\rangle = (-1)^{|\alpha|}\langle f, \tilde{K}_h * D^\alpha\varphi\rangle.$$

We have $\tilde{K}_h * D^\alpha\varphi = D^\alpha(\tilde{K}_h * \varphi)$. Hence,

$$\begin{aligned}\langle D^\alpha(K_h * f), \varphi\rangle &= (-1)^{|\alpha|}\langle f, D^\alpha(\tilde{K}_h * \varphi)\rangle \\ &= \langle D^\alpha f, \tilde{K}_h * \varphi\rangle \\ &= \langle K_h * D^\alpha f, \varphi\rangle,\end{aligned}$$

and thus the corollary is proved.

CHAPTER 2

FUNCTIONS WITH GENERALIZED DERIVATIVES

In this chapter, the classes of functions $W_p^l(U)$ introduced by Sobolev are studied. The investigation of these classes is based on certain integral representations of a function by its derivatives of order l. Deducing such representations, we use the properties of linear differential operators satisfying the complete integrability condition. In Section 1, integral representations of the Sobolev type are constructed for starlike domains. Section 2 studies integral representations of the same type for a wider class of domains. They are used to study the classes $W_p^l(U)$. First, the representations of local character are constructed. The final form of the representations we are interested in is obtained in Section 4. At the same time, we establish in Section 2 some properties of the basic class of domains—the class J, for which the properties of the functional spaces of $W_p^l(U)$ are investigated. In Section 3, estimates are established for integrals of the potential type, which are later used in Section 4 to establish the relations between the classes $W_p^l(U)$ and the classes $L_q(U)$ and $C(U)$. Section 4 studies the classes $W_p^l(U)$. In Section 5 we give the general theorem about differentiability almost everywhere of functions of the classes $W_p^l(U)$ and some of its corollaries. Besides, some theorems about the behaviour of functions of the classes W_p^l on almost all k-dimensional planes of the space $\mathbf{R}^n$ for $l \leqslant k \leqslant n$ are proved.

§1 Sobolev-Type Integral Representations

1.1. Preliminary Remarks

When studying properties of functions with generalized derivatives, one essentially uses formulae expressing the value of a function in terms of some integrals. These integrals contain derivatives of fixed order, or more generally, values of some differential operators of a given function. The importance of such integral representations was first shown by Sobolev, who, in particular, has constructed integral representations of a function in terms of its derivatives of a given order r.

Let us first introduce an auxiliary notion. Let U by an open set in $\mathbf{R}^n$. Suppose that a function $F(x, y)$ of variables $x \in U, y \in U$ is assigned which is defined for any $x, y \in U$ such that $x \neq y$. Then $F(x, y)$ is said to be a function of the type $|x-y|^r$ if, for every x, $F(x, y)$ as a function of y belongs to the class C^∞ in the domain $U \backslash \{x\}$, with the estimates

$$|F(x, y)| \leqslant C|x-y|^r \tag{1.1}$$

being valid, and for every multiindex α,

$$|D_y^\alpha F(x, y)| \leqslant C_\alpha |x-y|^{r-|\alpha|}. \tag{1.2}$$

Differentiation on the left-hand side of (1.2) is held by the variable y, and the magnitudes C and C_α in (1.1) and (1.2) are finite constants independent of x.

The simplest integral representation of those we are interested in is the identity of the form:

$$f(x) = \int_U K_0(x, y) f(y)\, dy + \sum_{i=1}^n \int_U K_i(x, y) \frac{\partial f}{\partial y_i}(y)\, dy. \tag{1.3}$$

Here U is an open set in $\mathbf{R}^n$, and $K_0, K_1, \ldots, K_n$ are some functions of the variables $x \in U, y \in U$.

Equality (1.3) is supposed to be valid for every function $f : U \to \mathbf{R}^n$ of the class C^1. Certain constraints are usually imposed upon the functions $K_0, K_1, \ldots, K_n$. Equality (1.3) is said to be the Sobolev integral representation if the functions $K_0, K_1, \ldots, K_n$ satisfy the following conditions.

A) The function $K_0(x, y)$ belongs to the class C^∞ as a function of the variable $z = (x, y) \in U \times U$; and for any fixed $x \in U$, the function $y \mapsto K_0(x, y)$ has compact support contained in U.

B) For $i = 1, 2, \ldots, n$, the function $K_i(x, y)$ is continuous by the totality of the variables (x, y) on the set of all pairs $(x, y) \in U \times U$ such that $x \neq y$. For every $x \in U$, there exists a compact set $A_x \subset U$ such that $K_i(x, y) = 0$ for $y \notin A_x$ for every $i = 1, 2, \ldots, n$.

C) For every $i = 1, 2, \ldots, n$, $K_i(x, y)$ is the function of the type $|x-y|^{1-n}$.

Let us first note that for an arbitrary open set U, the integral representation (1.3) satisfying the Conditions A, B, and C may not exist. This follows, for instance, from the fact that some results which will be obtained by means of this representation for an arbitrary domain U are wrong, in general.

U is said to be a domain of the type S if it admits at least one integral representation of the Sobolev type of the form (1.3), where the functions $K_0, K_1, \ldots, K_n$ satisfy the above Conditions A, B, C.

It goes without saying that one may also study integral representations of the form (1.3), in which the functions K_i, $i = 0, 1, \ldots, n$ satisfy the conditions differing from those given here. Such integral representations are used in investigations in the theory of spaces of functions with generalized derivatives. They are beyond the scope of this book.

We may reformulate the problem of constructing an integral representation of the form (1.3) by using the notion of a generalized function. Let us do it as follows. We have to obtain the functions $K_0(x,y)$, $K_1(x,y), \dots, K_n(x,y)$ satisfying the equation

$$K_0(x,y) - \sum_{i=1}^{n} \frac{\partial}{\partial y_i} K_i(x,y) = \delta(y-x). \tag{1.4}$$

Here $\delta(x)$ denotes, as above, the Dirac function. Hence it is clear that, to a great extent, the solution of this problem is defined arbitrarily. Namely, let $K_0, K_1, \dots, K_n$ be a system of functions satisfying equation (1.4), and let $H_0(x,y), H_1(x,y), H_2(x,y), \dots, H_n(x,y)$ be functions such that

$$H_0(x,y) - \sum_{i=1}^{n} \frac{\partial H_i}{\partial y_i}(x,y) = 0. \tag{1.5}$$

The system of functions satisfying this condition is constructed easily—$H_1, H_2, \dots, H_n$ may be assigned arbitrarily, but they must have derivatives on the left-hand side of (1.5), and then the function H_0 may be obtained by them from equality (1.5). It is obvious that the functions $K_0' = K_0 + H_0$, $K_1' = K_1 + H_1, \dots, K_n' = K_n + H_n$ then also satisfy equation (1.4).

There are many ways of constructing integral representations of the form (1.3). The large arbitrariness in defining such a representation explains the variety of the ways to construct them. The requirement that the Conditions A, B, and C should be satisfied does not restrict the freedom of choice very significantly. Below, we show the way to construct an integral representation of the form (1.3) so that the Conditions A, B, and C hold for some rather wide class of domains in $\mathbf{R}^n$, namely the domains of the class J introduced by John [98].

If for an open set $U \subset \mathbf{R}^n$, the integral relation of the form (1.3) holds, then by using some formal transformations, one can obtain from it various integral representations. As an example, let us show how one can obtain, starting from (1.3), the integral representation of a function $f \in C^l(U)$ by means of its derivatives of the order l.

Let U be an open set of the class S in $\mathbf{R}^n$, let $K_0, K_1, \dots, K_n$ be functions satisfying the above Conditions A, B, and C and such that for every function $f \in C^1(U)$ equality (1.3) is valid. Let $f : U \to \mathbf{R}^k$ be a function of the class $C^l(U)$. Let us assign arbitrary points $x, \xi \in U$, and let us assume that

$$\varphi(x,\xi) = \sum_{|\alpha| \leqslant l-1} \frac{(x-\xi)^\alpha}{\alpha!} D^\alpha f(\xi). \tag{1.6}$$

For every ξ, $\varphi(x,\xi)$ as the function of the variable x is the Taylor polynomial of order $l-1$ of the function f at the point ξ. Let us apply the method which is classical for the case $n=1$; namely, for a fixed value of x, we shall differentiate $\varphi(x,\xi)$ by the variables ξ_i, which are components of the point ξ. In the case $n=1$, as is known (see any textbook on mathematical analysis), for such differentiation all the derivatives of the function f whose order is less that l are reduced, and the final result only contains the derivative of the order l. An analogous situation takes place in the general case. Denote by δ_i the n-dimensional multiindex $(\delta_{i1},\delta_{i2},\dots,\delta_{in})$, where δ_{ij} is the Kronecker symbol, i.e., $\delta_{ii}=1$, $\delta_{ij}=0$ for $i\neq j$. We have

$$\frac{\partial\varphi}{\partial\xi_i}(x,\xi)=\sum_{|\alpha|\leqslant l-1\ \alpha\geqslant\delta_i}\left[-\frac{(\delta x-\xi)^{\alpha-\delta_i}}{(\alpha-\delta_i)!}\right]D^\alpha f(\xi)+\sum_{|\alpha|\leqslant l-1}\frac{(x-\xi)^\alpha}{\alpha!}D^{\alpha+\delta_i}f(\xi). \tag{1.7}$$

Now let us note that the conditions: $|\alpha|\leqslant l-1,\alpha\geqslant\delta_i$, are equivalent to the following one: $\alpha=\beta+\delta_i$, where $|\beta|\leqslant l-2$. Due to this, the first sum on the right-hand side of (1.7) is equal to

$$-\sum_{|\beta|\leqslant l-2}\frac{(x-\xi)^\beta}{\beta!}D^{\beta+\delta_i}f(\xi).$$

By substituting this expression into (1.7) and by replacing β by α, we obtain that

$$\frac{\partial\varphi}{\partial\xi_i}(x,\xi)=\sum_{|\alpha|=l-1}\frac{(x-\xi)^{|\alpha|}}{\alpha!}D^{\alpha+\delta_i}f(\xi). \tag{1.8}$$

On the right-hand side of (1.8), only the derivatives of order l of the function f are contained.

Now let us apply the integral representation (1.3) to the function $\varphi(x,\xi)$ as well as to the function of the variable ξ (so far, x is considered to be some parameter). We obtain

$$\varphi(x,\xi)=\int_U\Big(K_0(\xi,y)\,\varphi(x,y)+\sum_{i=1}^n K_i(\xi,y)\frac{\partial\varphi}{\partial y_i}(x,y)\Big)\,dy. \tag{1.9}$$

Note that for $\xi=x$ in the expression $\sum_{|\alpha|\leqslant l-1}\frac{(x-\xi)^\alpha}{\alpha!}D^\alpha f(\xi)$, all the summands except the one corresponding to the value $\alpha=0$ vanish. The latter summand equals $f(x)$ so that $\varphi(x,x)=f(x)$. By putting $\xi=x$ in (1.9) and by substituting expressions for the derivatives $\frac{\partial\varphi}{\partial y_i}(x,y)$ according to the formula (1.8), we obtain

$$f(x)=\int_U K_0(x,y)\sum_{|\alpha|\leqslant l-1}\frac{(x-y)^\alpha}{\alpha!}D^\alpha f(y)\,dy+\sum_{|\alpha|=l}\int_U K_\alpha(x,y)D^\alpha f(y)\,dy, \tag{1.10}$$

where

$$K_\alpha(x,y) = \sum_{\alpha_i \neq o} K_i(x,y)\frac{(x-y)^{\alpha-\delta_i}}{(\alpha-\delta_i)!} \tag{1.11}$$

(the summation is held by all α for which $|\alpha| = l$ and $\alpha_i > 0$). Let us transform the first integral on the right-hand side of (1.10) by using the fact that for fixed x, $K_0(x,y)$ is a test function in U of the class C^∞ of the variable y. As a result, we obtain

$$\begin{aligned}\int_U K_0(x,y) \sum_{|\alpha|\leqslant l-1} \frac{(x-y)^\alpha}{\alpha!} D^\alpha f(y)\,dy \\ = \int_U f(y) \sum_{|\alpha|\leqslant l-1} (-1)^{|\alpha|} D_y^\alpha [\frac{(y-x)^\alpha}{\alpha!} K_0(x,y)]\,dy.\end{aligned}$$

We set

$$K_0^l(x,y) = \sum_{|\alpha|\leqslant l-1} (-1)^{|\alpha|} D_y^\alpha \left[\frac{(x-y)^\alpha}{\alpha!} K_0(x,y)\right]. \tag{1.12}$$

Equality (1.10) acquires the final form:

$$f(x) = \int_U K_0^l(x,y)\,f(y)\,dy + \sum_{|\alpha|=l} \int_U K_\alpha(x,y) D^\alpha f(y)\,dy. \tag{1.13}$$

The functions K_0^l and K_α in this equality are expressed in terms of the formulae (1.12) and (1.11), respectively. Let us note some of their properties. First let us prove the following simple proposition.

Lemma 1.1. *If a function $F(x,y)$ of the variables $x \in U$, $y \in U$, $x \neq y$, is a function of the type $|x-y|^r$, then for every n-dimensional multiindex α, $(x-y)^\alpha \cdot F(x,y)$ is the function of the type $|x-y|^{r+|\alpha|}$.*

Proof. Let $F(x,y)$ be a function of the type $|x-y|^r$. Then it is obvious that for any multiindex β, the derivative $D^\beta F(x,y)$ is the function of the type $|x-y|^{r-|\beta|}$. We have: $|(x-y)^\alpha \cdot F(x,y)| \leqslant C|x-y|^{r+|\alpha|}$. Then, for $i = 1, 2, \ldots, n$,

$$\frac{\partial}{\partial y_i}[(x-y)^\alpha F(x,y)] = -\alpha_i (x-y)^{\alpha-\delta_i} F(x,y) + (x-y)^\alpha \frac{\partial}{\partial y_i} F(x,y).$$

Hence we conclude that

$$\left|\frac{\partial}{\partial y_i}[(x-y)^\alpha \cdot F(x,y)]\right| \leqslant (\alpha_i \cdot C_0 + C_{\delta_i})\,|x-y|^{|\alpha|+r-1}.$$

It is necessary to prove that for any multiindex β,

$$|D^\beta[(x-y)^\alpha F(x,y)]| \leqslant M_\beta |x-y|^{|\alpha|+r-|\beta|}, \tag{1.14}$$

where $M = \text{const} < \infty$. In the case $|\beta| = 1$, this is proved for any r and α. Suppose that inequality (1.14) is proved for $|\beta| = l$ for any r and α, and let $|\beta| = l + 1$. Let β_i be some of the components of the multiindex β which are distinct from zero. Then we have:

$$D_y^\beta[(x-y)^\alpha \cdot F(x,y)]$$
$$= D_y^{\beta-\delta_i}[-\alpha_i(x-y)^{\alpha-\delta_i} \cdot F(x,y)] + D_y^{\beta-\delta_i}[(x-y)^\alpha \frac{\partial}{\partial y_i} F(x,y)]. \tag{1.15}$$

Here $|\beta - \delta_i| = l, F(x,y)$ is the function of the type $|x-y|^r$, $\frac{\partial F}{\partial y_i}(x,y)$ is the function of the type $|x-y|^{r-1}$. Each of the summands on the right-hand side of equality (1.15) admits, due to the induction hypothesis, the estimate of the form (1.14). Hence the lemma obviously follows.

Considering the integral representation (1.13), we obtain that the following statements are valid.

A$_l$) The function $K_0^l(x,y)$ belongs to the class C^∞ in $U \times U$, and for every $x \in U$, $y \mapsto K_0^l(x,y)$ is a test function of the variable y in U.

This obviously follows from the expression $K_0^l(x,y)$ in terms of $K_0(x,y)$ and from the fact that $K_0(x,y)$ possesses similar properties.

B$_l$) Each of the functions $K_\alpha(x,y)$ is continuous on the set of all $(x,y) \in U \times U$ such that $x \neq y$; and for every $x \in U$, the function $y \mapsto K_\alpha(x,y)$ is compactly supported in U.

This fact follows directly from Condition B(1) for the integral representation (1.3) and for the expressions of the function $K_\alpha(x,y)$ by means of the functions $K_i(x,y)$.

C$_l$) $K_\alpha(x,y)$ are the functions of the type $|x-y|^{l-n}$ for $x \in U$, $y \in U$, $x \neq y$.

The given proposition is the corollary of Lemma 1.1 and of representation (1.11) of the function K_α.

Let us note that in the deduction of formula (1.13), we only used the Property A of the function K_0 in the integral representation (1.3). Then it was only necessary that each of the integrals in (1.3) have certain value, no matter what the function f of the class C^l might be.

1.2. Integral Representations in a Curvilinear Cone

First let us construct the integral representation by giving value to a function $f \in C^1(U)$ at a fixed point x of the domain U. In Subsection 1.4, we shall show how an integral representation of the form (1.3) valid for all $x \in U$ may be constructed according to the above integral representation.

Every continuous mapping $\xi : [a,b] \to \mathbf{R}^n$ is said to be a path or a parametrized curve in the space $\mathbf{R}^n$. The path $\xi : [a,b] \to \mathbf{R}^n$ is called absolutely continuous if each of its components $\xi_1, \xi_2, \ldots, \xi_n$ is absolutely continuous. We say that the path ξ lies in a set $A \subset \mathbf{R}^n$ if for any $t \in [a,b]$, $\xi(t) \in A$.

Suppose that assigned are a function $f : U \to \mathbf{R}^k$ of the class C^1 (here U is an open set in $\mathbf{R}^n$) and an absolutely continuous path $\xi : [a,b] \to \mathbf{R}^n$ lying in U. The set $\xi([a,b])$ is compact; therefore f satisfies the Lipschitz condition on $\xi([a,b])$, that is

$$|f(x'') - f(x')| \leqslant L|x'' - x'|$$

for any $x', x'' \in \xi([a,b])$, where $L = \text{const} < \infty$. In particular, we obtain that $|f[\xi(t')] - f[\xi(t'')]| \leqslant L|\xi(t') - \xi(t'')|$ for any $t', t'' \in [a,b]$. This allows us to state that the function $f[\xi(t)]$ is absolutely continuous. We have

$$\frac{d}{dt}(f[\xi(t)]) = \sum_{i=1}^{n} \frac{\partial f}{\partial x_i}[\xi(t)]\xi_i'(t),$$

and for any $t_1, t_2 \in [a,b]$,

$$f[\xi(t_2)] - f[\xi(t_1)] = \int_{t_1}^{t_2} \sum_{i=1}^{n} \frac{\partial f}{\partial x_i}[\xi(t)]\xi_i'(t)\,dt. \tag{1.16}$$

The deduction of integral representation of a function given below is based on some general idea that may be used in other similar problems. It is as follows. A family of paths is constructed depending on a parameter $h \in \mathbf{R}$ such that the beginning of each path is at the given point x. Applying formula (1.16) to the case $\xi(t) = \xi(t,h)$, $a \leqslant t \leqslant b$, where the function $\xi(t,h)$ is such that $\xi(a,h) = x$ for all h, we obtain

$$f(x) = f[\xi(b,h)] - \int_{a}^{b} \sum_{i=1}^{n} \frac{\partial f}{\partial x_i}[\xi(t,h)]\xi_i'(t,h)\,dt. \tag{1.17}$$

Let us multiply both parts of equality (1.17) by the function $w(h)$ of the parameter h, such that $\int w(h)\,dh = 1$ and let us integrate by h. As a result, we obtain the relation

$$f(x) = \int f[\xi(b,h)]\,w(h)\,dh - \int \left(\int_{a}^{b} \sum_{i-1}^{n} \frac{\partial f}{\partial x_i}[\xi(t,h)]\xi_i'(t,h)\,dt \right) w(h)\,dh.$$

The desired formula is obtained by transforming integrals on the right-hand side of the given equality.

Let us assign an arbitrary function $f : U \to \mathbf{R}^k$ of the class C^1. Let there be a point $x \in U$ and a path $\xi : [0,1] \to U$ such that $\xi(0) = x$ and for any $t_1, t_2 \in [0,1]$,

$$|\xi(t_1) - \xi(t_2)| \leqslant K|t_1 - t_2|, \tag{1.18}$$

where $K = \text{const} < \infty$. Let S be a compact set in $\mathbf{R}^n$. Let us define a family of paths $\xi_h(t)$, where $0 \leqslant t \leqslant 1$, $h \in S$, setting $\xi_h(t) = \xi(t) + th$. The family of paths ξ_h thus defined shades some set T. Let us denote by S_t an image of the set S with respect to the homothetical mapping $h \mapsto \xi(t) + th$. It is obvious that $T = \cup_{t\in[0,1]} S_t$. The set T is called a curvilinear cone. T turns into a direct circular cone if the function $\xi(t)$ has the form $\xi(t) = x + lt$ and if S is an $(n-1)$-dimensional ball lying in a hyperplane orthogonal to the vector l.

Below it will be supposed that the set T constructed above is contained in the open set U, and the compact set S, by means of which T was constructed, is the support of some bounded measurable function w such that

$$\int_{\mathbf{R}^n} w(y)\,dy = 1. \tag{1.19}$$

Below, some additional restrictions will be imposed upon the function w. Let us denote by T_τ, where $0 < \tau \leqslant 1$, a set shaded by the curve $\xi_h(t), 0 \leqslant t < \tau$, $h \in S$. We have: $T_\tau = \cup_{0\leqslant t\leqslant \tau} S_t$. The set T_τ is the image of the compact set $S \times [0,\tau] \subset \mathbf{R}^{n+1}$ with respect to the continuous mapping $(h,t) \mapsto \xi(t) + th$, and consequently, it is compact.

Let us establish some estimates of sizes of the set T_τ. Suppose that the inclusion

$$S \subset B(0,r) \tag{1.20}$$

holds. According to the condition, $|\xi(t_2) - \xi(t_1)| \leqslant K|t_2 - t_1|$ for any $t_1, t_2 \in [0,1]$. Let $y \in T_\tau$. Then $y = \xi(t) + th$, where $h \in S$ and $t \in [0,\tau]$. Hence, $|y - x| \leqslant |\xi(t) - \xi(0)| + t|h| \leqslant (K + |h|)t \leqslant (K + r)\tau$, and due to the arbitrariness of $y \in T_\tau$, we obtain that

$$T_\tau \subset B[x, \tau(r+K)]. \tag{1.21}$$

Let $f : U \to \mathbf{R}^k$ be a function of the class C^1. Setting $\xi_h(t) = \xi(t) + th$, we obtain

$$f(x) = f(\tau h + \xi(\tau)) - \int_0^\tau \sum_{j=1}^n \frac{\partial f}{\partial x_j}(th + \xi(t))\,(h_j + \xi_j'(t))\,dt. \tag{1.22}$$

Let us multiply both parts of equality (1.22) by $w(h)$, and let us integrate the result by the ball $B_0 = B(0,r) \supset S$. Taking (1.19) into account, we obtain

$$\begin{aligned} f(x) = &\int_{B(0,r)} f(\tau h + \xi(\tau)) w(h)\,dh \\ &- \int_{B(0,r)} w(h) \left\{ \int_0^\tau \sum_{j=1}^n \frac{\partial f}{\partial x_j}(th + \xi(t))(h_j + \xi_j'(t))\,dt \right\} dh. \end{aligned} \tag{1.23}$$

It should be noted that

$$\int_{B(0,r)} f(\tau h + \xi(t))w(h)\,dh = \int_{B(\xi(\tau),\tau r)} f(z)w\left(\frac{z-\xi(\tau)}{\tau}\right)\frac{dz}{\tau^n}.$$

Since the support of w is compact, the whole set U may be considered to be the integration domain here.

Note that the support of the function $z \mapsto w(\frac{z-\xi(t)}{\tau})$ is contained in the set T_τ.

If, according to the Fubini theorem, we change the integration order on the right-hand side of (1.23), it yields

$$f(x) = \int_U f(y)w\left(\frac{y-\xi(\tau)}{\tau}\right)\frac{dy}{\tau^n} - \int_0^\tau \left\{\int_{B(0,r)} \sum_{j=1}^n \frac{\partial f}{\partial x_j}(th+\xi(t))\left(h_j - \xi_j'(t)\right)w(h)\,dh\right\}dt. \quad (1.24)$$

Application of the Fubini theorem is justified here, since all the functions under consideration are bounded and measurable. After that, in the second summand on the right-hand side of (1.24), let us replace the integration variable, assuming $th + \xi(t) = y$. This results in the integral

$$\int_0^\tau \left(\int_{B(\xi(t),rt)} \sum_{j=1}^n \frac{\partial f}{\partial x_j}(y)\left(y_j - \xi_j(t) + t\xi_j'(t)\right)\, w\left(\frac{y-\xi(t)}{t}\right)\frac{dy}{t^{n+1}}\right)dt.$$

The integrand function vanishes outside of the set $S_t \subset U$ for every fixed t; therefore the entire set U may be considered to be the integration domain in the inner integral. By applying the Fubini theorem once more, we obtain

$$f(x) = \int_U f(y)w\left(\frac{y-\xi(\tau)}{\tau}\right)\frac{dy}{\tau^n} + \int_U \sum_{j=1}^n \frac{\partial f}{\partial x_j}(y)H_{\tau,j}(y-x,x)\,dy, \quad (1.25)$$

where

$$H_{\tau,j}(y-x,x) = \int_0^\tau (y_j + \xi_j(t) - t\xi_j'(t))w\left(\frac{y-\xi(t)}{t}\right)\frac{dt}{t^{n+1}}. \quad (1.26)$$

Equality (1.25) is the desired integral representation. However, the second application of the Fubini theorem demands some foundation, since in this case the integrand turned out to be unbounded. The desired justification is given in Lemma 1.2. First, let us introduce some notations. Let

$$\alpha(t) = \frac{\xi(t)-\xi(0)}{t} = \frac{\xi(t)-x}{t}, \qquad \beta(t) = \alpha(t) - \xi'(t). \quad (1.27)$$

By using these notations, we can rewrite the expression for $H_{t,j}(z,x)$ as follows:

$$\begin{aligned}H_{\tau,j}(z,x) &= \int_0^\tau [z_j - t\beta_j(t)]\, w\,[\frac{z}{t} - \alpha(t)]\frac{dt}{t^{n+1}}\\ &= \int_{1/\tau}^{\infty} [tz_j - \beta_j\left(\frac{1}{t}\right)]\, w\,[zt - \alpha\left(\frac{1}{t}\right)]t^{n-2}dt.\end{aligned} \tag{1.28}$$

Lemma 1.2. *Let $F : U \to \mathbf{R}^k$ be a continuous function. Let*

$$I_j(F) = \int_{B(0,r)} \left\{\int_0^\tau F(ht + \xi(t))\,(\xi_j'(t) + h_j)\, dt\right\} w(h)\, dh, \tag{1.29}$$

where $\xi(t)$ and $w(h)$ have the above meaning. Then the equality

$$I_j(F) = -\int_U F(y) H_{\tau,j}(y-x,x)\, dy$$

is valid, where the function $H_{\tau,j}(y,x)$ is described by equalities (1.28). For every $y \neq 0$, the inequality

$$|H_{\tau,j}(y,x)| \leqslant 3M(K+r)^n/(n-1)|y|^{n-1} \tag{1.30}$$

holds, where $M = \sup|w(h)|$, r and K are defined from the relations (1.18) and (1.20). The support of the function $y \mapsto H_{\tau,j}(y-x,x)$ is contained in the set T_τ.

Proof. By applying the Fubini theorem, we obtain

$$I_j(F) = \int_0^\tau \left\{\int_S F(\xi(t) + ht)[\xi_j'(t) + h_j]\, w(h)\, dh\right\} dt.$$

Let us perform the change of the integration variable in the inner integral by setting $\xi(t) + th = z$. We have: $h = \frac{z-\xi(t)}{t}$, $dh = \frac{1}{t^n}\, dz$, whence

$$I_j(F) = \int_0^\tau \left\{\int_{S_t} F(z)[t\xi_j'(t) - \xi_j(t) + z_j]\, w\left(\frac{z-\xi(t)}{t}\right)\frac{dz}{t^{n+1}}\right\} dt. \tag{1.31}$$

Since the integrand vanishes outside of $S_t \subset U$, the entire set U may be considered to be the integration domain. Let us assign arbitrary $\varepsilon > 0$, and let $I_j(F,\varepsilon)$ be the integral obtained if the outer integration in (1.31) is performed over the segment $[\varepsilon,\tau]$. We have $I_j(F) = \lim_{\varepsilon\to 0} I_j(F,\varepsilon)$. Let us apply the Fubini theorem to the integral expressing $I_j(F,\varepsilon)$. Since in this integral t is separated from zero by the positive constant ε, no problem arises due to unboundedness of the integrand, and we obtain

$$I_j(F,\varepsilon) = -\int_U F(y) H_{\tau,j}(y-x,x,\varepsilon)\, dy, \tag{1.32}$$

where

$$H_{\tau,j}(z,x,\varepsilon)=\int_\varepsilon^\tau (z_j+x_j-\xi_j(t)+t\xi_j'(t))\,w\left(\frac{z+x-\xi(t)}{t}\right)\frac{dt}{t^{n+1}}. \quad (1.33)$$

In the integral of (1.33), let us perform the change of the variable by setting $s=\frac{1}{t}$. Taking into account that $x_j-\xi_j(t)+t\xi_j'(t)=-t\beta_j(t), x-\xi(t)=t\alpha(t)$, we obtain

$$H_{\tau,j}(z,x,\varepsilon)=\int_{1/t}^{1/\varepsilon}[sz_j-\beta_j(\frac{1}{s})]\,w(sz-\alpha(\frac{1}{s}))\,s^{u-2}\,ds. \quad (1.34)$$

We have $|\alpha(\frac{1}{s})|\leqslant K$ for all s. If $z\neq 0$ and $s\geqslant (K+r)/|z|$, then $|sz-\alpha(\frac{1}{s})|\geqslant (K+r)-K=r$; therefore, if $z\neq 0$ and $s\geqslant (K+r)/|z|$, then $w(sz-\alpha(\frac{1}{s}))=0$. Thus, the integrand in (1.34) vanishes for sufficiently large values of s. Hence it follows that for every $z\neq 0$, $\lim_{\varepsilon\to 0}H_{\tau,j}(z,x,\varepsilon)$ exists and is finite, so the integral of (1.28) converges for every $z\neq 0$.

We have $|w(x)|\leqslant M$ for all $x\in\mathbf{R}^n$. Taking into account that the integrand in (1.34) vanishes for $s\geqslant (K+r)/|z|$ (provided $z\neq 0$), we obtain the inequality

$$\begin{aligned}|H_{\tau,j}(z,x,\varepsilon)| &\leqslant M\int_0^{(r+K)/|z|}(s|z|+2K)\,s^{n-2}ds\\ &= M\frac{(r+K)^n}{n|z|^{n-1}}+MK\frac{(r+K)^{n-1}}{(n-1)|z|^{n-1}}\\ &\leqslant \frac{3M(r+K)^n}{|z|^{n-1}}. \end{aligned}\quad (1.35)$$

Taking the limit for $\varepsilon\to 0$, we obtain that for any $z\neq 0$,

$$|H_{\tau,j}(z,x)|\leqslant\frac{3M(r+K)^n}{(n-1)|z|^{n-1}},$$

and thus estimate (1.30) is proved.

From equality (1.33) it is clear that if $H_{\tau,j}(y-x,x,\varepsilon)\neq 0$, then $w(\frac{y-\xi(t)}{t})\neq 0$ for at least one value of $t\in[\varepsilon,\tau]$, and therefore $y=\xi(t)+th$, where $h\in S$, that is, $y\in T_\tau$. Hence it follows that the support of the function $y\to H_{\tau,j}(y-x,x,\varepsilon)$ is contained in T_τ; therefore, the support of the function $y\mapsto H_{\tau,j}(y-x,x,\varepsilon)$ is also contained in T_τ.

The integrand in (1.32) vanishes outside of the compact set T_τ. Since F is continuous, it is bounded on T_τ. Hence, taking estimate (1.35) into account, we conclude that the integrand in (1.32) is majorized by the integrable function $\frac{C}{|z-x|^{n-1}}$. Due to the Lebesgue dominated convergence theorem, we obtain

$$\begin{aligned}I_j(F)&=\lim_{\varepsilon\to 0}I_j(F,\varepsilon)\\ &=-\int_U\lim_{\varepsilon\to 0}(F(z)H_{\tau,j}(z-x,x,\varepsilon))\,dz\\ &=-\int_U F(z)H_{\tau,j}(z-x,x)\,dz.\end{aligned}$$

The lemma is proved.

By setting $F = \frac{\partial f}{\partial x_j}$ in (1.29) and by using Lemma 1.2, we obtain (1.25).

The function w in our constructions satisfies some rather weak conditions only. This prevents us from making any conclusions relative to derivatives of the function $H_{\tau,j}(y-x,x)$.

Lemma 1.3. *Let ξ be as above, and let the function w be defined as follows: $w(h) = \frac{1}{r^n} w_0(\frac{h}{r})$, where $w_0 : \mathbf{R}^n \to \mathbf{R}$ is a function of the class C^∞ such that $S(w_0) \subset B(0,1)$ and*

$$\int_{\mathbf{R}^n} w_0(h)\, dh = 1,$$

and let the function $H_{\tau,j}(y,x)$ be defined in terms of w by formula (1.26). Then the function $z \mapsto H_{\tau,j}(z-x,x)$ belongs to the class C^∞ in the domain $\mathbf{R}^n \setminus \{x\}$, and the estimates

$$|H_{\tau,j}(y,x)| \leqslant M_0/|y|^{n-1},$$
$$|D_y^\alpha H_{\tau,j}(y,x)| \leqslant M_\alpha/|y|^{n+|\alpha|-1} \tag{1.36}$$

are valid, where

$$M_\alpha = \|w_0\|_{|\alpha|} \left(1 + \frac{K}{r}\right)^{n+|\alpha|}. \tag{1.37}$$

Proof. Let us make use of the second integral representation of the function $H_{\tau,j}$. Let us assign arbitrary $\varepsilon > 0$ and let $|y| > \varepsilon$. Let us set $L = (r+K)/\varepsilon$. Then for $t > L$, $|yt - \alpha(\frac{1}{t})| > r$; therefore, $w[yt - \alpha(\frac{1}{t})] = 0$, hence it follows that here

$$H_{\tau,j}(y,x) = \int_{1/t}^{L} \left[ty_j - \beta_j\left(\frac{1}{t}\right)\right] w\left[yt - \alpha\left(\frac{1}{t}\right)\right] t^{n-2} dt. \tag{1.38}$$

According to the condition, w is a function of the class C^∞. Therefore, the integrand in (1.38) is a function of the class C^∞ of the variable y. Hence it follows that $H_{\tau,j}(y,x)$ is a function of the class C^∞ in the domain $|y| \geqslant \varepsilon$ for any $\varepsilon > 0$; that is, $y \mapsto H_{\tau,j}(y,x)$ is a function of the class C^∞ in the domain $\mathbf{R}^n \setminus \{0\}$. Let us assign an arbitrary differentiation operator $D^\alpha = D_1^{\alpha_1} D_2^{\alpha_2} \dots D_n^{\alpha_n}$. We have

$$D_y^\alpha H_{\tau,j}(y,x) = \int_{1/t}^{L} D_y^\alpha \left\{\left[ty_j - \beta_j\left(\tfrac{1}{t}\right)\right] w\left[yt - \alpha\left(\tfrac{1}{t}\right)\right]\right\} t^{n-2} dt. \tag{1.39}$$

Then, for $\alpha_j > 0$,

$$D_y^\alpha \left\{\left[ty_j - \beta_j\left(\tfrac{1}{t}\right)\right] w\left[yt - \alpha\left(\tfrac{1}{t}\right)\right]\right\}$$
$$= \alpha_j t^{|\alpha|} (D^{\alpha-\delta_j} w)\left(yt - \alpha\left(\tfrac{1}{t}\right)\right) + \left(ty_j - \beta_j\left(\tfrac{1}{t}\right)\right) t^{|\alpha|} (D^\alpha w)\left(yt - \alpha\left(\tfrac{1}{t}\right)\right).$$

In the case $\alpha_j = 0$,

$$D_y^\alpha \{[ty_j - B_j(\tfrac{1}{t})]\, w\, [yt - \alpha(\tfrac{1}{t})]\} = [ty_j - \beta_j(\tfrac{1}{t})]\, t^{|\alpha|}(D^\alpha w)\, [yt - \alpha(\tfrac{1}{t})].$$

For all $x \in \mathbf{R}^n$, we have

$$|D^\alpha w(x)| = \frac{1}{r^{n+|\alpha|}} \left|(D^\alpha w_0)\left(\frac{x}{r}\right)\right| \leqslant \frac{\|w_0\|_{|\alpha|}}{r^{n+|\alpha|}}.$$

For every $j = 1, 2, \ldots, n$, $|\beta_j(1/t)| \leqslant 2K$, $|y_j| \leqslant y$. If in the integral of (1.39) we substitute the upper limit by the magnitude $(r + R)/|y|$, and the lower limit—by zero, we obtain after some obvious transformations,

$$|D_y^\alpha H_{\tau,j}(y, x)| \leqslant \frac{M_\alpha}{|y|^{n+|\alpha|-1}},$$

where

$$M_\alpha = \|w\|_{|\alpha|} \left(\frac{r + K}{r}\right)^{n+|\alpha|}.$$

The lemma is proved.

To complete the subsection, let us briefly describe another way of obtaining formula (1.25), leaving it to the reader to work out the details. The main idea of this method is widely used in the investigations on the theory of functions with generalized derivatives (see, for instance, [35]). Let us assign an arbitrary function w of the class $C_0^\infty(\mathbf{R}^n)$ such that its support is contained in the ball $B(0, r)$ and

$$\int_{\mathbf{R}^n} w(x)\, dx = 1.$$

Let $\xi(t)$ have the above sense, and let $f : U \to \mathbf{R}$ be a function of the class $C^1(U)$. For $t \in [0, 1]$, we put

$$f(x, t) = \frac{1}{t^n} \int_U f(y) w\left(\frac{y - \xi(t)}{t}\right) dt.$$

For $t \to 0$, $f(x, t) \to f(x)$; consequently,

$$f(x) = f(x, \tau) - \int_0^\tau \frac{\partial f}{\partial t}(x, t)\, dt. \tag{1.40}$$

After differentiation by the parameter t of the integral representing the function $f(x, t)$, we obtain

$$\begin{aligned} \frac{\partial f}{\partial t}(x, t) = {} & \int_U f(y) \sum_{i=1}^n \frac{\partial w}{\partial x_i}\left(\frac{y - \xi(t)}{t}\right) \left(\frac{\xi_i(t) - y_i - t\xi_i'(t)}{t^{n+2}}\right) dy \\ & - \int_U f(y) \frac{n}{t^{n+1}} w\left(\frac{y - \xi(t)}{t}\right) dy. \end{aligned} \tag{1.41}$$

We have:

$$\frac{\partial w}{\partial x_i}\left(\frac{y-\xi(t)}{t}\right) = t\frac{\partial}{\partial y_i}\left[w\left(\frac{y-\xi(t)}{t}\right)\right],$$

whence

$$\frac{\partial f}{\partial t}(x,t) = \int_U f(y)\sum_{i=1}^{n}\frac{\partial}{\partial y_i}\left[w\left(\frac{y-\xi(t)}{t}\right)\right]\frac{\xi_i(t)-y_i-t\xi_i'(t)}{t^{n+1}}\,dy$$
$$-\int_U f(y)\frac{n}{t^{n+1}}w\left(\frac{y-\xi(t)}{t}\right)dy.$$

We transform the former integral by the integration formula by parts, assuming that the support of the function $y\mapsto w(\frac{y-\xi(t)}{t})$ is contained in U for every t.

$$\frac{\partial f}{\partial t}(x,t) = \int_U -w\left(\frac{y-\xi(t)}{t}\right)\sum_{i=1}^{n}\frac{\partial}{\partial y_i}\left[f(y)\frac{\xi_i(t)-y_i-t\xi_i'(t)}{t^{n+1}}\right]dy$$
$$-\int_U f(y)\frac{n}{t^{n+1}}w\left(\frac{y-\xi(t)}{t}\right)dt \tag{1.42}$$
$$=\int_U\sum_{i=1}^{n}\frac{\partial f}{\partial x_i}(y)\frac{t\xi_i'(t)-\xi_i(t)+y}{t^{n+1}}\cdot w\left(\frac{y-\xi(t)}{t}\right)dy.$$

By substituting the obtained expression for $\frac{\partial f}{\partial t}(x,t)$ into (1.41), we obtain

$$f(x)=\int_U w\left(\frac{x-\xi(\tau)}{\tau^n}\right)f(\tau)-\sum_{i=1}^{n}\int_U\frac{\partial f}{\partial x_i}(y)H_i(x,y,\tau)\,dy, \tag{1.43}$$

where

$$H_i(x,y,\tau)=\int_0^\tau\frac{-t\xi_i'(t)+\xi_i(t)}{t^{n+1}}w\left(\frac{y-\xi(t)}{t}\right)dt. \tag{1.44}$$

Thus, the integral representation (1.25) is obtained. The advantage of this method to obtain formula (1.25) is that is can be performed for considerably weaker assumptions on the function f. The requirement $w\in C_0^\infty(\mathbf{R}^n)$ may be removed by the subsequent limit transition.

The magnitude $f(x,t)$ may also be represented as follows:

$$f(x,t)=\int_U f(yt+\xi(t))\,w(y)\,dy.$$

Hence (under the assumption that $f\in C^1(U)$), we obtain

$$\frac{\partial f}{\partial t}(x,t)=\int_U\sum_{i=1}^{n}\frac{\partial f}{\partial x_i}(yt+\xi(t))\,[y_i+\xi_i'(t)]\;w(y)\,dy.$$

By substituting this expression in (1.40), we obtain equality (1.29).

1.3. Domains of the Class J

The integral representation obtained in the previous subsection was preliminary. Here we define the class of domains in $\mathbf{R}^n$ for which the final representation will be constructed.

For arbitrary points $x, y \in \mathbf{R}^n$, $[x, y]$ here denotes the segment connecting the points x and y, i.e., the set of all points $z \in \mathbf{R}^n$ of the form $z = (1-t)x+ty$, where $0 \leqslant t \leqslant 1$.

A domain $U \subset \mathbf{R}^n$ is said to belong to the class $J(r, R)$, where $0 < r \leqslant R < \infty$ if the following conditions are valid. There exists a point $a \in U$ such that any other point $x \in U$ may be connected in U with a point a by a rectifiable curve $x(s)$ (the parameter s, $0 \leqslant s \leqslant l$, denotes the length of the arc) such that its length $l \leqslant R, x(0) = x, x(l) = a$, and for every $s \in [0, l]$, the inequality

$$\rho(x(s), \partial U) \geqslant \frac{rs}{l} \tag{1.45}$$

holds. The point a mentioned in this definition is said to be the marked point of the domain U. U is said to be a domain of the class J if $U \in J(r, R)$ for some r and R.

The class of domains $J(r, R)$ was introduced and named "the domains with inner radius r and outer radius R" by John [97].

A domain $U \in \mathbf{R}^n$ is said to be a starlike domain with respect to a ball $B(a, r) \subset U$ if for every point $x \in B(a, r)$ and for any point $y \in U$, the segment $[x, y] \subset U$.

Every convex open set is obviously a starlike domain with respect to a ball.

Let U be a starlike domain in $\mathbf{R}^n$ with respect to a ball $B(a, r)$ and let it be contained in a ball $B(a, r)$. Then U belongs to the class $J(r, R)$. In this case, for an arbitrary point $x \in U$, the curve $x(s)$ defined by the equality $x(s) = (1 - s/l)x + (s/l)a$, where $l = |a - x|$ satisfies all the conditions of the definition for a domain of the class $J(r, R)$.

Domains which are starlike with respect to a ball have rather special structure. The main goal of this subsection is to show that the class J contains some rather wide classes of domains which are characterized by natural geometric conditions.

Let U be a domain of the class $J(r, R)$. Putting $s = l$ in inequality (1.45) and taking into account that $x(l) = a$, we obtain that $\rho(a, \partial U) \geqslant r$, and consequently, U contains the ball $B(a, r)$. From the definition it also follows that if $U \in J(r, R)$, then for every point $x \in U$, $|x - a| \leqslant l \leqslant R$; therefore, $U \subset B(a, R)$.

Let us somewhat modify the initial definition of a domain of the class J as it was done in paper [78].

Lemma 1.4. *An open domain U in $\mathbf{R}^n$ belongs to the class J iff there exist a point $a \in U$ and the numbers $L < \infty$ and $\alpha \geqslant 0, \alpha \leqslant 1$, such that for every point x belonging to the closure $\bar{U}$ of the domain U, there exists a rectifiable* *curve $x(s)$ (the parameter s is the length of the arc, $s \in [0, l]$) whose length l does not exceed L, such that $x(0) = x$, $x(l) = a$, and for all $s \in [0, l]$ the inequality*

$$\rho[x(s), \partial U] \geqslant \alpha s$$

is valid.

Proof. Let U be a domain of the class J. Then $U \in J(r, R)$ for some r and R, $0 < r \leqslant R < \infty$. Let a be the marked point of U. According to the definition, for an arbitrary point $x \in U$, there exists a rectifiable curve $x(s)$, $0 \leqslant s \leqslant l$ (the parameter s is the length of the arc), such that $x(0) = x$, $x(l) = a, l \leqslant R$, and for all $s \in [0, l]$,

$$\rho[x(s), \partial U] \geqslant \frac{rs}{l} \geqslant \left(\frac{r}{R}\right) s.$$

We obtain that for every point $x \in U$, the condition of the lemma is satisfied with the constants $\alpha = r/R$, $L = R$. Let x be an arbitrary boundary point of U. Let us construct a sequence (x_m), $m = 1, 2, \ldots$, of points of the domain U, which converges to x for $m \to \infty$. For every m, let us construct a rectifiable curve $x_m(s)$, $s \in [0, l_m]$ where $l_m \leqslant R$, $x_m(0) = x_m$, $x_m(l_m) = a$, and the parameter s is the length of the arc; also,

$$\rho[x_m(s), \partial U] \geqslant \alpha s$$

for all $s \in [0, l_m]$ for every m. The family of vector-functions $x_m : [0, l_m] \to \mathbf{R}^n$ is equipotentially uniformly continuous because for any s', $s'' \in [0, l_m]$,

$$|x_m(s') - x_m(s'')| \leqslant |(s' - s'')|$$

and is bounded (at least, because they all lie in the domain U, which is bounded). Due to this, from the sequence $(x_m(s))$, one can extract a subsequence for which l_m tends to some limit l_0 and $x_m(s)$ converges to some function $x_0 : [0, l_0] \to \mathbf{R}^n$ uniformly on every segment $[0, l]$ where $0 < l < l_0$. To make notations less cumbersome, let us assume that the initial sequence $(x_m(s))$ satisfies all these conditions. Let us note that $|a - x_m| \leqslant l_m$ and $\lim_{m\to\infty} |a - x_m| = |a - x| \geqslant r$, whence it follows that $l_0 \geqslant r$. For the limit function $x_0(s)$, we have $x_0(0) = x$, $x_0(l_0) = a$, and $\rho[x_0(s), \partial U] \geqslant \alpha s$ for all $s \in [0, l]$. However, here the parameter s for the limit function x_0 is, in general, not the length of the arc. For every m, we have $|x_m(s') - x_m(s'')| \leqslant |s' - s''|$ for any s' and s''. Hence, by taking the limit, we obtain that for the limit curve $x_0(s)$, $|x_0(s') - x_0(s'')| \leqslant |s' - s''|$ for any $s', s'' \in [0, l_0]$. Let $y_0(\sigma) = x_0[s(\sigma)]$ be the parametrization of the curve $x_0(s)$, where the parameter σ is the length of the arc, $\sigma \in [0, \lambda]$, $\lambda \leqslant l_0$. For every $s \in [0, l]$, the length of the arc $[x_0(0), x_0(s)]$ of the given curve is, by definition, the lowest upper bound of the sums

$$\sum_{i=1}^{k} |x_0(s_i) - x_0(s_{i-1})| \leqslant \sum_{i=1}^{k} |s_i - s_{i-1}| = s_k - s_0 = s$$

(here $s_0 = 0 < s_1 < s_2 < \cdots < s_k = s$). Hence we obtain that $s(\sigma) \geqslant \sigma$ for any σ. We have: $y(0) = x$, $y(\lambda) = a$, and $\rho[y(\sigma), \partial U] = \rho[x[s(\sigma)], \partial U] \geqslant \alpha \cdot s(\sigma) \geqslant \alpha\sigma$. Thus, the necessity of lemma is completely proved.

Suppose that for the domain U the condition of the lemma is satisfied. Denote by R the lowest of the values of L such that any point $x \in \bar{U}$ may be connected with the point a by a rectifiable curve $x(s)$, $0 \leqslant s \leqslant l$ (the parameter s is the length of the arc) so that $\rho[x(s), \partial U] \geqslant \alpha \cdot s$ for all $s \in [0, l]$ and $l \leqslant L$. Let us put $r = \alpha R$. For every curve $x(s)$, $0 \leqslant s \leqslant l$, of the above type, we have $\rho(a, \partial U) = \rho[x(l), \partial U] \geqslant \alpha \cdot L$; and since l may be taken arbitrarily close to R, we obtain that $\rho(a, \partial U) \geqslant r$. Let us take an arbitrary point $x \in \bar{U}$ and let us construct a rectifiable curve $x(s)$, $s \in [0, l], l \leqslant R$ (as usual, s is the length of the arc), such that $x(0) = x, x(l) = a$, and for all $s \in [0, l]$, $\rho[x(s), \partial U] \geqslant \alpha s = (r/R)s$. The definition of a domain of the class $J(r, R)$, however, demands that a stronger inequality $\rho[x(s), \partial U] \geqslant (r/l)s$ should be satisfied. We shall ensure that the definition is fulfilled if we prolong the curve $x(s)$ so that its length becomes equal to R. To do this, let us assign an arbitrary vector e such that $|e| = 1$, and let us complete the definition of the function x by setting for $s \in [l, R]$ $x(s) = a + e\varphi(s)$, where the real function $\varphi(s)$ is chosen so that $|\varphi(s)| \leqslant \alpha \cdot (R - s)$ for all $s \in [l, R]$, $s \in [l, R]$, $\varphi(l) = 0$. The parameter s is the length of the arc in the segment $[l, R]$ as well.

To construct φ, let us assign (first, an arbitrary) number β such that $0 < \beta < 1$, and let E be the set of all points s of the form $s = s_m = R - \beta^m(R - l)$, where $m = 0, 1, 2, \ldots$. Let us put $\varphi(s) = \varphi(s, E) = \inf_{m \geqslant 0}\{|s - s_m|\}$. It is easy to verify that for any $s', s'' \in [l, R]$, $|\varphi(s') - \varphi(s'')| \leqslant |s' - s''|$, so that the function φ satisfies the Lipschitz condition, consequently, it is absolutely continuous. Let us consider a separate segment $[s_m, s_{m+1}]$. Let $\sigma_m = (1/2)[s_m + s_{m+1}]$. Then for $s \in [s_m, \sigma_m]$, $\varphi(s) = s - s_m$, and for $s \in [\sigma_m, s_{m+1}]$, $\varphi(s) = s_{m+1} - s$. Hence it is clear that $\varphi'(s) = \pm 1$ in each of the intervals (s_m, σ_m) and (σ_m, s_{m+1}). Since the interval $[s_m, s_{m+1}]$ is arbitrary, $\varphi'(s) = \pm 1$ almost everywhere in $[l, R]$; consequently, s is the length of the arc for the curve $x(s)$ in the segment $[l, R]$ as well. For $s \in [s_m, s_{m+1}]$, we have: $|\varphi(s)| \leqslant \frac{1}{2}(s_{m+1} - s_m) = \frac{1}{2}\beta^m(1 - \beta) \cdot (R - l)$. Let us now choose β so that the inequality $(1 - \beta)/2 < \alpha$ is valid. Then for every $s \in [s_m, s_{m+1}]$, we have $|\varphi(s)| \leqslant \alpha \cdot (R - s_{m+1}) \leqslant \alpha \cdot (R - s)$ since m is arbitrary; consequently, we obtain that $|\varphi(s)| \leqslant \alpha(R - s)$ for all $s \in [l, R]$. For every $s \in [l, R]$, we have $\rho[x(s), \partial U] \geqslant \rho(a, \partial U) - \rho(a, x(s)) = \alpha \cdot R - |\varphi(s)| \geqslant \alpha \cdot s$. Thus, having assigned an arbitrary point $x \in \bar{U}$, we constructed by it a curve $x(s), 0 \leqslant s \leqslant l$, where $l = R$, the parameter s is the length of the arc, $x(0) = x$, $x(l) = a$, and $\rho[x(s), \partial U] \geqslant rs/l$ for all $s \in [0, l]$. Thus we have obtained that U is a domain of the class $J(r, R)$. This completes the proof of the lemma.

Remark. *In the proof of the lemma we have incidentally proved the following. If U is a domain of the class $J(r, R)$ with the marked point a, then for every point x belonging to the closure of U not only for points $x \in U$, as is necessary according to the definition of a domain of the class $J(r, R)$, there exists a* *rectifiable curve* $x(s)$, $0 \leqslant s \leqslant l$ *(the parameter* s *is the length of the arc), such that* $x(0) = x$, $x(l) = a$, *and for all* $s \in [0, l]$, $\rho[x(s), \partial U] \geqslant rs/l$.

Let $A \subset \mathbf{R}^n$. A mapping $f : A \to \mathbf{R}^n$ is said to be a Lipschitz homeomorphism if there exists a constant L such that $1 \leqslant L < \infty$, and for any $x_1, x_2 \in A$, the inequalities

$$\frac{1}{L}|x_1 - x_2| \leqslant |f(x_1) - f(x_2)| \leqslant L|x_1 - x_2| \tag{1.46}$$

are valid.

The domain U in the space $\mathbf{R}^n$ is said to be a domain with Lipschitz boundary or, in other words, U is a domain of the class Lip if for every point $x_0 \in \partial U$ there exist a neighbourhood V and a Lipschitz homeomorphism $\varphi : V \to \mathbf{R}^n$ such that the set $\varphi(V)$ is the cube $Q(0, 1)$, the set $V \cap \partial U$ is mapped by the function φ onto the section of this cube by the plane $x_n = 0$, and $f(V \cap U)$ is the set of the points of this cube for which $x_n > 0$.

The notion of a domain with Lipschitz boundary is often described in another way than the one given above. Let us give this definition. Let U be a domain in $\mathbf{R}^n$. Then U is said to be a domain of the class Lip* if for every one of its boundary points x_0, one can find a neighbourhood V and a Cartesian orthogonal system of coordinates $(t_1, t_2, \ldots, t_n)$ in the space $\mathbf{R}^n$, such that the following Conditions A, B, and C are satisfied.

A) The set V in this system of coordinates is defined by a system of inequalities $a_i < t_i < b_i$, $i = 1, 2, \ldots, n$.

B) The set $V \cap \partial U$ is defined by the equation $t_n = f(t_1, t_2, \ldots, t_{n-1})$, where the function f is defined in an $(n-1)$-dimensional rectangle $(a_1, b_1) \times (a_2, b_2) \times \cdots \times (a_{n-1}, b_{n-1}) = V_0$, satisfies the Lipschitz condition, and f is such that $a_n < f(t_1, t_2, \ldots, t_{n-1}) < b_n$ for all $(t_1, t_2, \ldots, t_{n-1}) \in V_0$.

C) $V \cap U$ is the set of all points $x \in V$ for which $t_n > f(t_1, t_2, \ldots, t_{n-1})$.

If U is a domain of the class Lip*, then U is also a domain of the class Lip, i.e., Lip* $\subset$ Lip. Indeed, let $U \in$ Lip*, and let x_0 be an arbitrary boundary point of U. Let us choose a Cartesian orthogonal system of coordinates and a neighbourhood V of a point x_0 so that all the conditions of the previous definition be valid. The neighbourhood V in this case may be chosen so that the following condition should also be fulfilled:

$$a_n - \delta/2 < f(t_1, t_2, \ldots, t_{n-1}) < a_n + \delta/2,$$

where $\delta =$ const, $\delta > 0$, and the rectangle V_0 is a cube. This can obviously be obtained if we diminish, if necessary, the neighbourhood V. Let $\eta > 0$ be the length of edge of the cube V_0. The point x_0 may be assumed to be the origin of the chosen Cartesian system of coordinates. Let W be a set of all points $x \in V$ for which the inequalities

$$f(t_1, t_2, \ldots, t_{n-1}) - \delta/2 < t_n < f(t_1, t_2, \ldots, t_{n-1}) + \delta/2$$

are valid. Let us denote by φ a mapping which associates to the point with the coordinates $(t_1, t_2, \ldots, t_n)$ the points $(x_1, x_2, \ldots, x_n)$, where $x_i = \frac{1}{\eta} \cdot t_i$ for $i = 1, 2, \ldots, n-1$ and $x_n = \frac{1}{\delta}[t_n - f(t_1, t_2, \ldots, t_{n-1})]$. It is easy to prove that φ is a Lipschitz homeomorphism of the neighbourhood W of the point x_0 onto the cube $Q(0,1)$ such that the set $\varphi(\partial U \cap W)$ is the section of the cube by the plane $x_n = 0$, and $\varphi(U \cap W)$ is the upper half of this cube, $\varphi(U \cap W) = \{x \in Q(0,1) | x_n > 0\}$. Since $x_0 \in \partial U$ is arbitrary, we thus have proved that U is a domain with Lipschitz boundary.

Let us remark that the literature by a domain with Lipschitz boundary implies domains of the class Lip* in the sense of the above definition. Due to this let us give an example showing that Lip $\neq$ Lip*.

Let $n \geqslant 2$. In the space $\mathbf{R}^n$, let us introduce a cylindrical system of coordinates $(x_1, \ldots, x_{n-2}, r, \theta)$, where $r \geqslant 0, \theta$ is arbitrary. Cartesian coordinates of a point are expressed by the formulae: $x_{n-1} = r\cos\theta$, $x_n = r\sin\theta$; the remaining ones are $x_1, x_2, \ldots, x_{n-2}$. Let φ be a mapping

$$(x_1, \ldots, x_{n-2}, r, \theta) \mapsto (x_1, \ldots, x_{n-2}, r, \theta + \varepsilon \ln \tfrac{1}{r})$$

of the space $\mathbf{R}^n$ onto itself. The plane $\theta = \text{const}$ by the mapping φ is transformed into an $(n-1)$-dimensional cylinder constructed on a spiral which in the plane $x_1 = x_2 = \cdots = x_{n-2} = 0$ makes an infinite set of turns around the origin of coordinates. Therefore, the halfspace $0 < \theta < \pi$ by the mapping φ is transformed into a domain U such that no neighbourhood of any point $x \in \partial U$ for which $x_{n-1} = x_n = 0$ is uniquely projected on any plane. Consequently, the boundary of this domain cannot be assigned by the equation $t_n = f(t_1, t_2, \ldots, t_{n-1})$ in any Cartesian system of coordinates. The mapping φ belongs to the class C^∞ in the domain where $r > 0$. In the cylindrical system of coordinates, the linear element of the space is expressed by the formula

$$ds^2 = dx_1^2 + \cdots + dx_{n-2}^2 + dr^2 + r^2 d\theta^2.$$

We have

$$\begin{aligned} d\varphi^2 &= dx_1^2 + \cdots + dx_{n-2}^2 + dr^2 + r^2(d\theta - \frac{\varepsilon}{r} dr)^2 \\ &= dx_1^2 + \cdots + dx_{n-2}^2 + (1+\varepsilon^2)\, dr^2 - 2r\varepsilon dr\, d\theta + r^2 d\theta^2. \end{aligned}$$

It is easy to verify that

$$\lambda_1(\varepsilon) \leqslant \frac{d\varphi^2}{ds^2} \leqslant \lambda_2(\varepsilon), \tag{1.47}$$

where

$$\begin{aligned} \lambda_1(\varepsilon) &= 1 + \frac{\varepsilon^2}{2} - \varepsilon\sqrt{1 + \frac{\varepsilon^2}{4}}, \\ \lambda_2(\varepsilon) &= 1 + \frac{\varepsilon^2}{2} + \varepsilon\sqrt{1 + \frac{\varepsilon^2}{4}}, \\ \lambda_1(\varepsilon) &\to 1, \qquad \lambda_2(\varepsilon) \to 1 \quad \text{for} \quad \varepsilon \to 0. \end{aligned}$$

Let x, y be two arbitrary points in $\mathbf{R}^n$ such that the segment $L = [x, y]$ contains no points of the plane $x_{n-1} = 0$, $x_n = 0$. The image of the segment L is the curve $\varphi(L)$ for which due to inequalities (1.47), we have

$$s(\varphi(L)) \leqslant \sqrt{\lambda_2(\varepsilon)}s(L) = \sqrt{\lambda_2(\varepsilon)}|y - x|.$$

Hence $|\varphi(y) - \varphi(x)| \leqslant \sqrt{\lambda_2(\varepsilon)}|y - x|$. By continuity, this inequality is valid for any x, y. Let the points x, y be such that the segment $L = [\varphi(x), \varphi(y)]$ contains no points of the plane $x_{n-1} = 0, x_n = 0$. From the inequalities (1.47), it follows that $|\varphi(y) - \varphi(x)| \geqslant \sqrt{\lambda_1(\varepsilon)}\varphi^{-1}(L) \geqslant \sqrt{\lambda_1(\varepsilon)}|x - y|$. Consequently, we obtain that for any $x, y \in \mathbf{R}^n$,

$$\frac{1}{L(\varepsilon)}|x - y| \leqslant |\varphi(x) - \varphi(y)| \leqslant L(\varepsilon)|x - y|,$$

where $L(\varepsilon) \to 0$ for $\varepsilon \to 0$. The image U of the halfspace $0 < \theta < \pi$, due to our definition, is a domain of the class Lip. At the same time, it is not a domain of the class Lip*, at least because for any point $x \in \partial U$ for which $x_{n-1} = x_n = 0$, none of its neighbourhoods on U can be represented by the equation $t_n = f(t_1, t_2, \dots, t_{n-1})$ in any Cartesian orthogonal system of coordinates.

$U \subset \mathbf{R}^n$ is said to be a domain of the class C^k, where $k \geqslant 1$ is an integer, if for each of its boundary points x_0, one can find a neighbourhood V which is an open rectangle $H = (a_1, b_1) \times (a_2, b_2) \times \cdots \times (a_n, b_n)$ in $\mathbf{R}^n$, such that the set $x \cap \partial U$ is represented by the equation $x_i = f(x_1, \dots, x_{i-1}, x_{i+1}, \dots, x_n)$, where f is a function of the class C^k. This function is defined on the corresponding rectangle H_i, which is a face of the rectangle H in $\mathbf{R}^{n-1}$, and such that $a_i < f(x_i, \dots, x_{i-1}, x_{i+1}, \dots, x_n) < b_i$ for all points $(x_1, \dots, x_{i-1}, x_{i+1}, \dots, x_n)$ of the rectangle H_i.

It is clear that if U is a domain of C^k, then U is also a domain of the class Lip*; consequently, it is a domain of the class Lip as well.

From the classical theorem on implicit functions, it follows that the condition that U is a domain of the class C^k is equivalent to the following. For every point $x_0 \in \partial U$, there exist a neighbourhood V and a diffeomorphism $\varphi : V \to \mathbf{R}^n$ of the class C^k (that is, the topological mapping of the class C^k such that its Jacobian $J(x, \varphi) = det\ \varphi'(x)$ vanishes nowhere), which satisfy the following conditions: the set $\varphi(V)$ is a cube $Q(0, 1)$, $\varphi(V \cap \partial U)$ is the section of this cube by a plane $x_n = 0$, and $\varphi(V \cap U)$ is the set of all points $x \in Q(0, 1)$ for which $x_n > 0$.

We see that the two ways of introducing the class of domains (which in the Lipschitz case lead to different results) prove to be equivalent for domains of the class C^k.

Lemma 1.5. (Invariance of the class J with respect to locally Lipschitz homeomorphisms). *Let U be a domain of the class J in $\mathbf{R}^n$, let $\varphi : U \to \mathbf{R}^n$ be* *a homeomorphic mapping of U in $\mathbf{R}^n$. Suppose that there exists a number $K \geqslant 1$, $K < \infty$ such that for every point $x_0 \in U$, one can find a neighbourhood $B(x_0, \delta) \subset U$ for which $(1/K)|x_1 - x_2| \leqslant |\varphi(x_1) - \varphi(x_2)| \leqslant K \cdot |x_1 - x_2|$ for any $x_1, x_2 \in B(x_0, \delta)$. Then $V = \varphi(U)$ is a domain of the class J.*

Remark. *The condition of the lemma is satisfied, in particular, if φ is a Lipschitz homeomorphism of the domain U.*

Proof. Let a be a marked point of the domain U, $\xi(s)$, $a \leqslant s \leqslant b$ is an arbitrary rectifiable curve lying in the domain U(the parameter s denotes the length of the arc), $\eta(s) = \varphi[\xi(s)]$. The set $\xi([a, b])$ is compact. If we cover it according to the Borel theorem by a finite number of neighbourhoods satisfying the condition given in the statement of the lemma, then we obtain that for every finite sequence of points $s_0 = a < s_1 < s_2 < \cdots < s_m = b$ for which the magnitude $max(s_i - s_{i-1})$ is sufficiently small

$$\begin{aligned}\sum_{i=1}^{m} |\eta(s_i) - \eta(s_{i-1})| &= \sum_{i=1}^{m} |\varphi[\xi(s_i)] - \varphi[\xi s_{i-1}]| \\ &\leqslant \sum_{i=1}^{m} K|\xi(s_i) - \xi(s_{i-1})| \\ &\leqslant K \sum_{i=1}^{m} (s_i - s_{i-1}) \\ &= K|b - a|.\end{aligned}$$

Taking the limit under the condition that $\max(s_i - s_{i-1}) \to 0$, we obtain that the curve η is rectifiable and its length exceeds the length of ξ by no more that K times.

The inverse mapping $\psi = \varphi^{-1}$ obviously satisfies the conditions of the lemma with the same constant K. Therefore, from what was proved above, it follows that each of the mappings φ and ψ transforms any rectifiable curve into the rectifiable one which is no more than K times greater than the length of the initial curve.

Let us put $b = \varphi(a)$. Let $y \in V, x = \psi(y)$. According to Lemma 1.4, there exists a rectifiable curve $x(s), 0 \leqslant s \leqslant l$ (s is the length of the arc), such that $x(0) = x$, $x(l) = a$, $l \leqslant L$ and for every $s \in [0, l]$, $\rho[x(s), \partial U] \geqslant \alpha \cdot s$ where the constants $\alpha > 0$ and $L < \infty$ only depend on the domain U. Suppose that $y(s) = \varphi[x(s)]$. The curve $y(s)$ is rectifiable, its length not exceeding $L_1 = KL$. We have $y(0) = y$, $y(l) = b$. Let $\sigma(s)$ be the length of the arc $[y(0), y(s)]$ of the curve $y(s)$. Then we have $\frac{1}{K}s \leqslant \sigma(s) \leqslant Ks$. Let us take an arbitrary point $s_0 \in [0, l]$. Let $\rho(s_0) = \rho[y(s_0), \partial V]$ and let z be a point of the set ∂V which is the nearest to $y(s_0)$ so that $|y(s_0) - z| = \rho(s_0)$. By the mapping ψ, the segment $[z, y(s)]$ is transformed into some curve T which connects the point $x(s_0)$ with a boundary point of U.

The length of the curve T is not less than $\rho(x(s_0), \partial U)$. On the other hand, applying the arguments given at the beginning of the proof for the segment $[y(s_0), z]$ and to the mapping $\varphi^{-1} = \psi$, we obtain that the length of T does not exceed $K\rho(s_0)$, consequently, $K\rho(s_0) \geqslant \rho(x(s_0), \partial U) \geqslant \alpha \cdot s_0$. Besides, we have $\sigma(s_0) \leqslant K s_0$. Finally, we obtain that

$$\rho(s_0) \geqslant \frac{\alpha}{K^2}\sigma(s_0).$$

Thus, we see that for the domain V all the conditions of Lemma 1.4 are valid with the constants $L_1 = LK$, $\alpha_1 = \alpha/K^2$. The lemma is proved.

Lemma 1.6. *If an open domain U is a union of a finite number of domains of the class J, then U is a domain of the class J.*

Proof. Let a domain U admit the representation $U = U_1 \cup U_2$, where U_1 and U_2 are domains of the class J. Let a_i, $i = 1, 2$ be marked points of the domains U_1 and U_2, $L_i > 0$, $\alpha_i \in (0, l]$, $i = 1, 2$ are constants such that every point $x \in U_i$ may be connected with a_i by a rectifiable curve $x(s)$, $0 \leqslant s \leqslant l \leqslant L_i$ (s is the length of the arc), so that $x(0) = x$, $x(l) = a_i$, and $\rho[x(s), \partial U_i] \geqslant \alpha_i \cdot s$ for all s.

The set U is connected; therefore, there exists a rectifiable curve $\xi(s)$, $0 \leqslant s \leqslant K$ (where the parameter s is the length of the arc), connecting the points a_2 and a_1 so that $\xi(0) = a_2$, $\xi(K) = a_1$. Let $\delta = \inf_{s\in[0,K]} \rho[\xi(s), \partial U]$. It is obvious that $\delta > 0$. We shall prove that for the domain U, the conditions of Lemma 1.4 hold with respect to the marked point a_1 for the proper values of the constants α and $L < \infty$, $0 < \alpha \leqslant 1$. Let $x \in U$. Suppose that $x \in U_2$. Then there exists a rectifiable curve $\eta(s)$, $0 \leqslant s \leqslant l$ (s is the length of the arc), such that $l \leqslant L_2$, $\eta(0) = x$, $\eta(l) = a_2$, and $\rho[\eta(s), \partial U] \geqslant \alpha_2 \cdot s$ for all $s \in [0, l]$. Let us set $x(s) = \eta(s)$ for $s \in [0, l]$, $x(s) = \xi(s - l)$ for $s \in [l, l + K]$. This defines the rectifiable curve connecting the point x with the point a_1. For $0 \leqslant s \leqslant l$, we have $\rho[x(s), \partial U] \geqslant \rho[x(s), \partial U_2] \geqslant \alpha_2 s$. But if $l \leqslant s \leqslant l + K$, then $\rho[(x(s), \partial U)] \geqslant \delta \geqslant \delta s/(l + K) = \alpha_0 s$, where $\alpha_0 = \delta/(l + K)$. The length of the curve $x(s)$ does not exceed $L_2 + K$. But if the point x belongs to U_1, then there exists a rectifiable curve $x(s)$, where $0 \leqslant s \leqslant l \leqslant L_1$, lying in U_2, such that $x(0) = x$, $x(l) = a_1$, and for all $s \in [0, l]$, the inequality $\rho[x(s), \partial U] \geqslant \rho[x(s), \partial U_1] \geqslant \alpha_1 s$ is valid. Consequently, we obtain that for the domain U, all the conditions of the lemma are valid with the constants $L = \max\{L_1, L_2 + K\}$ and $\alpha = \min\{\alpha_1, \alpha_2, \alpha_0\}$. Thus, we have proved that $U \in J$.

Suppose that the lemma is true for the case when U is a union of m domains of the class J, and let $U = \cup_{k=1}^{m+1} U_k$, where each of the domains U_k belongs to the class J. Since U is a connected open set, then the domain U_{m+1} is intersected by at least one domain U_k, where $1 \leqslant k \leqslant m$. Let $U_{m+1} \cap U_{k_0} \neq \emptyset$. Then according to what was proved above, $U_{m+1} \cap U_{k_0}$ is the domain of the class J, and we obtain that U is the union of m domains $U_1, \ldots, U_{k_0-1}, U_{k_0} \cap$

$U_{m+1}, \dots, U_m$, each of them belonging to the class J. Therefore, due to the inductive hypothesis, $U \in J$. The lemma is proved.

As a corollary of the lemma, we obtain the following result.

Lemma 1.7. *Every bounded domain with Lipschitz boundary in the space* $\mathbf{R}^n$ *is a domain of the class* J.

Proof. Let U be a bounded domain with Lipschitz boundary. Then, by definition, every point $x \in \bar{U}$ has a neighbourhood V such that the set $V \cap U$ is transformed into a cube by Lipschitz mapping. Every cube is obviously a domain of the class J; therefore, due to Lemma 1.5, $V \cap U$ is a domain of the class J. Every point $X \in U$ has a neighbourhood which is a ball. The set $\overline{U}$ is compact, and every point $x \in \overline{U}$ has a neighbourhood V such that $V \cap U$ is a domain of the class J. Due to the Borel theorem, $\bar{U}$ is covered by a finite number of neighbourhoods of such kind. Let $\{V_1, V_2, \dots, V_m\}$ be an open covering of the set U such that $U_i = V_i \cap U$ is a domain of the class J for every $i = 1, 2, \dots, m$. We obviously have: $U = \cup_{i=1}^{m} U_i$; therefore, due to Lemma 1.6, U is a domain of the class J. Q. E. D.

Corollary. *Every domain* U *of the class* C^k, *where* $k \geqslant 1$ *in the space* $\mathbf{R}^n$, *belongs to the class* J.

1.4. Integral Representations of Smooth Functions in Domains of the Class J

Let us assign an arbitrary domain U of the class J in the space $\mathbf{R}^n$, and let $f : U \to \mathbf{R}^k$ be a function of the class C^1. Using the results of Subsection 1.2, we shall construct a representation of the function f in terms of the derivatives $\frac{\partial f}{\partial x_1}, \frac{\partial f}{\partial x_2}, \dots, \frac{\partial f}{\partial x_n}$. Let $U \in J(r, R)$, where, as usual, $0 < r \leqslant R < \infty$, and let a be the marked point of the domain U. We have: $B(a, r) \subset U$.

For every point $x \in U$, there exists a curve $x(s), 0 \leqslant s \leqslant l$ (s is the length of the arc), where $l \leqslant R$, such that $x(0) = x$, $x(l) = a$, and $\rho(x(s), \partial U) \geqslant rs/l$ for all $s \in [0, l]$. Let us assign a function w such that $S(w) \subset B(0, r)$, and let us construct a curvilinear cone T which is a set of all points $y = x(t) + th$, where $h \in S(w)$. Having constructed, as in Subsection 1.2, the integral representation of the function f in this cone, we seem to obtain the desired. However, an unpleasant detail arises: the dependence on x of the functions in the obtained representation turns out to be hardly controllable. Passing to a new point x, we have to construct a new curve $x(s)$ each time. Eventually, it turns out that we cannot even guarantee the measurability of the obtained functions $K_i(x, y)$. When there exists a way of a unique construction of the curve $x(s)$ by the point x (below, we consider some specific cases when such situation takes place), the construction of Subsection 1.2 directly yields the desired integral representation. However, we do not know how to do it in the general case. Below, we shall act as follows. Although in general, we cannot give the regular way of constructing a path connecting x with the point a and satisfying the conditions we need, the problem can be solved locally for small neighbourhood of the point x. The desired integral representation is obtained by pasting together separate local representations by means of partition of unity.

For $x \in U$, let $\rho(x) = \rho(x, \partial U)$. Let us define some numbers p, q from the condition

$$q/p = r/2R, \qquad 2p + q = 1. \tag{1.48}$$

Obviously, $0 < p < 1/2$, $0 < q < 1$.

Below, u denotes a point of the domain U. We assume that either $u = a$ or $|u - a| \geqslant r/2$, as for the rest: the point u is arbitrary. The first step of further calculations is to construct an integral representation which is valid for all x lying in some neighbourhood of the point u. Let us construct some neighbourhood B_u of the point u and a path $\eta : [0,1] \to \mathbf{R}^n$ connecting x with the point a in the domain U. If $u = a$, then we assume $B_u = B(a, r/2)$, and $\eta(t) = a$ for all $t \in [0,1]$. If $|u - a| \geqslant r/2$, then, due to the definition of a domain of the class $J(r, R)$, there exists a rectifiable curve $x(s), 0 \leqslant s \leqslant l$ (s is the length of the arc), such that $l \leqslant R$, $x(0) = u, x(l) = a$, and $\rho[x(s)] \geqslant rs/l$ for all $s \in [0, l]$. Let us put

$$r' = \min\{q\rho(u), l\},\ B_u = B(a, r'), \tag{1.49}$$

and for $t \in [0,1]$ let $\eta(t) = x(lt)$. The function η is absolutely continuous in $[0,1]$, besides $|\eta'(t)| = l|x'(lt)| \leqslant l \leqslant R$ for almost all $t \in [0,1]$. For any $t \in [0,1]$ we have: $\rho[\eta(t), \partial U] = \rho[x(lt), \partial U] \geqslant (r/l)lt = rt$.

The closure of the ball B_u for all cases is contained in U.

For $x \in B_u$, we put

$$\xi(t) = \xi(t, x, u) = \eta(t) + (1 - t)(x - u). \tag{1.50}$$

Below, the symbol u is omitted whenever no misunderstanding may occur. For $h \in \mathbf{R}^n$, we set

$$\xi_h(t, x, u) \equiv \xi_h(t, x) = \xi(t, x) + ht.$$

Let us show that if $x \in \bar{B}_u$ and $|h| < r/2$, then $\xi_h(t, x)$ belongs to U for all $t \in [0,1]$. If $u = a$, then $B_u = B(a, r/2)$, and for any $x \in B_u$ and $h \in B(a, r/2)$, we have

$$|\xi_h(t, x) - a| = |(1 - t)(x - a) + th| \leqslant (1 - t)|x - a| + t|h| \leqslant r/2 < r,$$

so in this case, $\xi_h(t, x) \in B(0, r) \subset U$ for all $t \in [0,1]$.

Let $|u - a| \geqslant r/2$. If $t \leqslant (p/l)\rho(u)$; then, since $|\eta(t) - u| = |\eta(t) - \eta(0)| \leqslant lt$, due to (1.48), we have

$$\begin{aligned} |\xi_h(t, x) - u| &\leqslant |\eta(t) - u| + t|h| + (1 - t)|x - u| < t(l + r/2) + (1 - t)r' \\ &\leqslant [p(1 + r/2l) + q]\rho(u) \leqslant (2p + q)\rho(u) = \rho(u); \end{aligned}$$

and consequently in this case, $\xi_h(t,x) \in U$. Suppose that $t > (p/l)\rho(u)$. Then

$$\begin{aligned}|\xi_h(t,x) - \eta(t)| &\leqslant t|h| + (1-t)|x-u| \\ &< tr/2 + q \cdot \rho(u) \\ &< t(r/2 + ql/p) \\ &= t[r/2 + (r/2)l/R] \leqslant tr,\end{aligned}$$

since $l \leqslant R$ and due to (1.48), $q/p = r/2R$. Since for all $t \in [0,1]$, $\rho[\eta(t), \partial U] \geqslant rt$, then from what was proved above, it follows that in this case, $\xi_h(t,x) \in U$. Q. E. D.

Now let us use the results of Subsection 1.2. Let us assign an arbitrary point $u \in U$, and let B_u be a neighbourhood of the point u defined as above. Let $r_1 = r/2$ and let the function $w \in L_1(\mathbf{R}^n)$ be such that $S(w) \subset B(0, r_1)$, and

$$\int_{\mathbf{R}^n} w(h)\, dh = 1.$$

It is reasonable to emphasize the case where the function w has the form

$$w(h) = \frac{1}{r_1^n} w_0\left(\frac{h}{r_1}\right) = \frac{2}{r^n} w_0\left(\frac{2h}{r}\right), \tag{1.51}$$

where w_0 is the function of the class $C_0^\infty[B(0,1)]$. If w is defined in this way, then the regular case applies.

Lemma 1.8. *Suppose that U is a domain of the class $J(r,R)$ and B_u is defined as above. Then for every function $f : U \to \mathbf{R}^k$ of the class C^1 for any $\tau \in (0,1]$ for all $x \in B_u$, the equality*

$$f(x) = \int_U f(z) w\left[\frac{z - \xi(\tau,x,u)}{\tau}\right] \frac{dz}{\tau^n} + \int_U \sum_{j=1}^n H_{\tau,j}(z-x,x,u) \frac{\partial f}{\partial x_j}(z)\, dz \tag{1.52}$$

is valid, where the functions $H_{\tau,j}(y,x,u)$ are defined by the function $\xi(t) = \xi(t,x,u) = \eta(x) + (1-t)(x-u)$ by equalities (1.26). For fixed $x \in B$, the supports of the functions

$$z \mapsto w\left(\frac{1}{\tau}[z - \xi(\tau,x,u)]\right), \qquad z \mapsto H_{\tau,j}(z-x,x,u)$$

are contained in the compact set $T_\tau(u)$, which is the image of the set $S(w) \times [0,t] \times \bar{B}_u$ for the mapping $(h,t,x) \mapsto \xi_h(t,x,u)$.

For every $y \neq 0$, the estimate

$$|H_{\tau,j}(y,u)| \leqslant \frac{M_0}{|y|^{n-1}} \tag{1.53}$$

is valid. In the regular case, the function $y \mapsto H_{\tau,j}(y,u)$ *belongs to the class* C^∞ *in the domain* $\mathbf{R}^n \setminus \{0\}$, *and for any* α,

$$|D_y^\alpha H_{\tau,j}(y,x,u)| \leqslant \frac{M_\alpha}{|y|^{n+|\alpha|-1}}. \tag{1.54}$$

Here

$$M_\alpha = \|w_0\|_{|\alpha|} \left(\frac{5R}{r}\right)^{n+|\alpha|}$$

for every α ($\alpha = 0$ *included), where the functions* w *and* w_0 *are connected by the relation (1.51).*

Proof. Equality (1.52), as was mentioned, is a direct corollary of equality (1.25). To obtain estimates (1.52) and (1.54), it is necessary to obtain the magnitude K such that for the function $\xi(t) = \xi(t,x) = \eta(t) + (1-t)(x-u)$ the inequality $|\xi(t_1) - \xi(t_2)| \leqslant K|t_1 - t_2|$ holds for any $t_1, t_2 \in [0,1]$. We have

$$\begin{aligned} |\xi(t_1,x) - \xi(t_2,x)| &= |\eta(t_1) - \eta(t_2)| + |t_1 - t_2| \cdot |x-u| \\ &\leqslant (l + r')\,|t_1 - t_2| \\ &\leqslant (R + qR)\,|t_1 - t_2| \\ &\leqslant 2R\,|t_1 - t_2|, \end{aligned}$$

so that one can take $K = 2R$. Now estimates (1.53) and (1.54) follow from the estimates of Lemma 1.3 (inequalities (1.36)). The lemma is proved.

Lemma 1.9. *If* $w(h) = \frac{2^n}{r^n} w_0(\frac{2h}{r})$, *where* $w_0 \in C_0^\infty[B(0,1)]$, *then the function*

$$\lambda : (x,z) \mapsto \frac{1}{t^n} w \left[\frac{z - \xi(\tau,x,u)}{\tau}\right]$$

belongs to the class C^∞, *and for every multiindex* α *with* $|\alpha| \geqslant 0$,

$$|D_z^\alpha \lambda(x,z)| \leqslant \frac{2^{n+|\alpha|}\|w_0\|_{|\alpha|}}{(r\tau)^{n+|\alpha|}}. \tag{1.55}$$

Proof. We have

$$\frac{1}{\tau^n} w\left(\frac{z - \xi(\tau,x,u)}{\tau}\right) = \frac{2^n}{(\tau r)^n} w_0 \left(\frac{2(z - \eta(\tau) - (1-\tau)x}{r\tau}\right).$$

Hence, it directly follows that λ is a function of the class C^∞. By means of induction by $K = |\alpha|$, it is easy to obtain estimate (1.55) as well. The lemma is proved.

Lemma 1.10. *For every $x \in B_u$, the functions $z \mapsto w(\frac{z-\xi(\tau,u,x)}{\tau})$, $z \mapsto H_{\tau,j}(z-x,x,u)$ vanish outside the ball $B(x,3R\tau)$.*

Proof. Let $K < \infty$ be such that $|\xi'(t)| \leqslant K$ for almost all $t \in [0,1]$ where $\xi(t)$ is defined by formula (1.50). Then due to (1.21), the set T_τ is contained in the ball $B(x, \tau(r_1+K))$. The functions mentioned in the lemma vanish outside of the set T_τ. In the present case, one can take $K = 2R$. We have $r_1 = r/2 \leqslant R$; therefore, $T_\tau \subset B(x,3R\tau)$. The lemma is proved.

Now let us construct integral representations defined in the entire domain. For every point $u \in U$, some of its neighbourhood B_u is defined. We have $u \in B_u$ so that the family of balls $(B_u)_{u\in U}$ forms an open covering of the set U. Let $(\lambda_m) m = 1,2,\dots$, be the partition of unity subordinate to the open covering $(B_u)_{u\in U}$. Let λ_m be such that $S(\lambda_m) \subset B_{u_m}$. Let us write out equality (1.52) for $u = u_m$ and let us multiply both parts of the obtained relation by $\lambda_m(x)$. We shall have:

$$\lambda_m(x)f(x) = \int_U f(z)\lambda_m(x)w\left(\frac{z-\xi(\tau,x,u_m)}{\tau}\right)\frac{dz}{\tau^n} - \int_U \sum_{j=1}^n \lambda_m(x)H_{\tau,j}(z-x,x,u_m)\frac{\partial f}{\partial x_j}(z)\,dz. \tag{1.56}$$

Equality (1.55) also remains valid for $x \notin B_{u_m}$, since for such x both of its parts vanish. Summing by m, we get

$$f(x) = \int_U H_0(x,z,\tau)\frac{dz}{\tau^n} + \int_U \sum_{j=1}^n H_{\tau,j}(z-x,x)\frac{\partial f}{\partial x_j}(z)\,dz, \tag{1.57}$$

where

$$H_0(x,z,\tau) = \sum_{m=1}^\infty \lambda_m(x)w\left[\frac{z-\xi(\tau,x,u_m)}{\tau}\right]dz,$$

$$H_{\tau,j}(z-x,x) = \sum_{m=1}^\infty \lambda_m(x)H_{\tau,j}(z-x,x,u_m). \tag{1.58}$$

For every point $x \in U$, there exists a neighbourhood in which only a finite number of summands is distinct from zero. Therefore, for every $x \in U$, each of the sums (1.57) is finite. Due to this, when passing from (1.56) to (1.57), no difficulties arise concerning the limit transition under the integral.

As a result of the above constructions, we can make the following statement.

Theorem 1.1. *If U is a domain of the class $J(r,R)$, then U belongs to the class S. Here, for every $\tau \in [0,1]$, the integral representation of the form (1.3) may be constructed so that the functions $K_0, K_1, K_2, \dots, K_n$ should satisfy the*

Conditions A, B, and C of Subsection 1.1, with the following conditions being satisfied as well:

I) Each of the functions $K_i(x,y)$, $i = 0,1,2,\dots,n$ vanishes if $|x-y| \geqslant 3R\cdot\tau$.

II) The estimates

$$|K_0(x,y) \leqslant \frac{C_0}{(\tau r)^n}, |D_y^\alpha K_0(x,y)| \leqslant \frac{C_\alpha}{(\tau r)^{n+|\alpha|}} \tag{1.59}$$

are valid, and for every $j = 1,2,\dots,n$,

$$|K_j(x,y)| \leqslant \frac{M_0}{|y-x|^{n-1}}, \tag{1.60}$$

$$|D_y^\alpha K_j(x,y) \leqslant \frac{M_\alpha}{|y-x|^{n-1+|\alpha|}} \tag{1.61}$$

for any multiindex α. The constant C_α, $|\alpha| \geqslant 0$ only depends on n and $|\alpha|$; $M_\alpha = D_\alpha(\frac{R}{r})^{n+|\alpha|}$, where D_α also depends on n and $|\alpha|$ only.

Proof. Let us show that $K_0(x,y) = H_0(x,y,\tau)$ and $K_j(x,y) = H_{\tau,j}(y-x,x)$, where H_0 and $H_{\tau,j}$ are defined by formulae (1.58), are the desired functions. For every point $x \in U$ there exists a neighbourhood V in which only a finite number of summands in each of the sums in (1.58) is distinct from zero. Hence it follows that the functions $y \mapsto K_j(x,y)$, $j = 1,2,\dots,n$, belong to the class C^∞ in the domain $U\backslash\{x\}$. Here

$$D_y^\alpha K_j(x,y) = \sum_{m=1}^{\infty} \lambda_m(x) D_y^\alpha H_{\tau,j}(y-x,x,u_m).$$

Therefore, estimates (1.60) and (1.61) are the corollary of estimates (1.53) and (1.54) and of the fact that $\lambda_m(x) \geqslant 0$ and $\sum \lambda_m(x) = 1$ for all x. Inequalities (1.59) are the corollary of (1.55).

From II) it follows, in particular, that $K_i(x,y)$ for $i = 1,2,\dots,n$ is a function of the type $|x-y|^{1-n}$, thus Condition C of Subsection 1.1 is satisfied.

For every point $x \in U$, there exists a neighbourhood V in which only a finite number of functions $\lambda_m(x)$ is distinct from zero. Let them be the functions $\lambda m_1(x), \lambda m_2(x), \dots, \lambda m_k(x)$. Let us set $A(x) = \cup_{i=1}^K T_\tau(u_{m_i})$. The set $A(x)$ is compact. For $x' \in V$ for every $y \notin A(x)$, $H_{\tau,j}(y-x',x',u_{m_i}) = 0$, whence it follows that $H_{\tau,j}(y-x,x') = 0$ for $x' \in V$ and $y \notin A(x)$. This proves that for the integral representation constructed here the Condition B of Subsection 1.1 is satisfied as well.

Lemma 1.9 yields that $K_0(x,y)$, as the function of the variables $(x,y) \in U \times U$, belongs to the class C^∞. For every x, $K_0(x,y)$ vanishes for $y \notin A(x)$ where $A(x)$ is the compact set constructed above. This implies that for every x, $K_0(x,y)$ as the function of y is a test function in U. Consequently, Condition A of Subsection 1.1 is also satisfied. This completes the proof of the theorem.

Let us especially note the case where in the representations under consideration, the parameter τ equals 1 everywhere. For the function $\xi(\tau, x, u)$ in equality (1.52) we have: $\xi(1, x, u) = a$ so that $\xi(1, x, u)$ does not depend on τ. Hence it follows that in the present case, $K_0(x, y) = w(y - a)$; therefore, we can formulate the following statement.

Corollary 1. *Let U be a domain of the class $J(r, R)$. Then for every function $f \in C^1$, the integral representation*

$$f(x) = \int_U f(y)w(y-a)\,dy + \sum_{i=1}^{n} \int_U K_i(x,y)\frac{\partial f}{\partial x_i}(y)\,dy \tag{1.62}$$

is valid. Here w is a function of the class C^∞ such that $S(w) \subset B(0, r/2)$,

$$\int_{\mathbf{R}^n} w(y) \cdot dy = 1,$$

and for all y, $|w(y)| \leqslant \frac{C}{r^n}$ where $C = const$. The integral representation (1.62) satisfies the Conditions A, B, and C of Subsection 1.1, and for the functions $K_0(x,y) = w(y-a)$ and $K_i(x,y)$, $i = 1, 2, \ldots, n$, estimates (1.59), (1.60), and (1.61) are valid. (Here, in (1.59), one should take $\tau = 1$.)

Corollary 2. *Let U be a domain of the class $J(r, R)$. Then for every $\tau > 0$, $\tau \leqslant 1$, one can take the functions $K_0^l(x,y)$, $K_\alpha(x,y)$, where α is an arbitrary multiindex of order l, such that for any function f of the class C^l the equality*

$$f(x) = \int_U K_0^l(x,y)f(y)\,dy + \int_U \sum_{|\alpha|=l} K_\alpha(x,y)D^\alpha f(y)\,dy.$$

is valid. Here the functions K_0^l and K_α satisfy the following conditions:

I) Each of the functions $K_0^l(x,y)$, $K_\alpha(x,y)$ vanishes if $y \notin B(x, 3R\tau)$.

II) For any $x, y \in U$,

$$|K_0^l(x,y)| \leqslant \frac{C}{(\tau\, r)^{n+l}}, \tag{1.63}$$

$$|K_\alpha(x,y)| \leqslant \frac{M_{l,0}}{|x-y|^{n-l}}, \tag{1.64}$$

$$\forall_\beta \ |D_y^\beta K_\alpha(x,y)| \leqslant \frac{M_{l,\beta}}{|x-y|^{n-l+|\beta|}}, \tag{1.65}$$

where

$$M_{l,0} \leqslant C_0 \left(\frac{R}{r}\right)^n, \qquad M_{l,\beta} \leqslant C_\beta \left(\frac{R}{r}\right)^{n+|\beta|},$$

and the constants $C_0 < \infty$, $C_\beta < \infty$ only depend on n and β.

III) For every point $x \in U$, one can find a neighbourhood V of the point x and a compact set $A(x)$ such that if $x' \in V$, then for any $y \notin A(x)$ $K_0^l(x', y) = 0$, and $K_\alpha(x', y) = 0$ for every α with $|\alpha| = l$.

To prove this statement it suffices to apply the conclusion, by means of which in Subsection 1.1 equality (1.13) was obtained from (1.3), to the integral representation of the theorem.

§2 Other Integral Representations

2.1. Sobolev-Type Integral Representations for Simple Domains

The main efforts made in Subsection 1.4 were aimed at obtaining integral representations of the (1.3) form for the domains whose boundary may have a rather irregular structure. However, if we only consider domains of rather simple structure, then all the above arguments are much more simplified. Let us give some examples.

1. A domain U in the space $\mathbf{R}^n$ is said to be a starlike domain with respect to a ball $B(a,r) \subset U$ if it is bounded and for any point $x \in U$ and for any point $y \in B(a,r)$, the segment $[x,y] \subset U$.

If the domain U is starlike with respect to the ball, then the formulae (1.25) of Subsection 1.2 allow us to directly obtain the integral representation of the form (1.3) without using the cumbersome constructions of Subsection 1.4.

Let us assign a function $w_0 \in C_0^\infty[B(0,1)]$ such that

$$\int_{\mathbf{R}^n} w_0(x)\,dx = 1,$$

and let us put $w(x) = \frac{1}{r^n}(\frac{x}{r})$. The function $w \in C_0^\infty[B(0,r)]$, and

$$\int_{\mathbf{R}^n} w(x)\,dx = 1.$$

Now let us use formula (1.25) and choose the path $\xi(t)$ as follows: we set $\xi(t) = ta + (1-t)x = x + t(a-x)$. In this case,

$$\begin{aligned}
\alpha(t) &= \frac{\xi(t)-\xi(0)}{t} = a - x,\\
\xi'(t) &= a - x,\\
\beta(t) &= \alpha(t) - \xi'(t) = 0.
\end{aligned}$$

The expression for the magnitude $H_{\tau,j}(z,x)$ takes the form

$$H_{\tau,j}(z,x) = z_j \int_0^\tau w\left(\frac{z}{t} + x - a\right)\frac{dt}{t^{n+1}} = z_j H(z,x,\tau),$$

where

$$H(z,x,\tau) = \int_0^\tau w\left(\frac{z}{t} + x - a\right)\frac{dt}{t^{n+1}} = \int_{1/\tau}^\infty w\,(zt + x - a)\,t^{n-1}dt. \quad (2.1)$$

Due to what was proved in Subsection 1.2, we obtain that the following proposition is valid.

Theorem 2.1. *Let U be a domain in $\mathbf{R}^n$ which is starlike with respect to a ball $B(a,r)$ and contained in a ball $B(a,R)$. Then for every function $f : U \to \mathbf{R}^k$ of the class C^1, the following integral representation of the form (1.3) takes place:*

$$f(x) = \int_U f(y)K_0(x,y)\,dy + \sum_{j=1}^{n} \int_U \frac{\partial f}{\partial x_j}(y)K_j(x,y)\,dy. \tag{2.2}$$

Here

$$K_0(x,y) = \frac{1}{t^n} w\left(\frac{y-x}{\tau} + x - a\right), \tag{2.3}$$

$$K_j(x,y) = (y_j - x_j)H(y-x,x,\tau), \tag{2.4}$$

$0 < \tau \leqslant 1$, $w \in C_0^\infty(\mathbf{R}^n)$, *and* $S(w) \subset B(0,r)$ *and* $\int_{\mathbf{R}^n} w(h)\,dh = 1$, $H(z,x,\tau)$ *is defined by formula (2.1). The integral representation (2.2) satisfies the Conditions A, B, and C of Subsection 1.1.*

Remark. For arbitrary functions K_i, $i = 0,1,2,\dots,n$, the estimates of Subsection 1.2 remain true.

The integral representation obtained in Theorem 2.1 has an important advantage which does not take place in the general case and is as follows. The function $H(z,x,\tau)$ defined by equality (2.4) admits the representation

$$H(z,x,\tau) = \Phi(z,x) + \psi(z,x,\tau),$$

where $\Phi(z,x)$ is a positively homogeneous function of the power $-n$ by the variable y, belonging to the class C^∞ on the set of all $(x,z) \in \mathbf{R}^n \times \mathbf{R}^n$ for which $z \neq 0$, $\psi(z,x,\tau)$ is a function of the class C^∞ in $\mathbf{R}^n \times \mathbf{R}^n$. We obtain such representation if we put

$$\Phi(z,x) = \int_0^\infty w(zt + x - a)\,t^{n-1}dt,$$

$$\psi(z,x,\tau) = -\int_0^{1/\tau} w(zt + x - a)\,t^{n-1}dt.$$

We only have to prove the homogeneity of Φ (as of the function of the variable y). Let $\lambda > 0$. We have

$$\Phi(\lambda y, x) = \int_0^\infty w(y\lambda t + x - a)t^{n-1}dt.$$

Performing the change of the variable $\lambda t = s$, we obtain $t = \frac{s}{\lambda}$, $dt = \frac{ds}{\lambda}$, whence

$$\Phi(\lambda y, x) = \frac{1}{\lambda^n} \int_0^\infty w(ys + x - a) s^{n-1} ds = \frac{1}{\lambda^n} \Phi(y, x).$$

Q. E. D.

The abovementioned property of the function $H(z, x, \tau)$ plays an important role in some problems. It is the foundation for the methods of estimating functions in terms of the values of differential operators. These methods are based on the Zygmund–Calderon theorem on singular integral operators (see [85], [86], [71]).

2. Below, we establish the integral representation which (formally being the corollary of (1.25)), however, has certain peculiarities not arising in the general case. In particular, that the summand on the right-hand side of the representation, which contains no derivative, has the form

$$\frac{1}{|G|} \int_G f(y)\, dy,$$

where $G \subset U$. The functions $K_i(x, y)$, $i = 1, 2, \ldots, n$, in the representation under consideration also have a simple geometric meaning. As in the previous case, we consider here domains of a much more restricted class than that considered in Section 1.

Let U be a bounded open domain in $\mathbf{R}^n$, let $G \subset U$ be an open set. U is said to be starlike with respect to G if for every point $y \in G$ and for any point $x \in U$, the segment $[x, y] \subset U$. The set G, in particular, may coincide with U; in this case, the condition that U is a starlike domain with respect to G simply means that the set U is convex.

Below, we need the following simple remark. Let there be a mapping $f : \mathbf{R}^n \to \mathbf{R}^n$. Then, for every set $A \subset \mathbf{R}^n$ the equality

$$\chi_A[f(x)] = \chi_{f^{-1}(A)}(x) \tag{2.5}$$

is valid.

Let us assign an arbitrary open domain $U \subset \mathbf{R}^n$ and an open set $G \subset U$ such that U is a domain that is starlike with respect to G. In particular, G may coincide with U. Let us assign arbitrary points $x \in U, y \in G$ and a parameter $\tau \in (0, 1]$. For $t \in (0, 1]$, let $G(x, t)$ be the set of all points z of the form $z = x + t(y - x)$, where $y \in G$. It is obvious that $G(x, t) \subset U$. The set $G(x, t)$ is obtained from G by the homothety with respect to x with the stretching coefficient t.

Let $f : U \to \mathbf{R}^n$ belong to the class $C^1(U)$. For $y \in G$, $x \in U$, we have

$$\begin{aligned} f(x + \tau(y - x)) - f(x) &= \int_0^\tau \frac{d}{dt} [f(x + t(y - x))]\, dt \\ &= \int_0^\tau \sum_{i=1}^n \frac{\partial f}{\partial x_i} [x + t(y - x)] (y_i - x_i)\, dt. \end{aligned}$$

Hence,

$$f(x) = f(x + \tau(y - x)) + \int_0^{\tau} \sum_{i=1}^{n} \frac{\partial f}{\partial x_i} [x + t(y - x)] (x_i - y_i)\, dt.$$

Let us integrate both parts of this equality by G. As a result, we obtain

$$|G| f(x) = \int_G f(x + \tau(y - x))\, dy$$
$$+ \int_G \left(\int_0^{\tau} \sum_{i=1}^{n} \frac{\partial f}{\partial x_i} [x + t(y - x)] (x_i - y_i)\, dt \right) dy.$$

In the former integral, we perform the change of the variable by setting $x + \tau(y - x) = z$, in the latter we change the integration order. The mapping $y \mapsto x + \tau(y - x)$ transforms G into $G(x, \tau)$. Due to this, the set $G(x, \tau)$ becomes the integration domain in the first integral after the change of variable. We have: $|G(x, \tau)| = \tau^n |G|$. As a result,

$$f(x) = \frac{1}{|G(x, \tau)|} \int_{G(x,\tau)} f(y)\, dy$$
$$+ \frac{1}{|G|} \int_0^{\tau} \left(\sum_{i=1}^{n} \int_G \frac{\partial f}{\partial x_i} [x + t(y - x)] (x_i - y_i)\, dy \right) dt.$$

In the inner integral let us perform a change of the integration variable by putting $x + t(y - x) = z$. As a result, the set $G(x, t)$ becomes the integration domain. For every function $F : U \to \mathbf{R}^n$, for $A \subset U$,

$$\int_A F(z)\, dz = \int_U \chi_A(y) F(y)\, dy.$$

Taking this fact into account and by applying the Fubini theorem, we once again obtain

$$f(x) = \frac{1}{|G(x, \tau)|} \int_{G(x,\tau)} f(y)\, dy$$
$$+ \frac{1}{|G|} \int_U \sum_{i=1}^{n} \frac{\partial f}{\partial x_i}(y)(x_i - y_i) \left(\int_0^{\tau} \chi_{G(x,t)}(y) \frac{dt}{t^{n+1}} \right) dy. \tag{2.6}$$

The second application of the Fubini theorem demands some grounding, which is realized by means of arguments similar to those used in the proof of Lemma 1.2. Let us transform the expression

$$\int_0^{\tau} \chi_{G(x,t)}(y) \frac{dt}{t^{n+1}}.$$

Let $h_t(y) = x + t(y-x)$. The inverse mapping h_t^{-1} is assigned by the formula $h_t^{-1}(y) = \frac{y-x}{t} + x$. Due to (2.5), we have $\chi_G(h_t^{-1}(y)) = \chi_{h_t(G)}(y) = \chi_{G(x,t)}(y)$, consequently,

$$\int_0^\tau \chi_{G(x,t)}(y)\frac{dt}{t^{n+1}} = \int_0^\tau \chi_G\left(\frac{y-x}{t} + x\right)\frac{dt}{t^{n+1}}.$$

Equality (2.6) allows us to conclude that for every function $f \in C^1(U)$, the equality

$$f(x) = \frac{1}{|G(x,\tau)|}\int_{G(x,\tau)} f(y)\,dy + \int_U \sum_{i=1}^n \frac{\partial f}{\partial x_i}(y)(x_i - y_i)K_\tau(x,y)\,dy \quad (2.7)$$

is valid where for $x \neq y$,

$$K_\tau(x,y) = \frac{1}{|G|}\int_0^\tau \chi_G\left(\frac{y-x}{t} + x\right)\frac{dt}{t^{n+1}}. \quad (2.8)$$

Putting $\tau = 1$, we obtain, in particular,

$$f(x) = \frac{1}{|G|}\int_G f(y)\,dy + \int_U \sum_{i=1}^n \frac{\partial f}{\partial x_i}(y)(x_i - y_i)K_1(x,y)\,dy, \quad (2.9)$$

where

$$K_1(x,y) = \frac{1}{|G|}\int_0^1 \chi_G\left(\frac{y-x}{t} + x\right)\frac{dt}{t^{n+1}}. \quad (2.10)$$

Equalities (2.7) and (2.9) are the integral representations whose deduction we intended to obtain.

Let us specially consider the case when the set G is convex. In integral (2.8), let us perform the change of the integration variable. Let us put $t = |y-x|/s$. As a result, we obtain

$$\begin{aligned} K_\tau(x,y) &= \frac{1}{|G|\,|y-x|^n}\int_{\frac{|y-x|}{\tau}}^\infty \chi_G\left(x + s\frac{y-x}{|y-x|}\right)s^{n-1}ds \\ &= \frac{1}{|G||y-x|^n}\Phi\left(x, \frac{y-x}{|y-x|}, \frac{|y-x|}{\tau}\right), \end{aligned} \quad (2.11)$$

where

$$\Phi(x,e,h) = \int_h^\infty \chi_G(x+se)s^{n-1}ds,$$

e is the unit vector in $\mathbf{R}^n$. The set $l(x,e)$ of all points $y = x + se$ where $s \geqslant 0$ is a ray starting from the point x and directed together with the vector e. If $l(x,e) \cap G = \emptyset$, then $\Phi(x,e,h) = 0$. Put $l(x,e) \cap G \neq \emptyset$. Then $l(x,e) \cap G$

is a segment. Let $\lambda_1(x,e) < \lambda_2(x,e)$ be the values of the parameter s which correspond to the ends of this segment. Then we have

$$\Phi(x,e,h) = \tfrac{1}{n}\left([\lambda_2(x,e)]^n - [\lambda_1(x,e)]^n\right) \quad \text{for} \quad h \leqslant \lambda_1(x,e),$$
$$\Phi(x,e,h) = \tfrac{1}{n}\left([\lambda_2(x,e)]^n - h^n\right) \quad \text{if} \quad \lambda_1(x,e) < h \leqslant \lambda_2(x,e),$$

and finally,

$$\Phi(x,e,h) = 0 \quad \text{for} \quad h > \lambda_2(x,e).$$

The integral representations (2.7) and (2.9) do not satisfy the Conditions A, B, and C of Subsection 1.1, so they cannot be the means of constructing the representation of a function in terms of its derivatives of higher orders. Nevertheless, they may be useful in certain problems, since by means of these representations, one can easily establish explicit values of constants in estimates obtained in terms of the integral representation of the form (1.3).

3. Let us consider the case when $U = \mathbf{R}^n$, and $f : \mathbf{R}^n \to \mathbf{R}^m$ is a compactly supported function of the class C^k, that is $f \in C_0^k(\mathbf{R}^n)$ where $k \geqslant 1$ is an integer. Here the integral representation of the function f in terms of its derivatives can be obtained in a simpler way than in the general case. Then the integral representation has a remarkable peculiarity: it has no summands containing the function itself in the integrand.

Let σ denote an $(n-1)$-dimensional Lebesgue measure, that is, the area of an $(n-1)$-dimensional sphere $S(0,1)$ in $\mathbf{R}^n$. Let $F \in L_1(\mathbf{R}^n)$. Then the equality

$$\int_{\mathbf{R}^n} F(x)\,dx = \int_0^\infty \left(\int_{S(0,1)} F(ru) r^{n-1} d\sigma(u) \right) dr \tag{2.12}$$

is valid (the representation of the Lebesgue integral in spherical coordinates).

Thus, let $f : \mathbf{R}^n \to \mathbf{R}^m$ be a function of the class $C_0^k(\mathbf{R}^n)$. Let us assign an arbitrary unit vector u, and let $\varphi(r) = f(x + ru)$. The function φ of the variable r has compact support in $\mathbf{R}$. Applying the formula of integration by parts, we obtain

$$\int_0^\infty \varphi^{(k)}(r) r^{k-1} dr = (-1)^k (k-1)!\varphi(0) = (-1)^k (k-1)! f(x).$$

We have

$$\varphi^{(k)}(r) = k! \sum_{|\alpha|=k} \frac{D^\alpha f(x+ru)}{\alpha!} u^\alpha.$$

Hence

$$f(x) = (-1)^k k \int_0^\infty \sum_{|\alpha|=k} \frac{D^\alpha f(x+ru)}{\alpha!} u^\alpha r^{k-1} dr.$$

Integrating this equality by the variable $u \in S(0,1)$ with respect to the measure σ, we obtain

$$\sigma_{n-1} f(x) = (-1)^k k \int_0^\infty \left(\int_{S(0,1)} \sum_{|\alpha|=k} \frac{D^\alpha f(x+ru)}{\alpha!} u^\alpha r^{k-1} d\sigma(u) \right) dr.$$

(Here $\sigma_{n-1} = \sigma(S(0,1))$ is the area of the $(n-1)$-dimensional sphere $S(0,1)$.) The function $F(z) = D^\alpha f(x+z) \frac{z^\alpha}{|z|^n}$ belongs to the class $L_1(\mathbf{R}^n)$, and due to (2.12)

$$\int_{\mathbf{R}^n} F(z)\, dz = \int_0^\infty \left(\int_{S(0,1)} D^\alpha f(x+ru) u^\alpha r^{k-1} d\sigma(u) \right) dr.$$

Hence,

$$\sigma_{n-1} f(x) = (-1)^k k \int_{\mathbf{R}^n} \sum_{|\alpha|=k} \frac{D^\alpha f(x+z)}{\alpha!} \frac{z^\alpha}{|z|^n} dz.$$

In the latter integral, let us perform the change of the integration variable putting $z + x = y$. Then $z = y - x$, and for every α such that $|\alpha| = k$, $(-1)^k (y-x)^\alpha = (x-y)^\alpha$. Finally we have

$$f(x) = \frac{k}{\sigma_{n-1}} \int_{\mathbf{R}^n} \sum_{|\alpha|=k} \frac{D^\alpha f(y)}{\alpha!} \frac{(x-y)^\alpha}{|x-y|^n} dy. \tag{2.13}$$

4. Let us show how by starting from (1.3) one can obtain the integral representation of a function f by means of a collection of its derivatives $\frac{\partial^{\lambda_1} f}{\partial x_1^{\lambda_1}}, \frac{\partial^{\lambda_2} f}{\partial x_2^{\lambda_2}}, \dots, \frac{\partial^{\lambda_n} f}{\partial x_n^{\lambda_n}}$. We shall do it in order to give an example of the application of the idea, being the foundation of the deduction of the integral representation of a function by its derivatives of order r. The deduction was given in Subsection 1.1. Such integral representations were first constructed by Ilyin, and they play an important role in the theory of anisotropic Sobolev spaces, i.e., in the theory of classes of functions with different differentiability order by different variables. Below, we do not touch upon this important theme, since its presentation causes great technical difficulties. The deduction given here differs from that given by Ilyin.

Thus, let a vector $\lambda = (\lambda_1, \lambda_2, \dots, \lambda_n)$ be given, where $\lambda_1 > 0, \lambda_2 > 0, \dots, \lambda_n > 0$ are integers. Let us put $l = \lambda_1 + \lambda_2 + \dots + \lambda_n - n + 1$. Let $\alpha = (\alpha_1, \alpha_2, \dots, \alpha_n)$ be a multiindex. Below, the recording $\alpha < \lambda$ implies that $\alpha_i < \lambda_i$ for every $i = 1, 2, \dots, n$. Let f be an arbitrary function of the class $C^l(U)$. For $x \in U$, $\xi \in U$, we set

$$\varphi(x,\xi) = \sum_{\alpha<\lambda} \frac{(x-\xi)^\alpha}{\alpha!} D^\alpha f(\xi).$$

If $\alpha = (\alpha_1, \alpha_2, \ldots, \alpha_n) < \lambda$, then for every $i = 1, 2, \ldots, n$, $\alpha_i \leqslant \lambda_i - 1$, whence it follows that $|\alpha| = \alpha_1 + \alpha_2 + \cdots + \alpha_n \leqslant |\lambda| - n = l - 1$. This yields that for every $i = 1, 2, \ldots, n$, the function $\varphi(x, \xi)$ belongs to the class C^1 as the function of the variable ξ. We have

$$\frac{\partial \varphi}{\partial \xi_i}(x, \xi) = \sum_{\alpha < \lambda,\, \alpha \geqslant \delta_i} \left[-\frac{(x-\xi)^{\alpha - \delta_i}}{(\alpha - \delta_i)!} D^\alpha f(\xi) \right] + \sum_{\alpha < \lambda} \frac{(x-\xi)^\alpha}{\alpha!} D^{\alpha + \delta_i} f(\xi). \tag{2.14}$$

The conditions $\alpha < \lambda$, $\alpha \geqslant \delta_i$ are equivalent to the condition $\alpha = \beta + \delta_i < \lambda$. Due to this, equality (2.14) may be rewritten as follows,

$$\frac{\partial \varphi}{\partial \xi_i}(x, \xi) = \sum_{\alpha + \delta_i < \lambda} \left[-\frac{(x-\xi)^{\alpha}}{\alpha!} D^{\alpha+\delta_i} f(\xi) \right] + \sum_{\alpha < \lambda} \frac{(x-\xi)^\alpha}{\alpha!} D^{\alpha + \delta_i} f(\xi).$$

All the summands on the right-hand side of the latter equality, for which $\alpha + \delta_i < \lambda$, are reduced. The summands, for which $\alpha < \lambda$ and the condition $\alpha + \delta_i < \lambda$ does not hold, remain. If α is such that $\alpha < \lambda$ and $\alpha + \delta_i < \lambda$ is not true, then obviously $\alpha_i = \lambda_i - 1$ and $\alpha_j < \lambda_j$ for $j \neq i$. Let us put $\alpha - (\lambda_i - 1)\delta_i = \gamma$. The differentiation operator D^γ does not contain differentiation by the variable x_i. For the given α,

$$\frac{(x-\xi)^\alpha}{\alpha!} = \frac{(x-\xi)^\gamma}{\gamma!} \frac{(x_i - \xi_i)^{\lambda_1 - 1}}{(\lambda_i - 1)!}.$$

Due to the above, the derivative $\frac{\partial \varphi}{\partial \xi_i}(x, \xi)$ acquires the form:

$$\frac{\partial \varphi}{\partial \xi_i}(x, \xi) = \sum_{\gamma < \lambda - \lambda_i \delta_i} \frac{(x-\xi)^\gamma}{\gamma!} \frac{(x_i - \xi_i)^{\lambda_i - 1}}{(\lambda_i - 1)!} D^\gamma \left(\frac{\partial^{\lambda_i} f}{\partial \xi_i^{\lambda_i}}(\xi) \right). \tag{2.15}$$

Now let us apply the integral representation (1.3) to the function $\varphi(x, \xi)$ as to the function of the variable ξ (so far, x is considered as some parameter). We obtain

$$\varphi(x, \xi) = \int\limits_U K_0(\xi, y) \varphi(x, y)\, dy + \sum_{i=1}^n \int_U K_i(\xi, y) \frac{\partial \varphi}{\partial y_i}(x, y)\, dy. \tag{2.16}$$

It is obvious that $\varphi(x, x) = f(x)$. By putting $\xi = x$ in (2.16), by substituting the expression for $\varphi(x, y)$ and by taking (2.15) into account, we obtain

$$f(x) = \int_U K_0(x, y) \sum_{\alpha < \lambda} \frac{(x-y)^\alpha}{\alpha!} D^\alpha f(y)\, dy$$
$$+ \sum_{i=1}^n \int_U K_i(x, y) \sum_{\gamma < \lambda - \lambda_i \delta_i} \frac{(x-y)^\gamma}{\gamma!} \frac{(x_i - y_i)^{\lambda_i - 1}}{(\lambda_i - 1)!} D^\gamma \left(\frac{\partial^{\lambda_i} f}{\partial y_i^{\lambda_i}}(y) \right) dy. \tag{2.17}$$

This is not the desired integral representation yet. We shall obtain the desired representation using the following proposition.

Lemma 2.1. *Let $F(x,y)$, $x \in U$, $y \in U$, be a function of the type $|x-y|^{r-n}$, where $r > 0$ and $\alpha = (\alpha_1, \alpha_2, \ldots, \alpha_n)$ is an n-dimensional multiindex such that $l = |\alpha| < r$. Suppose that $x_0 \in U$ is such that the function $y \mapsto F(x_0, y)$ is compactly supported in U. Then for every function $u(x) \in C^l(U)$, the equality*

$$\int_U F(x_0,y) D_y^\alpha u(y)\, dy = (-1)^l \int_U D_y^\alpha F(x_0,y) u(y)\, dy \tag{2.18}$$

is valid.

Proof. We shall prove the statement of the lemma by induction by $l = |\alpha|$. Let $l = 1$. Then $D_y^\alpha = \frac{\partial}{\partial y_i}$. For $y = (y_1, y_2, \ldots, y_n) \in \mathbf{R}^n$, let $\hat{y}_i$ denote a point $\mathbf{R}^{n-1}$ which is obtained by crossing out the ith coordinate of the vector y, and let U_t be the section of U by the plane $y_i = t$. According to the Fubini theorem,

$$\int_U F(x_0,y) \frac{\partial u}{\partial y_i}(y)\, dy = \int_{-\infty}^{+\infty} \left(\int_{U_t} F(x_0,y) \frac{\partial u}{\partial y_i}(y)\, d\hat{y}_i \right) dt.$$

For $t \neq x_{0i}$, the magnitude $F(x_0, y)$ as a function of the variable $\hat{y}_i$ in the domain U_t is a test function of the class C^∞. Hence, by means of integration by parts, we obtain that for every such t,

$$\int_{U_t} F(x_0,y) \frac{\partial u}{\partial y_i}(y)\, d\hat{y}_i = -\int_{U_t} \frac{\partial F}{\partial y_i}(x_0,y) u(y) d\hat{y}_i.$$

The latter equality is valid for every $t \neq x_{0i}$, that is, for almost all t, whence it follows that

$$\int_{-\infty}^{+\infty} \left(\int_{U_t} F(x_0,y) \frac{\partial u}{\partial y_i}(y) d\hat{y}_i \right) dt = -\int_{-\infty}^{+\infty} \left(\int_{U_t} \frac{\partial F}{\partial y_i}(x_0,y)\, u(y)\, d\hat{y}_i \right) dt. \tag{2.19}$$

The function $y \mapsto F(x_0, y)$ vanishes outside of some compact set $a \subset U$; therefore, the derivative $\frac{\partial F}{\partial y_i}(x_0, y)$ also vanishes outside of A. The derivative $\frac{\partial F}{\partial y_i}(x, y)$ is the function of the type $|x-y|^{r-1-n}$. The function $u(x)$ is continuous, therefore it is bounded on A. Hence it follows that for all $y \in U$,

$$\left| \frac{\partial F}{\partial y_i}(x_0,y)\, u(y) \right| \leqslant C|x_0 - y|^{r-1-n}.$$

Since $r - 1 > 0$ and the function $\frac{\partial F}{\partial y_i}(x_0, y)\, u(y)$ equals zero for $y \notin A$, it follows from the latter estimate that it is integrable by U. According to the Fubini theorem, it is possible to state that the integral on the right-hand side of equality (2.19) is equal to

$$-\int_U \frac{\partial F}{\partial y_i}(x_0,y)\, u(y)\, dy,$$

and for the case $l = 1$, the lemma is proved.

Suppose that lemma is valid for the case $|\alpha| = l$, and let $|\alpha| = l + 1 < r$. Let α_i be one of the components of the multiindex α which are distinct from zero. According to what was proved above,

$$\int_U F_0(x,y) D_y^\alpha u(y)\, dy = \int_U F_0(x,y) \frac{\partial}{\partial y_i} D_y^{\alpha-\delta_i} u(y)\, dy$$
$$= -\int_U \frac{\partial F}{\partial y_i}(x_0,y) D_y^{\alpha-\delta_i} u(y)\, dy.$$

The function $\frac{\partial F}{\partial y_i}(x,y)$ is the function of the type $|x-y|^{r-1-n}$ and for $x = x_0$, as the function y, it has compact support in U. We have: $|\alpha - \delta_i| = |\alpha| - 1 = l$; therefore, according to the inductive admission,

$$-\int_U \frac{\partial F}{\partial y_i}(x_0,y) D_y^{\alpha-\delta_i} u(y)\, dy = (-1)^{l+1} \int_U D_y^{\alpha-\delta_i} \left(\frac{\partial F}{\partial y_i}(x_0,y) \right) u(y)\, dy$$
$$= (-1)^{|\alpha|} \int_U D_y^\alpha F(x_0,y) u(y)\, dy.$$

The lemma is proved.

Let us now bring expression (2.17) to the final form. In the first integral in (2.17), the function $y \mapsto K_0(x,y)$ belongs to the class $C_0^\infty(U)$. Applying the rule of integration by parts, we obtain

$$\int_U K_0(x,y) \sum_{\alpha<\lambda} \frac{(x-y)^\alpha}{\alpha!} D^\alpha f(y)\, dy = \int_U K_\lambda(x,y) f(y)\, dy,$$

where

$$K_\lambda(x,y) = \sum_{\alpha<\lambda} (-1)^{|\alpha|} D_y^\alpha \left[\frac{(x-y)^\alpha}{\alpha!} K_0(x,y) \right]. \tag{2.20}$$

Then, due to Lemma 1.1, $K_i(x,y)(x-y)^\gamma (x_i - y_i)^{\lambda_i - 1}$ is the function of the type $|x-y|^{|\gamma|+\lambda_i-n}$. Applying Lemma 2.1, we obtain that for every $\gamma < \lambda - \lambda_i \delta_i$,

$$\int_U K_i(x,y)(x-y)^\gamma (x_i-y_i)^{\lambda_i-1} D^\gamma \left(\frac{\partial^{\lambda_i} f}{\partial y_i^{\lambda_i}} \right)(y)\, dy$$
$$= (-1)^{|\gamma|} \int_U D_y^\gamma \left[K_i(x,y)(x-y)^\gamma (x_i-y_i)^{\lambda_i-1} \right] \frac{\partial^{\lambda_i} f}{\partial y_i^{\lambda_i}}(y)\, dy.$$

Eventually, we obtain the following integral relation:

$$f(x) = \int_U K_\lambda(x,y) f(y)\, dy + \sum_{i=1}^n \int_U K_{\lambda,i}(x,y) \frac{\partial^{\lambda_i} f}{\partial y_i^{\lambda_i}}(y)\, dy, \tag{2.21}$$

where $K_\lambda(x,y)$ is defined by equality (2.20), and $K_{\lambda,i}(x,y)$ is expressed by the formula

$$K_{\lambda,i}(x,y) = \left(\sum_{\gamma<\lambda-\lambda_i\sigma_i} (-1)^{|\gamma|} D_y^\gamma \left[\frac{(x-y)^\gamma}{\gamma!} K_i(x,y) \right] \right) \frac{(x_i-y_i)^{\lambda_i-1}}{(\lambda_i-1)!}.$$

5. Let us construct integral representations similar to those considered above for the functions of the class C^1 on a sphere of radius r in the space $\mathbf{R}^{n+1}$.

First, let us make some remarks concerning the integration on a sphere. Let there be a sphere $S(a,r) = \{x \in \mathbf{R}^{n+1} \,|\, |x-a| = r\}$. Let us assign an arbitrary point $x \in S(a,r)$. Let us put $x - a = z$ and let $x' = a - z$ be the point of sphere $S(a,r)$, which is diametrically opposite to x. Denote by $\Gamma(x)$ the section of the sphere $S(a,r)$ by the hyperplane $\{y \in \mathbf{R}^{n+1} | \langle y-a, z\rangle = 0\}$. This hyperplane passes through the point a and is orthogonal to the straight line xx'. Let σ and τ be measures in $S(a,r)$ and $\Gamma(x)$, respectively, such that for an arbitrary Borel set $E \subset S(a,r)$, $\sigma(E)$ is the n-dimensional area of E; and, similarly, for any Borel set $E \subset \Gamma(x)$, $\tau(E)$ is the $(n-1)$-dimensional area of E. Let y be a point of the sphere $S(a,r)$ distinct from x and x'. Let us draw through the points x, x', and y a two-dimensional plane P. The straight line xx' divides the plane P into two half-planes, and let P' be the half-plane containing the point y. The half-plane P' intersects the $(n-1)$-dimensional sphere $\Gamma(x)$ at the unique point u. Let φ be the angle between the vectors $y-a$ and $u-a$, which has the sign "+" if y lies on the same side of $\Gamma(x)$ as the point x; otherwise, it has the sign "−". (Formally, $\varphi = \arcsin \frac{\langle x-a, y-a\rangle}{r^2}$). The point $u \in \Gamma(x)$ and the number $\varphi \in (-\frac{\pi}{2}, \frac{\pi}{2})$ are said to be coordinates of y in the spherical system of coordinates with the pole x. We shall write $y = (u, \varphi)$. Due to this, instead of $f(y)$, we write $f(u,\varphi)$. Let $(u, \frac{\pi}{2}) = x$, $(u, -\frac{\pi}{2}) = x'$ for every $u \in \Gamma(x)$.

Let $f : S(a,r) \to \mathbf{R}$ be a function integrable with respect to the measure σ. Then the equality

$$\int_{S(a,r)} f(y)\, d\sigma(y) = \int_{\Gamma(x)} \left\{ \int_{-\pi/2}^{\pi/2} f(u,\varphi) r \cos^{n-1}\varphi\, d\varphi \right\} d\tau(u) \qquad (2.22)$$

is valid.

The function $f : S(a,r) \to \mathbf{R}$, where $S(a,r)$ is a sphere in $\mathbf{R}^{n+1}$, is said to belong to the class C^k if for every point $x \in S(a,r)$, there exists a neighborhood V of this point in $\mathbf{R}^{n+1}$ and a function $\tilde{f}$ of the class $C^k(V)$ such that for all $x \in V \cap S(a,r)$, $\tilde{f}(x) = f(x)$. This notion also admits other equivalent definitions which are not going to be analyzed here.

If $f : S(a,r) \to \mathbf{R}$ belongs to the class C^1, then at every point $x \in S(a,r)$,a vector $\nabla f(x)$ is defined which is the gradient of the function f at the point x. If $\varphi(t)$, $\alpha < t < \beta$, is an arbitrary smooth curve lying on the sphere $S(a,r)$, then for all $t \in (\alpha, \beta)$, $\frac{d}{dt} f[\varphi(t)] = \langle (\nabla f)(\varphi(t)), \varphi'(t) \rangle$. The vector $\nabla f(x)$ lies in the tangent plane of the sphere $S(a,r)$ at the point x, and the vector function ∇f is continuous.

Let there be a function $f : S(a,r) \to \mathbf{R}$ of the class C^1. Let us assign an arbitrary point $x \in S(a,r)$ and let us introduce on $S(a,r)$ the spherical system of coordinates with the pole x. Let $y = (u, \varphi)$, where $u \in \Gamma(x)$ and $\frac{-\pi}{2} < \varphi < \frac{\pi}{2}$. We have:

$$f(x) = f(y) + \int_{\varphi}^{\pi/2} \frac{\partial f}{\partial \theta}(u,\theta)\, d\theta.$$

Integrating by y over the sphere $S(a,r)$, we obtain

$$\begin{aligned} r^n w_n f(x) &= \int_{S(a,r)} f(y)\, dy \\ &+ \int_{\Gamma(x)} \left\{ r \int_{-\pi/2}^{\pi/2} \cos^{n-1}\varphi \left\{ \int_{\varphi}^{\pi/2} \frac{\partial f}{\partial \theta}(u,\theta)\, d\theta \right\} d\varphi \right\} d\tau(u). \end{aligned} \tag{2.23}$$

Denote by $\nu(x,y)$ a vector lying in the intersection of the tangent plane of the sphere $S(a,y)$ with the two-dimensional plane passing through the points a, x, and y and such that $|\nu(x,y)| = 1$ and $\langle \nu(x,y), x - a \rangle > 0$. The vector $\nu(x,y)$ is defined at every point y for which the coordinate $\varphi \neq \pm\frac{\pi}{2}$ and the equality

$$\frac{\partial f}{\partial \theta}(u,\theta) = \frac{1}{r} \langle \nabla f(y), \nu(x,y) \rangle \tag{2.24}$$

is valid. For an arbitrary continuous function $F(u,\varphi)$, we have

$$\begin{aligned} &\int_{-\pi/2}^{\pi/2} \cos^{n-1}\varphi \left(\int_{\varphi}^{\pi/2} F(u,\theta)\, d\theta \right) d\varphi \\ &= \int_{-\pi/2}^{\pi/2} \left(\int_{\varphi}^{\pi/2} F(u,\theta) d\theta \right) d \left(\int_{-\pi/2}^{\varphi} \cos^{n-1}\theta\, d\theta \right) \\ &= \int_{-\pi/2}^{\pi/2} \left(\int_{-\pi/2}^{\varphi} \cos^{n-1}\theta\, d\theta \right) F(u,\varphi)\, d\varphi. \end{aligned} \tag{2.25}$$

For $y = (u, \varphi)$, we set

$$\lambda(x,y) = \frac{1}{r \cos^{n-1}\varphi} \int_{-\pi/2}^{\varphi} \cos^{n-1}\theta\, d\theta.$$

From equality (2.25), taking into account the introduced notation, we obtain that

$$\int_{\Gamma(x)} \left\{ \int_{-\pi/2}^{\pi/2} \cos^{n-1}\varphi \left(\int_{\varphi}^{\pi/2} F(u,\theta)\, d\theta \right) d\varphi \right\} d\tau(u)$$

$$= \int_{\Gamma(x)} \left(\int_{-\pi/2}^{\pi/2} \lambda(x,u,\varphi) r \cos^{n-1}\varphi F(u,\varphi) d\varphi \right) d\tau(u)$$

$$= \int_{S(a,r)} \lambda(x,y) F(y) d\sigma(y).$$

By putting $F(y) = \langle \nabla f(y), \nu(x,y) \rangle$ in this equality, due to (2.23) and (2.24), we obtain

$$f(x) = \frac{1}{r^n w_n} \int_{S(a,r)} f(y)\, dy + \int_{S(a,r)} \lambda(x,y) \langle \nabla f(y), \nu(x,y) \rangle d\sigma(y). \quad (2.26)$$

Equality (2.26) is the desired integral representation.

2.2. Differential Operators with the Complete Integrability Condition

Let U be an open set in $\mathbf{R}^n$. We shall consider here vector functions defined in U and taking values in an m-dimensional complex space $\mathbf{C}^m$. Suppose that in U, there are assigned matrix functions A_i, $i = 1, 2, \ldots, n$, and for every n, A_i is a square matrix of the mth order whose elements are complex functions of the class C^1. Let $y : U \to \mathbf{C}^m$ be a function which has continuous partial first-order derivatives in U. Then we set

$$(L_i y)(x) = \frac{\partial y}{\partial x_i}(x) - A_i(x)\, y(x). \quad (2.27)$$

Equalities (2.27) define some system of differential operators $L_1, L_2, \ldots, L_n$. The system of operators of the form (2.27) is said to be completely integrable if the system of differential equations

$$\frac{\partial y}{\partial x_i}(x) - A_i(x)\, y(x) = 0 \quad (2.28)$$

is completely integrable, that is, if it satisfies the following condition. For every point $a \in U$ and for any vector $b \in \mathbf{C}^m$, there exists a (unique) solution of system (2.28) satisfying the Cauchy condition $y(a) = b$.

Below, we construct integral representations of a function through values of differential operators for this function of the form (2.27), under the assumption that the condition of complete integrability is satisfied. By means of these values, various integral representations may be constructed. The examples will be given below.

First let us establish some properties of completely integrable systems of differential operators of the form (2.27). In particular, let us obtain some necessary and sufficient conditions of complete integrability of such a system. The result given below is a special case of the classical Frobenius theorem (see, for instance, [36]).

Below, we assume that assigned are an open set U in $\mathbf{R}^n$ and a system of linear differential operators $L_1, L_2, \ldots, L_n$ of the form (2.27) which are defined in U.

Lemma 2.2. *If the system of equations (2.28) is completely integrable, then for all $x \in G$ for any $j, k = 1, 2, \ldots, n$, the equality*

$$(L_k A_j)(x) - (L_j A_k)(x) = 0 \tag{2.29}$$

holds.

Remark 1. *Equality (2.29) has the following explicit form:*

$$\frac{\partial A_j}{\partial x_k}(x) - \frac{\partial A_k}{\partial x_j}(x) + A_j(x) A_k(x) - A_k(x)\, A_j(x) = 0.$$

Remark 2. *The condition (2.29) is equivalent to the following one: the operators L_j and L_k commute for any j and k, that is, for every function $y \in C^2$,*

$$L_j(L_k y) = L_k(L_j y).$$

Proof. Suppose that the system of equations (2.28) is completely integrable. Let us take an arbitrary point $x_0 \in U$ and a vector $y_0 \in \mathbf{C}^m$. Then, according to the definition of the complete integrability condition, there exists the solution $y : U \to \mathbf{C}^m$ of system (2.28) such that $y(x_0) = y_0$. For any $j, k = 1, 2, \ldots, n$, we have:

$$\begin{aligned} \frac{\partial y}{\partial x_j}(x) &= A_j(x)\, y(x), \\ \frac{\partial y}{\partial x_k}(x) &= A_k(x)\, y(x). \end{aligned} \tag{2.30}$$

Since $A_j(x)$ and $y(x)$ are the functions of the class C^1, then from (2.30) it follows that the derivatives $\frac{\partial y}{\partial x_j}$ and $\frac{\partial y}{\partial x_k}$ are also the functions of the class C^1. By differentiating the former equality in (2.30) by x_k and the latter by x_j, and by equating the derivatives $\frac{\partial^2 y}{\partial x_j \partial x_k}$ and $\frac{\partial^2 y}{\partial x_k \partial x_j}$, we obtain:

$$\frac{\partial A_j}{\partial x_k}(x)\, y(x) + A_j(x) \frac{\partial y}{\partial x_k} = \frac{\partial A_k}{\partial x_j}(x)\, y(x) + A_k(x) \frac{\partial y}{\partial x_j}(x).$$

By expressing the derivatives $\frac{\partial y}{\partial x_k}$ and $\frac{\partial y}{\partial x_j}$ from the equalities (2.30), we have

$$\left(\frac{\partial A_j}{\partial x_k} + A_j(x)\,A_k(x)\right)\,y(x) = \left(\frac{\partial A_k}{\partial x_j} + A_k(x)A_j(x)\right)y(x).$$

By setting $x = x_0$, we then obtain

$$\left[\frac{\partial A_j}{\partial x_k}(x_0) - \frac{\partial A_k}{\partial x_j}(x_0) + A_j(x_0)\,A_k(x_0) - A_k(x_0)\,A_j(x_0)\right]y_0 = 0,$$

that is,

$$[(L_k A_j)(x_0) - (L_j A_k)(x_0)]\;y_0 = 0.$$

Due to the arbitrariness of $y_0 \in \mathbf{C}^m$ and $x_0 \in G$ from what was proved above, it follows that equality (2.29) is valid for all $x \in G$. The lemma is proved.

Lemma 2.2 yields some necessary conditions of complete integrability of system (2.28). The goal of further arguments is to show that if the domain G is simply-connected, then these conditions are sufficient as well.

Lemma 2.3. *Let there be given an open set G in $\mathbf{R}^m$, an open set H in $\mathbf{R}^s$ and a mapping $\varphi : H \to G$ of the class C^2. Suppose that in the set G, matrix functions $A_j(x)$, $j = 1, 2, \ldots, n$, are assigned, where $A_j(x)$ for every x is an $m \times m$ matrix whose elements are complex functions of the class C^1. For $t \in H$, we set*

$$B_p(t) = \sum_{k=1}^{m} \frac{\partial \varphi_k}{\partial t_p} A_k[\varphi(t)], \quad p = 1, 2, \ldots, s.$$

Then, if for any $j, l = 1, 2, \ldots, n$,

$$\frac{\partial A_j}{\partial x_l} - \frac{\partial A_l}{\partial x_j} + A_j A_l - A_l A_j \equiv 0$$

in G, then

$$\frac{\partial B_p}{\partial t_q} - \frac{\partial B_q}{\partial t_p} + B_p B_q - B_q B_p \equiv 0 \tag{2.31}$$

in the set H.

Proof. The fact that equality (2.31) is valid is established by simple calculations which we leave to the reader to carry out.

Remark. *Under the conditions of the lemma, let $u : G \to \mathbf{R}^m$ be the solution of the system of equations (2.28). Then the function $v(t) = u[\varphi(t)]$ is the solution of the system of equations*

$$\frac{\partial v}{\partial t_k} = B_k(t)\,v(t).$$

Lemma 2.4. *Let H be a rectangle $(\alpha, \beta) \times (\gamma, \delta)$ in the plane $\mathbf{R}^2$. Suppose that in H the system of equations*

$$\frac{\partial y}{\partial t}(t,u) = A(t,u)\, y(t,u), \tag{2.32}$$

$$\frac{\partial y}{\partial u}(t,u) = B(t,u)\, y(t,u) \tag{2.33}$$

is given, where $A(t,u)$ and $B(t,u)$ are $m \times m$ matrices whose elements are complex functions, y is a mapping of H into $\mathbf{C}^m$. So, if

$$\frac{\partial A}{\partial t} - \frac{\partial B}{\partial t} + AB - BA \equiv 0 \tag{2.34}$$

in the domain H, then the system of equations (2.32), (2.33) is completely integrable.

Proof. Let us take an arbitrary point $(t_0, u_0) \in H$ and a vector $y_0 \in \mathbf{C}^m$. It should be proved that there exists a solution y of the system (2.32), (2.33) for which $y(t_0, u_0) = y_0$. First let us consider a system of ordinary equations:

$$z'(t) = A(t, u_0)\, z(t). \tag{2.35}$$

Let z_0 be the solution of the system (2.35) in the interval (α, β), which satisfies the condition $z(t_0) = y_0$. According to the classical results of the theory of ordinary differential equations such a solution z_0 does exist. Let y_t be the solution in the interval (γ, δ) of the system of ordinary equations

$$y'(u) = B(t,u)\, y(u), \tag{2.36}$$

which satisfies the condition $y(u_0) = z_0(t)$. Such a solution does exist. Let us set $y(t,u) = y_t(u)$. Thus, some function y is defined in H. Let us prove that it is the solution of the system (2.32), (2.33).

From (2.36) it obviously follows that the function y satisfies the equation (2.33) in H. From the classical results on the dependence of the solution of a system of ordinary differential equations on a parameter, it follows that the function y is continuous in H and has the derivative $\frac{\partial y}{\partial t}$. This derivative is continuous and satisfies the equation

$$\frac{\partial}{\partial u}\left(\frac{\partial y}{\partial t}(t,u)\right) = B(t,u)\frac{\partial y}{\partial t}(t,u) + \frac{\partial B}{\partial t}(t,u)\, y(t,u), \tag{2.37}$$

which is formally obtained if we differentiate both parts of the equality (2.33) by t and if we change the derivative $\frac{\partial^2 y}{\partial t \partial u}$ for $\frac{\partial^2 y}{\partial u \partial t}$. Let us arbitrarily fix $t \in (\alpha, \beta)$, and let us set

$$w(u) = \frac{\partial y}{\partial t}(t,u) - A(t,u)\, y(t,u).$$

We have $y(t, u_0) = z_0(t)$. Therefore, due to (2.35), $w(u_0) = 0$. Then

$$w'(u) = \frac{\partial}{\partial u}\left(\frac{\partial y}{\partial t}(t,u)\right) - \frac{\partial A}{\partial u}(t,u)\, y(t,u) - A(t,u)\frac{\partial y}{\partial u}(t,u).$$

By substituting here the expression for the derivative $\frac{\partial}{\partial u}(\frac{\partial y}{\partial t})$ from (2.37) and for the derivative $\frac{\partial y}{\partial u}$ from (2.33), we obtain

$$w'(u) = \left[\frac{\partial B}{\partial t}(t,u) - \frac{\partial A}{\partial u}(t,u) - A(t,u)\right] y(t,u) + B(t,u)\frac{\partial y}{\partial u}(t,u).$$

Due to (2.34),

$$\frac{\partial B}{\partial t} - \frac{\partial A}{\partial u} - AB = -BA,$$

whence

$$w'(u) = B(t,u)\left[\frac{\partial y}{\partial t}(t,u) - A(t,u)y(t,u)\right] = B(t,u)w(u).$$

Thus the function w is the solution of some homogeneous linear differential equation in the interval (γ, δ). Since $w(u_0) = 0$, it follows that $w(u) = 0$ for all $u \in (\gamma, \delta)$, that is,

$$\frac{\partial y}{\partial t}(t,u) - A(t,u)y(t,u) = 0$$

for all $u \in (\gamma, \delta)$. Since $t \in (\alpha, \beta)$ is arbitrary, this proves that the function y satisfies in H the equation (2.32) as well.

The function y is continuous. From Eqs. (2.32) and (2.33), it follows that the derivatives $\frac{\partial y}{\partial t}$ and $\frac{\partial y}{\partial u}$ are continuous,i.e., $y \in C^1$.

The lemma is proved.

Let us make some simple remarks about domains in $\mathbf{R}^n$.

Let A be an arbitrary set in $\mathbf{R}^n$. A path in the set A is any continuous mapping $\xi : [0,1] \to \mathbf{R}^n$ such that for all $t \in [0,1]$, the point $\xi(t) \in A$. The path $\xi : [0,1] \to \mathbf{R}^n$ is said to connect the points p and q if $\xi(0) = p$, $\xi(1) = q$.

Let $\xi : [0,1] \to \mathbf{R}^n, \eta : [0,1] \to \mathbf{R}^n$ be two arbitrary paths in the set $A \subset \mathbf{R}^n$ which connect the given points p and q. The paths ξ and η are said to be homotopic in A if there exists a continuous mapping ς of the square $[0,1] \times [0,1]$ in $\mathbf{R}^n$ such that $\varsigma(t,\tau) \in A$ for all $(t,\tau) \in [0,1] \times [0,1]$ and for any $t \in [0,1]$, $\varsigma(t,0) = \xi(t)$, $\varsigma(t,1) = \eta(t)$ and $\varsigma(0,\tau) = p$, $\varsigma(1,\tau) = q$ for all $\tau \in [0,1]$. The mapping ς having the above properties is said to be the deformation of the path ξ to the path η in the set A.

Let U be an open set in $\mathbf{R}^n$. If U is connected, for any two points $p, q \in U$ there exists a path connecting the points p and q in U. (This proposition may be considered to be the definition of connectedness of an open set in $\mathbf{R}^n$.)

The open set $U \subset \mathbf{R}^n$ is called simply connected if it is connected and if any two paths in U, connecting arbitrary points $p, q \in U$, are homotopic in U.

The path $\xi : [0,1] \to \mathbf{R}^n$ is said to be smooth if ξ is a function of the class C^∞. Let ξ and η be two smooth paths in the open set $U \subset \mathbf{R}^n$ which connect the points $p, q \in U$. Suppose that ξ and η are homotopic in U, and let $\varsigma : [0,1] \times [0,1] \to U$ be a deformation of the path ξ into the path η. The mapping ς is said to be the smooth deformation of ξ into η if ς belongs to the class C^∞ (that is, ς is the contraction on the square $[0,1] \times [0,1]$ of a function of the class C^∞. This function is defined on some open set $H \subset \mathbf{R}^2$ containing the square $[0,1] \times [0,1]$).

Lemma 2.5. *Let U be an open set in $\mathbf{R}^n$, $p, q \in U$. Then, if there exists a path $\xi : [0,1] \to \mathbf{R}^n$ lying in U and connecting the points p and q, then for every $\varepsilon > 0$, there exists a smooth path $\eta : [0,1] \to \mathbf{R}^n$ in the set U, connecting the points p and q such that $|\xi(t) - \eta(t)| < \varepsilon$ for all $t \in [0,1]$.*

Let ξ and η be smooth paths in the set U which connect the points $p, q \in U$. If ξ and η are homotopic in U, then there exists a smooth deformation in the set U of the path ξ to the path η.

Proof. Let $\xi : [0,1] \to \mathbf{R}^n$ be a path in the set U, $\xi(0) = p$, $\xi(1) = q$. Let $K = \xi([0,1])$. The set K is compact; therefore, there exists $\delta > 0$ such that the set $U_\delta(K)$ is contained in U. Let us assign an arbitrary $\varepsilon > 0$. Let $\varepsilon_1 = \min\{\frac{\varepsilon}{2}, \frac{\delta}{2}\}$. Let us construct a mapping $\eta_0 : [0,1] \to \mathbf{R}^n$ of the class C^∞ such that $|\eta(t) - \eta_0(t)| < \varepsilon_1$ for all $t \in [0,1]$. Let us put $\eta(t) = [p - \eta_0(0)](1-t) + [q - \eta_0(1)]t + \eta_0(t)$. It is obvious that η is a smooth path in $\mathbf{R}^n$. We have: $\eta(0) = p, \eta(1) = q$ so that the path η connects the points p and q, and $|\eta(t) - \xi(t)| \leqslant |p - \eta_0(0)|(1-t) + |q - \eta_0(1)|t + |\eta_0(t) - \xi(t)| < (1-t)\varepsilon_1 + t\varepsilon_1 + \varepsilon_1 = 2\varepsilon_1$. Since $\varepsilon_1 < \frac{\delta}{2}$, it follows that $\rho[\eta(t), K] < \delta$ for all $t \in [0,1]$; therefore $\eta(t) \in U$ for all $t \in [0,1]$, so that η is the path in U. Besides, we also have: $|\eta(t) - \xi(t)| < \varepsilon$ for all $t \in [0,1]$.

Let ξ and η be arbitrary smooth paths in U such that $\xi(0) = \eta(0) = p$, $\xi(1) = \eta(1) = q$. Suppose that the paths ξ and η are homotopic in U. Let ς be a deformation in the set U of the path ξ to a path η. Let us put $K = \varsigma([0,1] \times [0,1])$. The set K is compact and $U \supset K$. Let us find $\delta > 0$ such that $U_\delta(K) \subset U$. Let $\varsigma_1 : [0,1] \times [0,1] \to \mathbf{R}^n$ be the mapping of the class C^∞ such that $|\varsigma(t,\tau) - \varsigma_1(t,\tau)| < \frac{\delta}{4}$ for any t and τ. The desired smooth deformation of the path ξ to the path η is obtained if we consecutively correct the function ς_1. Let us set

$$\varsigma_2(t,\tau) = [p - \varsigma_1(0,\tau)]\,(1-t) + [q - \varsigma_1(1,\tau)]\,t + \varsigma_1(t,\tau).$$

It is obvious that $\varsigma_2(0,\tau) = p$, $\varsigma_2(1,\tau) = q$ for all $\tau \in [0,1]$, and $|\varsigma_2(t,\tau) - \varsigma(t,\tau)| < \frac{\delta}{2}$. Then we set $\varsigma_0(t,\tau) = [\xi(t) - \varsigma_2(t,0)](1-\tau) + [\eta(t) - \varsigma_2(t,1)]\tau + \varsigma_2(t,\tau)$. We have: $\varsigma_0(t,0) = \xi(t), \varsigma_0(t,1) = \eta(t)$ so that ς_0 is the deformation of the path ξ to the path η. It is also easy to verify that $|\varsigma_0(t,\tau) - \varsigma(t,\tau)| < \delta$;

therefore, $\varsigma_0(t,\tau) \in U$ for all t and τ. Thus, we have constructed the smooth deformation of the path ξ to η in the domain U. The lemma is proved.

Theorem 2.1. *Let U be a domain in $\mathbf{R}^n$. Suppose that in U there is assigned a system of differential equations of the form (2.28). Then, if for all $x \in U$, equality (2.29) is valid and the domain U is simply connected, then system (2.28) is completely integrable in the domain U. Here for any $a \in G$ and $b \in \mathbf{C}^m$, the solution of system (2.28) which satisfies the condition $y(a) = b$ is unique.*

Proof. Let all conditions of the theorem be satisfied. Let us assign an arbitrary point $a \in U$ and a vector $b \in \mathbf{C}^m$. First let us prove that the solution of system (2.28) which satisfies the conditions $y(a) = b$ is unique if it does exist. Suppose that the function $y(x)$ is the solution of system (2.28), besides $y(a) = b$. Let us take an arbitrary point $x \in G$, and let $\xi : [0,1] \to G$ be a smooth path connecting the points a and x. Let us put $v(t) = y[\xi(t)]$. We have: $v'(t) = \sum_{j=1}^n \frac{\partial y}{\partial x_j}[\xi(t)]\xi_j'(t)$. Since y is the solution of system (2.29), $\frac{\partial y}{\partial x_j} = A_j y$. Let us set

$$A_\xi(t) = \sum_{j=1}^n \xi_j'(t) A_j[\xi(t)]. \tag{2.38}$$

Then we obtain

$$v'(t) = A_\xi(t)v(t).$$

Thus, the function $v(t) = y[\xi(t)]$ is the solution in the segment $[0,1]$ of the linear system of differential equations, and $v(0) = b$. Due to the known results of the theory of ordinary differential equations, such a solution is unique if it does exist. Consequently, we obtain that the value $v(1) = y[\xi(1)] = y(x)$ is uniquely defined, and the statement of the theorem concerning the uniqueness of the solution of system (2.28) is proved.

Now let us prove the existence of the solution $y(x)$ of system (2.28) such that $y(a) = b$, where $a \in G$ and $b \in \mathbf{C}^m$ are arbitrary. Let us take an arbitrary point $x \in G$ and let $\xi : [0,1] \to G$ be a smooth path connecting in G the points a and x. Let $A_\xi(t)$ be the matrix function defined according to formula (2.38). Let us consider the linear system of ordinary differential equations

$$\frac{\partial v}{\partial t} = A_\xi(t)v(t). \tag{2.39}$$

As was shown above, if y is the solution of system (2.28), then the function $y[\xi(t)]$ is the solution of system (2.39). Denote by $v_\xi(t)$ the solution of system (2.39) defined in the interval $[0,1]$ and such that $v(0) = b$. Due to the classical results of the theory of ordinary differential equations, the solution of system (2.39) which satisfies this condition exists and is unique.

We shall establish that the magnitude $v_\xi(1)$ does not depend on the choice of the path ξ in G which connects the points a and x. By setting $v_\xi(1) = y(x)$, we obtain, due to arbitrariness of x, some function y defined in G. This function is the desired solution of the system (2.28), as it will be shown below.

Let η be another arbitrary smooth path in the set G which connects the points a and x. Let us set $A_\eta(t) = \sum_{j=1}^n \eta_j'(t)A_j[\eta(t)]$, and let v_η be the solution in [0,1] of the system

$$\frac{dv}{dt} = A_\eta(t)v(t),$$

which satisfies the Cauchy condition $v(0) = b$. It is necessary to prove that $v_\xi(1) = v_\eta(1)$.

Since, according to the condition, the domain G is simply connected, the paths ξ and η are homotopic in G; therefore, due to Lemma 2.5, there exists a smooth deformation $\varphi : [0,1] \times [0,1] \to \mathbf{R}^n$ of the path ξ to the path η in the domain G. We have: $\varphi(t,0) = \xi(t)$, $\varphi(t,1) = \eta(t)$ for any $t \in [0,1]$, $\varphi(0,u) = a$, $\varphi(1,u) = x$ for all $u \in [0,1]$. The mapping φ belongs to the class C^∞. It may be considered to be defined on some rectangle $H = (\alpha,\beta) \times (\gamma,\delta) \supset [0,1] \times [0,1]$, and $\varphi(t,\tau) \in G$ for all $(t,\tau) \in H$. Let us put

$$A(t,u) = \sum_{j=1}^n \frac{\partial \varphi_j}{\partial t}(t,u)A_j[\varphi(t,u)], \tag{2.40}$$

$$B(t,u) = \sum_{j=1}^n \frac{\partial \varphi_j}{\partial u}(t,u)A_j[\varphi(t,u)]. \tag{2.41}$$

Due to Lemma 2.3, for any $(t,u) \in H$,

$$\frac{\partial A}{\partial u}(t,u) - \frac{\partial B}{\partial t}(t,u) + A(t,u)B(t,u) - B(t,u)A(t,u) = 0.$$

According to Lemma 2.4, it therefore follows that the system of equations

$$\frac{\partial z}{\partial t}(t,u) = A(t,u)z(t,u), \tag{2.42}$$

$$\frac{\partial z}{\partial u}(t,u) = B(t,u)z(t,u) \tag{2.43}$$

in the domain H is completely integrable. Let $z(t,u)$ be the solution of this system, such that $z(0,0) = b$.

We have $\varphi(0,u) = a$, $\varphi(1,u) = x$ for all $u \in [0,1]$. Hence it follows that $B(0,u) = B(1,u) = 0$ for $u \in [0,1]$; therefore, due to (2.43), $\frac{\partial z}{\partial u}(0,u) = \frac{\partial z}{\partial u}(1,u) = 0$ for $u \in [0,1]$. This yields that the magnitudes $z(0,u)$ and $z(1,u)$ are constant in the interval $[0,1]$. In particular, we obtain that $z(0,0) =$

$z(0,1) = b$. Further, let us note that $A(t,0) = A_\xi(t), A(t,1) = A_\eta(t)$. Due to (2.42), we obtain that $z(t,0) = v_\xi(t), z(t,1) = v_\eta(t)$. Hence it follows that $v_\xi(1) = z(1,0) = z(1,1) = v_\eta(1)$; thus, the coincidence of values of the functions v_ξ and v_η at the point $t = 1$ is proved.

We obtain that the vector $v_\xi(1)$ does not depend on the choice of the smooth path ξ connecting the points a and x in the domain G. Let us put $v_\xi(1) = y(x)$. Let us prove that the function $y : G \to \mathbf{C}^m$ thus defined is the solution of system (2.28) and $y(a) = b$.

Let $x = a$. Let us set $\xi(t) = a$ for all $t \in [0,1]$. Obviously, $\xi'(t) = 0$; therefore, $A_\xi(t) \equiv 0$, whence it follows that $v_\xi(t)$ in the segment $[0,1]$ is constant, and therefore, $y(a) = v_\xi(1) = b$.

Now let us establish the continuity of y. Let us take an arbitrary point $x_0 \in G$. Let us construct a smooth path ξ in the domain G for which $\xi(0) = a$, $\xi(1) = x_0$. The set $K = \xi([0,1])$ is compact; therefore, there exists $\delta > 0$ such that for every $x \in K$, $\rho(x, \partial U) \geqslant \delta$. Let us take an arbitrary point $x \in B(x_0, \delta)$ and let

$$\xi_x(t) = \xi(t) + t(x - x_0).$$

Obviously, ξ_x is the smooth path connecting the points a and x. Let us define the function v_{ξ_x} which is the solution of the system of equations

$$v'(t) = A_{\xi_x}(t)v(t) = \sum_{j=1}^{n} [\xi_j'(t) + x_j - x_{0j}] \times A_j[\xi(t) + t(x - x_0)]v(t), \quad (2.44)$$

which satisfies the condition $v(0) = b$. We have: $v_{\xi_x}(1) = y(x)$. The right-hand side of (2.44) is continuous with respect to x for $x \in B(x_0, \delta)$. Hence it follows that the function y is continuous in the ball $B(x_0, \delta)$. In particular, it is continuous at the point x_0. Since $x_0 \in G$ is arbitrary, this proves that y is a function continuous in G.

Let us assign arbitrary $j = 1, 2, \ldots, n$ and let us set $\varphi(t, u) = \xi(t) + tue_i$, where $|u| < \delta$, and ξ, δ, and x_0 have the above meaning. By the function φ let us define matrix and vector functions A, B according to formulae (2.40) and (2.41). System (2.42), (2.43) corresponding to the given A and B is completely integrable. Let $z(t, u)$ be its solution satisfying the condition $z(0,0) = b$. We have $B(t, 0) = 0$, whence it follows that $z(0, u)$ is constant for $u \in [0,1]$, and therefore, $z(0, u) = b$ for any $u \in [0,1]$. For every fixed u, the function $v(t) = z(t, u)$ is the solution of the system of ordinary differential equations

$$v'(t) = A_{\varphi_u}(t)v(t),$$

where $\varphi_u(t) = \varphi(t, u)$, because it is obvious that $A_{\varphi_u}(t) = A(t, u)$. Hence it follows that $z(1, u) = y[\varphi(1, u)] = y(x_0 + ul_j)$. This yields that the function $u \mapsto y(x_0 + ue_j)$ is differentiable in the interval $(-\delta, \delta)$. We have:

$$\frac{\partial y}{\partial u}(x_0) = \frac{\partial z}{\partial u}(1,0) = B(1,0)z(1,0).$$

It remains to note that $z(1,0) = y(x_0)$, $B(1,0) = A_j(x_0)$, and we obtain that the equality

$$\frac{\partial y}{\partial xj}(x_0) = A_j(x_0)y(x_0)$$

is valid. The theorem is proved.

Remark. If the system of equations (2.28) is completely integrable and the matrix functions A_j belong to the class C^k, then any solution of system (1.2) is the function of the class C^{k+1}.

2.3. Integral Representations of a Function in Terms of a System of Differential Operators with the Complete Integrability Condition

First let us establish some properties of differential operators that we are going to deal with.

Let us assign an arbitrary domain $G \subset \mathbf{R}^n$ and in it let us take a completely integrable system of differential equations

$$\frac{\partial y}{\partial x_j} = A_j(x)y(x), \tag{2.45}$$

$j = 1, 2, \ldots, n$, where A_j are matrices of the type $m \times m$ whose elements are complex functions of the class C^1 in the domain G.

Together with system (2.45), let us consider the following system of equations with respect to matrix functions,

$$\frac{\partial Y}{\partial x_i} = A_i(x)Y(x), \quad i = 1, 2, \ldots, n, \tag{2.46}$$

where $Y(x)$ is an $m \times m$ matrix for every $x \in G$. If $Y(x)$ is the solution of system (2.46), then every column of the matrix $Y(x)$ considered as a vector function with the values in $\mathbf{C}^m$ is the solution of system (2.45), whence it follows that system (2.46) is completely integrable; that is, for every point $a \in G$ and for any square matrix Y_0 of the order m, there exists a solution of system (2.46) such that $Y(a) = Y_0$. Indeed, let $b^{(i)}$ be the ith column of the matrix Y_0, let $y^{(i)}$ be the solution of system (2.45) for which $y^{(i)}(a) = b^{(i)}$. Let $Y(x)$ be the matrix whose ith column is $y^{(i)}$. Then $Y(x)$ is the solution of system (2.46) and $Y(a) = Y_0$.

If Y is the solution of system (2.46) and h is an arbitrary vector in $\mathbf{C}^m$, then the vector function $Y(x)h$ is the solution of system (2.45).

Let ξ be an arbitrary point of the domain G. Let us denote by $\Omega(x, \xi)$ the solution $Y(x)$ of the matrix equation (2.46) such that $Y(\xi) = I$, where I is the unit matrix of the order m. The matrix function $\Omega(x, \xi)$ is said to be the fundamental matrix of system (2.45).

Let us note some properties of the matrix function $\Omega(x, \xi)$.

If b is a vector in $\mathbf{C}^m$, then $y(x) = \Omega(x, \xi)b$ is the solution of system (2.45), satisfying the condition $y(\xi) = b$. Similarly, if Y_0 is a square matrix of the order m whose elements are complex numbers, then $Y(x) = \Omega(x, \xi)Y_0$ is the solution of system (2.46) such that $Y(\xi) = Y_0$.

From the definition of Ω, it directly follows that

$$\Omega(x, x) = I \tag{2.47}$$

for every $x \in G$.

Let x, y, z be three arbitrary points of the domain G. Let us put $Y_1(x) = \Omega(x, y)\Omega(y, z)$, $Y_2(x) = \Omega(x, z)$. The matrix functions Y_1 and Y_2 are the solution of system (2.46). Here $Y_1(y) = I$, $\Omega(y, z) = \Omega(y, z)$, and also $Y_2(y) = \Omega(y, z)$; that is, $Y_1(y) = Y_2(y)$. Hence it follows that $Y_1(x) = Y_2(x)$ for all x, and thus, for any $x, y, z \in G$, the equality

$$\Omega(x, y)\Omega(y, z) = \Omega(x, z) \tag{2.48}$$

is valid. In particular, by letting $z = x$, we obtain

$$\Omega(x, y)\Omega(y, x) = I. \tag{2.49}$$

From (2.49), it follows that the matrix $\Omega(x, y)$ is nondegenerate for any $x, y \in G$.

Let us take an arbitrary point $a \in G$. By setting $z = a$ in (2.48) and by taking (2.49) into account, we obtain that

$$\Omega(x, y) = Y(x)\,[Y(y)]^{-1} \tag{2.50}$$

for all $x \in G$, where $Y(x) = \Omega(x, a)$.

Let the matrix function $Y(x)$ be the solution of system (2.46) such that $\det Y(x) \neq 0$ for all $x \in G$. The function $Y(x) = \Omega(x, a)$, for instance, where Ω is the fundamental matrix of system (2.45), satisfies this condition. The operator of multiplication to the left of the vector function $y : G \to \mathbf{C}^m$ by the matrix function Y is also denoted by Y. Let us show that the equalities

$$\frac{\partial}{\partial x_j} = Y^{-1} \circ L_j \circ Y, \tag{2.51}$$

$$L_j = Y \circ \frac{\partial}{\partial x_j} \circ Y^{-1} \tag{2.52}$$

are valid. Actually, we have

$$\begin{aligned} L_j(Y(x)y) &= \frac{\partial}{\partial x_j}(Y(x)y) - A_j(x)Y(x)y = \frac{\partial Y}{\partial x_j}(x)y \\ &+ Y(x)\frac{\partial y}{\partial x_j} - A_j(x)Y(x)y = Y(x)\frac{\partial y}{\partial x_j}, \end{aligned}$$

because $\frac{\partial Y}{\partial x_j} - A_j Y = 0$. Hence,

$$Y(x)^{-1} L_j \left[Y(x) y(x) \right] = \frac{\partial y}{\partial x_j}(x);$$

thus, equality (2.51) is proved. Equality (2.52) obviously follows from (2.51).

Let U be an open set of the class S in the space $\mathbf{R}^n$, let

$$L_j y = \frac{\partial y}{\partial x_j} - A_j y, \quad j = 1, 2, \ldots, n$$

be a system of differential operators with the complete integrability condition. Let us show how, with the help of the equality (1.3), one can construct the integral representation of a function f through the magnitudes $L_1 f, L_2 f, \ldots, L_n f$. Equality (1.3) is valid for scalar functions. Let $f : U \to \mathbf{R}^m$ be a vector function of the class $C^1(U)$. By applying equality (1.3) to each of the components $f_1, f_2, \ldots, f_m$ of the function f, we obtain that equality (1.3) is also valid for vector-functions. Let $Y(x)$ be a matrix function which is the solution of system (2.46) and such that $\det Y(x) \neq 0$ for all $x \in U$. Then we have:

$$L_j f(x) = Y(x) \frac{\partial}{\partial x_j} \left[(Y(x))^{-1} f(x) \right].$$

Let us put $Y(x)^{-1} f(x) = g(x)$. By applying equality (1.3) to the vector function $g(x)$, we obtain

$$\begin{aligned} g(x) &= [Y(x)]^{-1} f(x) \\ &= \int_U K_0(x,y) [Y(y)]^{-1} f(y) dy + \sum_{i=1}^{n} \int_U K_i(x,y) \frac{\partial}{\partial y_i} \left([Y(y)]^{-1} f(y) \right) dy. \end{aligned}$$

Hence,

$$\begin{aligned} f(x) = &\int_U K_0(x,y) Y(x) [Y(y)]^{-1} f(y) dy \\ &+ \sum_{i=1}^{n} \int_U K_i(x,y) \left(Y(x) [Y(y)]^{-1} \right) Y(y) \frac{\partial}{\partial y_i} \left((Y(y))^{-1} f(y) \right) dy. \end{aligned}$$

We have $Y(x)[Y(y)]^{-1} = \Omega(x,y)$, where $\Omega(x,y)$ is the fundamental matrix of the system of differential operators $L_1, L_2, \ldots, L_n$. Hence we obtain the desired integral representation

$$f(x) = \int_U K_0(x,y) \Omega(x,y) f(y) dy + \sum_{i=1}^{n} \int_U K_i(x,y) \Omega(x,y) (L_i f)(y) dy. \quad (2.53)$$

2.4. Integral Representations for the Deformation Tensor and for the Tensor of Conformal Deformation

The formula (2.53) obtained in the previous subsection may be used to construct integral representations of a function in terms of differential operators of a more general type. The corresponding examples may be found in the author's papers [62], [64]. In particular, in the Russian version of this book, the integral representation of a function in terms of its derivatives of the order r was constructed by just this method. The deduction of this representation given in Subsection 1.1 was obtained as a result of the union of separate stages of argumentation given in the Russian edition. The general scheme of application of formula (2.53) to the construction of integral representations is as follows. Suppose that we have to construct the integral representation of a function $y : G \to \mathbf{R}^m$, where G is an open set in $\mathbf{R}^n$, by means of the magnitude $P(x, D)y$, where $P(x, D)$ is a linear differential operator of arbitrary power with coefficients depending on x ($P(x, D)y$ is a vector in $\mathbf{R}^k$). To do this, let us construct some completely integrable system of differential operators $L_j = \frac{\partial}{\partial x_j} - A_j(x)$, $j = 1, 2, \ldots, n$ having the following property. By the vector-function y, one may define some vector function f which is obtained by adding to the components of y with the magnitudes expressed linearly by these components themselves and by their derivatives, such that the equalities

$$(L_j f)(x) = \frac{\partial f}{\partial x_j}(x) - A_j(x)f = R_j(x, D)P(x, D)y \tag{2.54}$$

are valid, where $R_j(x, D)$, $j = 1, 2, \ldots, n$ are some linear differential operators. For the vector function f, let us construct the integral representation of the form (2.53) by means of the magnitudes $L_1 f, L_2 f, \ldots, L_n f$. If we take the components of f which coincide with the components of the vector function y, then we obtain some representation of the vector function $y(x)$. However, this representation is not the desired one, because on the right-hand side instead of the integrand $P(x, D)y$ we shall have the expression $R(x, D)P(x, D)y$. The desired representation is obtained as a result of the additional transformation of integrals according to the formula of integration by parts similar to the one carried out in Example 4 of Subsection 2.1.

Let us consider some examples illustrating the realization of the above idea for concrete cases.

I. Let us assign an arbitrary open set $U \subset \mathbf{R}^n$. For an arbitrary vector function $v : U \to \mathbf{R}^n$ of the class C^1, let $v'(x)$ denote the Jacobi matrix of the mapping v at the point x. If A is a square matrix, then let A^* be the transposed matrix of A, and $\operatorname{tr} A$ be the trace of A. We set

$$(Q_1 v)(x) = \tfrac{1}{2}\left[v'(x) + (v'(x))^*\right]. \tag{2.55}$$

The matrix function $Q_1 v$ is said to be the deformation tensor corresponding to the vector field v.

For simplicity, let us use the following notations. Differentiation by the variable x_i is denoted by the index i placed after a comma, so that $u_{,i} = \frac{\partial u}{\partial x_i}$, $v_{k,j} = \frac{\partial v_k}{\partial x_j}$, etc. Below, δ_{ij} denotes Kronecker's symbol, i.e., $\delta_{ij} = 1$ for $i = j$, $\delta_{ij} = 0$ for $i \neq j$, $i, j = 1, 2, \ldots, n$. In these notations for the elements of the matrix $Q_1 v$, we obtain the following expression,

$$((Q_1 v)(x))_{ij} = \frac{1}{2}\left(\frac{\partial v_i}{\partial x_j}(x) + \frac{\partial v_j}{\partial x_i}(x)\right) = \frac{1}{2}(v_{i,j} + v_{j,i}). \tag{2.56}$$

Suppose that v is a vector function of the class $C^2(U)$. Let us set

$$\tfrac{1}{2}(v_{i,j} + v_{j,i}) = q_{ij} = (Q_1 v)_{ij}, \tag{2.57}$$

$$\tfrac{1}{2}(v_{i,j} - v_{j,i}) = \eta_{ij}. \tag{2.58}$$

Then

$$\frac{\partial v_i}{\partial x_j} = q_{ij} + \eta_{ij} \tag{2.59}$$

for any $i, j = 1, 2, \ldots, n$. By differentiating term by term the equality $v_{i,k} + v_{k,i} = 2q_{ik}$ by x_j, and the equality $v_{j,k} + v_{k,j} = 2q_{jk}$ by x_i, and by subtracting the obtained results from each other, we have

$$\tfrac{1}{2}(v_{i,jk} - v_{j,ik}) = q_{ik,j} - q_{jk,i},$$

whence

$$\frac{\partial \eta_{ij}}{\partial x_k} = q_{ik,j} - q_{jk,i}. \tag{2.60}$$

Now let us assign some system of differential equations with respect to unknown functions, $v_1, v_2, \ldots, v_n$, η_{ij}, $i, j = 1, 2, \ldots, n$, $i < j$. This system arises if we put $q_{ij} = 0$. Taking into account that only the values η_{ij} are considered for which $i < j$, we obtain

$$\frac{\partial v_i}{\partial x_j} = \begin{cases} -\eta_{ji} & \text{for } j < i, \\ 0 & \text{for } j = i, \\ \eta_{ij} & \text{for } j > i, \end{cases} \tag{2.61}$$

$$\frac{\partial \eta_{ij}}{\partial x_k} = 0. \tag{2.62}$$

System (2.61), (2.62) is equivalent to the system of equations

$$v_{i,j} + v_{j,i} = 0, \quad i, j = 1, 2, \ldots, n \tag{2.63}$$

in the following sense. If v_i, $i = 1, 2, \ldots, n$, η_{ij}, $1 \leqslant i < j \leqslant n$ is the solution of system (2.61), (2.62), then the functions $v_1, v_2, \ldots, v_n$ satisfy the system of equations (2.63). On the contrary, if $v_1, v_2, \dots, v_n$ is the solution of system (2.63), then the set of functions v_i, $i = 1, 2, \dots, n$, $\eta_{ij} = \frac{1}{2}(v_{i,j} - v_{j,i})$, $1 \leqslant i < j \leqslant n$, is the solution of system (2.61), (2.62). Let us show that the system of equations (2.61), (2.62) is completely integrable. The simplest way to do this is by solving this system. From equations (2.62), it follows that the functions η_{ij} are constant. Let $\eta_{ij} = A_{ij}$. Let $A_{ii} = 0$, and let $A_{ij} = -A_{ji}$ for $j < i$. Then, from (2.61), we obtain

$$v_i(x) = a_i + \sum_{j=1}^{n} A_{ij} x_j, \tag{2.64}$$

where a_i, $i = 1, 2, \dots, n$, are constants. On the contrary, if v_i is expressed as above and if $\eta_{ij} = A_{ij}$, with the matrix (A_{ij}) being skew-symmetric, then the system of functions v_i, η_{ij}, $1 \leqslant i < j \leqslant n$ is the solution of system (2.61), (2.62). The solution of the Cauchy problem

$$\begin{aligned} \eta_{ij}(a) &= h_{ij}, \quad 1 \leqslant i < j \leqslant n, \\ v_i(a) &= h_i, \quad\ \ i = 1, 2, \dots, n \end{aligned} \tag{2.65}$$

is given by the formula

$$v_i(x) = h_i + \sum_{j=1}^{n} h_{ij}\,(x_j - a_j), \qquad \eta_{ij}(x) = h_{ij}, \tag{2.66}$$

where we set $h_{ii} = 0$ and $h_{ij} = -h_{ji}$ for any i, j. Complete integrability of system (2.61), (2.62) is established.

By using the results of Subsection 2.3, let us now construct the desired integral representation. Let $x \in \mathbf{R}^n$. The system of magnitudes $v_i(x)$, $i = 1, 2, \dots, n$, $\eta_{kl}(x)$, $k, l = 1, 2, \dots, n$, $k < l$, is considered as a vector in the space $\mathbf{R}^m$, where $m = n + n(n-1)/2 = n(n+1)/2$. The first n components of an arbitrary vector $h \in \mathbf{R}^m$ are numbered by indices $i = 1, 2, \dots, n$, the next $n(n-1)/2$ components are numbered by pairs of indices k, l, where $1 \leqslant k < l \leqslant n$. Let us define the operators L_j as follows. For the vector function $y(x) = (v_i(x), \eta_{kl}(x))$, the vector $z(x) = L_j y(x)$ has the components

$$\begin{aligned} z_i &= \frac{\partial v_i}{\partial x_j} - \eta_{ij} \quad \text{for } i < j, \qquad z_j = \frac{\partial v_j}{\partial x_j}, \\ z_i &= \frac{\partial v_i}{\partial x_j} + \eta_{ij} \quad \text{if } i > j, \\ z_{kl} &= \frac{\partial \eta_{kl}}{\partial x_j}. \end{aligned}$$

The system of equations $L_j y(x) = 0$, $j = 1, 2, \dots, n$, obviously coincides with system (2.61), (2.62). Let there be a vector function $v(x) =$

$(v_1(x), v_2(x), \dots, v_n(x))$ of the class C^2. Let us construct with it the vector function $f(x)$ with values in $\mathbf{R}^m$, by setting $f(x) = (v_i(x), \eta_{j,k}(x))$, $i,j,k = 1,2,\dots,n$, $j < k$, where η_{jk} is defined by f by equalities (2.58). Taking into account equalities (2.59) and (2.60), we obtain that for every $j = 1,2,\dots,n$,

$$L_j f(x) = (q_{ij}(x), q_{kj,l} - q_{lj,k}), \tag{2.67}$$

$i,k,l = 1,2,\dots,n$, $k < l$. We see that the operators L_j and $P(x,D) = Q_1$ are interconnected by the relation of the form (2.54).

Let $\theta(x,y)$ be the fundamental matrix of system (2.61), (2.62). Then for every vector $h = (h_i, h_{kl}) \in \mathbf{R}^m$, $i,k,l = 1,2,\dots,n$, $k < l$, $\theta(x,y)h$ is the solution of system (2.61), (2.62) satisfying the condition (2.65); consequently, it is expressed by equalities (2.66). Let $f(x)$ be the vector $(v_i(x), \eta_{kl}(x)) \in \mathbf{R}^m$, where η_{kl} is expressed in terms of the components of v according to (2.58). Then the functions $L_j f$ are expressed according to formula (2.67). To apply formula (2.53), we have to obtain the vector $\theta(x,y)L_j f(y) \in \mathbf{R}^m$. Since we only want to obtain the integral representation of the vector v_i, we only have to take the components $\theta_i(x,y)L_j f(y)$, $i = 1,2,\dots,n$, of the vector $z = \theta(x,y)L_j f(y)$. We have

$$\theta_i(x,y)L_j f(y) = q_{ij}(y) + \sum_{l=1}^{n} (q_{ij,l} - q_{lj,i})(x_l - y_l).$$

In the same way, we obtain that

$$\theta_i(x,y)f(y) = v_i(y) + \sum_{k=1}^{n} \frac{1}{2}\left(\frac{\partial v_i}{\partial y_k} - \frac{\partial v_k}{\partial y_i}\right)(x_k - y_k).$$

By applying formula (2.53), we now obtain the following integral representation for the component v_i of the vector function v:

$$\begin{aligned} v_i(x) = \int_G K_0(x,y)\left\{v_i(y) + \sum_{k=1}^{n}\left[\frac{\partial v_i}{\partial y_k}(y) - \frac{\partial v_k}{\partial y_i}(y)\right](x_k - y_k)\right\}dy \\ + \sum_{j=1}^{n}\int_G K_j(x,y)\left\{q_{ij}(y) + \sum_{k=1}^{n}(q_{ij,k}(y) - q_{kj,i}(y))(x_k - y_k)\right\}dy. \end{aligned} \tag{2.68}$$

This formula is the preliminary one. The final result is obtained from (2.68) by integration by parts. To do this, Lemma 2.1 is used. As a result of obvious transformations, we have:

$$v_i(x) = \int_G \sum_{k=1}^{n} K_{0ik}(x,y)\, v_k(y)\, dy + \int_G \sum_{j=1}^{n}\sum_{k=1}^{n} K_{jik}(x,y)\, q_{ik}(y)\, dy, \tag{2.69}$$

where for every $j = 0, 1, 2, \ldots, n$,

$$K_{jik}(x,y) = \left(K_j(x,y) - \sum_{l=1}^{n} \frac{\partial}{\partial y_l} \left[K_j(x,y)(x_l - y_l) \right] \right) \delta_{ik} + \frac{\partial}{\partial y_i} \left[K_j(x,y)(x_k - y_k) \right],$$

$$q_{ik}(x) = ((Q_1 v)(x))_{ik}, \quad i, k = 1, 2, \ldots, n. \tag{2.70}$$

From the properties of the functions $K_j(x,y)$ in the integral representation (1.3), it follows that for the function K_{jik} the following statements are valid.

For any i, k, the function $K_{0ik}(x,y)$ belongs to the class C^∞ in $G \times G$, and for any $x \in G$, $y \mapsto K_{0ik}(x,y)$ is a compactly supported function in the domain G.

For every $j = 1, 2, \ldots, n$, $K_{jik}(x,y)$ is a function of the type $|x - y|^{1-n}$. Here, for every point x, one can show a neighbourhood $U(x) \subset G$ of the point x and a compact set $A \subset G$ such that if $x' \in U(x)$, then $K_{jik}(x', y)$ vanishes if $y \notin A$.

If we start from integral representations of the form (1.3) constructed in Subsection 2.1, then we obtain the representation of the form (2.69) in which the functions $K_{jik}(x,y)$ possess some additional properties.

Let G be a domain of the class S in $\mathbf{R}^n$, and let $v : G \to \mathbf{R}^n$ be a vector function of the class C^3. Let us put

$$Q_2 v(x) = \tfrac{1}{2} \left[v'(x) + (v'(x))^* \right] - \tfrac{1}{n} \operatorname{div} v(x) I_n, \tag{2.71}$$

where I_n is the unit matrix of the order n. The matrix function $Q_2 v$ is called the tensor of conformal deformation of the vector field $v(x)$.

By using shortened notations introduced in Subsection I when studying the deformation tensor of a vector field, we obtain the following expression for the elements of the matrix $(Q_2 v)(x)$:

$$(Q_2 v)_{ij} = \tfrac{1}{2} (v_{i,j} + v_{j,i}) - \tfrac{1}{n} \operatorname{div} v \delta_{ij}, \tag{2.72}$$

where, as usual, δ_{ij} is Kronecker's symbol, $\delta_{ij} = 0$ for $i \neq j$, $\delta_{ii} = 1$.

The system of equations

$$Q_2 v = 0,$$

in the case $n = 2$, turns into the Cauchy–Riemann system of equations. Below, we set $n \geqslant 3$. Let v be a vector function of the class C^3.

Let

$$\begin{aligned} q_{ij}(x) = (Q_2 v(x))_{ij} &= \tfrac{1}{2} (v_{i,j}(x) + v_{j,i}(x)) - \tfrac{1}{n} \operatorname{div} v \delta_{ij}, \\ \eta_{ij}(x) &= \tfrac{1}{2} (v_{i,j} - v_{j,i}), \quad i, j = 1, 2, \ldots, n, \end{aligned} \tag{2.73}$$

$$w = \tfrac{1}{n} \operatorname{div} v, \tag{2.74}$$

$$\varsigma_i = w_{,i} \tag{2.75}$$

First let us show that the first derivatives of the functions v_i, η_{ij}, w, and ς_i are linearly expressed in terms of the functions themselves and in terms of the derivatives of different orders of the function q_{ij}.

First of all, we have

$$v_{i,k} = \eta_{ik} + \delta_{ik} w + q_{ik}, \tag{2.76}$$

$$w_{,k} = \varsigma_k. \tag{2.77}$$

Now let us construct expressions for the derivatives of the functions η_{ij}, $i \neq j$. Let $k \neq i$, $k \neq j$. Thus we have: $v_{i,k} + v_{k,i} = 2q_{ik}$, $v_{j,k} + v_{k,j} = 2q_{jk}$. By differentiating term by term the first equality by x_j, the second one by x_i, and by subtracting the obtained equalities, we have

$$\tfrac{1}{2}\left(v_{i,jk} - v_{j,ik}\right) = q_{ik,j} - q_{jk,i},$$

that is, for $k \neq i$, $k \neq j$, $i \neq j$,

$$\eta_{ij,k} = q_{ik,j} - q_{jk,i}.$$

Then, for $i \neq j$,

$$\eta_{ij,j} = \tfrac{1}{2}\left(v_{i,jj} - v_{j,ij}\right) = \tfrac{1}{2}\left(v_{i,jj} + v_{j,ij} - v_{j,ij} - v_{j,ij}\right) = q_{ij,j} - v_{j,ji}.$$

We have $v_{j,j} = q_{jj} + w$. Hence, we obtain that

$$\eta_{ij,j} = -\varsigma_i + q_{ij,j} - q_{jj,i}.$$

If we exchange i and j and take into account that $\eta_{ij} = -\eta_{ji}$, then we obtain

$$\eta_{ij,i} = \varsigma_j - q_{ij,i} + q_{ii,j}.$$

The final expression for the derivative $\eta_{ij,k}$ may be represented by the formula

$$\eta_{ij,k} = \delta_{ik}\varsigma_j - \delta_{jk}\varsigma_i + q_{ik,j} - q_{jk,i}, \tag{2.78}$$

which is valid for any i, j, k.

Now let us give the expression for the derivative $\varsigma_{j,k}$. We have: $\varsigma_{j,k} = w_{,jk}$. Let $j \neq k$. Let us fix $i \neq j$, $i \neq k$. For every i, $w = v_{i,i} - q_{ii}$. Let $j \neq k$. Then

$$\begin{aligned} w_{,jk} &= v_{i,ijk} - q_{ii,jk} = v_{i,jik} - q_{ii,jk} \\ &= (2q_{ij} - v_{j,i})_{,ik} - q_{ii,jk} \\ &= 2q_{ij,ik} - v_{j,kii} - q_{ii,jk}. \end{aligned}$$

By interchanging j and k, we obtain

$$w_{,jk} = 2q_{ik,ij} - v_{k,jii} - q_{ii,jk}.$$

Summing up the obtained expressions for $w_{,jk}$, we have

$$\varsigma_{j,k} = w_{j,k} = q_{ij,ik} + q_{ik,ij} - q_{kj,ii} - q_{ii,jk}. \tag{2.79}$$

This expression is sufficient for us. We can get rid of some asymmetry in formula (2.79) which arises due to the presence of the index $i \neq j, \ \ i \neq k$. To do this, let us perform in (2.79) the summation by i. Taking into account that the right-hand side of (2.79) vanishes for $i = j$ and for $i = k$, we obtain:

$$\varsigma_{j,k} = \frac{1}{n-2}\left(\sum_{i=1}^{n} q_{ij,ik} + \sum_{i=1}^{n} q_{ik,ij} - \sum_{i=1}^{n} q_{kj,ii}\right). \tag{2.80}$$

Now let us consider the case $j = k$. We have for $i \neq j$,

$$\varsigma_{j,j} = v_{i,ijj} - q_{ii,jj} = v_{i,jij} - q_{ii,jj}.$$

By interchanging i and j, we further obtain

$$\varsigma_{i,i} = v_{j,iij} - q_{jj,ii}.$$

Summing up the given equalities term by term, we have for $i \neq j$,

$$\varsigma_{j,j} + \varsigma_{i,i} = 2q_{ij,ij} - q_{ii,jj} - q_{jj,ii}.$$

Summing this by j and taking into account that in the case $j = i$ the right-hand side of the latter equality vanishes, we obtain

$$\operatorname{div}\varsigma + (n-2)\varsigma_{i,i} = 2\sum_{j=1}^{n} q_{ij,ij} - \sum_{j=1}^{n} q_{ii,jj}.$$

Summing it by i, we obtain

$$(n-1)\operatorname{div}\varsigma = \sum_{i=1}^{n}\sum_{j=1}^{n} q_{ij,ij}.$$

The two latter equalities yield the following representation for $\varsigma_{i,i}$:

$$\varsigma_{i,i} = \frac{2}{n-2}\sum_{j=1}^{n} q_{ij,ij} - \frac{1}{n-2}\sum_{j=1}^{n} q_{ii,jj} - \frac{1}{(n-2)(n-1)}\sum_{i=1}^{n}\sum_{j=1}^{n} q_{ij,ij}. \tag{2.81}$$

Equalities (2.80) and (2.81) are embraced by the following unique formula

$$\varsigma_{j,k} = \frac{1}{n-2}\left(\sum_{i=1}^{n} q_{ij,ik} + \sum_{i=1}^{n} q_{ik,ij} - \sum_{i=1}^{n} q_{kj,ii}\right) + \frac{\delta_{jk}}{(n-2)(n-1)}\sum_{i=1}^{n}\sum_{l=1}^{n} q_{il,il}. \tag{2.82}$$

Let us denote by H_n a vector space whose elements are all possible systems of numbers X of the form: $X = (v_i, \eta_{jk}, w, \varsigma_l)$, where $i, j, k, l = 1, 2, \ldots, n$, and $\eta_{jk} = -\eta_{kj}$ for any j and k. The dimensionality of H_n equals $n + \frac{n(n-1)}{2} + 1 + n = \frac{(n+1)(n+2)}{2}$. Let us consider the system:

$$\begin{aligned}
v_{i,s} &= \eta_{i,s} + w\delta_{i,s},\\
\eta_{jk,s} &= -\varsigma_j\delta_{ks} + \varsigma_k\delta_{js},\\
w_{,s} &= \varsigma_s,\\
\varsigma_{l,s} &= 0,
\end{aligned} \tag{2.83}$$

where $i, j, k, l = 1, 2, \ldots, n$, $s = 1, 2, \ldots, n$. In the shortened form, this system may be written down as follows:

$$L_s X \equiv \frac{\partial X}{\partial x_s} - A_s X \equiv 0, \quad s = 1, 2, \ldots, n, \tag{2.84}$$

where $X = (v_i, \eta_{jk}, w, \varsigma_l)$, $i, j, k, l = 1, 2, \ldots, n$, is the vector function with values in H_n,

$$L_s X = (v_{i,s} - \eta_{i,s} - w\delta_{i,s}, \eta_{jk,s} + \varsigma_j\delta_{ks} - \varsigma_k\delta_{js}, w_{,s} - \varsigma_s, \varsigma_{l,s})$$

for any vector $X \in H_n$. Let there be a vector function $v : G \to \mathbf{R}^n$ of the class C^3. Suppose that the functions η_{ij}, w, and ς_k are defined by v by equalities (2.73), (2.74), and (2.75). Let $Q_2 v = (q_{ij})$, $i, j = 1, 2, \ldots, n$. Due to equalities (2.76), (2.77), (2.78), and (2.82), the vector $L_s f$, where $f = (v_i, \eta_{jk}, w, \varsigma_l)$, is expressed as follows:

$$\begin{aligned}
L_s f = &(q_{is}, q_{js,k} - q_{ks,j}, 0, \frac{1}{n-2}\sum_{\alpha=1}^{n}(q_{\alpha l,s\alpha} + q_{\alpha s,l\alpha} - q_{ls,\alpha\alpha})\\
&+ \frac{\delta_{ls}}{(n-2)(n-1)}\sum_{\alpha=1}^{n}\sum_{\beta=1}^{n} q_{\alpha\beta,\alpha\beta}).
\end{aligned} \tag{2.85}$$

Hence it is clear that $L_s f = T_s(D)(Q_2 f)$, that is, for the operators $L_s = \frac{\partial}{\partial x_s} - A_s$, the relation of the form (2.54) is valid.

Let us show that the system of differential operators $L_s = \frac{\partial}{\partial x_s} - A_s$ is completely integrable. As in the previous example, we shall do it by explicit construction of the solution of system (2.83). Let us assign an arbitrary point $y \in \mathbf{R}^n$, $y = (y_1, y_2, \dots, y_n)$, and systems of numbers a_i, A_{jk}, a_0 and b_l, $i, j, k, l = 1, 2, \dots, n$, $A_{jk} = -A_{kj}$ for any j, k. We shall look for the solution of system (2.83) such that

$$v_i(y) = a_i, \qquad \eta_{jk}(y) = A_{jk}, \qquad \varsigma_l(y) = b_l. \tag{2.86}$$

The latter of the equations (2.83) allows us to state that the functions ς_l are constant; therefore, $\varsigma_l(y) \equiv b_l$. Hence, it follows that

$$w(x) = a_0 + \sum_{i=1}^{n} b_i(x_i - y_i).$$

According to the second equation of (2.83), we have

$$\eta_{jk,j} = b_k, \qquad \eta_{jk,k} = -b_j, \qquad \eta_{jk,l} = 0$$

for $l \neq j$, $l \neq k$. Hence,

$$\eta_{jk}(x) = A_{jk} + b_k(x_j - y_j) - b_j(x_k - y_k).$$

Finally, the first of the equations in (2.83) yields

$$v_{i,j} = A_{ij} + b_j(x_i - y_i) - b_i(x_j - y_j) \quad \text{for } i \neq j,$$

$$v_{i,i} = a_0 + \sum_{i=1}^{n} b_i(x_i - y_i).$$

By integrating the given relations, we finally obtain the explicit expression for the solution of system (2.83):

$$v_i = a_i + \sum_{j=1}^{n} A_{ij}(x_j - y_j) + [a_0 + \langle b, x - y \rangle](x_i - y_i) - b_i |x - y|^2/2.$$

In the vector recording, the latter equality takes the form:

$$v(x) = a + A(x - y) + [a_0 + \langle b, x - y \rangle](x - y) - b|x - y|^2/2. \tag{2.87}$$

Here A is a skew-symmetric matrix, a, b are vectors in $\mathbf{R}^n$, and a_0 is a real number.

Now we can write down the desired integral representation. Let $v : G \mapsto \mathbf{R}^n$ belong to the class C^3, let $f(x) = (v_i(x), \eta_{jk}(x), w(x), \varsigma_l(x))$ be the vector function with values in H, where η_{jk}, w, and ς_l are expressed by the components of the vector function v by equalities (2.73), (2.74), and (2.75). The vector function $\theta(x, y) f(y)$ is the solution of $X(x)$ of system (2.83), satisfying the condition $X(x) = f(y)$. Taking into account equality (2.87), we conclude that the first n components of the vector function $\theta(x,y)f(y)$ are expressed as follows:

$$\theta_i(x,y)f(y) = v_i(y) + \sum_{j=1}^{n} \eta_{ij}(y)(x_j - y_j) + (x_i - y_i)[w(y) + \langle \varsigma(y), x - y\rangle] - \varsigma_i(y)|x-y|^2/2. \tag{2.88}$$

Since we intend to obtain only the integral representation of the vector function $v(x)$ which forms the first n components of the vector function $f(x)$, the other components of the vector function $\theta(x,y)f(y)$ are beyond our interest. Similarly, taking into account equality (2.85), we obtain the representation for the first i components of the vector function $\theta(x,y)(L_s f)(y)$. We do not give the final result, since it is cumbersome.

By applying the integral representation (2.53), we obtain

$$v(x) = \int_U K_0(x,y)\tilde{\theta}(x,y)f(y)dy + \sum_{i=1}^{n} \int_U K_i(x,y)\tilde{\theta}(x,y)T_s(D)(Q_2 v)(y)dy.$$

Here $\tilde{\theta}(x,y)$ implies the rectangle matrix formed by the first n lines of the matrix $\theta(x,y)$. The obtained representation is the desired one. To obtain the desired representation from it, it is necessary to perform additional integration by parts, which is done just as in the previous example.

§3 Estimates for Potential-Type Integrals

3.1. Preliminary Information

Let us assign an arbitrary bounded open set U in $\mathbf{R}^n$. Let $K(x,y)$ be a function of variables $x \in R^n$, $y \in R^n$, which is defined for any $x \in U$, $y \in U$ such that $x \neq y$. We assume K to satisfy the following conditions:

A) $K(x,y)$ is continuous in $U \times U$ for $x \neq y$.

B) There exist numbers $L < \infty$ and $l < n$, $l > 0$ such that

$$|K(x,y)| \leqslant L|x-y|^{l-n}$$

for any $(x,y) \in U \times U$, where $x \neq y$.

C) There exists a number $\tau > 0$ such that $d(U) \geqslant \tau$ and $K(x,y) = 0$ if $|x-y| \geqslant \tau$.

If the function K satisfies the above Conditions A, B, C, then, for short, K is said to be a normal kernel. The number τ in the Condition C is called the finiteness radius of K, the number $n - l$ is the singularity order of K.

Let $K(x, y)$ be a normal kernel in the bounded open set U, and let $t \geqslant 1$ be such that $(n - l)t < n$.

Denote by $M_t(K)$ the lowest of the numbers $M \leqslant \infty$ such that

$$\left(\int_U |K(x, y)|^t dy\right)^{1/t} \leqslant M, \quad \left(\int_U |K(x, y)|^t dx\right)^{1/t} \leqslant M \quad (3.1)$$

for any $x \in U$, $y \in U$.

Let us give some estimates for the magnitude $M_t(K)$. For every x, we have

$$\int_U |K(x, y)|^t dy \leqslant M^t \int_{|x-y| \leqslant \tau} |x - y|^{(l-n)t} dy.$$

The latter integral can easily be calculated by the transition to spherical coordinates. We have for $0 \leqslant \gamma \leqslant n$,

$$\int_{|z| \leqslant d} \frac{dz}{|z|^{n-\gamma}} = \omega_n \int_0^d \frac{r^{n-1}\, dr}{r^{n-\gamma}} = \omega_n \frac{d^\gamma}{\gamma}$$

(ω_n is the area of the unit sphere). In our case, $\gamma = n - t(n - l)$, we obtain

$$\int_U |K(x, y)|^t dy \leqslant \omega_n M^t \frac{\tau^{n-t(n-l)}}{n - t(n - l)}.$$

Similarly, we estimate the latter of the integrals in (3.1). We eventually obtain

$$M_t(K) \leqslant C\tau^{n/t-n+l}, \quad (3.2)$$

where $C = Lw_n{}^{1/t}[n - t(n - l)]^{-1/t}$.

Let U be a bounded open set in $\mathbf{R}^n$, let K be a normal kernel in U; μ is an arbitrary measure in U such that K is μ-integrable. We set

$$(P_K\mu)(x) = \int_U K(x, y)\, \mu(dy) \quad (3.3)$$

for every x for which the integral on the right-hand side is defined and finite.

Let us consider the iterated integrals

$$\int_U \left(\int_U |K(x, y)|\, |\mu|\, (dy)\right) dx, \quad (3.4)$$

$$\int_U \left(\int_U |K(x, y)|\, dx\right) |\mu|\, (dy). \quad (3.5)$$

Due to the Tonelli theorem, if at least one of these integrals is finite, then the other is also finite, and these integrals equal each other. From (3.2), the finiteness of integral (3.5) obviously follows. Consequently, integral (3.4) is finite as well. This allows us to conclude that the magnitude $(P_{|K|}|\mu|)(x) = \int_U |K(x, y)||\mu|(dy)$ is defined and finite for almost all x, and the function $(P_{|K|}|\mu|)$ thus obtained is integrable. Hence it follows that the magnitude

$(P_K\mu)(x)$ is defined and finite for almost all x. The function $P_K\mu$ is integrable. The Fubini theorem yields that for any bounded measurable function φ,

$$\int_U \varphi(x)(P_K\mu)(x)\,dx = \int_U \left(\int_U K(x, y)\varphi(x)\,dx\right)\mu(dy). \tag{3.6}$$

Similarly, if $u \in L_1(U)$, then we suppose that

$$(P_K u)(x) = \int_U K(x, y)\,u(y)\,dy.$$

Formally, $P_K u = P_K\mu$, where the measure μ is the indefinite integral of the function u.

The functions $P_K\mu$ and $P_K u$, where μ is a measure, u is an integrable function, are called the potentials of the measure μ or of the function $u(x)$ with respect to the kernel K.

Let us also describe some special approximation of the function K. Let us first assign a function $\lambda : \mathbf{R}^n \to \mathbf{R}$ of the class C^∞ such that $\lambda(x) = 1$ for $|x| \leqslant 1/2$, $\lambda(x) = 0$ for $|x| > 1$, and $0 \leqslant \lambda(x) \leqslant 1$ for all x. Let $h > 0$. We set

$$K_h(x, y) = K(x, y)\,\lambda\left(\frac{x-y}{h}\right),$$

$$E_h(x, y) = K(x, y) - K_h(x, y) \quad \text{for } x \neq y, \; E_h(x, x) = 0.$$

We have

$$K(x, y) = K_h(x, y) + E_h(x, y). \tag{3.7}$$

Here the function K_h satisfies the same Conditions A, B, C as the function K, and the function E_h is bounded and continuous. The finiteness radius of K_h in this case equals $\min\{h, r\}$.

3.2. Lemma on the Compactness of Integral Operators

Lemma 3.1. *Let U be a bounded open set in $\mathbf{R}^n$, and let $K : U \times U \to R$ be a bounded continuous function. Then for any $p \geqslant 1$, $q \geqslant 1$, the operators P_K maps $L_p(U)$ into $L_q(U)$, and in the case $p > 1$, $q > 1$, the operator P_K is completely continuous.*

Proof. Let $M < \infty$ be such that $|K(x,y)| \leqslant M$ for any $(x, y) \in U \times U$. Let us take an arbitrary function $u \in L_p(U)$. Due to boundedness of U, the function u is integrable on U. Let $x_0 \in U$. For $x \to x_0$, $K(x,y)u(y) \to K(x_0,y)u(y)$ for all $y \in U$. Here $|K(x,y)u(y)| \leqslant M|u(y)|$ for all x. According to the Lebesgue theorem about majorant convergence, we conclude that $(P_K u)(x) \to (P_K u)(x_0)$ for $x \to x_0$; therefore, the function $P_K u$ is continuous.

Let $A \subset U$ be measurable. For all $x \in U$, we have for $u \in L_p(U)$,

$$\left|\int_A K(x, y)u(y)\,dy\right| \leqslant M \int_A |u(y)|\,dy \leqslant M|A|^{1-1/p}\|u\|_{L_p(U)}. \quad (3.8)$$

By setting $A = U$, we obtain, in particular, that if $u \in L_p(U)$, where $p \geqslant 1$, then the function $P_K u$ is bounded. Thus, P_K maps $L_p(U)$ into the set of all bounded, continuous functions on U; and therefore, P_K maps $L_p(U)$ into $L_q(U)$ for any $p \geqslant 1$, $q \geqslant 1$.

Let us now prove that if $p > 1$ and $q \geqslant 1$, then P_K is a completely continuous mapping of $L_p(U)$ into $L_q(U)$. Let (u_m), $m = 1, 2, \ldots$, be an arbitrary sequence of functions from $L_p(U)$ weakly converging to zero in $L_p(U)$. It is necessary to prove that in this case, $\|P_K u_m\|_{L_q(U)} \to 0$ for $m \to \infty$.

Since $u_m \to 0$ is weak, the sequence of norms $\|u_m\|_{L_p(U)}$ is bounded. Let $\|u_m\|_{L_p(U)} \leqslant L = \text{const} < \infty$ for all m. Let us arbitrarily assign $h > 0$. Let V_h be a set of all $x \in U$ such that $\rho(x, \partial U) \geqslant h$, and $W_h = U \setminus V_h$. The set V_h is compact, and $|W_h| \to 0$ for $h \to 0$. Let us put $v_m = P_K u_m$, and let

$$\xi_m(x) = \int_{V_h} K(x, y)\,u_m(x)\,dx, \qquad \eta_m(x) = \int_{W_h} K(x, y)\,u_m(x)\,dx.$$

Then $v_m = \xi_m + \eta_m$ for every m.

Applying inequality (3.8), we obtain

$$|\eta_m(x)| \leqslant M\|U_m\|_{L_p(U)}|W_h|^{1-1/p} \leqslant LM|W_h|^{1-1/p}.$$

Let us arbitrarily assign $\varepsilon > 0$ and choose $h > 0$ so that

$$\|\eta_m\|_{L_q(U)} < \varepsilon/3 \quad (3.9)$$

for all m. Due to the previous inequality, this is possible, since $|W_h| \to 0$ for $h \to 0$.

For every x, the function of the variable y, which is equal to $K(x, y)$ for $y \in V_h$ and is equal to zero for $y \notin V_h$, obviously belongs to the space $L_{p'}(U)$ conjugated to $Lp(U)$, and therefore,

$$\xi_m(x) = \int_{V_h} K(x, y)\,u_m(y)\,dy \to 0$$

for $m \to \infty$ for all $x \in U$. Due to inequality (3.8), we have

$$|\xi_m(x)| \leqslant ML|U|^{1-1/p} = M_1$$

for all $x \in U$. Then we obtain

$$\|\xi_m\|_{L_q(U)} = \left(\int_U |\xi_m(x)|^q\,dx\right)^{1/q} \leqslant \left(\int_{W_h} |\xi_m(x)|^q\,dx\right)^{1/q} + \left(\int_{V_h} |\xi_m(x)|^q\,dx\right)^{1/q} \leqslant M_1|W_h|^{1/q} + \left(\int_{V_h} |\xi_m(x)|^q\,dx\right)^{1/q}. \quad (3.10)$$

Here h is supposed to be chosen so that

$$M_1 |W_h|^{1/q} < \varepsilon/3. \tag{3.11}$$

Now let us note that since the set $V_h \times V_h$ is compact, then the function $K(x,y)$ is uniformly continuous on $V_h \times V_h$. Let w be its continuity modulus on $V_h \times V_h$. Then for any $x_1, x_2 \in V_h$, we have

$$|\xi_m(x_1) - \xi_m(x_2)| \leqslant \int_{V_h} |K(x_1, y) - K(x_2, y)|\,|u_m(y)|\,dy \leqslant$$
$$\leqslant L\omega(|x_1 - x_2|)\,|U|^{1-1/p}.$$

This implies that the family of functions (ξ_m) is uniformly equicontinuous on the set V_h. Since $\xi_m(x) \to 0$ is pointwise on V_h, $\xi_m(x) \to 0$ uniformly on V_h, so

$$\left(\int_{V_h} |\xi_m(x)|^q\,dx\right)^{1/q} \to 0$$

for $m \to \infty$. Let us find m_0 such that for $m \geqslant m_0$,

$$\left(\int_{V_h} |\xi_m(x)|^q\,dx\right)^{1/q} < \varepsilon/3. \tag{3.12}$$

For $m \geqslant m_0$, due to inequalities (3.10), (3.11), and (3.12), we have

$$\|\xi_m\|_{L_q(U)} < \varepsilon/3 + \varepsilon/3 = \frac{2}{3}\varepsilon.$$

Hence, due to (3.9), it follows that for $m > m_0$,

$$\|v_m\|_{L_q(U)} \leqslant \|\xi_m\|_{L_q(U)} + \|\eta_m\|_{L_q(U)} < \frac{2}{3}\varepsilon + \frac{1}{3}\varepsilon = \varepsilon.$$

Due to the arbitrariness of $\varepsilon > 0$, we thus obtain that $\|v_m\|_{L_q(U)} \to 0$ for $m \to 0$.

The lemma is proved.

3.3. Basic Inequalities

1) The case $lp > n$.

Lemma 3.2. *Let numbers l and p be such that $0 < l < n$, $lp > n$, $p \geqslant 1$. Suppose that a bounded open set U in $\mathbf{R}^n$ and a kernel K which satisfies the Condition A, B, and C of Subsection 3.1 are given. Let $u \in L_p(U)$. Then P_K is a continuous bounded function in U, and for all $x \in U$, the inequality*

$$|(P_K u)(x)| \leqslant M_{p'}(K)\|u\|_{L_p(U)} \leqslant cL\tau^{l-n/p}\|u\|_{L_p(U)} \tag{3.13}$$

is valid, where $p' = p/(p-1)$, $c = \text{const}$, τ is the radius of finiteness of the kernel K.

Proof. We have $p'(n-l) < n$. Applying the Hölder inequality, we obtain

$$|(P_K u)(x)| \leqslant \left(\int\limits_U |u(y)|^p dy\right)^{1/p} \left(\int\limits_U |K(x, y)|^{p'} dy\right)^{1/p'}$$
$$= \|u\|_{L_p(U)} M_{p'}(K).$$

This proves the boundedness of the function $P_K u$. Due to (3.2), we have $M_{p'}(K) \leqslant Lc\tau^{n/p'-n+l} = cL\tau^{l-n/p}$, and estimate (3.13) is established.

To prove the continuity of the function $P_K u$, let us use the representation $K = K_h + E_h$ constructed in Subsection 3.2. We have

$$(P_K u)(x) = (P_{K_h} u)(x) + (P_{E_h} u)(x). \tag{3.14}$$

For simplicity, let $h < \tau$. Since the function E_h is bounded and continuous, the function $P_{K_h} u$ is continuous. According to the proof above,

$$|(P_{K_h} u)(x)| \leqslant \|u\|_{L_p} M_{p'}(K_h) \leqslant cL \|u\|_{L_p} h^{l-n/p}. \tag{3.15}$$

From (3.13) and (3.15) it follows that for $h \to 0$ $(P_{K_h} u)(x) \to (P_K u)(x)$ uniformly in U. Hence the continuity of $P_K u$ follows.

The lemma is proved.

2) The case $lp \leqslant n$.

Lemma 3.3. *Let numbers l and p be such that $0 < l < n$, $lp \leqslant n$, $p \geqslant 1$, and let U be a bounded open set in $\mathbf{R}^n$; K is a normal kernel in the domain U with singularity order $n - l$ and with finiteness radius τ. Then the function $P_K u$ belongs to the class $L_q(U)$ for any $q \geqslant p$ for which*

$$\gamma = 1/q - 1/p + l/n > 0. \tag{3.16}$$

Here, if m is described by the equality $1/m = 1 - 1/p + 1/q$, then for any such q, we have

$$\|P_K u\|_{L_q(U)} \leqslant M_m(K) \|u\|_{L_p(U)} \leqslant cL\tau^{n\gamma} \|u\|_{L_p(U)}, \tag{3.17}$$

where c depends on l, n, p and q, and L is the constant of the condition B of the definition of a normal kernel.

In the case where $p > 1$, $q > 1$, with p and q satisfying condition (3.16), the operator P_K maps $L_p(U)$ into $L_q(U)$ completely continuously (i.e., is compact).

If μ is an arbitrary measure in U such that $|\mu|(U) < \infty$, then $P_K \mu \in L_q(U)$ for every $q \geqslant 1$ such that

$$\gamma = 1/q - 1 + l/n > 0. \tag{3.18}$$

Here for any such q,

$$\|P_K \mu\|_{L_q(U)} \leqslant M_q(K) |\mu|(U) \leqslant cL\tau^{n\gamma} |\mu|(U). \tag{3.19}$$

Proof. First let us consider the potential of a function of the class $L_p(U)$, where $p > 1$. The arguments are based on the use of the Hölder inequality.

$$\left|\int ABC\,dx\right| \leqslant \left(\int |A|^r dx\right)^{1/r}\left(\int |B|^s dx\right)^{1/s}\left(\int |C|^t dx\right)^{1/t}, \tag{3.20}$$

where r, s, t are positive numbers such that $1/r + 1/s + 1/t = 1$.

Let p, q, and l satisfy all the conditions of the lemma, besides $p < q$. Let us define r, s, and t so that $1/r = 1/p - 1/q$, $1/s = 1 - 1/p$, $t = q$. It is obvious that r, s, and t are positive, also $1/r + 1/s + 1/t = 1$. Let us put $A = |u(y)|^\alpha$, $B = |K(x,y)|^\beta$, $C = |u(y)|^{1-\alpha}|K(x,y)|^{1-\beta}$, where $\alpha = 1 - p/q$, $\beta = q/(s+q)$. Obviously, $0 < \alpha < 1$, $0 < \beta < 1$, $ABC = |u(y)|\ |K(x,y)|$. Hence, applying inequality (3.20), we obtain

$$\begin{aligned}|(P_K u)(x)| &\leqslant \int_U |K(x, y)|\,|u(y)|\,dy = \int_U ABC\,dy \\ &\leqslant \left(\int_U |u(y)|^{\alpha r} dy\right)^{1/r}\left(\int_U |K(x, y)|^{\beta s} dy\right)^{1/s} \\ &\times\left(\int_U |u(y)|^{(1-\alpha)q}|K(x, y)|^{(1-\beta)q} dy\right)^{1/q}.\end{aligned}$$

We have $\alpha r = p$, $(1-\alpha)q = p$, $m = \beta s = (1-\beta)q = qs/(s+q)$. Further, $1/m = 1/s + 1/q = 1 - 1/p + 1/q > 1 - l/n$, whence $n > (n-l)m$. Taking this fact into account, we obtain

$$\begin{aligned}|(P_K u)(x)| &\leqslant (\|u\|_{L_p(U)})^{1-p/q}\left(\int_U |K(x, y)|^m dy\right)^{1-1/p} \\ &\times\left(\int_U |u(y)|^p |K(x, y)|^m dy\right)^{1/q}.\end{aligned}$$

By raising both parts of this inequality to the degree q and by integrating it term by term, we have

$$\begin{aligned}\int_U |(P_K u)(x)|^q dx &\leqslant (\|u\|_{L_p(U)})^{q-p}\left(\int_U |K(x, y)|^m dy\right)^{q/s} \\ &\times\int_U\left(\int_U |K(x, y)|^m |u(y)|^p dx\right)dy.\end{aligned} \tag{3.21}$$

Since $m < n/(n-l)$, then

$$\left(\int_U |K(x, y)|^m dy\right) \leqslant (M_m(K))^m < \infty. \tag{3.22}$$

Applying the Fubini theorem, we obtain

$$\begin{aligned}\int_U\left(\int_U |K(x, y)|^m |u(y)|^p dy\right)dx &= \int_U |u(y)|^p\left(\int_U |K(x, y)|^m dx\right)dy \\ &\leqslant \|u\|^p_{L_p}(M_m(K))^m.\end{aligned} \tag{3.23}$$

By substituting estimates (3.22) and (3.23) into (3.21), we obtain the inequality

$$\int_U |(P_K u)(x)|^q dx \leqslant \|u\|^{q-p}_{L_p(U)} M_m(K)^{mq/s} \|u\|^p_{L_p(U)} (M_m(K))^m.$$

Hence

$$\|P_K u\|_{L_q(U)} \leqslant \|u\|_{L_p(U)} M_m(K)^{m(1/s+1/q)}.$$

It remains to note that $m(1/s+1/q) = 1$, and the first inequality of the lemma is proved. By limit transition for $q \to p+0$, we obtain that it is valid for $q = p$ as well.

Let $p > 1$, $q > 1$, and let p and q satisfy inequality (3.16). Let us prove that in this case the operator P_K from $L_p(U)$ into $L_q(U)$ is compact. To do this, let us use the representation $K = K_h + E_h$ constructed in Subsection 3.1. We have $P_K = P_{K_h} + P_{E_h}$. The function E_h is bounded and continuous on the set $U \times U$. So, according to Lemma 3.1, the operator P_{E_h} is compact. For $h < \tau$, due to what was proved above, we have

$$\|P_{K_h} u\|_{L_q(U)} \leqslant ch^{n\gamma} \|u\|_{L_p(U)},$$

consequently, the norm of P_{K_h} (since P_{K_h} is the operator acting from $L_p(U)$ to $L_q(U)$) does not exceed $ch^{n\gamma}$ and may be chosen to be arbitrarily small. Thus, the operator P_K may be represented as a sum of the compact operator and an operator with an arbitrarily small norm. Hence, the compactness of P_K follows.

Now let us obtain the inequality of the lemma concerning the potential of measure in U. Let μ be a measure in U such that $|\mu|(U) < \infty$. Let us arbitrarily assign $q > 1$ such that inequality (3.18) holds. Then $q < n/(n-l)$. Let us put $p = q/(q-1)$. It is easy to see that $lp > n$. Let us take an arbitrary function $v \in L_p(U)$. We obtain

$$\left| \int_U v(x)(P_K\mu)(x)\, dx \right| \leqslant \int_U |v(x)| \left(\int_U |K(x,y)|\, d|\mu|(y) \right) dx$$
$$= \int_U \left(\int_U |K(x,y)|\, |v(x)|\, dx \right) d|\mu|(y).$$

According to Lemma 3.2, the inner integral does not exceed $M_{p/(p-1)}(K)\, \|v\|_{L_p}$; consequently, we obtain that for every function $v \in L_p(U)$,

$$\left| \int_U v(x)(P_K\mu)(x)\, dx \right| \leqslant |\mu|(U)\, M_{p/(p-1)} \|v\|_{L_p}.$$

Hence it follows that $P_K\mu \in L_q(U)$, besides,

$$\|P_K\mu\|_{L_q(U)} \leqslant M_q(K)\, |\mu|\, (U).$$

This completes the proof of the lemma.

§4 Classes of Functions with Generalized Derivatives

4.1. Definition and the Simplest Properties

Let $U \in \mathbf{R}^n$ be an open set and let f be a generalized function in U. Then f is said to belong to the class $W^l_{p,\mathrm{loc}}(U)$ where $p \geqslant 1$, $l \geqslant 1$; besides, l is an integer if for every n-dimensional multiindex α such that $|\alpha| = l$, the generalized function $D^\alpha f$ is a function of the class $L_{p,\mathrm{loc}}(U)$.

If for any multiindex α for which $|\alpha| = l$, the generalized function $D^\alpha f$ is a measure in U, then f is said to belong to the class $\overline{W}^l_{1,\mathrm{loc}}(U)$.

The identity $D^\alpha(D^\beta f) = D^{\alpha+\beta}(f)$, which is valid for generalized functions, allows us to conclude that if f is a generalized function of the class $W^l_{p,\mathrm{loc}}(U)$ and if there is given an integer k such that $1 \leqslant k < l$, then all its derivatives of the order k belong to the class $W^{l-k}_{p,\mathrm{loc}}(U)$. Just the same, if $f \in \overline{W}^l_{1,\mathrm{loc}}(U)$, then all the derivatives of the order k, where $1 \leqslant k < l$, of the generalized function f belong to the class $\overline{W}^{l-k}_{1,\mathrm{loc}}(U)$.

If $1 \leqslant p_1 < p_2$, then obviously $L_{p_2,\mathrm{loc}}(U) \subset L_{p_1,\mathrm{loc}}(U)$. Hence it follows that for $1 \leqslant p_1 < p_2$ for any l,

$$W^l_{p_2,\mathrm{loc}}(U) \subset W^l_{p_1,\mathrm{loc}}(U). \tag{4.1}$$

Every function of the class $L_{p,\mathrm{loc}}(U)$ as a generalized function is naturally identified with some measure which is its indefinite integral. Due to this, for every l for any $p \geqslant 1$, we have the inclusion

$$W^l_{p,\mathrm{loc}}(U) \subset \overline{W}^l_{1,\mathrm{loc}}(U). \tag{4.2}$$

If $f \in \overline{W}^1_{1,\mathrm{loc}}(U)$, then f is said to locally be a function of bounded variation in the set U.

The symbol $W^l_p(U)$ denotes the totality of all generalized functions $f \in W^l_{p,\mathrm{loc}}(U)$ all of whose derivatives of the order l are functions of the class $L_p(U)$. By $\overline{W}^l_1(U)$ we denote the set of all $f \in \overline{W}^l_{1,\mathrm{loc}}(U)$ for which every derivative of the order l is a measure in U such that its complete variation $|D^\alpha f|(U)$ on the set U is defined and finite.

4.2. Integral Representations for Elements of the Space $\overline{W}^l_{1,\mathrm{loc}}$

The main tool to investigate the spaces $\overline{W}^l_{1,\mathrm{loc}}$, $W^l_{p,\mathrm{loc}}$ are some of the integral representations which are the generalization of representations obtained in Subsection 2.4 for smooth functions.

Theorem 4.1. *Let U be an arbitrary domain of the class $J(r, R)$ in the space $\mathbf{R}^n$ and let f be a generalized function of the class $\overline{W}^l_{1,\mathrm{loc}}(U)$. Then for every n-dimensional multiindex α such that $0 \leqslant |\alpha| = k < l$ and for any $\tau \in [0,1]$, the derivative $D^\alpha f$ admits the representation*

$$D^\alpha f(x) = \int\limits_U \zeta_\tau^{(\alpha)}(x, z) f(z)\, dz + \int\limits_U \sum_{|\gamma|=l} H_{\tau,\gamma}^{(\alpha)}(x, z) D^\gamma f(z)\, dz. \quad (4.3)$$

Here $\zeta_\tau^{(\alpha)}(x,z)$ is the function of the class C^∞ with respect to variables (x,z), and $H_{\tau,\gamma}^{(\alpha)}(x,z)$ is the function of the class C^∞ with respect to (x,z) in the domain where $x \neq z$. For every point $x_0 \in U$, there exist a neighbourhood $V \subset U$ of the point x_0 and a compact set $T \subset U$ such that for $x \in V$ and $z \notin T$, $\zeta_\tau^{(\alpha)}(x,z) = 0$ and $H_{\tau,\gamma}^{(\alpha)}(x,z) = 0$. For $z \notin B(x, 3R\tau)$, if $x \in U$, then $H_{\tau,\gamma}^{(\alpha)}(x,z) = 0$ and $\zeta_\tau^{(\alpha)}(x,z) = 0$. The functions $\zeta_\tau^{(\alpha)}$ and $H_{\tau,\gamma}^{(\alpha)}$ admit the following estimates:

$$|\zeta_\tau^{(\alpha)}(x, z)| \leqslant c_1/\tau^{n+|\alpha|},$$
$$|H_{\tau,\gamma}^{(\alpha)}(x, z)| \leqslant c_2/|z - x|^{n-l+|\alpha|},$$

where the constants c_1, c_2 only depend on n, l, k, and on the domain U. In the case $\tau = 1$, the function $\zeta_\tau^{(\alpha)}(x,z)$ admits the representation:

$$\zeta_1^{(\alpha)}(x, z) = \sum_{|\lambda|<l-|\alpha|} x^\lambda \varphi_\lambda^\alpha(z),$$

where φ_λ^α are the functions of the class C^∞ which are test functions in U.

Remark 1. The first summand on the right-hand side of (4.3) is meant here as a result of the action of the generalized function f upon the function $z \mapsto \zeta_\tau^{(\alpha)}(x,z)$ of the class $C_0^\infty(U)$.

Remark 2. The meaning of equality (4.3) is as follows. The generalized function $D^\alpha f$ coincides with the function assigned by the expression on the right-hand side of this equality.

Proof. First let us obtain the integral formula similar to formula (2.47) of Lemma 2.11. Let us assign an arbitrary point $\xi \in U$ so that either $\xi = a$ or $|\xi - a| \geqslant r/2$, and let $B_\xi = B(a,r)$ in the case $\xi = a$ and $B_\xi = B(\xi, r')$, where r' is described by formula (2.26) in the case $|\xi - a| \geqslant r/2$. Let us define, as Subsection 2.3 shows, the path $\theta : [0,1] \to \mathbf{R}^n$ connecting in U the point ξ with the point a, and just as in subsection 2.3, let us set

$$\varphi(t, x, \xi) = \theta(t) + (1-t)(x-\xi).$$

Let us arbitrarily choose a function $w_0 \in C_0^\infty(\mathbf{R}^n)$ such that $S(w_0) \subset B(0,1)$, and the integral of ω_0 by $\mathbf{R}^n$ equals 1, and let $\omega(h) = (2/r)^n \omega_0(2h/r)$, and $H \subset B(0, r/2)$ is the support of the function ω. Let $T_\tau(\xi)$, where $0 < \tau \leqslant 1$, denote the set crossed out by the curve $\varphi_h(t,x) = ht + \varphi(t,x)$, $0 \leqslant t \leqslant \tau$ when h runs through the set H, x stands for the set $\overline{B}_\xi$. As was said in Subsection 2.3, the set $T_\tau(\xi)$ is compact and contained in U. Let $\delta > 0$ be such that for all $z \in T_\tau(\xi)$, $\rho(z, \partial U) > \delta$. Let $G = U_\delta(T_\tau(\xi))$. Let f_h be a mean function for f with respect to a kernel K chosen arbitrarily. Below, we set $0 < h < \delta/2$. Then the function f_h is defined on the set G and belongs to the class C^∞. By applying the formula of Lemma 2.11, we obtain that for all

$x \in B$,

$$D^{\alpha} f_h(x) = \int_G \zeta_{\tau}^{(\alpha)}(x, z, \xi) f_h(z)\, dz + \int_G \sum_{|\gamma|=l} H_{\tau,\gamma}^{(\alpha)}(x, z, \xi)(D^{\gamma} f_h)(z)\, dz.$$

Let us introduce some notations. We set

$$I_{0,h}(x, \xi) = \int_G \zeta_{\tau}^{(\alpha)}(x, z, \xi) f_h(z)\, dz,$$

$$I_0(x, \xi) = \int_G \zeta_{\tau}^{(\alpha)}(x, z, \xi) f(z)\, dz,$$

$$I_{\gamma,h}(x, \xi) = \int_G H_{\tau,\gamma}^{(\alpha)}(x, z, \xi) D^{\gamma} f_h(z)\, dz,$$

$$I_{\gamma}(x, \xi) = \int_G H_{\tau,\gamma}^{(\alpha)}(x, z, \xi) D^{\gamma} f(z)\, dz.$$

For every h, we have

$$D^{\alpha} f_h(x) = I_{0,h}(x, \xi) + \sum_{|\gamma|=l} I_{\gamma,h}(x, \xi),$$

$$\int_{B_\xi} D^{\alpha} f_h(x) \varphi(x)\, dx = \int_{B_\xi} \left[I_{0,h}(x, \xi) + \sum_{|\gamma|=l} I_{\gamma,h}(x, \xi) \right] \varphi(x)\, dx. \tag{4.7}$$

Let us take an arbitrary function $\varphi \in C_0^{\infty}(B_{\xi})$. For $h \to 0$, due to the properties of mean functions given in Subsection 2.1,

$$\int_{B_\xi} \varphi(x) D^{\alpha} f_h(x)\, dx \to \int_{B_\xi} D^{\alpha} f(x) \varphi(x)\, dx. \tag{4.8}$$

Let us show that

$$\int_{B_\xi} I_{0,h}(x, \xi) \varphi(x)\, dx \to \int_{B_\xi} I_0(x, \xi) \varphi(x)\, dx, \tag{4.9}$$

and for every γ such that $|\gamma| = l$,

$$\int_{B_\xi} I_{\gamma,h}(x, \xi) \varphi(x)\, dx \to \int_{B_\xi} I_{\gamma}(x, \xi) \varphi(x)\, dx. \tag{4.10}$$

The set G lies strictly inside U; therefore, f is in G a finite-order derivative of some continuous function F, $\langle f, \theta \rangle = (-1)^{|\gamma|} \langle F, D^{\gamma}\theta \rangle$ for any function $\theta \in C_0^{\infty}(U)$ such that $S(\theta) \subset G$.

Let us put $G_1 = U_{\delta/2}(T_r(\xi))$. Then $\overline{G}_1 \subset G$, and for every function $\theta \in C_0^{\infty}$ such that $S(\theta) \subset G_1$, we have

$$\begin{aligned}\langle K_h * f, \theta\rangle &= \langle f, \widetilde{K}_h * \theta\rangle \\ &= (-1)^{|\nu|}\langle F, D^\nu(\widetilde{K}_h * \theta)\rangle = (-1)^{|\nu|}\langle F, \widetilde{K}_h * D^\nu\theta\rangle \\ &= (-1)^{|\nu|}\langle K_h * F, D^\nu\theta\rangle.\end{aligned}$$

By setting $\theta(z) = \zeta_\tau^{(\alpha)}(x, z, \xi)$, we obtain that

$$I_{0,h}(x, \xi) = \int_G (-1)^{|\nu|} F_h(z) D_z^\nu \zeta_\tau^{(\alpha)}(x, z, \xi)\, dz.$$

Due to the properties of mean functions given in Subsection 2.1 of Chapter 1, $F_h \to F$ for $h \to 0$ uniformly in G_1. According to the classical theorems about integrals depending on parameter it hence follows that uniformly with respect to x,

$$\begin{aligned}I_{0,h}(x, \xi) &\to \int_G (-1)^{|\nu|} F(z) D_z^\nu \zeta_\tau^{(\alpha)}(x, z, \xi)\, dz \\ &= \int_G D^\nu F(z) \zeta_\tau^{(\alpha)}(x, z, \xi)\, dz = \int_G f(z) \xi_\tau^{(\alpha)}(x, z, \xi)\, dz = I_0(x, \xi).\end{aligned}$$

Hence it follows that

$$\int_{B_\xi} I_{0,h}(x, \xi) \varphi(x)\, dx \to \int_{B_\xi} I_0(x, \xi) \varphi(x)\, dx$$

for $h \to 0$. Thus, relation (4.9) is proved.

Now let us transform the expression

$$\int_{B_\xi} I_{\gamma,h}(x, \xi) \varphi(x)\, dx.$$

By applying the Fubini theorem, we obtain

$$\int_{B_\xi} I_{\gamma,h}(x, \xi) \varphi(x)\, dx = \int_G \Phi_\gamma(z) D^\gamma f_h(z)\, dz,$$

where

$$\Phi_\gamma(z) = \int_{B_\xi} H_{\tau,\gamma}^{(\alpha)}(x, z, \xi) \varphi(x)\, dx.$$

The function Φ_γ is continuous due to the properties of integrals of the potential type which were proved in §3. It vanishes for $z \in T_r(\xi)$, and for $h \to 0$,

$$\int_G \Phi_\gamma(z) D^\gamma f_h(z)\, dz \to \int_G \Phi_\gamma(z) D^\gamma f(z)\, dz \tag{4.11}$$

due to what was proved in §3 of Chapter 1. The derivative $D^\gamma f$ is a measure in U, and the first integral in (4.11) with respect to this measure is the integral of the continuous function Φ_γ. Applying the Fubini theorem, we conclude that

$$\int_G \Phi_\gamma(z) D^\gamma f(z)\,dz = \int_{B_\xi}\left(\int_G H^{(\alpha)}_{\tau,\gamma}(x, z, \xi) D^\gamma f(z)\,dz\right)\varphi(x)\,dx$$
$$= \int_{B_\xi} I_\gamma(x, \xi)\varphi(x)\,dx.$$

Hence, the validity of relation (4.10) follows as well.

Taking the limit for $h \to 0$ in equality (4.7), according to relations (4.8), (4.9), and (4.10), we conclude that

$$\int_{B_\xi} \varphi(x) D^\alpha f(x)\,dx = \int_{B_\xi}\left[I_0(x, \xi) + \sum_{|\gamma|=l} I_\gamma(x, \xi)\right]\varphi(x)\,dx$$

for any $\varphi \in C_0^\infty$ such that $S(\varphi) \in B_\xi$. This implies that the generalized function $D^\alpha f$ coincides in the ball B_ξ with the function $I_0(x,\xi) + \sum_{|\gamma|=l} I_\gamma(x,\xi)$, i.e., the function

$$x \to \int_G \zeta^{(\alpha)}_\tau(x, z, \xi) f(z)\,dz + \int_G \sum_{|\gamma|=l} H^{(\alpha)}_{\tau,\gamma}(x, z, \xi) D^\gamma f(z)\,dz$$

in the ball B_ξ is the derivative $D^\alpha f$ of the generalized function f. We see that this derivative in the ball B_ξ is the usual function. By putting $\alpha = 0$, we obtain, in particular, that the generalized function f is a function in the ball B_ξ.

So far, we have obtained some local integral representation of the generalized function f. To obtain from it an integral representation valid for the entire domain U, let us use the partition of unity.

The sets B_ξ form an open covering of the set U. Indeed, let us take an arbitrary point $x_0 \in U$. If $|x_0 - a| < r/2$, then $x \in B_a$. But if $|x_0 - a| \geqslant r/2$, then B_{x_0} is the set of the family of balls (B_ξ) which contains x_0. Let (λ_m), $m = 1, 2, \ldots$, be the partition of unity which is subordinate to the open covering B_ξ. For every m, let ξ_m be such that the support of λ_m is contained in the ball B_{ξ_m}. For every m, let us define some function F_m by setting for $x \in B_{\xi_m}$,

$$F_m(x) = I_0(x, \xi_m) + \sum I_\gamma(x, \xi_m).$$

For $x \notin B_{\xi_m}$ let us set $F_m(x) = 0$. Then the generalized function $\lambda_m(x)D^\alpha f(x)$ coincides with the function $\lambda_m F_m$. Indeed, for any $\varphi \in C_0^\infty$,

$$\langle \lambda_m D^\alpha f, \varphi\rangle = \langle D^\alpha f, \lambda_m\varphi\rangle = \int_{B_{\xi_m}} \lambda_m \varphi F_m\,dx,$$

since $S(\lambda_m\varphi) \subset B_{\xi_m}$. In the latter integral as the integration domain, one may take the entire domain U.

Let us put $F(x) = \sum_m \lambda_m(x) F_m(x)$ for all x for which the sum is defined. For every point $x_0 \in U$, there exists a neighbourhood V intersecting the supports of only a finite number of functions λ_m. Let them be the functions

$\lambda_{m_1}, \lambda_{m_2}, \dots, \lambda_{m_k}$. For $x \in V$,

$$\sum_m \lambda_m(x) F_m(x) = \sum_{j=1}^{k} \lambda_{m_j}(x) F_{m_j}(x).$$

Hence it follows that the function F is defined almost everywhere and is integrable on V. Since $x_0 \in U$ was taken arbitrarily, it follows that $F(x)$ is defined almost everywhere in U and is a locally integrable function in U.

For arbitrary $\varphi \in C_0^\infty$, we have

$$\langle D^\alpha f, \varphi\rangle = \sum_m \langle \lambda_m D^\alpha f, \varphi\rangle = \int_U \sum_m \lambda_m F_m \varphi \, dx = \int_U F\varphi \, dx,$$

therefore, $F = D^\alpha f$ everywhere in U.

According to what was proved above for all $x \in B_{\xi_m}$,

$$\lambda_m(x) F_m(x) = \int_G \lambda_m(x) \zeta_\tau^{(\alpha)}(x, z, \xi_m) f(z)\, dz + \int_G \lambda_m(x) \sum_{|\gamma|=l} H_{\tau,\gamma}^{(\alpha)}(x, z, \xi_m) D^\gamma f(z)\, dz. \tag{4.12}$$

The right-hand side of this equality does not change if we take the entire set U as the integration domain. Since for $x \in B_{\xi_m}$, $\lambda_m(x) = 0$, equality (4.12) is valid for $x \in B_{\xi_m}$, too. Summing it up by m, we obtain

$$F(x) = \int_G \zeta_\tau^{(\alpha)}(x, z) f(z)\, dz + \int_G \sum_{|\gamma|=l} H_{\tau,\gamma}^{(\alpha)}(x, z) D^\gamma f(z)\, dz,$$

where

$$\zeta_\tau^{(\alpha)}(x, z) = \sum_{m \in N} \lambda_m(x) \zeta_\tau^{(\alpha)}(x, z, \xi_m),$$

$$H_{\tau,\gamma}^{(\alpha)}(x, z) = \sum_{m \in N} \lambda_m(x) H_{\tau,\gamma}^{(\alpha)}(x, z, \xi_m).$$

Let us take an arbitrary point $x_0 \in U$ and let us find its neighbourhood V in which only a finite number of functions λ_m are distinct from zero. Let them be the functions $\lambda_{m_1}, \lambda_{m_2}, \dots, \lambda_{m_k}$. Let $T = \cup_{j=1}^{k} T_\tau(\xi_{m_j})$. It is obvious that T is compact, $T \subset U$, and for $x \in V$, $z \notin T$, each of the functions $\xi_\tau^{(\alpha)}$, $H_{\tau,\gamma}^{(\alpha)}$ vanishes. Note that for every $j = 1, 2, \dots, k$, $x \in B_{\xi_{m_k}}$; therefore, $T_\tau(\xi_{m_k})$ is contained in the ball $B(x, 3R\tau)$. Hence it follows that the functions $\zeta_\tau^{(\alpha)}(x, z)$ and $H_{\tau,\gamma}^{(\alpha)}(x, z)$ vanish if z lies outside of this ball.

Estimates (4.4) and (4.5) are the obvious corollary of analogous estimates for the functions $\zeta_\tau^{(\alpha)}(x, z, \xi)$ and $H_{\tau,\gamma}^{(\alpha)}(x, z, \xi)$ contained in Lemma 2.11.

In the case $\tau = 1$, the function $\zeta_\tau^{(\alpha)}(x, z, \xi)$ is expressed by formula (2.56); that is,

$$\zeta_1^{(\alpha)}(x, z, \xi) = \sum_{|\beta|<l} \frac{(-1)^{|\beta|}}{\beta!} D_z^\beta [D_x^\alpha (x-z)^\beta \omega(z-a)].$$

In particular, we see that if $\tau = 1$, then the function $\zeta_1^{(\alpha)}$ is independent of ξ. This implies that

$$\zeta_1^{(\alpha)}(x, z) = \sum_{|\beta|<l} \frac{(-1)^\beta}{\beta} D_z^\beta [(D_x^\alpha (x-z)^\beta)\omega(z-a)].$$

By removing the parentheses, we obtain

$$\zeta_1^{(\alpha)}(x, z) = \sum_{|\lambda|<l-|\alpha|} x^\lambda \psi_\lambda(z).$$

This completes the proof of the lemma.

4.3. The Imbedding Theorem

Let us assign an arbitrary domain U of the class J in the space $\mathbf{R}^n$. Let $U \in J(R, r)$ for some r, R such that $0 < r \leqslant R < \infty$, and let a be a marked point of the domain U. We shall now give some corollaries of integral representations obtained in Subsection 4.2. We shall only use the representation for which the parameter τ equals 1.

Let $f : U \to \mathbf{R}$ be a function of the class $W_p^l(U)$. We set

$$\|f\|_{L_p^l(U)} = \left(\int_U \left(\sum_{|\gamma|=l} \frac{l!}{\gamma!} [D^\alpha f(z)]^2\right)^{\frac{p}{2}} dz\right)^{\frac{1}{p}}. \tag{4.13}$$

The correspondence $f \to \|f\|_{L_p^l(U)}$ is some seminorm on the vector space $W_p^l(U)$.

By applying the formula of Theorem 4.1 for $\tau = 1$ and $\alpha = 0$, we obtain that for every function $f \in W_p^l(U)$, the equality

$$f(x) = \int_U \zeta(x, z) f(z)\, dz + \int_U \sum_{|\gamma|=l} H_\gamma(x, z) D^\gamma f(z)\, dz \tag{4.14}$$

is valid. Let us set

$$\int_U \zeta(x, z) f(z)\, dz = (\Pi_0 f)(x).$$

This describes some operator Π_0 on the set W_p^l. We have

$$\zeta(x, z) = \sum_{|\beta|<l} x^\beta \psi_\beta(z),$$

where ψ_β are functions of the class C^∞ whose supports are contained in the ball $B(a, r/2)$. Hence it is clear that $\Pi_0 f$ is a polynomial whose degree does not exceed $l-1$. Let us note that if the function f is a polynomial of degree not exceeding $l-1$, then the second integral in equality (4.14) vanishes, and we obtain that in this case, $f = \Pi_0 f$. Thus, the operator maps $W_p^l(U)$ to a finite-dimensional vector space $P_{l-1,n}$ of polynomials with degree not exceeding $l-1$ of n variables. If $f \in P_{l-1}, n$, then $\Pi_0 f = f$.

Integral representation (4.14) and all that was said above about the operator Π_0 remains valid in the case of the space $\overline{W}_1^l(U)$ as well.

Theorem 4.2. *Let U be a domain of a class $J(R,r)$ in $\mathbf{R}^n$, and let $f : U \to \mathbf{R}$ be a function of the class $W_p^l(U)$. Then, if $lp \leqslant n$ and $q \geqslant p$ is such that $\gamma = 1/q - 1/p + 1/n > 0$, the function f belongs to the class $L_q(U)$, and the estimate*

$$\|f - \Pi_0 f\|_{L_q(U)} \leqslant C \|f\|_{L_p^l(U)} \tag{4.15}$$

is also valid, where C is a constant. C only depends on n, l, p, q, and on the constants R and r. If $lp > n$, then the function f is continuous. In this case,

$$\|f - \Pi_0 f\|_{C(U)} \leqslant C \|f\|_{L_p^l(U)}. \tag{4.16}$$

The constant C here also depends on n, l, p, and on the magnitudes r and R.

Proof. The statement of the theorem is the direct corollary of equality (4.14) and Lemmas 3.2 and 3.3 about integrals of the potential type. Let us set $f - \Pi_0 f = g$, and let

$$g_\gamma = \int_U H_\gamma(x, z) D^\gamma f(z)\, dz = (P_{H_\gamma} D^\gamma f)(x).$$

The function H_γ due to Theorem 4.1 is defined and continuous on the set of all $(x,z) \in U \times U$ such that $x \neq z$ and

$$H(x, z) < M/|x-z|^{n-l},$$

where the constant M has the form $M = C(R/r)^n$, and C only depends on n. Let $lp \leqslant n$ and $q \geqslant p$ be such that $\gamma = 1/q - 1/p + l/n > 0$. Then, due to Lemma 3.3, $g_\gamma \in L_q(U)$; and, in addition, implies

$$\|g_\gamma\|_{L_q(U)} \leqslant C \|D^\gamma f\|_{L_p(U)}.$$

We have $g = \sum_{|\gamma|=l} g_\gamma$. Hence, we conclude that $g \in L_q(U)$, and also

$$\|g\|_{L_q(U)} \leqslant C \sum_{|\gamma|=l} \|D^\gamma f\|_{L_p(U)}. \tag{4.17}$$

For any finite system of nonnegative numbers $z_1, z_2, \ldots, z_m$ for any $p \geqslant 1$ we have

$$\frac{z_1 + z_2 + \cdots + z_m}{m} \leqslant \left(\frac{z_1^p + z_2^p + \cdots + z_m^p}{m} \right)^{\frac{1}{p}}.$$

Using this elementary inequality, we obtain

$$\sum_{|\gamma|=l} \| D^\gamma f \|_{L_p(U)} = \sum_{|\gamma|=l} \left(\int_U | D^\gamma f |^p dx \right)^{\frac{1}{p}} \leqslant C \left(\int_U \sum_{|\gamma|=l} | D^\gamma f |^p dx \right)^{\frac{1}{p}}. \tag{4.18}$$

It remains to be noted that

$$\left(\sum_{|\gamma|=l} | D^\gamma f |^p \right)^{\frac{1}{p}} \leqslant C \left(\sum_{|\gamma|=l} \frac{l!}{\gamma!} | D^\gamma f |^2 \right)^{\frac{1}{2}}, \tag{4.19}$$

where $C = \text{const}$. Comparing inequalities (4.17), (4.18), and (4.19), we obtain inequality (4.15).

In the case $lp > n$ in the above notations, by applying Lemma 3.2, we obtain that each of the functions g_γ is continuous, and that

$$\| g_\gamma \|_{C(U)} \leqslant C \| D^\gamma f \|_{L_p(U)}.$$

Hence, similar to the previous case, we obtain

$$\| g \|_{C(U)} \leqslant C \| f \|_{L_p^l(U)}.$$

The theorem is proved.

4.4. Corollaries of Theorem 4.2. Normalization of the Spaces $W_p^l(U)$

Here U denotes an arbitrary domain of the J type in the space $\mathbf{R}^n$. If f and g are two arbitrary functions of the class $W_p^l(U)$, then for any numbers $\alpha, \beta \in \mathbf{R}$, the function $\alpha f + \beta g$ also belongs to the class $W_p^l(U)$ so that the set $W_p^l(U)$ is a vector space. Let us now introduce some norm in this space.

Denote by $P_{l-1,n}$ a set of all polynomials of n variables whose degree does not exceed $l-1$. The set $P_{l-1,n}$ may naturally be considered to be a vector space. This is a finite-dimensional space. Every polynomial $\varphi \in P_{l-1,n}$ may be written as

$$\varphi(x) = \sum_{|\alpha| \leqslant l-1} a_\alpha x^\alpha. \tag{4.20}$$

The dimension of the space $P_{l-1,n}$ is equal to the number of n-dimensional multi-indices α such that $|\alpha| \leqslant l-1$. The polynomials $x \mapsto x^\alpha$, where $|\alpha| \leqslant l-1$ form the basis in the vector space $P_{l-1,n}$. In this vector space, one may introduce a norm. This can be done in various ways. Let $\varphi \in P_{l-1,n}$ and let it be assigned by equality (4.20). Let

$$N(\varphi) = \left(\sum_{|\alpha| \leqslant l-1} \frac{|\alpha|!}{\alpha!} a_\alpha^2 \right)^{1/2}. \tag{4.21}$$

The function N thus defined on $P_{l-1,n}$ is a norm in $P_{l-1,n}$. Let then

$$N_C(\varphi) = \sup_{x \in U} | \varphi(x) |, \tag{4.22}$$

$$N_{L_p}(\varphi) = \| \varphi \|_{L_p(U)}. \tag{4.23}$$

It can easily be verified that the functions N_C and N_{L_p} are also the norms in $P_{l-1,n}$. It is most difficult here to verify that if $N_{L_p}(\varphi) = 0$ or $N_C(\varphi) = 0$, then $\varphi \equiv 0$, that is, that all the coefficients of the polynomial φ are equal to zero. The validity of this statement follows from the fact that if the polynomial vanishes on any open subset $\mathbf{R}^n$, then all its coefficients equal zero. Due to finite dimensionality of the space $P_{l-1,n}$, any two norms in them are equivalent. However, in some cases, it may turn out that the choice of this or that special norm gives certain advantages. Herein the symbol $\|\cdot\|_{P_{l-1,n}}$ denotes an arbitrary norm in $P_{l-1,n}$. We only fix the choice of it when necessary.

Let there be a linear mapping $\Pi : W_p^l(U) \to P_{l-1,n}$. The mapping Π is said to be projective if for every $f \in P_{l-1,n}$, $\Pi f = f$. If Π is a projective operator from $W_p^l(U)$ to $P_{l-1,n}$, then for every f, we have

$$(\Pi f)(x) = \sum_{|\alpha| \leqslant l-1} x^\alpha l_\alpha(f), \tag{4.24}$$

where l_α are linear functionals in $W_p^l(U)$. The functionals l_α satisfy the orthogonality conditions

$$l_\alpha(x^\beta) = \begin{cases} 1, & \text{for } \alpha = \beta, \\ 0, & \text{for } \alpha \neq \beta, \end{cases} \tag{4.25}$$

for every β with $|\beta| \leqslant l-1$. Conversely, if a system of linear functionals is assigned on $W_p^l(U)$, satisfying the orthogonality conditions (4.25), then equality (4.24) defines the projective operator acting from $W_p^l(U)$ into $P_{l-1,n}$.

Above, we defined some special projective operator $\Pi_0 : W_p^l(U) \to P_{l-1,n}$. This operator admits the representation

$$(\Pi_0 f)(x) = \sum_{|\alpha| < l-1} x^\alpha \int_U \psi_\alpha(z) f(z)\, dz,$$

where ψ_α are the functions of the class $C_0^\infty(U)$.

The orthogonality conditions (4.25) in this case acquire the form

$$\int_U z^\beta \psi_\alpha(z)\, dz = \begin{cases} 1, & \text{for } \beta = \alpha, \\ 0, & \text{for } \beta \neq \alpha, \end{cases}$$

where $|\beta| \leqslant l-1$.

According to Theorem 4.2, in the case $lp \leqslant n$, we have

$$\|f - \Pi_0 f\|_{L_q(U)} \leqslant C \|f\|_{L_p^l(U)} \tag{4.26}$$

for any $q < np/(n - lp)$, and

$$\|f - \Pi_0 f\|_{C(U)} \leqslant C \|f\|_{L_p^l(U)}$$

in the case $lp > n$.

Let Π be a projective operator acting from $W_p^l(U)$ into $P_{l-1,n}$. For an arbitrary function $f \in W_p^l$, we set

$$\nu_\Pi(f) = \|\Pi f\|_{P_{l-1,n}} + \|f\|_{L_p^l(U)}. \tag{4.28}$$

The magnitude ν_Π is the norm in $W_p^l(U)$. Let us verify this statement. We have 1) $\nu_\Pi(f_1 + f_2) \leqslant \nu_\Pi(f_1) + \nu_\Pi(f_2)$, 2) $\nu_\Pi(\lambda f) = |\lambda|\nu_\Pi(f)$, which is obvious. It is also clear that if $f = 0$, then $\nu_\Pi(f) = 0$. Suppose that for the function f, $\nu_\Pi(f) = 0$. Then $\|f\|_{L_p^l(U)} = 0$; therefore, due to inequalities (4.26) and (4.27), the difference $f - \Pi_0 f$ equals zero, i.e. $f \in P_{l-1,n}$. Hence it follows that $\Pi f = f$. The condition allows us to conclude that $f \equiv 0$. Below, ν_Π is said to be the norm described by the projective operator Π. Note that in the definition of the norm ν_Π, there is some arbitrariness concerning the choice of the norm in $P_{l-1,n}$. Obviously, if we substitute a norm in $P_{l-1,n}$ by another one, we obtain equivalent norms in $W_p^l(U)$.

Now for $f \in W_p^l(U)$, we set

$$\|f\|_{W_p^l(U)} = \nu_{\Pi_0}(f) = \|\Pi_0 f\|_{P_{l-1,n}} + \|f\|_{L_p^l(U)}. \tag{4.29}$$

In other words, the magnitude $\|f\|_{W_p^l(U)}$ is the norm $\nu_\Pi(f)$ corresponding to the case where Π is a special projective operator Π_0. As for the norm in $W_p^l(U)$, we imply the norm described by equality (4.29) every time if no exception is made.

Let us note that for any choice of the projective operator $\Pi : W_p^l(U) \to P_{l-1,n}$, it is continuous with respect to the norm ν_Π.

Let us consider some norms arising if we modify the second summand in the right-hand part of (4.28). Let a system of numbers (u_α) be given where α runs through a set of all n-dimensional multi-indices α such that $|\alpha| = 1$. Then there exist the numbers K_1, K_2, L_1, L_2 such that $0 < K_1 < K_2 < \infty$, $0 < L_1 < L_2 < \infty$, and

$$K_1^p \sum_{|\alpha|=l} |u_\alpha|^p \leqslant \left(\sum \frac{l!}{\alpha!} u_\alpha^2\right)^{p/2} \leqslant K_2^p \sum_{|\alpha|=l} |u_\alpha|^p, \tag{4.30}$$

$$L_1 \sum_{|\alpha|=l} |u_\alpha| \leqslant \left(\sum_{|\alpha|=l} |u_\alpha|^p\right)^{\frac{1}{p}} \leqslant L_2 \sum_{|\alpha|=l} |u_\alpha|. \tag{4.31}$$

In this case the magnitudes K_1 and K_2, L_1 and L_2 only depend on l, n, and p.

By setting $u_\alpha = D^\alpha f$ in (4.30) and by integrating by U, we obtain

$$K_1 \sum_{|\alpha|=l} \|D^\alpha f\|_{L_p(U)}^p \leqslant (\|f\|_{L_p^l(U)})^p \leqslant K_2 \sum_{|\alpha|=l} \|D^\alpha f\|_{L_p(U)}^p.$$

By putting $u_\alpha = \|D^\alpha f\|_{L_p(U)}$ and by applying inequality (4.31), we obtain

$$M_1 \sum_{|\alpha|=l} \|D^\alpha f\|_{L_p(U)} \leqslant \|f\|_{L_p^l(U)} \leqslant M_2 \sum_{|\alpha|=l} \|D^\alpha f\|_{L_p(U)}, \tag{4.32}$$

where $M_1 = K_1L_1$, $M_2 = K_2L_2$, $0 < M_1 < M_2 < \infty$. Inequality (4.32), in particular, allows us to conclude that the magnitude

$$\|\Pi f\|_{P_{l-1,n}} + \sum_{|\alpha|=l} \|D^\alpha f\|_{L_p(U)} \tag{4.33}$$

is a norm in $W_p^l(U)$ equivalent to norm (4.28).

From Theorem 4.2, the following theorem directly follows.

Theorem 4.3. *If a domain U in $\mathbf{R}^n$ belongs to the class J, then for any l, p, and n for every function $f \in W_p^l(U)$, the following estimates are valid. In the case $lp \leqslant n$,*

$$\|f\|_{L_q(U)} \leqslant C\|f\|_{W_p^l(U)}$$

for any $q < np/(n-lp)$, $q > 1$, and in the case $lp > n$,

$$\|f\|_{C(U)} \leqslant C\|f\|_{W_p^l(U)}.$$

The constants C only depend on l, p, n, q, and on the domain U.

Proof. From Theorem 4.2, we obtain

$$\|f\|_{L_q(U)} \leqslant \|\Pi_0 f\|_{L_q(U)} + C\|f\|_{L_p^l(U)} = N_{L_q}(\Pi_0 f) + C\|f\|_{L_p^l(U)}$$

for $lp \leqslant n$, and for $lp > n$,

$$\|f\|_{C(U)} \leqslant \|\Pi_0 f\|_{C(U)} + S\|f\|_{L_p^l(U)} = N_C(\Pi_0 f) + S\|f\|_{L_p^l(U)}.$$

According to what was said above about the equivalence of different norms in $P_{l-1,n}$, the theorem is proved.

Theorem 4.4. *If U is a domain of the class J, then the space $W_p^l(U)$ provided with norm (4.29) is complete.*

Proof. Let (f_m), $m = 1, 2, \ldots$, be an arbitrary sequence of functions from $W_p^l(U)$ which is fundamental with respect to norm (4.29). We are to prove that there exists a function $f \in W_p^l(U)$ such that $\|f_m - f\|_{W_p^l(U)} \to 0$ for $m \to \infty$.

For every function $\varphi \in W_p^l(U)$ for the multiindex α with $|\alpha| = l$, we have

$$\|D^\alpha \varphi\|_{L_p(U)} \leqslant C\|\varphi\|_{L_p^l(U)},$$

where $C = \text{const}$. Hence, we obtain that for any $m, k \geqslant 1$,

$$\|D^\alpha f_m - D^\alpha f_k\|_{L_p(U)} \leqslant C\|f_m - f_k\|_{L_p^l(U)} \leqslant C\|f_m - f_k\|_{W_p^l(U)}$$

and therefore, $\|D^\alpha f_m - D^\alpha f_k\|_{L_p(U)} \to 0$ for $m \to \infty$, $k \to \infty$. Due to the completeness of the space $L_p(U)$, it follows that for every multiindex α with $\|\alpha\| = l$, there exists a function f_α such that $\|D^\alpha f_m - f_\alpha\|_{L_p(U)}$ for $m \to \infty$. We also have

$$\|\Pi_0 f_m - \Pi_0 f_k\|_{P_{l-1,n}} \leqslant \|f_m - f_k\|_{W_p^l(U)}$$

for any m and k, whence it follows that $\|\Pi_0 f_m - \Pi_0 f_k\| \to 0$ for $m \to \infty$ and $k \to \infty$. Hence, it follows that the sequence of polynomials $(\Pi_0 f_m)$, $m = 1, 2, \ldots$, converges to some polynomial ψ whose degree does not exceed $(l-1)$. Since any norm in $P_{l-1,n}$ is equivalent to any of the norms N_C, N_{L_p}, then the polynomials $\Pi_0 f_m$ uniformly converge to ψ. Let us set $F_m = f_m - \Pi_0 f_m$. According to Theorem 4.2, for any m and k in the case $lp \leqslant n$, we have

$$\|F_m - F_k\|_{L_p(U)} \leqslant C\|f_m - f_k\|_{L_p^l} \leqslant C\|f_m - f_k\|_{W_p^l},$$

and for $lp > n$, the analogous inequality holds for the norm in $C(U)$. This allows us to conclude that for $m \to \infty$, $k \to \infty$, $\|F_m - F_k\|L_p(U) \to 0$ (in the case $lp \leqslant n$); therefore, due to the completeness of the space $L_p(U)$, the sequence of functions (F_m) converges in $L_p(U)$ to some function F. Then the sequence of functions (f_m) converges in $L_p(U)$ to the function $f = F + \psi$. In the case $lp > n$, the functions (f_m) uniformly converge to the function f. We have

$$(\Pi_0 f_m)(x) = \sum_{|\alpha| \leqslant l-1} x^\alpha \int_U \Psi_\alpha(z) f_m(z)\, dz,$$

where ψ_α are the functions of the class C^∞. Hence, it follows that for $m \to \infty$, $\Pi_0 f_m(x) \to \Pi_0 f(x)$; therefore, $\psi = \Pi_0 f(x)$.

Thus, we have constructed the function f and a set of functions $\{f_\alpha\}$ where $|\alpha| = l$ such that for $m \to \infty$,

$$\|\Pi_0 f_m - \Pi_0 f\|_{P_{l-1,n}} \to 0 \tag{4.34}$$

and

$$\|D^\alpha f_m - f_\alpha\|_{L_p(U)} \to 0. \tag{4.35}$$

Let us show that $f_\alpha = D^\alpha f$ for every α. Indeed, let us assign an arbitrary function $\theta \in C_0^\infty(U)$. For every m, we have

$$\int_U (D^\alpha\theta)(x) f_m(x)\, dx = (-1)^l \int_U \theta(x) D^\alpha f_m(x)\, dx.$$

Taking the limit for $m \to \infty$ in this equality, we obtain

$$\int_U D^\alpha\theta(x) f(x)\, dx = (-1)^l \int_U \theta(x) f_\alpha(x)\, dx.$$

Due to the arbitrariness of $\theta \in C_0^\infty(U)$, this proves that $f_\alpha = D^\alpha f$. According to relations (4.34) and (4.35), it follows, due to equivalence of norms (4.29) and (4.33), that $\|f_m - f\|_{W_p^l(U)} \to 0$ for $m \to \infty$. This completes the proof of the theorem.

Let us investigate the correlation of norms in the space $W_p^l(U)$ which are obtained for the different choice of the operators Π_1 and Π_2.

Theorem 4.5. *Let the projective operators Π_1 and Π_2 be given. Then the norms ν_{Π_1} and ν_{Π_2} in the space $W_p^1(U)$ are equivalent iff the operator Π_2 is bounded with respect to the norm ν_{Π_1}.*

Proof. Suppose that the norms ν_{Π_1} and ν_{Π_2} are equivalent. Then there exists a constant $K < \infty$ such that $\nu_{\Pi_2}(f) \leqslant K\nu_{\Pi_1}(f)$ for any function $f \in W_p^l(U)$. Taking into account the expression for the norm ν_{Π_2}, we thus obtain $\|\Pi_2 f\|_{P_{l-1,n}} \leqslant K\nu_{\Pi_1}(f)$. Thus, the necessity of the theorem is proved.

Let us prove the sufficiency. Suppose that the operator Π_2 is bounded with respect to the norm ν_{Π_1}. This implies that there exists a number $K < \infty$ such that $\|\Pi_2 f\|_{P_{l-1,n}} \leqslant K\nu_{\Pi_1}(f)$. Hence, we obtain

$$\nu_{\Pi_2}(f) = \|\Pi_2 f\| + \|f\|_{L_p^l} \leqslant K\nu_{\Pi_1}(f) + \|f\|_{L_p^l}$$
$$\leqslant K\|\Pi_1 f\| + (K+1)\|f\|_{L_p^l} \leqslant (K+1)\,\nu_{\Pi_1}(f).$$

Now let us estimate the norm ν_{Π_1} via ν_{Π_2}. Note that since $\Pi_1 f, \Pi_2 f \in P_{l-1,n}$, then $\Pi_2\Pi_1 f = \Pi_1 f$ and $\Pi_1\Pi_2 f = \Pi_2 f$. For every function $f \in W_p^l$, we have $\Pi_1(f - \Pi_1 f) = \Pi_1 f - \Pi_1 f = 0$, therefore, $\nu_{\Pi_1}(f - \Pi_1 f) = \|f\|_{L_p^l}$. Now we have

$$\|(\Pi_2 f - \Pi_1 f)\|_{P_{l-1,n}} = \|\Pi_2(f - \Pi_1 f)\|_{P_{l-1,n}} \leqslant K\nu_{\Pi_1}(f - \Pi_1 f)$$
$$= K\|f - \Pi_1 f\|_{L_p^l} = K\|f\|_{L_p^l}.$$

Hence,

$$\nu_{\Pi_1}(f) = \|\Pi_1 f\|_{P_{l-1,n}} + \|f\|_{L_p^l} \leqslant \|\Pi_2 f\|_{P_{l-1,n}} + \|\Pi_2 f - \Pi_1 f\|_{P_{l-1,n}}$$
$$+ \|f\|_{L_p^l} \leqslant \|\Pi_2 f\|_{P_{l-1,n}} + (K+1)\|f\|_{L_p^l} \leqslant (K+1)\,\nu_{\Pi_2}(f).$$

This completes the proof of the theorem.

Corollary. *Let there be a system of functions (φ_α) of the class $C_0^\infty(U)$ where α runs through the set of all multiindices such that $\|\alpha\| \leqslant l-1$. Suppose that the functions φ_α satisfy the orthogonality conditions*

$$\int_U z^\beta \varphi_\alpha(z)\,dz = \begin{cases} 1, & \text{for } \beta = \alpha, \\ 0, & \text{for } \beta \neq \alpha, \end{cases}$$

where $|\beta| \leqslant l-1$. Let us define the operator Π by setting for $f \in W_p^l(U)$,

$$(\Pi f)(x) = \sum_{|\alpha|\leqslant l-1} x^\alpha \int_U f(z)\varphi_\alpha(z)\,d\bar{z}.$$

The norm ν_Π in W_p^l defined by means of this operator is equivalent to the norm ν_{Π_0} which corresponds to the special projective operator.

Indeed, for every α with $|\alpha| \leqslant l-1$, the linear functional

$$l_\alpha : f \to \int_U \varphi_\alpha(z) f(z)\,dz$$

is obviously bounded in $L_q(U)$ and in $C(U)$, and therefore, due to Theorem 4.3, it is bounded with respect to the norm $\|f\|_{W_p^l(U)}$. Hence, the boundedness of the operator Π follows. This proves the corollary.

The space $P_{l-1,n}$ is a subspace of the space $W_p^l(U)$. The factor space $W_p^l(U)/P_{l-1,n}$ is denoted by the symbol $L_p^l(U)$. The elements of the set $L_p^l(U)$ are classes of functions from $W_p^l(U)$. In this case, two functions f and g from $W_p^l(U)$ belong to the same element of the space $L_p^l(U)$ iff the difference $f-g$ is a polynomial with degree not exceeding $l-1$. Let u be an arbitrary element of the space $L_p^l(U)$. Let us choose arbitrary $f \in u$ and set

$$\|u\|_{L_p^l(U)} = \|f\|_{L_p^l(U)}.$$

It is obvious that the magnitude $\|u\|_{L_p^l(U)}$ is independent of the choice of $f \in u$. This describes some norm in $L_p^l(U)$.

Let us assign an arbitrary projective operator $\Pi : W_p^l(U) \to P_{l-1,n}$ which is continuous with respect to norm (4.29). The operator Π makes it possible to uniquely associate to every element u of the space $L_p^l(U)$ some element $\Pi^* u$ of the space $W_p^l(U)$. Namely, let $f \in u$. Let us put $\Pi^* u = f - \Pi f$. The value of $\Pi^* u$ does not depend on the choice of $f \in u$. Indeed, let g be another arbitrary element of the set u. Then $g = f + h$, where $h \in P_{l-1,n}$; therefore, $g - \Pi g = f - \Pi f + h - \Pi h = f - \Pi f$, because $h = \Pi h$. The set $\Pi^*(L_p^l(U))$ coincides with the totality of all $f \in W_p^l(U)$ such that $\Pi f = 0$. Actually, we suppose that $f \in W_p^l(U)$ is such that $\Pi f = 0$, and let $u \in L_p^l(U)$ be the element of the space $L_p^l(U)$ to which f belongs. Then $\Pi^* u = f - \Pi f = f$ so that $f \in \Pi^*(L_p^l(U))$. Conversely, let $f \in \Pi^*(L_p^l(U))$. Then $f = g - \Pi g$ for some $g \in W_p^l(U)$. Hence, $\Pi f = \Pi g - \Pi(\Pi(g))$. Since $\Pi g \in P_{l-1,n}$, then $\Pi(\Pi g) = \Pi(g)$, and we obtain that $\Pi f = \Pi g - \Pi g = 0$.

The mapping $u \to \Pi^* u$ is the isometric imbedding of $L_p^l(U)$ into $W_p^l(U)$. Indeed, let $f = \Pi^* u$. Then $f \in u$ and $\Pi f = 0$. Hence, we conclude that

$$\|f\|_{W_p^l(U)} = \|f\|_{L_p^l(U)} = \|u\|_{L_p^l(U)}.$$

Note that the result remains valid if we replace the norm $\|\cdot\|_{W_p^l(U)}$ by any other norm ν_P, where P is a projective operator from $W_p^l(U)$ to $P_{l-1,n}(U)$.

Due to continuity of the operator Π, the set $\Pi^*(L_p^l(U))$ is a closed subset of $W_p^l(U)$. Since the operator Π^* is isometric, it follows that the space $L_p^l(U)$ is complete.

Let $f \in W_p^l(U)$. Then the function f belongs to the class $L_p(U)$, as follows from Theorem 4.3. Let us consider its derivative $D^\alpha f$, where $|\alpha| = r < l$. The function $D^\alpha f$ belongs to the class $W_p^{l-r}(U)$, therefore $D^\alpha f \in L_p(U)$. Consequently, we obtain that for every function $f \in W_p^l(U)$, the magnitude

$$N_{l,p}(f) = \sum_{|\alpha| \leq l} \| D^\alpha f \|_{L_p(U)}$$

is defined and finite.

Lemma 4.1. *A function $N_{l,p}$ is a norm in $W_p^l(U)$ equivalent to the main one.*

Proof. $N_{l,p}$ is obviously a norm. The operator $\Pi_0(f)$ is obviously bounded in $L_p(U)$; consequently, we obtain $\|\Pi_0 f\|_{P_{l-1,n}} \leq K\|f\|_{L_p(U)}$, where $K < \infty$, $K = \text{const}$. Hence, by taking inequality (4.32) into account, we obtain that

$$\|f\|_{W_p^l(U)} \leq K\|f\|_{L_p(U)} + M_2 \sum_{|\alpha|=l} \|D^\alpha f\|_{L_p(U)} \leq C N_{l,p}(f),$$

where the constant $C = \max\{K, M_2\}$.

To complete the proof, we have to estimate the magnitude $N_{l,p}(f)$ in terms of $\|f\|_{W_p^l(U)}$. Let α be such that $|\alpha| = r < l$. The function $D^\alpha f$ belongs to the class $W_p^{l-r}(U)$. Obviously, due to Theorem 4.3, we have

$$\|D^\alpha f\|_{L_p(U)} \leq C\|D^\alpha f\|_{W_p^{l-r}(U)} = C\,(\|\tilde{\Pi}_0 f\|_{P_{l-r-1,n}} + \|D^\alpha f\|_{L_p^{l-r}(U)}).$$

Here Π_0 is the operator of projecting $W_p^{l-r}(U)$ to a space of polynomials whose degree does not exceed $l - r - 1$. We have

$$(\tilde{\Pi}_0 D^\alpha f)(x) = \sum_{|\beta| < l-r} x^\beta \int_U \psi_\beta(z) D^\alpha f(z)\, dz.$$

Here the functions ψ_β belong to the class $C_0^\infty(U)$, therefore

$$\int_U \psi_\beta(z) D^\alpha f(z)\, dz = (-1)^r \int_U D^\alpha \psi_\beta(z) f(z)\, dz.$$

Hence, it is clear that the operator $f \mapsto \tilde{\Pi}_0 D^\alpha f$ is continuous in $L_p(U)$; therefore

$$\|\tilde{\Pi}_0 D^\alpha f\|_{P_{r-l-1,n}} \leq \|f\|_{L_p(U)}.$$

Let us also note that

$$\|D^\alpha f\|_{L_p^{l-r}(U)} \leq \|f\|_{L_p^l(U)},$$

and we obtain that

$$\|D^\alpha f\|_{L_p(U)} \leq C\,(\|f\|_{L_p(U)} + \|f\|_{L_p^l(U)}).$$

From the above-said, the estimate

$$N_l^p(f) \leqslant C(\|f\|_{L_p(U)} + \|f\|_{L_p^l(U)}).$$

follows.

It remains to note that due to Theorem 4.3, $\|f\|_{L_p(U)} \leqslant C\|f\|_{W_p^l(U)}$, whence it follows that the magnitude $N_{l,p}(f)$ is estimated in terms of $\|f\|_{W_p^l(U)}$. This completes the proof of the lemma.

4.5. Approximation of Functions from W_p^l by Smooth Functions

Lemma 4.2. *Let f be a function of the class $W_{p,\mathrm{loc}}^l(U)$, where U is an open set in $\mathbf{R}^n$. Let us assign an arbitrary averaging kernel K and set $f_h = K_h * f$. Then, for every compact set $A \subset U$ for every multiindex α such that $|\alpha| = l$,*

$$\|D^\alpha f_h - D^\alpha f\|_{L_p(A)} \to 0$$

for $h \to 0$.

Proof. The statement of the lemma obviously follows from the fact that $D^\alpha f_h = K_h * D^\alpha f$ and from Theorem 2.2 of Chapter 1.

Corollary. *Let $f \in W_{p,\mathrm{loc}}^l(U)$. Then for every compact set $A \subset U$, any α with $|\alpha| \leqslant l\, \|D^\alpha f_h - D^\alpha f\|_{L_p(A)} \to 0$ for $h \to 0$.*

If $|\alpha| = l$, then there is nothing to prove. In the case $|\alpha| = r < l$, the desired result follows from the fact that if $f \in W_{p,\mathrm{loc}}^l(U)$, then $f \in W_{p,\mathrm{loc}}^r(U)$ for any $r < l$.

Lemma 4.3. *Let U be a bounded open set in $\mathbf{R}^n$. Then there exists a function $\eta : \mathbf{R}^n \to \mathbf{R}$ of the class $C^\infty(\mathbf{R}^n)$ such that for all $x \in U$, $0 < \eta(x) < 1/2\rho(x, \partial U)$, $|\nabla\eta(x)| < 1/2$ and $\eta(x) = 0$ if $x \notin U$.*

Proof. For a point $x \in U$, denote by U_x a ball $B(x, \delta)$, where $\delta = 1/2\rho$ $(x, \partial U)$. The sets U_x obviously form an open covering of the set U. Let (λ_m), $m \in N$, be the partition of unity corresponding to this covering. The functions λ_m are nonnegative, with each of them belonging to the class $C_0^\infty(\mathbf{R}^n)$, and for all $x \in U$, $\sum_{m=1}^\infty \lambda_m(x) = 1$. Hence it follows, in particular, that at every point $x \in U$, $\lambda_m(x) > 0$ for at least one m. For $x \notin U$, $\lambda_m(x) = 0$ for all m. Let $U_{x_m} = B(x_m, \delta_m)$ be the set of the covering $\{U_x\}$ which contains $S(\lambda_m)$.

For an arbitrary function $\varphi \in C_0^\infty(\mathbf{R}^n)$, we set for $r \geqslant 0$,

$$\|\varphi\|_r = \sup_{|\alpha| \leqslant r,\, x \in R^n} |D^\alpha \varphi(x)|.$$

Obviously, $\|\varphi\|_r < \infty$ for any function $\varphi \in C_0^\infty(\mathbf{R}^n)$ and $\|\varphi\|_0 \leqslant \|\varphi\|_1 \leqslant \cdots \leqslant \|\varphi\|_r \leqslant \ldots$.

It is obvious that for all $x \in U_{x_m}$, $\rho(x, \partial U) \geqslant 1/2\delta_m$. Let us put $K_m = \|\lambda_m\|_m$ and let $\alpha_m > 0$ be such that $K_m \alpha_m < 2^{-m}$, $\alpha_m < 5^{-m}\delta_m$, $\|\lambda_m\|_1 \alpha_m < 3^{-m}/\sqrt{n}$. Let us prove that the function $\eta = \sum_{m=1}^{\infty} \alpha_m \lambda_m$ is the desired one. Indeed, for every $r \geqslant 1$, we have for $m \geqslant r$,

$$\|\alpha_m \lambda_m\|_r \leqslant \|\alpha_m \lambda_m\|_m = \alpha_m K_m < {}^1/_{2^m}.$$

Hence, it follows that the series of derivatives of the order r of functions $\alpha_m \lambda_m$ is uniformly converging; therefore, the function η belongs to the class C^r for every $r \geqslant 0$. Since $r \geqslant 0$ is arbitrary here, we consequently obtain that $\eta \in C^\infty(\mathbf{R}^n)$. For $x \notin U$, it is obvious that $\eta(x) = 0$. For every point $x \in U$, there exists $m \geqslant 1$ such that for some m $\alpha_m \lambda_m(x) > 0$; consequently, $\eta(x) > 0$ for all $x \in U$. For every $x \in U_{x_m}$,

$$\alpha_m \lambda_m(x) < \left(\frac{1}{5}\right)^m \delta_m < 2\left(\frac{1}{5}\right)^m \rho(x, \partial U).$$

For $x \notin U_{x_m}$, $\lambda_m(x) = 0$; therefore, for such x, the latter inequality is also valid. Summing up by m, we obtain

$$\eta(x) = \sum_{m=1}^{\infty} \alpha_m \lambda_m(x) < 2 \sum_{m=1}^{\infty} \left(\frac{1}{5}\right)^m \rho(x, \partial U) = \frac{1}{2} \rho(x, \partial U)$$

for all $x \in U$. Finally, due to the inequality $\alpha_m \|\lambda_m\|_1 < 3^{-m}/\sqrt{n}$, we conclude that

$$\left| \frac{\partial \eta}{\partial x_i}(x) \right| < \sum_{m=1}^{\infty} 3^{-m}/\sqrt{n} = 1/2\sqrt{n},$$

whence it follows that $|\nabla \eta(x)| \leqslant 1/2$ for all $x \in U$.

Lemma is proved.

Theorem 4.6. *Let U be a bounded open set in $\mathbf{R}^n$. Suppose that $f : U \to R$ is a function of the class $L_p(U)$ such that all its generalized derivatives $D^\alpha f$, where $|\alpha| \leqslant l$ belong to the class $L_p(U)$. Then there exists a sequence (f_m), $m = 1, 2, \ldots$, of functions of the class $C^\infty(U)$ such that for $m \to \infty$, $\|D^\alpha f_m - D^a f\|_{L_p(U)} \to 0$ for every α with $|\alpha| \leqslant l$.*

Proof. We obtain the desired sequence of functions f_m if we properly modify the averaging operation. Let η be a functions constructed in Lemma 4.3. Let us assign a function $\sigma : \mathbf{R} \to \mathbf{R}$ belonging to the class $C^\infty(\mathbf{R})$ and satisfying the following conditions: the function σ is nondecreasing, $\sigma(t) = 0$ for $t \leqslant 0$, $\sigma(t) = 1$ for $t \geqslant 2$, $0 < \sigma(t) < t$ for $t > 0$, and $\sigma'(t) < 1$ for all t. We leave it to the reader, as an easy exercise, to construct the function σ satisfying all these conditions. Now for $m = 1, 2, \ldots$, we set

$$h_m(x) = \frac{1}{m^l} \sigma[m\eta(x)].$$

The function h_m belongs to the class $C^\infty(\mathbf{R}^n)$, $h_m(x) = 0$ for $x \notin U$, and for every $x \in U$,

$$0 < h_m(x) < \frac{1}{m^l}(m\eta(x)) = \frac{1}{m^{l-1}}\eta(x) < \frac{1}{2}\rho(x, \partial U).$$

For every x,

$$\frac{\partial h_m}{\partial x_i}(x) = \frac{1}{m^{l-1}}\sigma'(m\eta(x))\frac{\partial \eta}{\partial x_i}(x),$$

whence we conclude that $|\nabla h_m(x)| < 1/2$ for all $x \in U$.

Let us denote by U_m the totality of all points $x \in U$ for which $\eta(x) > 2/m$. For $x \in U_m$, $h_m(x) = 1/m^l = \text{const}$. The sequence of sets (U_m) is increasing, and $\bigcup_{m=1}^{\infty} U_m = U$. Now let us note, and this is very essential hereafter, that the derivatives of the order r, where $r \leqslant l$, of the functions h_m are bounded in U by a magnitude independent of m; that is, there exists a constant $M < \infty$ such that for any α with $|\alpha| \leqslant l$ for all $x \in U$ and for all m, the inequality

$$|D^\alpha h_m(x)| < M$$

holds.

Indeed, let $|\alpha| = r \leqslant l$. Then the equality

$$D^\alpha h_m(x) = \frac{1}{m^{l-r}}\sigma^{(r)}(m\eta(x))A_1^r(x) + \\ + \frac{1}{m^{l-r+1}}\sigma^{(r-1)}(m\eta(x))A_2^r(x) + \ldots + \frac{1}{m^{l-1}}\sigma'(m\eta(x))A_r^r(x) \tag{4.36}$$

holds, where each of the functions A_i^r is some polynomial of derivatives of the function η. The validity of equality (4.36) is easily verified by means of induction by r. The derivatives of the function η are the bounded functions, since η is a function of the class $C_0^\infty(\mathbf{R}^n)$. Hence, the boundedness of the functions A_i^r follows. The derivatives of the function σ are bounded functions, too. Hence, the boundedness of the functions $D^\alpha h_m$ for $|\alpha| \leqslant l$ obviously follows.

Let us assign an arbitrary averaging kernel $K \geqslant 0$, and for $x \in U$ let us set

$$f_m(x) = \frac{1}{[h_m(x)]^n}\int_U f(y)K\left(\frac{y-x}{h_m(x)}\right)dy. \tag{4.37}$$

Let $f_m^* = K_{\alpha_m} * f$, where $\alpha_m = 1/m^l$. On the set U_m, $h_m(x) = 1/m^l$; therefore, $f_m(x) = f_m * (x)$ for all $x \in U$. Let us show that the function f_m belongs to the class $C^\infty(U)$. For fixed $x \in U$, the function $y \mapsto K[(y-x)/h_m(x)]$ vanishes outside of the ball $B(x, h_m(x))$. Since $h_m(x) < 1/2\rho(x, \partial U)$, this ball is contained in U together with the closure. Let us fix an arbitrary point $x_0 \in U$. Put $\delta = \rho(x_0, \partial U)$, and let a point x be such that $|x - x_0| < \delta/9$. Then

$$\rho(x, \partial U) \leqslant |x - x_0| + \rho(x_0, \partial U) < (10/9)\delta,$$

therefore, $h_m(x) < 1/2\rho(x, \partial U) < (5/9)\delta$. Hence, it follows that if $|x - x_0| < \delta/9$, then the ball $B(x, h_m(x))$ is contained in the ball B_0 with the centre x_0 and with radius $(2/3)\delta$, so for $|x - x_0| < \delta/9$,

$$f_m(x) = \frac{1}{[h_m(x)]^n} \int_{B_0} f(y) K\left(\frac{y-x}{h_m(x)}\right) dy. \tag{4.38}$$

The function h_m in the ball B_0 is bounded from below by a constant $\gamma > 0$, and, due to the classical theorems about integrals depending on parameter, from equality (4.38), it is clear that the function f_m has continuous derivatives of all orders for all $x \in B_0 = B(x_0, 2/3\delta)$. Since $x_0 \in U$ is an arbitrary point, this proves that $f_m \in C^\infty(U)$ for every m.

Now let us obtain some estimates for all derivatives of the functions f_m. The representation of the function f_m by formula (4.37) makes it easy to obtain that f_m belongs to the class C^∞. However, it is difficult to establish some estimates for derivatives of the function f_m proceeding from this expression. Let us perform the change of the integration variable in the integral of (4.37) by setting $(y-x)/h_m(x) = z$. We obtain $y = x + zh_m(x)$, whence

$$f_m(x) = \int_{B(0,1)} f(x + zh_m(x)) K(z)\, dz. \tag{4.39}$$

Let us first show that the derivatives of the function f_m by x may be calculated by means of the same formulae as in the case where f is the function of the class C^∞. Let us again assign an arbitrary point $x_0 \in U$. Put $\rho(x_0, \partial U) = \delta$. For $|x - x_0| < \delta/9$ and $|z| < 1$, the point $x + zh_m(x)$ lies in the ball $B_0 = B(x_0, 2/3\delta)$. Let (g_k), $k = 1, 2, \ldots$, be a sequence of functions of the class C^∞ such that for any α with $|\alpha| \leqslant l$, $\|D^\alpha g_k - D^\alpha f\|_{L_p(B_0)} \to 0$ for $k \to \infty$. Such a sequence of functions (g_k) may be obtained, for instance, if we apply the averaging operation to the function f. Let

$$f_{m,k}(x) = \int_{B(0,1)} g_k(x + zh_m(x)) K(z)\, dz. \tag{4.40}$$

Let $|\alpha| = l$. Then we have

$$D^\alpha f_{m,k}(x) = \int_{B(0,1)} D_x^\alpha (g_k(x + zh_m(x))) K(z)\, dz \tag{4.41}$$

and

$$D_x^\alpha (g_k(x + zh_m(x))) = \sum_{\beta < \alpha} (D^\beta g_k)(x + zh_m(x)) P_\beta^\alpha(x, z), \tag{4.42}$$

where the factor $P_\beta^\alpha(x, z)$ is a polynomial with respect to derivatives with the order not exceeding $|\alpha|$ of the function h_m and to components of the vector z. The validity of equality (4.42) may easily be verified via induction by the number $r = |\alpha|$.

Let us first show that for $k \to \infty$, $D^\alpha f_{m,k}(x) \to D^\alpha f_m(x)$ for every x such that $|x - x_0| < \delta/9$. Representation (4.38) proves to be convenient here. If in equality (4.40) we perform the change of integration variable by the formula $x + zh_m(x) = y$, then we obtain

$$f_{m,k}(x) = \int_{B_0} g_k(y) \frac{1}{[h_m(x)]^n} K\left(\frac{y-x}{h_m(x)}\right) dy,$$

where $B_0 = B(x_0, 2/3\delta)$. Hence,

$$D^\alpha f_{m,k}(x) = \int_{B_0} g_k(y) D_x^\alpha \left\{ \frac{1}{[h_m(x)]^n} K\left(\frac{y-x}{h_m(x)}\right) \right\} dy.$$

The factor for $g_k(y)$ in the integral is a bounded function. Since $g_k \to f$ in $L_p(B_0)$, it follows that for $k \to \infty$,

$$D^\alpha f_{m,k}(x) \to \int_{B_0} f(y) D_x^\alpha \left\{ \frac{1}{[h_m(x)]^n} K\left(\frac{y-x}{h_m(x)}\right) \right\} dy.$$

The latter integral equals $D^\alpha f_m(x)$. This proves that $D^\alpha f_{m,k}(x) \to D^\alpha f_m(x)$ for all $x \in B(x_0, \delta/9)$.

Let us put for $x \in B(x_0, \delta/9)$ and $|z| < 1$,

$$\psi_\alpha(x, z) = \sum_{\beta \leqslant \alpha} (D^\beta f)[x + zh_m(x)] P_\beta^\alpha(x, z),$$

$$u_\alpha(x) = \int_{B(0,1)} \psi_\alpha(x, z) K(z)\, dz.$$

Since the derivatives of the function h_m are bounded functions, from equality (4.40), we obtain

$$|D^\alpha f_{m,k}(x) - u_\alpha(x)| \leqslant C \sum_{\beta \leqslant \alpha} \int_{B(0,1)} |D^\beta g_k(x + zh_m(x)) - D^\beta f(x + zh_m(x))| K(z)\, dz,$$

where $C = \text{const}$ is independent of k. The integral in the sum in the right-hand side is equal to

$$\frac{1}{[h_m(x)]^n} \int_{B_0} |D^\beta g_k(y) - D^\beta f(y)| K\left(\frac{y-x}{h_m(x)}\right) dy,$$

and since $D^\beta g_k \to D^\beta f$ in $L_p(B_0)$ for $k \to \infty$, this integral tends to zero. Hence we conclude that

$$D^\alpha f_{m,k}(x) \to u_\alpha(x)$$

for $k \to \infty$ uniformly in B_0; therefore, $u_\alpha(x) = D^\alpha f_m(x)$ for all $x \in B(x_0, \delta/9)$. The point $x_0 \in U$ is an arbitrary one; consequently, we obtain that derivatives whose order does not exceed l of the function f_m may be obtained from expression (4.39) if we formally differentiate the integrand by x just as if f were a sufficiently smooth function.

The main idea of the proof is as follows. According to the condition, the derivatives $D^\alpha f$ where $|\alpha| \leqslant l$ are all integrable in U in degree p. Let us assign some inner subdomain V of the domain U so that the integrals of

$|D^\alpha f(x)|^p$ on the boundary band $U\setminus V$ be sufficiently small. On the set V, $\|D^\alpha f_m - D^\alpha f\| \to 0$ for $m \to \infty$, due to the fact that for sufficiently large m, f_m coincides on the set V with the averaging of the function f_m. The integrals of $|D^\alpha f_m|^p$ on the boundary band $U\setminus V$ are also small. This is so because the derivatives of the functions f_m are bounded from above uniformly with respect to m. Comparing these facts, we obtain

$$\|D^\alpha f_m - D^\alpha f\|_{L_p(U)} \to 0$$

for $m \to \infty$. For $h > 0$ let $V_h = \{x \in U \mid \rho(x, \partial U) \geqslant h$, $G_h = U\setminus V_h$. For $h \to 0$ for all α with $|\alpha| \leqslant l$,

$$\int_{G_h} |D^\alpha f(x)|^p dx \to 0.$$

Let us arbitrarily assign $\varepsilon > 0$ and let us fix the value of $h > 0$ such that

$$\int_{G_{2h}} |D^\alpha f(x)|^p dx < \varepsilon$$

for all α such that $|\alpha| \leqslant l$.

The sequence of open sets (U_m), $m = 1, 2, \dots$, is the increasing one, their union coincides with U. Due to compactness of the set V_h, it follows that there exists m_1 such that for $m \geqslant m_1$, $U_m \supset V_h$. On the set U_m, $f_m = f_m^*$, where f_m^* is a mean function; therefore, due to Lemma 4.2, provided that $|\alpha| \leqslant l$,

$$\|D^\alpha f_m - D^\alpha f\|_{L_p(V_h)} \to 0$$

for $m \to \infty$.

Let m_2 be such that for $m \geqslant m_2$,

$$\|D^\alpha f_m - D^\alpha f\|_{L_p(V_h)} \leqslant \varepsilon^{1/p} \tag{4.44}$$

for all α for which $|\alpha| \leqslant l$, $m_0 = \max\{m_1, m_2\}$.

Now let us estimate the integrals of the functions $D^\alpha f_m$, where $|\alpha| \leqslant l$ on the boundary band G_h. According to what was proved above, we have

$$D^\alpha f_m(x) = \int_{B(0,1)} \sum_{\beta \leqslant \alpha} (D^\beta f)[x + zh_m(x)] P_\beta^\alpha K(z)\, dz,$$

where the factors P_β^α are polynomials relative to the components of the vector z and to derivatives of the order not exceeding l of the function h_m. The sequence of functions (h_m), $m = 1, 2, \dots$, is constructed so that the above derivatives are, uniformly in m, bounded,

$$|D^\alpha f_m(x)| \leqslant C \sum_{\beta \leqslant \alpha} \int_{B(0,1)} |(D^\beta f)[x + zh_m(x)]| K(z)\, dz. \tag{4.45}$$

Let us show that the mapping $\theta_z : x \mapsto x + zh_m(x)$, where $|z| < 1$, is the diffeomorphism of the set U into itself. First, let us note that for $x \in U$, $\theta_z(x) \in U$. Then, simple calculations show that the Jacobian of the mapping

θ_z equals $1+\langle z, \nabla h_m(x)\rangle$. Since, according to the condition, $|\nabla h_m(x)| < 1/2$, we obtain that $3/2 > J(x, \theta_z) > 1/2$ for every $x \in U$. Let us show that θ_z is a one-to-one mapping. Let x_1 and x_2 be two arbitrary points of the set U, also, $x_1 \neq x_2$. We set $\rho(x_1, \partial U) \geqslant \rho(x_2, \partial U)$. Let $\rho(x_1, \partial U) = \delta$, and let $|x_1 - x_2| < \delta$. The ball $B(x_1, \delta)$ is contained in U, and the points x_1 and x_2 are contained in it. We have

$$h_m(x_2) - h_m(x_1) = \int_0^1 h_m'[x_1 + t(x_2 - x_1)](x_2 - x_1)\,dt. \tag{4.46}$$

The integrand makes sense here due to convexity of the ball $B(x_1, \delta)$. Since $|\nabla h_m(x)| < 1/2$ for all x, we obtain $|h_m(x_2) - h_m(x_1)| \leqslant (1/2)|x_2 - x_1|$ from equality (4.46). Hence, we conclude that in the case under consideration,

$$|\theta_z(x_1) - \theta_z(x_2)| \geqslant |x_1 - x_2| - |h_m(x_1) - h_m(x_2)| \geqslant {}^1/_2 |x_1 - x_2|$$

(therefore, $\theta_z(x_1) \neq \theta_z(x_2)$). Suppose that $|x_1 - x_2| \geqslant \delta$. Then we obtain

$$|\theta_z(x_1) - \theta_z(x_2)| = |x_1 + zh_m(x_1) - x_2 - zh_m(x_2)|$$

$$\geqslant |x_1 - x_2| - h_m(x_1) - h_m(x_2) > |x_1 - x_2| - {}^1/_2\rho(x_1, \partial U)$$

$$- {}^1/_2\rho(x_2, \partial U) > 0,$$

so that in this case $\theta_z(x_1) \neq \theta_z(x_2)$ as well. Thus, the fact that θ_z is a one-to-one mapping is proved.

Now let us turn to inequality (4.45). Let us denote by ν_α a number of summands on the right-hand side of this inequality and obtain

$$|D^\alpha f_m(x)|^p \leqslant C^p \nu_\alpha^{p-1} \sum_{\beta < \alpha} \left\{ \int_{B(0,1)} |(D^\beta f)[\theta_z(x)]| K(z)\,dz \right\}^p$$

$$\leqslant C^p \nu_\alpha^{p-1} \sum_{\beta \leqslant \alpha} \int_{B(0,1)} |(D^\alpha f)[\theta_z(x)]|^p K(z)\,dz. \tag{4.47}$$

Let us integrate both sides of this inequality by the domain G_h. We obtain

$$\int_{G_h} \left(\int_{B(0,1)} |(D^\alpha f)[\theta_z(x)]|^p K(z)\,dz \right) dx$$

$$\leqslant \int_{B(0,1)} \left(2 \int_{G_h} |(D^\alpha f)[\theta_z(x)]|^p J(x, \theta_z)\,dx \right) K(z)\,dz.$$

Due to the theorem about the change of variable, the inner integral here is equal to

$$\int_{\theta_z(G_h)} |(D^\alpha f)(y)|^p\,dy \leqslant \int_{G_{3h}} |(D^\alpha f)(y)|^p\,dy < \varepsilon. \tag{4.48}$$

The former inequality here follows from the fact that $\theta_z(G_h) \subset G_{2h}$. Taking into account that

$$\int K(z)\,dz = 1,$$

and by comparing inequalities (4.47) and (4.48), we obtain

$$\int_{G_h} |D^\alpha f_m(x)|^p\, dx \leqslant C^p \nu_\alpha^p \varepsilon.$$

By comparing inequalities (4.43), (4.44), and (4.49), we obtain for every $m \geqslant m_0$,

$$\int_U |D^\alpha f_m - D^\alpha f|^p\, dx \leqslant \int_{G_h} |D^\alpha f_m - D^\alpha f|^p\, dx$$

$$+ \int_{V_h} |D^\alpha f_m - D^\alpha f|^p\, dx \leqslant 2^{p-1} \int_{G_h} |D^\alpha f_m|^p\, dx$$

$$+ 2^{p-1} \int_{G_h} |D^\alpha f|^p\, dx + \varepsilon < C_1 \varepsilon,$$

where C_1 is the constant independent of ε. Since $\varepsilon > 0$ is arbitrary, this proves that

$$\|D^\alpha f_m - D^\alpha f\|_{L_p(U)} \to 0$$

for $m \to \infty$. The theorem is proved.

Corollary. *Let U be a domain of the class J in the space $\mathbf{R}^n$. Then for every function $f \in W_p^l(U)$, there exists a sequence of functions (f_m), $m = 1, 2, \ldots$, of the class $C^\infty(U)$ such that $\|f_m - f\|_{W_p^l(U)} \to 0$ for $m \to \infty$.*

Proof. The magnitude

$$N_{l,p}(f) = \sum_{|\alpha| \leqslant l} \|D^\alpha f\|_{L_p(U)}$$

is a norm in $W_p^l(U)$ which is equivalent to the main one. According to the theorem, there exists a sequence (f_m), $m = 1, 2, \ldots$, of functions from $C(U)$ such that

$$N_{l,p}(f_m - f) \to 0$$

for $m \to \infty$. This proves the given statement.

4.6. Change of Variables for Functions with Generalized Derivatives

Let U be an open set in $\mathbf{R}^n$. A mapping $\varphi : U \to \mathbf{R}^n$ is said to be a diffeomorphism of the class C^r where $r \geqslant 1$ if the following conditions are satisfied:

1) the mapping φ is a one-to-one mapping;
2) for all $x \in U$, the Jacobian

$$J(x, \varphi) = \det \left(\frac{\partial \varphi_i}{\partial x_j}(x) \right)$$

of the mapping φ is distinct from zero.

Provided conditions 1) and 2) are satisfied, φ maps the open set U onto some open set $V \subset \mathbf{R}^n$. The inverse mapping φ^{-1} is also a diffeomorphism here.

For an arbitrary linear mapping $T : \mathbf{R}^n \to \mathbf{R}^n$, we set

$$|T| = \sup_{|\xi|=1} |T_\xi|.$$

Let a diffeomorphism $\varphi : U \to \mathbf{R}^n$ be given of the class C^1. φ is said to be a diffeomorphism of the class $C^1(\delta, K)$, where $\delta > 0$, $K > 0$, if for all $x \in U$, $J(x, \varphi) \geqslant \delta$ and $|\varphi'(x)| \leqslant K$.

Lemma 4.4. *Let U be a domain of the J type in the space $\mathbf{R}^n$ and let $\varphi : U \to \mathbf{R}^n$ be an arbitrary diffeomorphism of the class $C^1(\delta, K)$. Then the set $V = \varphi(U)$ is also a domain of the J type.*

Proof. Let $\psi = \varphi^{-1}$. Let us show that ψ belongs to the class $C^1(\delta_1, K_1)$ for some $\delta_1 > 0$, $K_1 > 0$. Let us take an arbitrary point $x \in U$ and let $y = \varphi(x)$, $T = \varphi'(x)$, $S = \psi'(y) = T^{-1}$. The mapping T is non-degenerate. Due to the classical results of algebra, there exist orthogonal transformations P and Q such that $T = PMQ$, where M is the transformation

$$(x_1, x_2, \ldots, x_n) \mapsto (\lambda_1 x_1, \lambda_2 x_2, \ldots, \lambda_n x_n),$$

where $0 < \lambda_1 \leqslant \lambda_2 \leqslant \cdots \leqslant \lambda_n$. It is obvious that

$$|\lambda_n| = T, \ \lambda_1\lambda_2 \ldots \lambda_n = |\det T| = |J(x, \varphi)|.$$

We have $S = Q^{-1}M^{-1}P^{-1}$. The transformations P^{-1} and Q^{-1} are orthogonal, and M^{-1} is the transformation

$$(x_1, x_2, \ldots, x_n) \mapsto (x_1/\lambda_1, x_2/\lambda_2, \ldots, x_n/\lambda_n).$$

Hence it follows that $|S| = 1/\lambda_1$, and $|\det S| = \lambda_1^{-1}\lambda_2^{-1} \ldots \lambda_n^{-1}$. This yields $|\det S| \geqslant \lambda_n^{-n} \geqslant K^{-n} = \delta_1$. Now, $\delta \leqslant |\det T| = \lambda_1\lambda_2 \ldots \lambda_n \leqslant \lambda_1\lambda_n^{n-1} \leqslant \lambda_1 K^{n-1}$, whence $\lambda_1 \geqslant \delta K^{1-n}$, therefore, $|S| \leqslant K^{n-1}/\delta = K_1$. The point $y \in V$ is an arbitrary one; this proves that $\psi \in C^1(\delta_1, K_1)$, where $\delta_1 = K^{-n}$, $K_1 = K^{n-1}/\delta$.

Let $x_0 \in U$, $\delta = \rho(x_0, \partial U)$, $B_0 = B(x_0, \delta)$. We put $G = \varphi[B(x_0, \delta)]$, $y_0 = \varphi(x_0)$. The set G is open, so there exists $\varepsilon > 0$ such that the ball $B(y_0, \varepsilon) \subset G$. Let $\delta_1 > 0$, $\delta_1 \leqslant \delta$ be such that for $|x - x_0| < \delta_1$, $|\varphi(x) - y_0| < \varepsilon$. Let us take arbitrary points $x_1, x_2 \in B(x_0, \delta)$. We put $y_1 = \varphi(x_1)$, $y_2 = \varphi(x_2)$. Then we obviously have

$$|\varphi(x_1) - \varphi(x_2)| \geqslant K|x_1 - x_2|$$

and

$$|x_1 - x_2| = |\psi(y_1) - \psi(y_2)| \leqslant K_1|y_1 - y_2| = |\varphi(x_1) - \varphi(x_2)|.$$

Hence it follows that for the mapping φ, all conditions of Lemma 2.5 are satisfied. Therefore, the desired statement is the direct corollary of Lemma 2.5.

Theorem 4.7. *Let U be a domain of the class J in the space $\mathbf{R}^n$ and let $\varphi : U \to \mathbf{R}^n$ be a diffeomorphism of the class C^l, $l \geqslant 1$, $V = \varphi(U)$. Suppose that φ belongs to the class C^l, $l \geqslant 1$, $V = \varphi(U)$. Suppose that φ belongs to the class $C^1(\delta, L)$, and let all the derivatives $D^\alpha\varphi$, where $|\alpha| \leqslant l$, be bounded functions. For an arbitrary function $f : V \to \mathbf{R}$, we set $\varphi^* f = f \circ \varphi$. If f is a function of the class $W_p^l(V)$, then the function $\varphi^* f$ belongs to the class $W_p^l(U)$, and the mapping $\varphi^* : W_p^l(V) \to W_p^l(U)$ thus defined is linear and continuous. Here, if $f \in W_p^l(V)$, then generalized derivatives of the function $\varphi^* f$ are expressed in terms of the derivatives of the function f and in terms of the derivatives of the components of the vector function φ by the same formulae as in the case of functions of the class C^∞.*

Proof. Let all conditions of the theorem be satisfied. Let us take an arbitrary function $f : V \to \mathbf{R}$. First, let us suppose that f belongs to the class $C^\infty(V)$. Then the function $\varphi^* f$ obviously belongs to the class $C^l(U)$. According to the classical results of analysis for every α such that $|\alpha| \leqslant l$, we have

$$D^\alpha(\varphi^* f) = \sum_{\gamma < \alpha} G_\gamma \varphi^* (D^\gamma f), \tag{4.50}$$

where G_γ is the polynomial with respect to the derivatives of the order not exceeding l of components of the vector function φ. From the condition of the theorem, it follows that the functions G_γ are bounded; consequently, we have

$$|D^\alpha(\varphi^* f)(x)| \leqslant M \sum_{\gamma \leqslant \alpha} |\varphi^*(D^\gamma f)|.$$

Hence it follows that for $|\alpha| \leqslant l$,

$$\|D^\alpha(\varphi^* f)\|_{L_p(U)} \leqslant M \sum_{\gamma < \alpha} \|\varphi^*(D^\gamma f)\|_{L_p(U)}. \tag{4.51}$$

Let F be an arbitrary function defined and integrable on the set V. For $x \in U$, we have

$$|(\varphi^* F)(x)| = |F[\varphi(x)]| \leqslant \frac{1}{\delta} |F[\varphi(x)]| \, |J(x, \varphi)|. \tag{4.52}$$

Due to the classical change of variables theorem, it follows that the function $\varphi^* F$ is integrable in U, and, in addition

$$\int_U |(\varphi^* F)(x)| \, dx \leqslant \frac{1}{\delta} \int_V |F(y)| \, dy.$$

Inequality (4.52) proves that if $f \in L_p(V)$, then $\varphi^* f \in L_p(U)$; also,

$$\|\varphi^* f\|_{L_p(U)} \leqslant \delta^{-1/p} \|f\|_{L_p(V)}. \tag{4.53}$$

From inequality (4.53) in particular, it follows that

$$\|\varphi^*(D^\gamma f)\|_{L_p(U)} \leqslant \delta^{-1/p} \|D^\gamma f\|_{L_p(V)}.$$

Hence, due to inequality (4.51), it follows that for $|\alpha| \leqslant l$,

$$\|D^{\alpha}(\varphi^{*}f)\|_{L_p(U)} \leqslant M\delta^{-1/p} \sum_{\gamma<\alpha} \|D^{\gamma}f\|_{L_p(V)}.$$

Summing it up by all α such that $|\alpha| \leqslant l$, we obtain that

$$N_{l,p}(\varphi^{*}f) = \sum_{|\alpha|<l} \|D^{\alpha}(\varphi^{*}f)\|_{L_p(U)} \leqslant C \sum_{|\alpha|\leqslant l} \|D^{\gamma}f\|_{L_p(V)} = CN_{l,p}(f).$$

So far, inequality (4.54) has been obtained for the functions of the class C^{∞} only. Suppose now that f is an arbitrary function of the class $W_p^l(V)$. Then, according to Theorem 4.6, there exists a sequence of functions (f_m), $m = 1, 2, \ldots$, of the class $C^{\infty}(V)$ such that for $m \to \infty$,

$$\|D^{\alpha}f_m - D^{\alpha}f\|_{L_p(V)} \to 0$$

for every α such that $|\alpha| \leqslant l$. Then the sequence of functions $(D^{\alpha}f_m)$, $m = 1, 2, \ldots$, for any such α, is the fundamental one in $L_p(V)$. From inequality (4.54), it follows that for all α, $|\alpha| \leqslant l$, the sequence of functions $D^{\alpha}(\varphi^{*}f_m)$ is also fundamental in $L_p(U)$; therefore, it converges in $L_p(U)$ to some function $g_{\alpha} \in L_p(U)$.

We have

$$\|\varphi^{*}f_m - \varphi^{*}f\|_{L_p(U)} \leqslant \delta^{-1/p}\|f_m - f\|_{L_p(V)},$$

whence it follows that the functions $\varphi^{*}f_m$ converge in $L_p(U)$ to the function $\varphi^{*}f$. Let us show that $g_{\alpha} = D^{\alpha}(\varphi^{*}f)$ for every α with $|\alpha| \leqslant l$. Let us assign an arbitrary function $\lambda \in C_0^{\infty}$. Then for every m, we have

$$\int_U D^{\alpha}\lambda(x)(\varphi^{*}f_m)(x)\,dx = (-1)^{|\alpha|}\int_U \lambda(x) D^{\alpha}(\varphi^{*}f_m)(x)\,dx \tag{4.55}$$

for any n-dimensional multiindex α with $|\alpha| \leqslant l$. The functions $D^{\alpha}\lambda$ are bounded. Taking the limit in equality (4.55) for $m \to \infty$, we obtain

$$\int_U D^{\alpha}\lambda(x)(\varphi^{*}f)(x)\,dx = (-1)^{|\alpha|}\int_U \lambda(x) g_{\alpha}(x)\,dx,$$

which proves that $g_{\alpha} = D^{\alpha}(\varphi^{*}f)$ and that the function $\varphi^{*}f$ belongs to the class W_p^l.

Due to (4.50), for every $m = 1, 2, \ldots$, the equality

$$D^{\alpha}(\varphi^{*}f_m) = \sum_{\gamma<\alpha} G_{\gamma}\varphi^{*}(D^{\gamma}f_m)$$

holds.

For $m \to \infty$, $D^{\alpha}(\varphi^{*}f_m) \to D^{\alpha}(\varphi^{*}f)$ in $L_p(U)$. The functions G_{γ} are continuous and bounded. For $m \to \infty$, $\varphi^{*}(D^{\gamma}f_m) \to \varphi^{*}(D^{\gamma}f)$ in the space $L_p(U)$. Hence it follows that the sum

$$\sum_{\gamma\leqslant\alpha} G_{\gamma}\varphi^{*}(D^{\gamma}f_m)$$

for $m \to \infty$ converges in $L_p(U)$ to the function

$$\sum_{\gamma \leqslant \alpha} G_\gamma \varphi^* (D^\gamma f).$$

Consequently, we obtain

$$D^\alpha (\varphi^* f) = \sum_{\gamma < \alpha} G_\gamma \varphi^* (D^\gamma f).$$

Thus, the derivatives of the function $\varphi^* f$ are expressed in terms of the derivatives of f by the same formulae as in the case of smooth functions.

This completes the proof of the theorem.

Corollary 1. *Let U be an open domain in $\mathbf{R}^n$, and let $\varphi : U \to V$ be a diffeomorphism of the class C^l where $l \geqslant 1$. Then for every function $f : V \to \mathbf{R}$ of the class $W^l_{p,\mathrm{loc}}(V)$, the function $\varphi^* f = f \circ \varphi$ belongs to the class $W^l_{p,\mathrm{loc}}(U)$. Here, the derivatives of the function $\varphi^* f$ whose order does not exceed l are expressed in terms of the derivatives of f according to the same formulae as in the case where $f \in C^l$.*

Proof. Let us take an arbitrary point $x_0 \in U$. Find $r > 0$ such that the closed ball $B(x_0, r)$ is contained in U. For all $x \in B(x_0, r)$, $|J(x, \varphi)| > 0$. Since the functions $\frac{\partial \varphi_i}{\partial x_j}$ and $J(\cdot, \varphi)$ are continuous and the set $B(x_0, r)$ is compact, there exist a constant $\delta > 0$ and a constant $K < \infty$ such that $|J(x, \varphi)| > \delta$ and $|\varphi'(x)| < K$ for all $x \in B(x_0, r)$ so that the restriction φ on the ball $B(x_0, r)$ is the mapping of the class $C^1(\delta, K)$. According to the theorem, it follows that the function $\varphi^* f$ is a function of the class W^l_p on the ball $B(x_0, r)$. Here the derivatives $D^\alpha(\varphi^* f)$ in the ball $B(x_0, r)$ are expressed by the same formulae as in the smooth case.

A point $x_0 \in U$ is arbitrary. Consequently, we obtain that for every point $x_0 \in U$, there exists a neighbourhood such that the restriction of the function $\varphi^* f$ upon it is a function of the class W^l_p.

This yields $\varphi^* f \in W^l_{p,\mathrm{loc}}(U)$. The corollary is proved.

4.7. Compactness of the Imbedding Operators

Theorem 4.8. *Let U be a domain of the class J in the space $\mathbf{R}^n$ and let numbers $p \geqslant 1$, $q \geqslant 1$, $l \geqslant 1$ be such that $lp \leqslant n$, $1 < q < np/(n - lp)$; l is an integer. Then the imbedding operator from $W^l_p(U)$ into $L_q(U)$ is compact (that is, every set of functions from $W^l_p(U)$ which is bounded in $W^l_p(U)$ is relatively compact in $L_q(U)$).*

Proof. The desired result is obtained by applying integral representation (4.3) for $r = 1$, $\alpha = 0$ for the functions of the class $W^l_p(U)$ and the results of Section 3. We have

$$f(x) = \int_U \zeta(x, z) f(z)\, dz + \sum_{|\gamma|=l} \int_U H_\gamma(x, z) D^\gamma f(z)\, dz$$
$$= (P_\zeta f)(x) + \sum_{|\gamma|=l} (P_{H_\gamma} D^\gamma f)(x).$$

Suppose that l, p, q satisfy all the conditions of the theorem. Due to Theorem 4.2, every bounded set in $W_p^l(U)$ is also bounded in $L_q(U)$. The function ζ is bounded and continuous in $U \times U$, and due to Lemma 3.2, the mapping $f \in L_q(U) \mapsto P_\zeta f \in L_q(U)$ is compact. Each of the functions $H_\gamma(x, z)$ satisfies all three conditions A), B), C) of Subsection 3.1; therefore, each of the operators P_{H_γ} compactly maps $L_p(U)$ into $L_q(U)$. By comparing the obtained results, we conclude that the operator

$$f \to P_\zeta f + \sum_{|\gamma|=l} P_{H_\gamma} D^\gamma f$$

is compact as the operator acting from $W_p^l(U)$ into $L_q(U)$.

The theorem is proved.

In the case $lp > n$, the space $W_p^l(U)$ is imbedded into $C(U)$. Compactness of imbedding remains valid in this case as well. To prove it, we need arguments other than those used in the case $lp \leqslant n$.

Lemma 4.5. *Let there be given a cube $Q = Q(a, r)$ and let $f \in W_p^1(Q)$, where $p > n$. Let us put*

$$\bar{f}(Q) = \sup_{x \in Q} f(x), \qquad \underline{f}(Q) = \inf_{x \in Q} f(x),$$

$$\operatorname*{osc}_Q f = \operatorname*{osc}_{x \in Q} f(x) = \bar{f}(Q) - \underline{f}(Q).$$

Then the estimate

$$\operatorname*{osc}_Q f \leqslant C r^{1-n/p} \|f\|_{L_p^1(Q)} \tag{4.56}$$

is valid where the constant C only depends on p and n.

Proof. First let us consider the case where Q is a cube $Q_0 = Q(0, 1)$. According to Theorem 4.2, f is continuous in Q_0. In this case, $l = 1$; therefore, the space $P_{l-1,n}$ coincides with the set of functions constant in Q_0. Due to Theorem 4.2, there exists an operator $\Pi_0 : W_p^1(Q_0) \to P_{0,n}$ such that for any function $f \in W_p^1(Q_0)$, the inequality $\|f - \Pi_0 f\|_{C(Q_0)} \leqslant \|f\|_{L_p^1(Q_0)}$ is valid. Since $\Pi_0 f = \text{const}$, we obtain

$$\operatorname*{osc}_{Q_0} f \leqslant C \|f\|_{L_p^1(Q_0)}.$$

Now let us consider the case where Q is an arbitrary cube $Q(a, r)$. Let us put $\varphi(t) = f(a + rt)$. Theorem 4.7 yields $\varphi \in W_p^1(Q_0)$, and also $\frac{\partial \varphi}{\partial t_i} = r \frac{\partial f}{\partial x_i}(a + rt)$ for all $t \in Q_0$. We have

$$\|\varphi\|_{L_p^1(Q_0)} = \left(\int\limits_{Q_0} \left\{\sum_{i=1}^{n} \left[\frac{\partial\varphi}{\partial t_i}(t)\right]^2\right\}^{\frac{p}{2}} dt\right)^{\frac{1}{p}}$$

$$= r\left(\int\limits_{Q(0,1)} \left\{\sum_{i=1}^{n} \left[\frac{\partial f}{\partial x_i}(a+rt)\right]^2\right\}^{\frac{p}{2}} dt\right)^{\frac{1}{p}}.$$

In the latter integral, we shall perform the change of the integration variable setting $a + rt = x$. By applying the formula of change of variables in the multiple integral, we obtain

$$\|\varphi\|_{L_p^1(Q_0)} = r^{1-\frac{n}{p}}\left(\int\limits_{Q(a,r)} \left\{\sum_{i=1}^{n} \left[\frac{\partial f}{\partial x_i}(x)\right]^2\right\}^{\frac{p}{2}}\right)^{\frac{1}{p}} = r^{1-\frac{n}{p}} \|f\|_{L_p^1[Q(a,r)]}.$$

Since obviously $\operatorname*{osc}_Q f = \operatorname*{osc}_{Q_0} \varphi$, due to (4.57), the inequality

$$\operatorname*{osc}_Q f \leqslant C r^{1-\frac{n}{p}} \|f\|_{L_p^1(Q)} \tag{4.57}$$

follows.

The lemma is proved.

Remark. The statement of the lemma remains valid if a cube is replaced by a ball in its formulation.

Corollary. *Let U be an arbitrary open set in $\mathbf{R}^n$. If a function f is continuous and belongs to the class $W_p^1(U)$, where $p > n$, then f locally satisfies in U the Hölder condition with the exponent $\alpha = 1 - n/p$.*

Proof. Let x be an arbitrary point of the set U. Let us find r such that the cube $Q(x,r) \subset U$. Let us take arbitrary points $x_1, x_2 \in Q(x,r)$. Let $\overline{Q}(\xi, h)$ be the smallest closed cube of those contained in $Q(x,r)$ and containing x_1 and x_2. Obviously, $|x_1 - x_2| \geqslant h$, and for sufficiently small $\varepsilon > 0$, the cube $Q(\xi, h+\varepsilon) \subset Q(a,r)$. Due to (4.56), we have

$$|f(x_2) - f(x_1)| \leqslant \operatorname*{osc}_{Q(\xi,h+\varepsilon)} f \leqslant C(h+\varepsilon)^{1-\frac{n}{p}} \|f\|_{L_p^1[Q(\xi,h+\varepsilon)]}$$

$$\leqslant C(h+\varepsilon)^{1-\frac{n}{p}} \|f\|_{L_p^1[Q(a,r)]}.$$

Directing ε to zero, we obtain

$$|f(x_2) - f(x_1)| \leqslant C h^{1-\frac{n}{p}} \|f\|_{L_p^1[Q(a,r)]} \leqslant C|x_2 - x_1|^{1-\frac{n}{p}} \|f\|_{L_p^1[Q(a,r)]}.$$

This proves the corollary

Remark. From the above arguments, it follows that if $f \in W_p^1(U)$ where $p > n$, then for any points x_1, x_2 for which there exists a cube Q such that

$x_1 \in Q$, $x_2 \in Q$, and $Q \subset U$, the inequality

$$|f(x_2) - f(x_1)| \leqslant C|x_2 - x_1|^{1-\frac{n}{p}} \|f\|_{L_p^1(U)}$$

is valid.

Lemma 4.6. *Let U be a domain of the class J in the space $\mathbf{R}^n$ and let $\mathcal{F}$ be a set of real functions defined in U. Suppose that there exist a constant $M < \infty$ and a nondecreasing function $\omega : [0,\infty) \to \mathbf{R}$ such that $\omega(0) = 0$, $\omega(t) > 0$ for $t > 0$, $\omega(t) \to 0$ for $t \to 0$, and*

$$\int_{+0} \frac{\omega(t)}{t} dt < \infty, \tag{4.58}$$

and for any function $f \in \mathcal{F}$ $|f(x)| \leqslant M$ for all $x \in U$. If the points $x_2, x_2 \in U$ are such that there exists a cube $Q \subset U$ containing the points x_1 and x_2, then

$$|f(x_2) - f(x_1)| \leqslant \omega(|x_2 - x_1|).$$

Then the set $\mathcal{F}$ is relatively compact in $C(U)$.

Proof. We have to prove that from any sequence (f_m), $m = 1, 2, \ldots$, of the function belonging to $\mathcal{F}$, one can extract a subsequence uniformly converging in U.

For $m \in \mathbf{N}$ let us set $V_m = \{x \in U | \rho(x, \partial U) \geqslant 1/m\}$. The sets V_m are compact, $V_m \subset V_{m+1}$ for every m and $\cup_{m=1}^{\infty} V_m = U$. Let us show that the family of functions $\mathcal{F}$ is equicontinuous on each of the sets V_m. Let us consider the relation $|f(x_2) - f(x_1)|/\omega(|x_2 - x_1|) = \delta(f, x_1, x_2)$, where $f \in \mathcal{F}$, $x_1, x_2 \in V_m$, $x_1 \neq x_2$. If $|x_1 - x_2| < (1/2m)\sqrt{n}$, then the cube $Q(x_1, (1/2m)\sqrt{n}) = Q$ is contained in U, and $x_1 \in Q$, $x_2 \in Q$. In this case, $\delta(f, x_1, x_2) \leqslant 1$. But if $|x_1 - x_2| \geqslant (1/2m)\sqrt{n}$, then $\delta(f, x_1, x_2) \leqslant 2M/\omega(1/2m\sqrt{n})$. Consequently, we obtain that $|f(x_2) - f(x_1)| \leqslant C\omega(|x_2 - x_1)$, where $C = \max\{1, 2M/\omega(1/2m\sqrt{n})\}$ for any $x_1, x_2 \in V_m$ and for any function $f \in \mathcal{F}$. This proves that $\mathcal{F}$ is equicontinuous on V_m.

Let a function $f : U \to \mathbf{R}$ be such that for any points $x_1, x_2 \in U$ for which there exists a cube $Q \subset U$ such that $x_1 \in Q$, $x_2 \in Q$, the inequality $|f(x_2) - f(x_1)| \leqslant \omega(|x_2 - x_1|)$ is valid. Let us estimate how much the value of $f(x)$ may change when the point x gets out of the set V_m. Since $U \in J$, there exists a number $\alpha \in (0,1)$ such that for any point $x \in U$, there exists a curve $x(s)$, $0 \leqslant s \leqslant l$ (the parameter s stands for the length of the arc) such that $x(0) = x$, $x(l) = a$ (a is the marked point of the domain U), and for all $s \in [0, l]$, $\rho(x(s), \partial U) \geqslant \alpha s$. Let $a \in V_m$ and let x be an arbitrary point of the domain U which does not belong to V_m. Let $x(s)$, $0 \leqslant s \leqslant l$, be a curve connecting x and a and satisfying the above conditions. Let s_m be the lowest of the values of s such that $x(s) \in V_m$. Let $x(s_m) = x_m$. Obviously, $\rho(x_m, \partial U) = 1/m$ and $\alpha s_m \leqslant \rho(x_m, \partial U)$ so that $s_m \leqslant 1/\alpha m$. Let us arbitrarily assign $t \in (0,1)$ and let us set $\sigma_\nu = s_m t^\nu$, $\nu = 0, 1, 2, \ldots$. For every ν, $\rho(x(\sigma_\nu), \partial U) \geqslant \alpha\sigma_\nu$; therefore, the cube $Q_\nu = Q[x(\sigma_\nu), \rho_\nu]$, where

$\rho_\nu = 2\alpha\sigma_\nu/\sqrt{n}$ is contained in U. The number t is chosen so that the point $x(\sigma_{\nu+1})$ belongs to the cube Q_ν for every ν. We have $|x(\sigma_{\nu+1}) - x(\sigma_\nu)| \leqslant |\sigma_{\nu+1} - \sigma_\nu| = s_m t^\nu(1-t) = \sigma_\nu(1-t)$, and the above condition will obviously hold if $1 - t < 2\alpha/\sqrt{n}$, i.e., if $t > 1 - 2\alpha/\sqrt{n}$. Let us put $t = 1 - \alpha/\sqrt{n}$. For every $\nu = 0, 1, 2, \dots,$

$$|f[x(\sigma_\nu)] - f[x(\sigma_{\nu+1})]| \leqslant \omega(\sigma_\nu - \sigma_{\nu+1}) = \omega[(1-t)s_m t^\nu].$$

Hence we conclude that

$$|f(x) - f(x_m)| \leqslant \sum_{\nu=0}^{\infty} |f[x(\sigma_\nu)] - f[x(\sigma_{\nu+1})]|$$

$$\leqslant \sum_{\nu=0}^{\infty} \omega[(1-t)s_m t^\nu] \leqslant \int_0^\infty \omega[(1-t)s_m t^{z-1}]\,dz. \qquad (4.59)$$

In the integral on the right-hand side, we perform the change of the integration variable by setting $(1-t)s_m t^{z-1} = u$. For $\eta > 0$, let

$$\psi(\eta) = \frac{1}{\ln t}\int_0^\eta \frac{\omega(u)}{u}\,du.$$

Obviously, $\psi(\eta) > 0$ for all $\eta > 0$, the function ψ is nondecreasing and $\psi(\eta) \to 0$ for $\eta \to 0$. The integral on the right-hand side of (4.59) equals $\psi[(1-t)s_m/t] \leqslant \psi[(1-t)/\alpha t m]$. As a result, we obtain

$$|f(x) - f(x_m)| \leqslant \psi(\beta/m),$$

where $\beta = (1-t)/\alpha t$.

Let (f_m) be an arbitrary sequence of functions from $\mathcal{F}$. The family $\mathcal{F}$ is equicontinuous on each of the sets V_m. By means of the usual diagonal process, we construct a sub-sequence f_{m_r}, $m_1 < m_2 < \cdots < m_r < \dots$, uniformly converging on each of the sets V_m. For every $x \in U$, there exists a limit $\lim_m f_{m_r}(x) = f_0(x)$. Let us prove that for $r \to \infty$, the functions f_{m_r} converge to f_0 uniformly in U. Note that if for the points $x_1, x_2 \in U$ there exists a cube $Q \subset U$ such that $x_1 \in Q$ and $x_2 \in Q$; then $|f_0(x_1) - f_0(x_2)| \leqslant \omega(|x_1 - x_2|)$. Let us arbitrarily assign $\varepsilon > 0$, and let us obtain by it $m_o \in \mathbf{N}$ such that $a \in V_{m_0}$ and $\psi(\beta/m_0) < \varepsilon/3$. Since for $r \to \infty$, $f_{m_r} \to f_0$ uniformly on the set V_{m_0}, there exists r_0 such that for $r \geqslant r_0 |f_{m_r}(x) - f_0(x)| < \varepsilon/3$ for all $x \in V_{m_0}$. Let $x \in V_{m_0}$. Let us construct a curve $x(s)$, $0 \leqslant s \leqslant l$ (s is the length of the arc) such that $x(l) = a$, $x(0) = x$, and $\rho[x(s), \partial U] \geqslant \alpha s$ for all $s \in [0, l]$. Let s_{m_0} be the lowest of the values of s such that $x(s) \in V_{m_0}$ and $x_{m_0} = x(s_{m_0})$. Then, according to what was proved above,

$$|f_{m_r}(x_{m_0}) - f_{m_r}(x)| \leqslant \psi(\beta/m_0), \; |f_0(x_{m_0}) - f_0(x)| \leqslant \psi(\beta/m_0),$$

therefore,

$$|f_{m_r}(x) - f_0(x)| \leqslant |f_{m_r}(x) - f_{m_r}(x_{m_0})| + |f_{m_r}(x_{m_0}) - f_0(x_{m_0})|$$
$$+ |f_0(x_{m_0}) - f_0(x)| < \varepsilon/3 + \varepsilon/3 + \varepsilon/3 = \varepsilon.$$

The fact that $\varepsilon > 0$ is arbitrary and that in order to obtain the latter inequality, we only needed that the inequality $r \geqslant r_0$ be valid, proves that $f_{m_r} \to f_0$ uniformly in U for $r \to \infty$.

The lemma is proved.

The function $\omega(t) = Ct^\alpha$, in particular, satisfies the condition of lemma (4.58). From what was proved above, the following proposition follows.

Theorem 4.9. *Let U be a domain of the class J in the space $\mathbf{R}^n$, $l \geqslant 1$, and $p \geqslant 1$ be such that $lp > n$, and l be an integer. Then the space $W_p^l(U)$ is compactly imbedded into $C(U)$.*

Proof. First let us consider the case $l = 1$, $p > n$. Let $\mathcal{F}$ be an arbitrary bounded set in $W_p^1(U)$. Then there exists a constant $M_1 < \infty$ such that $\|f\|_{W_p^1(U)} < M_1$ for every function $f \in \mathcal{F}$. According to Theorem 4.3, it follows that there exists a constant $M_2 < \infty$ such that if $f \in \mathcal{F}$, then $|f(x)| \leqslant M_2$ for all $x \in U$. Due to Lemma 4.5, if $f \in \mathcal{F}$ and the points x_1, x_2 are such that there exists a cube $Q = Q(a, r)$ for which $Q \subset U$, $x_1 \in Q$, $x_2 \in Q$, then $|f(x_2) - f(x_1)| \leqslant CM_1|x_2 - x_1|^{1-n/p}$. According to Lemma 4.6, it follows that the set of functions $\mathcal{F}$ is relatively compact in $C(U)$.

For the case $l = 1$, $p > n$, the theorem is proved.

Let $l > 1$, $lp > n$. First let us suppose that $(l-1)p \leqslant n$. Then

$$\frac{np}{n - (l-1)p} = \frac{n}{n/p - l + 1} > n,$$

since $lp > n$. Let us assign an arbitrary q such that $n < q < np/[n - (l-1)p]$. If the function $f \in W_p^l(U)$, then its first derivatives belong to the class $W_p^{l-1}(U)$. Here, if $\mathcal{F}$ is a bounded set in $W_p^l(U)$, then due to Theorem 4.3, the derivatives of functions belonging to $\mathcal{F}$ form a bounded set in $L_q(U)$. Hence, it follows that every bounded set in $W_p^l(U)$ is bounded in $W_q^1(U)$; therefore, it is relatively compact in $C(U)$.

If $(l-1)p > n$, then the first derivatives of the functions $f \in \mathcal{F}$ belong to the class $C(U)$, and if $\mathcal{F}$ is a bounded set of functions from $W_p^l(U)$, then their derivatives form a bounded set in the space $C(U)$, and therefore $\mathcal{F}$ is a bounded set of the space $C^1(U)$ having the natural norm $\|f\|_{C^1(U)} = \|f\|_{C(U)} + \sup_{x \in U} |f'(x)|$. Hence, it obviously follows that $\mathcal{F}$ is relatively compact in $C(U)$.

The theorem is proved.

4.8. Estimates with a Small Coefficient for the Norm in L_p^l

Lemma 4.7. *Let u be a function of the class $L_p(\mathbf{R}^n)$ where $p > 1$. For $h > 0$, we set*

$$(\theta_h u)(x) = \int_{B(x,h)} u(y)\,dy.$$

Then

$$|(\theta_h u)(x)| \leqslant (\sigma_n h^n)^{1-\frac{1}{p}} \|u\|_{L_p(R^n)} \tag{4.60}$$

for all $x \in \mathbf{R}^n$ and for any $q \geqslant p$, $\theta_h u \in L_q(\mathbf{R}^n)$, and also

$$\|\theta_h u\|_{L_q(R^n)} \leqslant (\sigma_n h^n)^{1+\frac{1}{q}-\frac{1}{p}} \|u\|_{L_p(R^n)}. \tag{4.61}$$

Proof. Let us first suppose that u is a continuous compactly supported function in $\mathbf{R}^n$. Then $\theta_h u$ is also a continuous compactly supported function in R^n. For $x \in \mathbf{R}^n$, $y \in \mathbf{R}^n$, we put $\chi(x,y) = 1$ for $|x-y| < h$, $\chi(x,y) = 0$ for $|x-y| \geqslant h$. Obviously, $[\chi(x,y)]^k = \chi(x,y)$, whatever the number $k > 0$ might be. We have

$$|(\theta_h u)(x)| = \left|\int_{R^n} \chi(x,y) u(y)\,dy\right| \leqslant \int_{R^n} \chi(x,y)|u(y)|\,dy.$$

Inequality (4.60) for the case $p = 1$ directly follows from the above inequality if we remark that $0 \leqslant \chi(x,y) \leqslant 1$ for any x, y. In the case $p > 1$, applying the Hölder inequality, we obtain

$$|(\theta_h u)(x)| \leqslant \left(\int_{R^n} [\chi(x,y)]^{\frac{p}{p-1}} dy\right)^{1-\frac{1}{p}} \left(\int_{R^n} |u(y)|^p\,dy\right)^{-\frac{1}{p}}$$

$$= (\sigma_n h^n)^{1-\frac{1}{p}} \|u\|_{L_p(R^n)},$$

and so inequality (4.60) for the case $u \in C_0(\mathbf{R}^n)$ is completely proved.

Let us prove (4.61). Suppose that $q > p > 1$ and $u \in C_0(\mathbf{R}^n)$. Let r, s, and t be such that $1/r + 1/s + 1/t = 1$, and let $\alpha \in (0,1)$. We have

$$|(\theta_h u)(x)| = \int_{R^n} \chi(x,y)\{\chi(x,y)|u(y)|^\alpha\}|u(y)|^{1-\alpha}\,dy.$$

Applying the Hölder inequality, we obtain

$$|(\theta_h u)(x)| \leqslant \left(\int_{R^n} \chi(x,y)\,dy\right)^{\frac{1}{r}} \left(\int_{R^n} \chi(x,y)|u(y)|^{\alpha s}\,dy\right)^{\frac{1}{s}}$$

$$\times \left(\int_{R^n} |u(y)|^{(1-\alpha)t}\,dy\right)^{\frac{1}{t}}.$$

The numbers r, s, t, and α are chosen so that the equalities $s = q$, $(1-\alpha)t = p$, $\alpha s = p$ be valid. Hence $\alpha = p/q$, $1/t = 1/p - 1/q$, $1/r = 1 - 1/p$. We have

$$\left(\int_{R^n} \chi(x,y)\,dy\right)^{\frac{1}{r}} \leqslant (\sigma_n h^n)^{1-\frac{1}{p}},$$

and as a result we obtain

$$|(\theta_h u)(x)|^q \leqslant (\sigma_n h^n)^{q\left(1-\frac{1}{p}\right)} \|u\|^{q-p}_{L_p(R^n)} \int_{R^n} \chi(x,y)|u(y)|^p\,dy.$$

By integrating both parts of the latter inequality and by applying the Fubini theorem, we obtain

$$(\|\theta_h u\|_{L_q(R^n)})^q \leqslant \sigma_n^{1+q-\frac{q}{p}} h^{nq\left(1-\frac{1}{p}\right)+n} \|u\|^q_{L_p(R^n)},$$

Thus, inequality (4.61) is proved for the case where $u \in C_0(\mathbf{R}^n)$ and $p > q > 1$. If $u \in C_0(\mathbf{R}^n)$, then both the function u and the function $\theta_h u$ are integrable to any power. Due to this, for the case $p \geqslant q \geqslant 1$, inequality (4.61) may be obtained from what was proved above by limit transition by p and q.

If u is an arbitrary function of the class $L_p(\mathbf{R}^n)$, then the desired inequality can easily be obtained from what was proved above by limit transition.

The lemma is proved.

Theorem 4.10. *Let U be a domain of the class J in the space $\mathbf{R}^n$ and let $f \in W_p^l(U)$. Let us assign an arbitrary n-dimensional multiindex α such that $|\alpha| < l$. Set $r = l - |\alpha|$. Then, if $rp \leqslant n$, the function $D^\alpha f$ belongs to the class $L_q U$ for any $q < np/(n-rp)$; and if t and q satisfy the inequalities $1 \leqslant t < q < np/(n-rp)$, the estimate $\forall \tau \in (0,1]$*

$$\|D^\alpha f\|_{L_q(U)} \leqslant \frac{C_1 \|f\|_{L_t(U)}}{\tau^{n\left(\frac{1}{t}-\frac{1}{q}\right)+|\alpha|}} + C_2 \tau^{r+n\left(\frac{1}{q}-\frac{1}{p}\right)} \|f\|_{L_p^l(U)}$$

holds where the constants C_1, C_2 only depend on the domain U and on the magnitudes n, l, p, q, $|\alpha|$.

Proof. Due to formula (4.3), we have

$$(D^\alpha f)(x) = \int_U \zeta_\tau^{(\alpha)}(x,z) f(z)\,dz + \sum_{|\gamma|=l} \int_U H_{\tau,\gamma}^{(\alpha)}(x,z) D^\gamma f(z)\,dz.$$

Let us put

$$\Phi_0(x) = \int_U \zeta_\tau^{(\alpha)}(x,z) f(z)\,dz, \qquad F_\gamma(x) = \int_U H_{\tau,\gamma}^{(\alpha)}(x,z) D^\gamma f(z)\,dz,$$

$$\Phi_1(x) = \sum_{|\gamma|=l} F_\gamma(x).$$

Let us estimate the norms in $L_q(U)$ of each of the functions Φ_0 and Φ_1 separately. We have $F_\gamma = P_K D^\gamma f$, where $K_\gamma = H_{\tau\gamma}^{(\alpha)}$. The function $K_\gamma(x,z)$

vanishes if $|x-z| \geqslant h = 3R\tau$, and for any x, z, we have

$$|K_\gamma(x,z)| \leqslant \frac{C}{|x-z|^{n-r}}.$$

According to Theorem 3.3, the estimate

$$\|F_\gamma\|_{L_q(U)} \leqslant Ch^{r+n(1/q-1/p)} \|D^\gamma f\|_{L_p(U)}$$

follows, whence we obtain

$$\|\Phi_1\|_{L_q(U)} \leqslant C\tau^{r+n(1/q-1/p)} \sum_{|\gamma|=l} \|D^\gamma f\|_{L_p(U)} \leqslant C_2 \tau^{r+n(1/q-1/p)} \|f\|_{L_p^l(U)}.$$

Now let us estimate the magnitude $\|\Phi_0\|_{L_q(U)}$. We have $|\zeta_\tau^{(d)}(x,z)| \leqslant C/\tau^{n+|d|}$ and $\zeta_\tau^{(\alpha)}(x,z) = 0$ for $|x-z| \geqslant h = 3R\tau$. Hence it follows that

$$|\Phi_0(x)| \leqslant \int_{R^n} |\zeta_\tau^{(\alpha)}(x,z)|\,|f(z)|\,dz$$

$$\leqslant \frac{C}{\tau^{n+|\alpha|}} \int_{B(x,h)\cap U} |f(z)|\,dz = \frac{C}{\tau^{n+|\alpha|}} (\theta_h|f|)(x),$$

therefore, due to Lemma 4.7,

$$\|\Phi_0\|_{L_q(U)} \leqslant \frac{C_1}{\tau^{n/t-n/q+|\alpha|}} \|f\|_{L_t(U)}.$$

The theorem is proved.

4.9. Functions of One Variable

Let us establish in what cases the function defined on the segment $I = (\alpha, \beta) \subset \mathbf{R}$ belongs to the class $W^1_{1,\mathrm{loc}}(I)$ or $W^1_{p,\mathrm{loc}}(I)$.

Let μ be a measure in the interval I. The function $f : I \to \mathbf{R}$ is said to be the generating one for the measure μ if f is continuous from the left at every point $x \in I$ and for any segment $[a,b)$ such that $\alpha < a < b < \beta$, $\mu([a,b)) = f(b) - f(a)$. Every measure μ in the segment I has at least one generating function. Indeed, let us assign an arbitrary point $x_0 \in I$ and set $f(x) = \mu([x_0,x))$ if $x_0 < x < \beta$, $f(x) = -\mu([x,x_0))$ in the case where $\alpha < x \leqslant x_0$, and finally $f(x_0) = 0$. It is easy to verify that the constructed function f is generating for the measure μ.

If f_1 and f_2 are two generating functions of the measure μ, then the difference $f_1 - f_2$ is identically constant in I. Actually, let x_1, x_2 be two arbitrary points of the segment I such that $x_1 < x_2$. Then $\mu([x_1,x_2)) = f_1(x_2) - f_1(x_1) = f_2(x_2) - f_2(x_1)$, whence we obtain that $f_1(x_2) - f_2(x_2) = f_1(x_1) - f_2(x_1)$. The arbitrariness of the points x_1, x_2 proves that $f_1 - f_2 = $ const in I.

If f is a generating function of the measure μ, then f is the function with bounded variation in every closed segment $[x_1,x_2] = I$. Indeed, let $\mu =$

$\mu^+ - \mu^-$ be the Jordan decomposition of the measure μ; h and g are generating functions of the measures μ^+ and μ^-. Due to nonnegativity of the measures μ^+ and μ^-, the functions h and g are nondecreasing. The difference $h - g$ is generating for the measure μ. According to the previous remark, we obtain that $f = h - g + C$; consequently, f is the difference of two nondecreasing functions; and therefore, it is the function with bounded variation in every segment $[x_1, x_2] \subset I$.

Lemma 4.8. *If $f : I \to \mathbf{R}$ is a generating function of a measure μ defined in a segment $I = (\alpha, \beta)$, then μ is the generalized derivative of the function f.*

Proof. Let $f : I \to \mathbf{R}$ be a generating function of a measure μ in a segment $I = (\alpha, \beta)$. Let us take an arbitrary function $\varphi \in C_0^\infty(\mathbf{R}^n)$ whose support is contained in the segment I. Then, applying the formula of part-by-part integration for the Stieltjes integral, we obtain due to the finiteness of φ,

$$\int_\alpha^\beta f(x)\varphi'(x)\,dx = -\int_\alpha^\beta \varphi(x)\,df(x).$$

However, the latter integral equals

$$-\int_I \varphi(x)\,\mu(dx).$$

Consequently, we obtain that for any function $\varphi \in C_0^\infty(I)$, the equality

$$\int_I f(x)\varphi'(x)\,dx = -\int_I \varphi(x)\,\mu(dx)$$

is valid. This completes the proof of the lemma.

Theorem 4.11. *A generalized function f assigned on a set $I = (\alpha, \beta) \subset \mathbf{R}$ belongs to the class $\overline{W}^1_{1,\mathrm{loc}}(I)$ iff it is the function with bounded variation on every segment $[x_1, x_2] \subset I$. If $\bar{f}$ is a generating function of the measure $\mu = df/dx$, then the difference $f - \bar{f}$ is constant in I.*

Proof. Suppose that $f \in \overline{W}^1_{1,\mathrm{loc}}(I)$. Let the measure μ be a derivative of df/dx and let the function f_0 be a generating function of the measure μ. Then, due to the previous lemma, the generalized derivative df_0/dx equals the measure μ. Hence it follows that the generalized derivative of the difference $f - f_0$ is identically equal to zero; therefore, the difference $f - f_0$ is the function constant in I. Thus, $f = f_0 + C$. This proves the necessity of the theorem.

Now let us suppose that the function $f : (\alpha, \beta) \to \mathbf{R}$ is the function with bounded variation in every segment $[x_1, x_2] \subset (\alpha, \beta)$. Then there exists a measure μ defined in segment I and such that for every segment $[x_1, x_2) \subset I$, the equality $\mu([x_1, x_2)) = f(x_2 - 0) - f(x_1 - 0)$.

Let us define by f some new function $\bar{f}$ setting $\bar{f} = f(x - 0)$ for every $x \in (\alpha, \beta)$. The function $\bar{f}$ differs from f at the points forming at the most a countable set; consequently, $\bar{f}(x) = f(x)$ almost everywhere in I. The function $\bar{f}$ is a generating function for the measure μ; therefore, $\mu = d\bar{f}/dx$. Since

$\bar{f}$ and f coincide almost everywhere, then they define the same generalized function. The measure μ is therefore also the derivative of the function f.

The theorem is proved.

Theorem 4.12. *A generalized function f assigned in an interval $I = (\alpha, \beta) \subset$ **R**, belongs to the class $W_1^1(I)$ iff f is an absolutely continuous function in the interval I. Besides, if $f : I \to$ **R** is an absolutely continuous function, then its usual derivative is at the same time the generalized derivative of the function f.*

Proof. Let $f \in W^1_{p,\mathrm{loc}}(I)$. Then f belongs to the class $\overline{W}^1_{1,\mathrm{loc}}(I)$ as well. The measure μ, which is the indefinite integral of the function df/dx, is a generalized derivative of f. Let us put for $x \in I$,

$$\bar{f}(x) = \int_{x_0}^{x} \frac{df}{dx}(t)\,dt.$$

The function $\bar{f}$ is a generating function of the measure μ. Consequently, the difference $f - \bar{f}$ (as a generalized function) is constant in I. The function $\bar{f}$ is absolutely continuous, and $f = \bar{f} + c$. The necessity of the theorem is proved.

Let us prove the sufficiency. Let f be an absolutely continuous function defined in the interval $I = (\alpha, \beta)$. Then for any interval $[x_1, x_2] \subset (\alpha, \beta)$,

$$f(x_2) - f(x_1) = \int_{x_1}^{x_2} f'(t)\,dt,$$

that is, f is a generating function of the indefinite integral of the function f', therefore, the indefinite integral of the function f' and the function f' itself are the generalized derivatives of the function f.

The theorem is proved.

4.10. Differential Description of Convex Functions

A set $U \subset \mathbf{R}^n$ is said to be convex if for any $x_1, x_2 \in U$, and for any $\lambda \in [0,1]$, the point $\lambda x_1 + (1-\lambda)x_2 \in U$.

Let U be a convex open set in the space $\mathbf{R}^n$. A function $f : U \to \mathbf{R}$ is called convex if for any $x_1, x_2 \in U$ and for any $\lambda \in [0,1]$, the inequality

$$f(\lambda x_1 + (1-\lambda)x_2) \leqslant \lambda f(x_1) + (1-\lambda) f(x_2)$$

is valid. If the function $f(x)$ has continuous second derivatives in the domain U, then f is convex in U iff the quadratic form

$$Q(x, \xi) = \sum_{i=1}^{n} \sum_{j=1}^{n} \frac{\partial^2 f}{\partial x_i \partial x_j} \xi_i \xi_j$$

is nonnegative for all $x \in U$.

Theorem 4.13. *Let $f(x)$ be an arbitrary locally summable function defined in a convex open set U of the space $\mathbf{R}^n$. Then the function f is convex in U iff for any vector $\xi = (\xi_1, \xi_2, \ldots, \xi_n)$, the generalized function*

$$Q(\xi) = \sum_{i=1}^{n} \sum_{j=1}^{n} \frac{\partial^2 f}{\partial x_i \partial x_j} \xi_i \xi_j$$

is nonnegative (i.e., is a nonnegative linear functional).

Proof. Let $\hat{U}_h$ be a set of all $x \in U$ which are at the distance exceeding h from $\mathbf{R}^n \backslash U$. The set $\hat{U}_h$ is convex, as can easily be verified. Let us assign an arbitrary nonnegative averaging kernel K.

First let us suppose that the function f is convex. Then the mean function $f_h = K_h * f$ is defined for all $x \in \hat{U}_h$ and is a convex function in $\hat{U}_h$; consequently, the quadratic form

$$Q_h(\xi, x) = \sum_{i,j=1}^{n} \frac{\partial^2 f_h}{\partial x_i \partial x_j} \xi_j \xi_i$$

is nonnegative for all $x \in \hat{U}_h$.

Let φ be a nonnegative function of the class $C_0^\infty(U)$. Then there exists $h_0 > 0$ such that for $0 < h \leqslant h_0$, $S(\varphi) \subset \hat{U}_h$. If $0 < h \leqslant h_0$, then

$$\int_{\hat{U}_{h_0}} \sum_{i,j=1}^{n} \frac{\partial^2 f_h}{\partial x_i \partial x_j} \xi_i \xi_j \varphi \, dx \geqslant 0.$$

Taking the limit for $h \to 0$, we obtain

$$\int_{U} \sum_{i,j=1}^{n} \frac{\partial^2 f}{\partial x_i \partial x_j} \xi_i \xi_j \varphi dx = \int_{\hat{U}_{h_0}} \sum_{i,j=1}^{n} \frac{\partial^2 f}{\partial x_i \partial x_j} \xi_i \xi_j \varphi dx \geqslant 0.$$

Since φ is an arbitrary nonnegative function from $C_0^\infty(U)$, the necessity of the theorem is proved.

Let us prove the sufficiency. If the generalized function

$$F(\xi) = \sum_{i,j=1}^{n} \frac{\partial^2 f}{\partial x_i \partial x_j} \xi_i \xi_j$$

is nonnegative, then the function

$$(K_h * F)(\xi) = \sum_{i,j=1}^{n} \frac{\partial^2 f_h}{\partial x_i \partial x_j} \xi_i \xi_j$$

is nonnegative, too. Hence it follows that the mean function f_h is convex for every h. Taking the limit for $h \to 0$, we obtain that the function f is convex.

The theorem is proved.

Corollary. *Let U be a convex set in $\mathbf{R}^n$, and let $f : U \to \mathbf{R}$ be a convex function. Then $f \in \overline{W}_1^2(U)$.*

Proof. For every vector ξ, the generalized function

$$Q(\xi, \xi) = \sum_{i,j=1}^{n} \frac{\partial^2 f}{\partial x_i \partial x_j} \xi_i \xi_j$$

is nonnegative; and consequently, $Q(\xi, \xi)$ is a measure on the δ-ring $\mathcal{B}_0(U)$. Hence it follows that for any vectors ξ and η, the generalized function

$$Q(\xi, \eta) = {}^1\!/_4 [Q(\xi+\eta, \xi+\eta) - Q(\xi-\eta, \xi-\eta)]$$

is a measure in the set U. By setting

$$\xi = e_i = \Big(0, \ldots, \underset{i}{1}, \ldots, 0\Big), \quad \eta = e_j = \Big(0, \ldots, \underset{j}{1}, \ldots, 0\Big),$$

we obtain that $\frac{\partial^2 f}{\partial x_i \partial x_j} = Q(e_i, e_j)$ is a measure in U. Q. E. D.

4.11. Functions Satisfying the Lipschitz Condition

Let $E \subset \mathbf{R}^n$ be an arbitrary set. A function $f : E \to \mathbf{R}$ is said to satisfy the Lipschitz condition with a constant $L > 0$, $L < \infty$ if for any $x_1, x_2 \in E$, the inequality

$$|f(x_1) - f(x_2)| \leqslant L|x_1 - x_2|$$

is valid. Suppose that f is a real function whose definition domain is some set $A \supset E$. Then f is said to satisfy the Lipschitz condition on the set E if the restriction of the function f on the set E satisfies the Lipschitz condition.

Theorem 4.14. *Let U be a convex open set in $\mathbf{R}^n$. A function $f : U \to \mathbf{R}$ is equivalent, in the sense of the theory of the Lebesque integral, to a function satisfying the Lipschitz condition with a constant L where $0 < L < \infty$ iff the function f belongs to the class $W_1^1(\mathbf{R}^n)$ and for almost all $x \in U$, the inequality*

$$|\nabla f(x)| = \sqrt{\sum_{i=1}^{n} \left|\frac{\partial f}{\partial x_i}(x)\right|^2} \leqslant L$$

is valid.

Proof. Let $f : U \to \mathbf{R}$ satisfy the Lipschitz condition with the constant L. For $h > 0$, let $\hat{U}_h$, as before, denote a set of all $x \in U$ such that $\rho(x) = \rho(x, \partial U) > h$. The set $\hat{U}_h$ is convex. Indeed, let x_1, x_2 be two arbitrary points of the set $\hat{U}_h$, $x = (1-t)x_1 + tx_2$, where $0 < t < 1$ is an arbitrary point of the segment connecting x_1 and x_2. Let $\rho_1 = \rho(x_1)$, $\rho_2 = \rho(x_2)$. We have: $\rho_1 > h$, $\rho_2 > h$. Let $\rho_0 = \min\{\rho_1, \rho_2\}$, $\rho_0 > 0$. Let us take an arbitrary point $y \in B(x, \rho_0)$ and let $z = y - x$, $y_1 = x_1 + z$, $y_2 = x_2 + z$. Then $y_1 \in U$, $y_2 \in U$, and $y = (1-t)y_1 + ty_2$, whence, due to the convexity of U, it follows that $y \in U$. Since $y \in B(x, \rho_0)$ is arbitrary, this proves that $B(x, \rho_0) \in U$. So, $\rho(x) \geqslant \rho_0 > h$, and therefore $x \in U$.

Let K be an averaging kernel. The function K is assumed to be nonnegative. Let us put $f_h = K_h * f$. For all $x \in \hat{U}_h$, we have

$$f_h(x) = \int_{|t|\leqslant 1} f(x+ht)K(t)dt.$$

Hence, we obtain that for any $x_1, x_2 \in \hat{U}_h$,

$$\begin{aligned}|f_h(x_1) - f_h(x_2)| &\leqslant \int_{|t|\leqslant 1} |f(x_1+ht) - f(x_2+ht)| \cdot K(t)dt \\ &\leqslant L|x_1 - x_2| \cdot \int_{|t|\leqslant 1} K(t)dt = L|x_1 - x_2|,\end{aligned}$$

that is, the function f_h satisfies the Lipschitz condition with the coefficient L. For an arbitrary point $x \in \hat{U}_h$ for every unit vector e for sufficiently small $t \neq 0$, the point $x + te \in \hat{U}_h$. Due to what was proved above, we have:

$$|f_h(x+te) - f_h(x)| \leqslant L|t|.$$

Hence,

$$\left|\frac{f_h(x+te) - f_h(x)}{t}\right| \leqslant L.$$

Taking the limit for $t \to 0$, we obtain

$$|\langle f_h(x), e\rangle| \leqslant L, \tag{4.62}$$

whence, due to arbitrariness of the unit vector e, it follows that $|\nabla f_h(x)| \leqslant L$ for all $x \in U$. In particular, we obtain that $|\frac{\partial f_h}{\partial x_i}(x)| \leqslant L$ for all $x \in \hat{U}_h$ and for any $i = 1, 2, \ldots, n$.

Let us assign an arbitrary function $\varphi \in C_0^\infty(U)$. Find $h_0 > 0$ such that $\hat{U}_h \supset S(\varphi)$ for $0 < h < h_0$. For $0 < h < h_0$, we have:

$$\begin{aligned}\left|\int_{\hat{U}_h} f_h(x)\frac{\partial \varphi}{\partial x_i}(x)dx\right| &= \left|\int_{\hat{U}_h} \frac{\partial f_h}{\partial x_i}(x)\varphi(x)dx\right| \\ &\leqslant L \cdot \int_{\hat{U}_h} |\varphi(x)|dx \leqslant L \cdot \int_U |\varphi(x)|dx.\end{aligned}$$

Taking the limit in this inequality for $h \to 0$, we obtain that for every $i = 1, 2, \ldots, n$ for any function $\varphi \in C_0^\infty(U)$,

$$\left|\int_U f(x)\frac{\partial \varphi}{\partial x_i}(x)dx\right| \leqslant L \cdot \|\varphi\|_{L_1(U)}.$$

Thus, the generalized derivative $\frac{\partial f}{\partial x_i}$ is a linear functional on $C_0^\infty(U)$ which is bounded in the norm $L_1(U)$. Since $C_0^\infty(U)$ is everywhere dense in $L_1(U)$, this functional is uniquely extended, with the norm on $L_1(U)$ being preserved. On the basis of the classical results with respect to the general form of linear functionals in $L_1(U)$, we obtain that for every $i = 1, 2, \ldots, n$, there exists a function $f_i \in L_\infty(U)$ such that $|f_i(x)| \leqslant L$ for all x and

$$\int_U f(x)\frac{\partial \varphi}{\partial x_i}(x)dx = -\int_U f_i(x)\varphi(x)\,dx$$

for any function $\varphi \in C_0^\infty(U)$. Hence it follows that $f_i = \frac{\partial f}{\partial x_i}$; therefore, $f \in W^1_{1,\mathrm{loc}}(U)$.

For $h \to 0$, $\frac{\partial f_h}{\partial x_i}(x) \to \frac{\partial f}{\partial x_i}(x)$ for almost all x. Hence it follows that $|\nabla f_h(x)| \to |\nabla f(x)|$ for almost all $x \in U$, too. Since $|\nabla f_h(x)| \leqslant L$ for all $x \in \hat{U}_h$, it follows that $|\nabla f(x)| \leqslant L$ for almost all $x \in U$. This proves the necessity of the theorem.

Let us prove the sufficiency. Suppose that the function $f : U \to \mathbf{R}$ belongs to the class $W^1_{1,loc}(U)$; besides, $|\nabla f(x)| \leqslant L$ for all $x \in U$. Let f_h be the mean functions, $f_h = K_h * f$, where the averaging kernel K is nonnegative. Let us assign an arbitrary unit vector $u = (u_1, u_2, \ldots, u_n)$. We have

$$v(x) = \langle \nabla f(x), u \rangle = \sum_{i=1}^{n} \frac{\partial f}{\partial x_i}(x) u_i.$$

Let $v_h(x) = \langle \nabla f_h(x), u \rangle = \sum_{i=1}^{n} \frac{\partial f_h}{\partial x_i}(x) u_i$. The function $v_h(x)$, as a result of this, is the averaging of the function v, and since $|v(x)| \leqslant L$ for almost all $x \in U$, then $|v_h(x)| \leqslant L$ for all $x \in \hat{U}_h$. Let us take arbitrary points $x_1, x_2 \in U$. Let h_0 be such that for $0 < h < h_0$, $x_1 \in \hat{U}_h$ and $x_2 \in \hat{U}_h$. We have

$$f_h(x_2) - f_h(x_1) = \int_0^1 \langle \nabla f_h[x_1 + t(x_2 - x_1)],\ x_2 - x_1 > dt,$$

whence we conclude that

$$|f_h(x_2) - f_h(x_1)| \leqslant L|x_2 - x_1|. \tag{4.63}$$

For $h \to 0$, almost everywhere $f_h(x) \to f(x)$. Inequality (4.63) yields that the family of functions f_h is uniformly equicontinuous in U. So, for $h \to 0$, the functions f_h converge to some function $\tilde{f}$ uniformly on every compact subset U. It is obvious that $\tilde{f}(x) = f(x)$ for almost all $x \in U$ and for any $x_1, x_2 \in U$,

$$|\tilde{f}(x_2) - \tilde{f}(x_1)| \leqslant L|x_2 - x_1|.$$

Thus, the sufficiency of the theorem is also proved.

§5 Theorem on the Differentiability Almost Everywhere

5.1. Definitions

As applications of theorems on differentiability almost everywhere of measures in the space $\mathbf{R}^n$ (the theorems were proved in Chapter I) let us give here some general theorems about differentiability almost everywhere of functions with generalized derivatives. First, let us introduce a general notion of differentiability.

Let $\mathcal{R}$ be a linear topological space whose elements are functions defined on a unit ball $B = \{x \,|\, |x| \leqslant 1\}$ of the space $\mathbf{R}^n$ and taking values in the space $\mathbf{R}^m$.

Let $U \subset \mathbf{R}^n$ be an open set, and let $f : U \to \mathbf{R}^m$ be an arbitrary mapping. Let us assign an arbitrary point $x_0 \in U$ and let $P_{x_0}(X)$ be a polynomial of a variable $X = (X_1, X_2, \dots, X_n)$ whose degree does not exceed r, where $r \geqslant 0$ is an integer. Here we suppose that the coefficients of the polynomial P_{x_0} are nothing but vectors in $\mathbf{R}^m$. The polynomial P_{x_0} is said to be a differential of the order r of a function f at a point x_0 in the sense of convergence in $\mathcal{R}$ if for sufficiently small $h > 0$, $h \leqslant h_0$, the function

$$\rho_h \colon X \mapsto \frac{1}{h^r}[f(x_0 + hX) - P_{x_0}(hX)]$$

belongs to the space $\mathcal{R}$ and for $h \to 0$ tends to zero in the sense of topology of the space $\mathcal{R}$. If at a point x_0, a function f has the differential of the order r in the sense of convergence in $\mathbf{R}$, then f is said to be r-multiple differentiable at the point x_0 in the sense of topology of $\mathcal{R}$.

The above definition does not exclude the case $r = 0$. A polynomial $R_r(X)$ of degree 0 is a vector in R^m. The differential of zero order in the sense of convergence in $\mathcal{R}$ is said to be the natural value of a function f at a point x_0 in the sense of convergence in $\mathcal{R}$. According to the above general definition, a vector $k \in \mathbf{R}^m$ is the natural value of a function f at a point x_0 in the sense of convergence in $\mathcal{R}$ if there exists h_0 such that for $0 < h < h_0$, the function $\rho_{0,h} : X \mapsto f(x_0 + hX) - k$ belongs to the class $\mathcal{R}$, for $h \to 0$ the function $\rho_{0,h}$, tending to zero in the topology of the space $\mathcal{R}$. If $f : U \to \mathbf{R}^m$ is a function of the class $L_{p,loc}(U)$, then due to Theorem 4.1 of Chapter 1 for almost all $x_0 \in U$, the magnitude $f(x_0)$ is the natural value at the point x_0 in the sense of convergence in $L_p(B)$.

Let us also consider a special case where $\mathcal{R} = M$ is a Banach space of bounded functions $\varphi : B \to \mathbf{R}^m$. Topology in M is described by means of the norm defined by the equality

$$\|\varphi\|_M = \sup_{x \in B} |\varphi(x)|.$$

The convergence in the sense of topology of the space M defined by this norm is the usual uniform convergence. Let us show that a polynomial P_{x_0} is a differential of the order r of a function f at a point x_0 in the sense of convergence in M iff the equality

$$f(x_0+X)=P_{x_0}(X)+o(|X|^r) \quad \text{for } X\to 0$$

is valid.

Actually, let us first prove the necessity of this condition. Let P_{x_0} be a differential of order r in the sense of convergence in M of a function f at a point x_0. Let us find h_0 such that for $0 < h < h_0$, the function

$$\rho_h : X \mapsto \frac{1}{h^r}[f(x_0+hX)-P_{x_0}(hX)]$$

belongs to the class M, and let

$$\theta(h)=\left\|\frac{1}{h^r}[f(x_0+hX)-P_{x_0}(hX)]\right\|_M .$$

Then $\theta(h)\to 0$ for $h\to 0$, and for all $X\in B=B(0,1)$, the inequality

$$\left|\frac{1}{h^r}[f(x_0+hX)-P_{x_0}(hX)]\right|\leqslant\theta(h)$$

holds for $0 < h < h_0$. Let us take an arbitrary vector $X\in\mathbf{R}^n$ such that $|X| < h_0/2$. Let us put $h = 2|X|$, $Y = (1/h)X$ so that $X = hY$. We have $|Y| < 1$, $h < h_0$, therefore

$$\left|\frac{1}{h^r}[f(x_0+hY)-P_{x_0}(hY)]\right|\leqslant\theta(h),$$

whence

$$|f(x_0+X)-P_{x_0}(X)|\leqslant|X|^r 2^r\theta(2|X|)=|X|^r\alpha(|X|),$$

where $\alpha(h)\to 0$ for $h\to 0$. This proves the necessity of the condition. Let us assume conversely that

$$f(x_0+X)=P_{x_0}(X)+o(X)\,|X|^r. \tag{5.1}$$

Let us assign $\varepsilon > 0$ and by means of this, let us obtain $\delta > 0$ such that for $|X| < \delta$,

$$\frac{|f(x_0+X)-P_{x_0}(X)|}{|X|^r}<\varepsilon.$$

Let $h < \delta$ and $X\in B$. We set $hX = Y$. Then $|Y|\leqslant h<\delta$, therefore

$$\frac{|f(x_0+hX)-P_{x_0}(hX)|}{h^r}=\frac{|f(x_0+Y)-P_{x_0}(Y)|}{h^r}$$

$$\leqslant\frac{|f(x_0+Y)-P_{x_0}(Y)|}{|Y|^r}<\varepsilon.$$

Hence

$$\frac{1}{h^r}\|f(x_0+hX)-P_{x_0}(hX)\|_M\leqslant\varepsilon.$$

Since $\varepsilon > 0$ is arbitrary,

$$\frac{1}{h^r}\|f(x_0+hX)-P_{x_0}(hX)\|_M \to 0$$

for $h \to 0$.

Thus, a function f is r-multiple differentiable at a point x_0 in the sense of convergence in M iff relation (5.1) is valid. As a rule, r-multiple differentiability of a function at a point implies only relation (5.1) to be valid. Therefore, if r-multiple differentiability of a function at a point is mentioned, we always imply differentiability in the sense of convergence in M.

From the abovesaid it follows, in particular, that a function $f : U \to \mathbf{R}^m$ has at a point $x_0 \in U$ the natural value in the sense of convergence in M if and only if f is continuous at the point x_0. Here $f(x_0)$ is the natural value at the point x_0.

Another important special case arises if we take, as $\mathcal{R}$, a space $\mathcal{F}$ of measurable functions $f : B \to R^m$, and if topology in $\mathcal{R}$ is introduced by means of the following metric

$$\rho(f, g) = \int_B \frac{|f(X)-g(X)|}{1+|f(X)-g(X)|}\, dX.$$

Convergence in the sense of this metric is nothing but convergence by a measure.

A differential in the sense of convergence in $\mathcal{F}$ is said to be an approximative differential. We leave it to the reader to see that the above definition of an approximative differential is equivalent to the traditional one (see, for instance, [70]).

Let $f : U \to \mathbf{R}^m$ be a function of the class $W^r_{p,\mathrm{loc}}(U)$, $p \geqslant 1$. Then for almost all $x_0 \in U$, the polynomial

$$P_{x_0}(X) = \sum_{|\alpha|\leqslant r} \frac{D^\alpha f(x_0)}{\alpha!} X^\alpha$$

is defined.

This polynomial is called a formal differential of the order r of a function f at a point x_0. We are also interested in the case where $f \in \overline{W}^r_{1,\mathrm{loc}}(U)$, i.e., in the case when the derivatives $D^\alpha f$ of the order $|\alpha| = r$ are measures.

In this case, the symbol $D^\alpha f(x_0)$ for $|\alpha| = r$ denotes the density of the measure $D^\alpha f$ at a point x_0.

In the following, a function $f : U \to \mathbf{R}$ is said to be exact at a point $x \in U$ if it is continuous at the point x in the sense of convergence in L_1. If f is exact at every point $x \in U$ at which its natural value is defined, then f is said to be exact in U. Every locally integrable in U function may be made exact in U if we change its values on a set of zero measure.

Here all the functions under consideration and their generalized derivatives are supposed to be exact in U.

5.2. Auxiliary Propositions

First let us prove an integral identity. Let $U \subset \mathbf{R}^n$ be an open set and let $x \in U$ and h be such that $0 < h < \rho(x, \partial U)$. Suppose that a function f of the class $C^\infty(U)$ is given. For $t \in (0, h)$, we put $\theta(t) = f(x + tX)$. Applying the Taylor formula with the residual term in the integral form, we obtain the following equality:

$$\theta(h) = \theta(t) + \frac{h-t}{1!}\theta'(t) + \ldots + \frac{(h-t)^r}{r!}\theta^{(r)}(t)$$

$$+ \frac{1}{(r-1)!}\int_t^h (h-\tau)^{r-1}[\theta^{(r)}(\tau) - \theta^{(r)}(t)]\,d\tau. \tag{5.2}$$

For every $t \geqslant 0$, we have

$$\theta^k(t) = \sum_{|\alpha|=k} \frac{k!}{\alpha!} D^\alpha f(x + tX) X^\alpha.$$

By substituting this expression into (5.2) and by making some obvious transformations, we obtain

$$f(x + hX) = \sum_{|\alpha| \leqslant r} \frac{(h-t)^{|\alpha|} X^\alpha}{\alpha!} D^\alpha f(x + tX)$$

$$+ r\int_t^h (h-\tau)^{r-1} \sum_{|\alpha|=r} \frac{X^\alpha}{\alpha!}[D^\alpha f(x + \tau X) - D^\alpha f(x + tX)]\,d\tau. \tag{5.3}$$

Suppose that a function $\varphi \in C_0^\infty$ is assigned such that $S(\varphi) \subset B(0,1)$. By multiplying both parts of equality (5.3) by $\varphi(x)$ and by integrating by X, we obtain

$$\int_{B(0,1)} \left[f(x + hX) - \sum_{|\alpha| \leqslant r} \frac{(h-t)^{|\alpha|} X^\alpha}{\alpha!} D^\alpha f(x + tX) \right] \varphi(X)\,dX$$

$$= r\int_t^h (h-\tau)^{r-1} \left(\int_{B(0,1)} \sum_{|\alpha|=r} \frac{X^\alpha}{\alpha!}[D^\alpha f(x+\tau X) - D^\alpha f(x+tX)] \right.$$

$$\left. \times \varphi(X)\,dX \right) d\tau. \tag{5.4}$$

When deducing equality (5.4), it was assumed that $f \in C^\infty$. In this case, the equality is also valid if $t = 0$. Our goal is to obtain equality (5.4) for the functions of the class $\overline{W}^r_{1,\text{loc}}$. In this case, the value $t = 0$ leads to certain difficulties.

Let there be a generalized function f defined in U. Let us assign an arbitrary averaging kernel K and let $f_\eta = K_\eta * f$. Let $\eta_0 > 0$ be such that $h + \eta_0 < \rho(x, \partial U)$. Then for $0 < \eta < \eta_0$, the function f_η is defined in the ball $B(x, h)$. Let us write out equality (5.4) for the function f. Let us show that for $\eta \to 0$,

$$\int_{B(0,1)} D^\alpha f_\eta(x+\tau X)\varphi(X)\,dX \to \int_{B(0,1)} D^\alpha f(x+\tau X)\varphi(X)\,dX, \quad (5.5)$$

with the convergence being uniform with respect to τ in the interval $[t,h]$, where $0<t<h$.

Actually, we have

$$\int_{B(0,1)} D^\alpha f_\eta(x+\tau X)\varphi(X)\,dX = \int_U D^\alpha f_\eta(z)\frac{1}{\tau^n}\varphi\left(\frac{z-x}{\tau}\right)dz$$

$$= \int_U D^\alpha f_\eta(z)\,\theta_\tau(z)\,dz = \int_U D^\alpha f(z)\,(\tilde{K}_\eta * \theta_\tau)(z)\,dz,$$

where $\tilde{K}(h)=K(-h)$, $\theta_\tau(z)=\frac{1}{\tau^n}\varphi(\frac{z-x}{\tau})$.

Let (τ_m) and (η_m) be such that for $m\to\infty$, $\eta_m\to 0$, for all m, $t\leqslant\tau_m\leqslant h$ and $\tau_m\to\tau_0\varepsilon[t,h]$. As can easily be verified, the functions $\tilde{K}_{\eta_m} * \theta_{\tau_m}$ for $m\to\infty$ then converge in C_0^∞ to the function θ_{τ_0}. Hence it follows that

$$\int_{B(0,1)} D^\alpha f_{\eta_m}(x+\tau_m X)\varphi(X)\,dX = \int_U D^\alpha f(z)\,(\tilde{K}_{\eta_m} * \theta_{\tau_m})(z)\,dz$$

$$\to \int_U D^\alpha f(z)\,\theta_{\tau_0}(z)\,dz = \int_{B(0,1)} D^\alpha f(x+\tau_0 X)\varphi(X)\,dX.$$

Since the sequences (η_m) and (τ_m) are arbitrary, it follows from what was proved above that the convergence in relation (5.5) is uniform with respect to τ for $\tau\in[t,h]$.

According to what was said above, equality (5.4) written for the function f_η in the limit is transformed into the relations concerning the generalized function f; consequently, equality (5.4) is valid for every generalized function f and for any function $\varphi\in C_0^\infty$ whose support is contained in the ball $B(0,1)$.

5.3. The Main Result

Theorem 5.1. *Let U be an open set in $\mathbf{R}^n$ and let $f: U\to R$ be a function of the class $W^r_{p,\mathrm{loc}}(U)$ (of the class $\overline{W}^r_{1,\mathrm{loc}}(U)$). Then for almost all $x\in U$, the formal differential of f at a point x is its differential in the sense of convergence in W^r_p (in $\overline{W}^r_1$, respectively).*

Proof. According to the above, the function f and all its derivatives are supposed to be exact in U. For $x\in U$ and $X\in B(0,1)$, we put

$$R_h(f,x,X)=\frac{1}{h^r}\left[f(x+hX)-\sum_{|\alpha|=r}\frac{D^\alpha f(x)}{\alpha!}h^{|\alpha|}X^\alpha\right].$$

We are to prove that for almost all $x\in U$, the magnitude

$$\|R_h(f,x,\cdot)\|_{W^r_p(B(0,1))}$$

where Π is the operator of the form

$$(\Pi u)(X) = \sum_{|\alpha| \leqslant l-1} X^\alpha \int_{B(0,1)} \varphi_\alpha(z) u(z) dz.$$

Here φ_α is a function of the class C_0^∞.

To prove the theorem, it suffices to show that the following two statements are valid:

A) Almost all points $x \in U$ are such that

$$\int_{B(0,1)} \varphi_\alpha(z) R_h(f, x, z) dz \to 0$$

for every α.

B) For almost all $x \in U$,

$$\|R(f, x, \cdot)\|_{L_p(B(0,1))} \to 0$$

for $h \to 0$.

Let us prove statement A. Suppose that x is a Lebesgue point of the function f and of each of its derivatives $D^\alpha f$ where $|\alpha| \leqslant r$. Then for any function $\varphi \in C_0^\infty$ such that $S(\varphi) \subset B(0,1)$ for $t \to 0$,

$$\int_{B(0,1)} D^\alpha f(x + tX) \varphi(X) dX \to D^\alpha f(x) \int_{B(0,1)} \varphi(X) dX.$$

Due to this, if we pass to the limit for $t \to 0$ in equality (5.4), we obtain

$$\int_{B(0,1)} R_h(f, x, X) \varphi(X) dX$$

$$= \frac{r}{h} \int_0^h \left(1 - \frac{\tau}{h}\right)^{r-1} \left(\int_{B(0,1)} \sum_{|\alpha|=r} \frac{X^\alpha}{\alpha!} [D^\alpha f(x + \tau X) - D^\alpha f(x)] \right.$$

$$\left. \times \varphi(X) dX \right) d\tau.$$

Due to the choice of the point x, the integral

$$\theta_\alpha(\tau) = \int_{B(0,1)} |D^\alpha f(x + \tau X) - D^\alpha f(x)| dX$$

tends to zero for $\tau \to 0$.

Let us assign arbitrary $\varepsilon > 0$. According to this, there exists $\delta > 0$ such that for $0 < \tau < \delta$, $|\theta_\alpha(\tau)| < \varepsilon$ for all α with $|\alpha| = r$. We have

$$\frac{r}{h}\int_0^h\left(1-\frac{\tau}{h}\right)^{r-1}\left(\int_{B(0,1)}\sum_{|\alpha|=r}\frac{X^\alpha}{\alpha!}[D^\alpha f(x+\tau X)-D^\alpha f(x)]\varphi(X)dX\right)d\tau$$

$$\leqslant\frac{r}{h}\int_0^h\left(1-\frac{\tau}{h}\right)^{r-1}\sum_{|\alpha|=r}M_\alpha\theta_\alpha(\tau)\,d\tau,$$

where $M_\alpha = \sup|\frac{X^\alpha}{\alpha!}\varphi(X)|$. Hence we conclude that for $0 < \tau < \delta$,

$$\left|\int_{B(0,1)}R_h(f,x,X)\varphi(X)\,dX\right|\leqslant\left(r\sum_{|\alpha|=r}M_\alpha\right)\varepsilon,$$

and since $\varepsilon > 0$ is arbitrary, then statement A is proved. All arguments in this case remain valid without changes for the case $f \in \overline{W}_1^r$ as well.

Now let us prove statement B. For $|\alpha| = r$, as follows from the formulae proved in the previous subsection (Theorem 4.7).

$$D_X^\alpha[R_h(f,x,X)] = D^\alpha f(x+hX) - D^\alpha f(x).$$

At every point $x \in U$ which is an L_p-point of each of the functions f, $D^\alpha f$

$$\|D^\alpha f(x+hX)-D^\alpha f(x)\|_{L_p(B(0,1))}\to 0$$

for $h \to 0$. Hence it follows that for any such point,

$$\|R_h(f,x,\cdot)\|_{L_p^r}\to 0$$

for $h \to 0$.

This proves the validity of statement B and completes the proof of the theorem.

5.4. Corollaries of the General Theorem on the Differentiability Almost Everywhere

Let us return to the general notion of a differential given in subsection 5.1. Suppose that a polynomial $P(X)$ of degree not exceeding l is the complete differential of a function $f : U \to \mathbf{R}^m$ at a point x_0 in the sense of convergence in the space M of bounded functions defined in the ball B_1. (The norm in M is defined according to the formula

$$\|f\|_M = \sup_{x\in B_1}|f(x)|\,).$$

In this case, a function f is said to be l-multiple differentiable at a point x_0 and P is said to be a differential of order l of a function f at a point x_0. The condition that P is a differential of order l of a function f at a point x_0, as shown in Subsection 5.1, is equivalent to the following one,

$$f(x_0+X) = P(X) + o(|X|^l) \quad \text{for } X \to 0.$$

The problem of differentiability almost everywhere of a function belonging to some class is one of the classical themes of the theory of functions of real variable. Theorem 5.1 provides some means to be effectively used to obtain various results of such kind. Let H be a set of functions contained in the space $W_p^l(U)$. According to Theorem 5.1 every function $u \in H$ is almost everywhere l-multiple differentiable in the sense of convergence in W_p^l. By using some special properties of functions of the class H, in many cases one can prove that the differential in the sense of convergence in W_p^l is at the same time the differential in some other sense. Let us give examples to illustrate the general ideas outlined above.

Theorem 5.2. *Let $u \in W_p^l(U)$, and $lp > n$. The u has a differential of order l for almost all $x \in U$; that is, for almost all $x \in U$, the following decomposition is valid:*

$$u(x+X) = \sum_{0 \leqslant |\alpha| \leqslant l} D^\alpha u(x) X^\alpha + o(|X|^l).$$

Proof. Let $P(x,X)$ be a formal differential of order l of the function u at the point x. Then, due to Theorem 5.1, for almost all $x \in U$, the function $X \mapsto P(x,X)$ is the complete differential of order l of the function u at the point x; that is, for almost all $x \in U$,

$$\left\|\frac{u(x+hX) - P(x,hX)}{h^l}\right\|_{W_p^l(B(0,1))} \to 0 \tag{5.6}$$

for $h \to 0$. Since $lp > n$, then due to Theorem 4.2 for such x,

$$\left\|\frac{u(x+hX) - P(x,hX)}{h^l}\right\|_{M(B(0,1))} \to 0,$$

that is, $P(x,X)$ is the usual differential of the function f at the point x. The theorem is proved.

Corollary. *Let $U \subset \mathbf{R}^n$ be an open set in $\mathbf{R}^n$. If a function $u : U \to \mathbf{R}$ satisfies the Lipschitz condition, then $u(x)$ is differentiable in U almost everywhere.*

Proof. Suppose U to be a ball. Then, due to Theorem 4.14, the function $u(x)$ belongs to the class $W_1^1(U)$, its derivatives in U being bounded. Therefore, $u \in W_p^1(U)$ for every $p > 1$. Due to the theorem, it follows that $u(x)$ is differentiable in U almost everywhere. In the general case, the set U may be represented as the union of an at the most countable set of balls. According to what was proved above, $u(x)$ is differentiable almost everywhere on each of these balls, therefore, it is differentiable on the set U. Q. E. D.

If in Theorem 5.2 we put $l = 1$, then we obtain the statement which may be considered to be a special case of one of Calderon's theorems [84].

Theorem 5.3. (Aleksandrov [1]). *Every convex function has the second differential almost everywhere.*

Proof. Due to Theorem 4.13 of this chapter, every convex function belongs to the class $\overline{W}_1^2$.

Let $P(X)$ be the complete second-order differential of the function u at a point $x_0 \in U$ in the sense of convergence in $\overline{W}_1^2$. Then, due to imbedding theorems, $P(X)$ is the complete differential in the sense of convergence in L_q for any $q < \frac{n}{n-2}$.

Let $P = P_0 + P_1 + P_2$, where P_i is a homogeneous polynomial of degree i. Then for $h \to 0$,

$$\left\|\frac{u(x_0+hX) - P(hX)}{h^2}\right\|_{L_q(B(0,1))} = \left\|\frac{u(x_0+hX) - P_0 - hP_1(X)}{h^2} - P_2(X)\right\|_{L_q(B(0,1))} \to 0,$$

Thus, the convex functions $\frac{u(x_0+hX)-P_0-hP_1(X)}{h^2}$ for $h \to 0$ converge in L_q to the function $P_2(X)$. Due to convexity, the convergence in L_q results in the uniform convergence, and consequently,

$$\left\|\frac{u(x_0+hX) - P(hX)}{h^2}\right\|_{M(B(0,1))} \to 0 \quad \text{for } h \to 0.$$

Thus, the complete differential in the sense of convergence in $\overline{W}_1^2$ is the complete differential in the sense of convergence in M. Theorem 5.3 is therefore the corollary of Theorem 5.1.

Let $u(x)$ be a function defined and continuous in an open set U. The function u is said to be monotone if for any open G such that $\overline{G} \subset U$, the equalities

$$\sup_{x\in G} u(x) = \sup_{x\in\partial G} u(x); \quad \inf_{x\in G} u(x) = \inf_{x\in\partial G} u(x)$$

hold.

Theorem 5.4. *Let $u : U \to \mathbf{R}$ be a continuous monotone function. Then, if $u \in W_n^1$, then $u(x)$ almost everywhere in U has the complete differential.*

Proof. According to Theorem 5.1, for almost all $x \in U$, there exists a function $P(x,X) = u(x) + \sum_{i=1}^n P_i(x)X_i$ such that

$$\left\|\frac{u(x_0+hX) - u(x_0)}{h} - \sum_{i=1}^n P_i(x_0)X_i\right\|_{W_n^1(B(0,1))} \to 0$$

for $h \to 0$. The functions

$$v_h : X \mapsto \frac{u(x_0+hX) - u(x_0)}{h}$$

are monotone in the ball $B(0,1)$. Their norms in $W_n^1(B(0,r)), r \leqslant 1/2$, for sufficiently small h, $0 < h \leqslant h_0$, are bounded in total. Hence it follows that for $0 < h \leqslant h_0$, the family of functions (v_h) is equicontinuous in the ball $B(0,1/2)$. So,

$$v_h(X) = \frac{u(x+hX) - u(x)}{h} \to \sum_{i=1}^{n} P_i(x)X_i$$

uniformly for $h \to 0$ for almost all $x \in U$. The theorem is proved.

As an example of the application of the results obtained here let us give the proof of Stepanov's theorem [69], which establishes the necessary and sufficient condition of the differentiability almost everywhere of a function defined on an arbitrary subset of $\mathbf{R}^n$.

First let us prove the following extension lemma.

Lemma 5.1. *Let there be a set $E \subset \mathbf{R}^n$ and a function $f : E \to \mathbf{R}$ satisfying the Lipschitz condition with a constant L. Then there exists a function $g : \mathbf{R}^n \to \mathbf{R}$ such that $g(x) = f(x)$ for all $x \in E$ and that g should satisfy the Lipschitz condition with the same constant L as f.*

Proof. For an arbitrary $x \in \mathbf{R}^n$, we set

$$g(x) = \inf_{\xi \in E} [f(\xi) + L|x - \xi|].$$

Let us prove that the function g is the desired one. Let $x \in E$. For every $\xi \in E$, the inequality

$$f(x) \leqslant f(\xi) + L|x - \xi|$$

is valid, for $\xi = x$ this inequality turning into equality. Hence it follows that for $x \in E$,

$$f(x) = \inf_{\xi \in E} [f(\xi) + L|x - \xi|],$$

that is, $g(x) = f(x)$ for any $x \in E$. Let x_1 and x_2 be arbitrary points in $\mathbf{R}^n$. For every $\xi \in E$, we have the inequality

$$g(x_1) \leqslant f(\xi) + L|x_1 - \xi| \leqslant f(\xi) + L|x_2 - \xi| + L|x_1 - x_2|. \tag{5.7}$$

Due to arbitrariness of $\xi \in E$ from the latter inequality, it follows that

$$g(x_1) \leqslant \inf_{\xi \in E} \{f(\xi) + L|x_2 - \xi| + L|x_1 - x_2|\} = g(x_2) + L|x_1 - x_2|.$$

By choosing an arbitrary point of the set E to stand for x_1 and by taking into account that $g(x) = f(x)$ for $x \in E$ from the inequality (5.7), we obtain, in particular, that $g(x) > -\infty$ for any $x \in \mathbf{R}^n$. Further, from inequality (5.7), it follows that

$$g(x_1) - g(x_2) \leqslant L|x_1 - x_2|.$$

Since x_1 and x_2 have the same rights, the inequality

$$g(x_2) - g(x_1) \leqslant L|x_2 - x_1|$$

is also valid, that is,

$$|g(x_1) - g(x_2)| \leqslant L|x_1 - x_2|.$$

The lemma is proved.

Let E be an arbitrary set in $\mathbf{R}^n$ and let $f : E \to \mathbf{R}$ be a function which is defined on E. Let $a \in E$ be a limit point of the set E. We set

$$L_f(a) = \overline{\lim_{x \to a}} \frac{|f(x) - f(a)|}{|x - a|}.$$

Let us introduce the following notations. We put

$$E_\rho = E \cap B(a, \rho), \qquad L_f(a, \rho) = \sup_{x \in E_\rho} \frac{|f(x) - f(a)|}{|x - a|}. \tag{5.8}$$

Obviously, $L_f(a) = \lim_{\rho \to 0} L_f(a, \rho)$.

Let a be a limit point of the set $E \subset \mathbf{R}^n$. The function $f : E \to \mathbf{R}$ is said to be differentiable at the point a by the set E if there exists a linear function $\lambda : \mathbf{R}^n \to \mathbf{R}$ such that

$$\frac{f(x) - f(a) - \lambda(x - a)}{|x - a|} \to 0$$

if $x \to a$ by the set E.

Theorem 5.5. (Stepanov [69]). *Let there be a set $E \subset \mathbf{R}^n$ and a function $f : E \to \mathbf{R}$. Then f is differentiable on the set E almost everywhere iff for almost all $x \in E$, the magnitude $L_f(x)$ is finite.*

Remark. This theorem does not imply the set E to be measurable.

Proof. The necessity of the condition of this theorem is obvious, so we only have to prove the sufficiency.

Thus, let the function $f : E \to \mathbf{R}$ be such that for almost all $x \in E$, $L_f(x) < \infty$. Let E' be the totality of all points $x \in E$ which are either not limit points of E or for which $L_f(x) = \infty$. Due to the condition of the theorem, E' is a set of zero measure. We put $E_0 = E \backslash E'$.

Let us denote by P_m a set of all points $x \in E_0$ for which

$$L_f(x, 1/m) \leqslant m.$$

Since for every point $x \in E_0$, $L_f(x) = \lim_{m\to\infty} L_f(x, 1/m) < \infty$, then every point $x \in E_0$ belongs to at least one of the sets P_m, that is

$$E_0 = \cup_{m=1}^{\infty} P_m.$$

Let $P_m = \cup_k P_{m,k}$ be the partition of the set P_m into a countable number of pairwise nonintersecting subsets, each of them having the diameter less than $1/m$. Such a partition may obviously be obtained if we first construct the partition of the space into pairwise nonintersecting cubes whose diameter is less than $1/m$, then we take the intersections of the set P_m with these cubes. The function f on each of the sets $P_{m,k}$ satisfies the Lipschitz condition with constant m.

Let us fix an arbitrary set $P_{m,k}$. Let us construct a function $g : \mathbf{R}^n \to \mathbf{R}$ such that $g(x) = f(x)$ for all $x \in P_{m,k}$, and that g satisfies the Lipschitz condition with constant m. Such a function g exists due to Lemma 5.1. Let $A_{m,k}$ be a set of points $x \in P_{m,k}$ which are not density points of the set $P_{m,k}$, let $B_{m,k}$ be the totality of all points $x \in P_{m,k}$ such that g is not differentiable at the point x. According to Corollary 3 of Theorem 4.1 (Chapter 1), $A_{m,k}$ is a set of zero measure. The corollary of Theorem 5.2 yields that $|B_{m,k}| = 0$ as well. Let us put $S_{m,k} = A_{m,k} \cup B_{m,k}$ and let $x_0 \in P_{m,k}$, and at the same time let $x_0 \notin S_{m,k}$. Then the function g is differentiable at the point x_0, and x_0 is the density point of $P_{m,k}$. Let λ be a differential of the function g at the point x_0, i.e., λ is a linear function such that

$$\frac{g(x) - g(x_0) - \lambda(x - x_0)}{|x - x_0|} \to 0$$

for $x \to 0$. Let us arbitrarily assign $\varepsilon > 0$, $\varepsilon < 1$ and let us obtain by it $\delta_1 > 0$ such that for $|x - x_0| < \delta_1$,

$$|g(x) - g(x_0) - \lambda(x - x_0)| \leqslant \varepsilon |x - x_0|.$$

Since x_0 is the density point of the set $P_{m,k}$, then there exists $\delta_2 > 0$ such that for $\rho < \delta_2$, the inequality

$$|P_{m,k} \cap B(x_0, \rho)| > \left(1 - \frac{\varepsilon^n}{m^n}\right) |B(x_0, \rho)| \tag{5.9}$$

is valid. Let us put $\delta_0 = \min\{\delta_1, \delta_2, \frac{1}{2m}\}$. Let x be a point of the set E such that $|x - x_0| = \rho < \delta_0$. We set $y = x_0 + (1 - \frac{\varepsilon}{m})(x - x_0)$. The point y lies in the ball $B = B(x_0, \rho)$, and the ball $B' = B(y, \frac{\varepsilon}{m}\rho)$ is contained in the ball B_ρ. Here $|B'| = \frac{\varepsilon^n}{m^n}|B_\rho|$. The ball B' contains the points of the set $P_{m,k} \cap B_\rho$. Indeed, otherwise we have: $|P_{m,k} \cap B_\rho| \leqslant |B_\rho| - |B'| = (1 - \frac{\varepsilon^n}{m^n})|B_\rho|$, which contradicts inequality (5.9). Let $z \in P_{m,k} \cap B_\rho$. Then

$$|z - x| \leqslant |z - y| + |y - x| \leqslant 2\rho\varepsilon/m. \tag{5.10}$$

Since $\rho < \delta_0 \leqslant \frac{1}{2m}$, then $|z - x| \leqslant 2\rho < \frac{1}{m}$, and since $z \in P_m$, then

$$|f(z) - f(x)| \leqslant m|z - x| \leqslant 2\rho\varepsilon. \tag{5.11}$$

Since λ is a differential of the function g at the point x_0, and since g satisfies the Lipschitz condition with coefficient m, the $n\lambda$ satisfies the Lipschitz condition with coefficient m as well; therefore, $|\lambda(x)| \leqslant m|x|$ for every vector $x \in \mathbf{R}^n$. We have: $x_0 \in P_{m,k}$, $z \in P_{m,k}$, and therefore, $g(x_0) = f(x_0)$, $g(z) = f(z)$. We have: $|f(x)-f(x_0)-\lambda(x-x_0)| \leqslant |f(x)-f(z)|+|\lambda(x-z)|+|f(z)-f(x_0)-\lambda(z-x_0)| = |f(x)-f(z)|+|\lambda(x-z)|+|g(z)-g(x_0)-\lambda(z-x_0)|$. Due to (5.11), $|f(x) - f(z)| \leqslant 2\varepsilon\rho$. From (5.10), it follows that $|\lambda(x - z)| \leqslant m|x-z| \leqslant 2\varepsilon\rho$. Since $|z-x_0| < \delta_1$, then $|g(z)-g(x_0)-\lambda(z-x_0)| \leqslant \varepsilon|z-x_0| \leqslant \varepsilon\rho$. Comparing these estimates, we obtain

$$|f(x) - f(x_0) - \lambda(x - x_0)| \leqslant 5\varepsilon\rho = 5\varepsilon|x - x_0|.$$

Since $\varepsilon > 0$, $\varepsilon > 1$ is arbitrary and since this inequality is valid if $|x-x_0| < \delta_0$, this proves that the function f is differentiable at the point x_0 by the set E.

Let $S = \cup_{m,k} S_{m,k}$. Then $|S| = 0$. Let $x \in E_0 \backslash S$. Then $x \in P_{m,k}$ for some m and k; therefore, the function f is differentiable at the point x by the set E.

This completes the proof of the theorem.

5.5. The Behaviour of Functions of the Class W_p^l on Almost All Planes of Smaller Dimensionality

Let P and Q be two k-dimensional planes in $\mathbf{R}^n$ where $1 \leqslant k \leqslant n-1$. P and Q are said to be parallel and are written as $P \| Q$ if P may be combined with Q by means of parallel transition. The relation $P \| Q$ is obviously the equivalence relation. The equivalence class by this relation in the set of all k-dimensional planes in $\mathbf{R}^n$ is called a k-dimensional direction in $\mathbf{R}^n$.

Let Γ be an arbitrary k-dimensional direction in $\mathbf{R}^n$. Let T be an $(n-k)$-dimensional plane which is completely orthogonal to some of the planes belonging to Γ. Then T is completely orthogonal to all the planes $P \in \Gamma$, and any plane P intersects T at the unique point which we denote by $\varkappa(P)$. Let E be an arbitrary subset of Γ. By $\varkappa(E)$, we denote the totality of all points $\varkappa(P)$ where $P \in E$. Let us note that the $(n-k)$-dimensional measure of the set $\varkappa(E)$ does not depend on the choice of the plane T. We denote it by $\mu_{n-k}(E)$. This or that condition is said to be valid for almost all planes belonging to the k-dimensional direction G if for the set E of planes that do not satisfy this condition, $\mu_{n-k}(E) = 0$.

Theorem 5.6. *Let U be an open set in $\mathbf{R}^n$ and let f be a function of the class $W^l_{p,loc}(U)$. Then for every k-dimensional direction Γ for almost all k-dimensional planes $P \in \Gamma$, the restriction of the function f on the set $P \cap U$ is the function of the class $W^l_p(P \cap U)$.*

Remark. In the theorem, the function f is supposed to be exact, i.e., it coincides with its natural values at every point where they are defined.

Proof. Without loss of generality, one may assume that the planes of the given k-dimensional direction are defined by the equations x_{k+i} = const, $i = 1, 2, \dots, n-k$. The general case may obviously be reduced to this by orthogonal transformations of coordinates. Let us introduce the following notations. For an arbitrary point $x = (x_1, x_2, \dots, x_n)$, we put $y = (x_1, \dots, x_k)$, $z = (x_{k+1}, \dots, x_n)$, and we write $x = (y, z)$. If a function f is given, then, in accordance with this, the magnitude $f(x)$ is denoted by $f(y, z)$ in the following.

Let us assign an arbitrary cube, $Q = (a_1, b_1) \times (a_2, b_2) \times \dots \times (a_n, b_n)$ whose closure is contained in U. Let us put

$$Q_1 = (a_1, b_1) \times \dots \times (a_k, b_k), \qquad Q_2 = (a_{k+1}, b_{k+1}) \times \dots \times (a_n, b_n).$$

We shall prove that for almost all $z \in Q_2$, the function $y \mapsto f(y, z)$ belongs to the class $W_p^l(Q_1)$. Let us assign an arbitrary averaging kernel K and let us put $f_h = K_h * f$. For sufficiently small h, $h < h_0$, $f_h(x)$ is defined for all $x \in Q$. For $h \to 0$, $\|f_h - f\|_{W_p^l(Q)} \to 0$. Let (h_m), $m = 1, 2, \dots$, be a decreasing sequence of values of h such that $0 < h_m < h_0$ for all m and $\|D^\alpha f_{h_m} - D^\alpha f\|_{L_p(Q)} < 1/2^m$ for all α with $|\alpha| \leqslant l$,

$$w_m(x) = \sum_{|\alpha| \leqslant l} |D^\alpha f_{h_m}(x) - D^\alpha f(x)|.$$

Then

$$\int_Q w_m(x)\, dx \leqslant C(l, n)/2^{mp},$$

where $C(l, n)$ is the number of n-dimensional multiindices such that $|\alpha| \leqslant l$. Let us set $w = \sum_{m=1}^{\infty} w_m$. The function w is integrable. Due to the Fubini theorem, for almost all $z \in Q_2$, the function $y \to w(y, z)$ is integrable by the cube Q_1. Let z be such that for it this fact is true. We have

$$\int_{Q_1} w(y, z)\, dy = \sum_{m=1}^{\infty} \int_{Q_1} w_m(y, z)\, dy.$$

Since the integral on the left-hand side is finite, then for this z,

$$\int_{Q_1} w_m(y, z)\, dy \to 0$$

for $m \to \infty$. Hence, we conclude that for the given z for $m \to \infty$,

$$\int_{Q_1} |D^\alpha f_{h_m}(y, z) - D^\alpha f(y, z)|^p\, dy \to 0$$

for any multiindex α with $|\alpha| \leqslant l$. We obtain that for the given z (setting $\alpha = 0$), the functions $y \mapsto f_{h_m}(y, z)$ converge in $L_p(Q_1)$ to $y \mapsto f(y, z)$, and the derivatives $D^\alpha f_{h_m}(y, z)$ converge in $L_p(Q_1)$ to the functions $D^\alpha f(y, z)$.

From the above, due to the completeness of the space W_p^l, it follows that the function $y \mapsto f(y, z)$ for the given z belongs to the class $W_p^l(Q_1)$.

Now the statement of the theorem easily follows from what was proved above if we represent U as a union of a countable number of cubes.

Corollary 1. *If f is a function of the class $W^1_{p,\mathrm{loc}}(U)$, then for every one-dimensional direction Γ for almost all $l \in \Gamma$, the restriction of f on the straight line l is an absolutely continuous function.*

5.6. The ACL-Classes

A function $\pi_j : \mathbf{R}^n \to \mathbf{R}^{n-1}$, $\pi_j(x_1, x_2, \dots, x_j, \dots, x_n) = (x_1, x_2, \dots, x_{j-1}, x_{j+1}, \dots, x_n) = \overline{x}_j$ is said to be the jth projection of $\mathbf{R}^n$ on $\mathbf{R}^{n-1}$. We write the point $x \in \mathbf{R}^n$ in the form $x = (\overline{x}_j, x_j)$. A function f, locally summable in the domain $G \subset \mathbf{R}^n$, belongs to the class $ACL(G)$ if for any $j = 1, 2, \dots, n$, the function $x_j \mapsto f(\overline{x}_j, x_j)$ for almost all $\overline{x}_j \in \pi_j(G)$ is absolutely continuous on any closed segment $[a, b]$ such that $\{\bar{x}_j\} \times [a, b] \subset G$.

If $f \in ACL(G)$, then f has almost everywhere in G the ordinary partial derivatives $\frac{\partial f}{\partial x_j}$. The function $f \in ACL_p(G)$ if $\frac{\partial f}{\partial x_j} \in L_p(G)$ for all $j = 1, 2, \dots, n$.

Theorem 5.7. *A function $f \in L_p^1(G)$ if and only if $f \in ACL_p(G)$.*

Proof. The necessity follows from the corollary of Theorem 5.6.

Sufficiency. If $\varphi \in C_0^\infty(G)$, then

$$\int_G f \frac{\partial \varphi}{\partial x_1}\, dx_1 = \int_{\pi_1^{-1}(G)} \left(\int_{-\infty}^{\infty} f(t, y_1) \frac{\partial \varphi}{\partial x_1}(t, y_1)\, dt \right) dy_1.$$

For almost all $y_1 \in \pi_1(G)$, the function $f(t_1, y_1)$ is absolutely continuous by t on every segment $[a, b] \in \pi_1^{-1}(y_1) \cap G$. Let us fix such a point y_1.

The set $\pi_1^{-1}(y_1) \cap G$ is the union of a countable number of the interval (α_j, β_j). The finiteness of the function φ and the absolute continuity of the function $f(t, y_1)$ by t allow us to apply the integration formula part by part:

$$\int_G f \frac{\partial \varphi}{\partial x_1}\, dx = - \int_{\pi_1^{-1}(G)} \left(\int_{-\infty}^{\infty} \frac{\partial f}{\partial x_1}(t, y_1)\, \varphi(t, y_1)\, dt \right) dy_1$$

$$= - \int_G \frac{\partial f}{\partial x_1}(x_1)\, \varphi(x)\, dx.$$

The theorem is proved.

CHAPTER 3

NONLINEAR CAPACITY

This chapter deals with the construction of the notion of nonlinear capacity which will be used to characterize sets of density points in the sense of Lebesgue for functions of the Sobolev classes and for convergence theorems (the analogies of theorems like those of Egorov, Luzin, etc.) for the spaces W_p^l. This characteristic is more refined than in the theory of L_p spaces. In these and in a number of related problems, the capacity plays a role similar to that of the measure for spaces of summable functions. The local character of properties of functions of the Sobolev classes in question allows us to investigate classes in $\mathbf{R}^n$. As is clear from the papers [82, 84], the use of Bessel transformations turned out to be a useful technical means for the spaces $W_p^l(\mathbf{R}^n)$. For Bessel transformations, the space $W_p^l(\mathbf{R}^n)$ is represented as the image of $L_p(\mathbf{R}^n)$ for integral transformation. A considerable part of the results and possible applications of the capacity concept for other scales of spaces, except the spaces W_p^l, cause us to give the invariant representation under the assumption that the classes of spaces under consideration were obtained from L_p by means of some integral transformation.

This concept of capacity originates from the papers [65,48].

In order to study boundary behaviour of the classes $W_p^l(G)$ in domains G of a Euclidean space, we introduce the notion of variational capacity (or conductivity, in the terminology of Maz'ya). This notion is introduced due to the absence of a suitable integral transformation which allows one to obtain $W_p^l(G)$ from a space of the L_p type.

For domains with smooth boundary, there exist good two-sided estimates between two types of capacity. Variational capacity will further be used to study mappings preserving the Sobolev classes, in the problem of extending the functions of these classes, and when proving imbedding theorems in domains with "bad" boundary.

§1 Capacity Induced by a Linear Positive Operator

1.1. Definition and the Simplest Properties

Let U be a domain in $\mathbf{R}^n$, and let V be a measurable set in $\mathbf{R}^m$. A continuous linear operator $T : L_p^+(U) \to L_{1,\mathrm{loc}}(V)$ is said to be positive if:

a) for any function $u \in L_p^+(U)$, the function Tu is nonnegative;

b) (preserving convergence) if the series $\sum_m u_m$, $u_m \in L_p^+(U)$, converges everywhere to the function $u \in L_p^+(U)$, then the series $\sum_m Tu_m$ converges everywhere to the function Tu.

The positive operator T is called a C-operator if for every nonnegative function $u \in C_0(U)$, the function Tu is continuous.

The positive operator T is called strongly positive if for every nonnegative function $u \in C_0(U)$, $u \not\equiv 0$, the function Tu is positive (vanishes nowhere).

To every positive operator $T : L_p^+(U) \to L_{1,loc}(U)$, we associate the notion of capacity.

A function $f \in L_p^+(U)$ is said to be admissible for the set $E \subset V$ and for the operator T if $(Tf)(x) \geqslant 1$ at every point $x \in E$.

The greatest lower boundary of the integral

$$\int\limits_U |f|^p \, dx,$$

taken on the set $M_{T,p}(E, V)$ of all functions admissible for the set E and for the positive operator T is called the (T,p)-capacity of the set E. Depending on the context, we use either the full notation

$$\mathrm{Cap}_{T,p}(E, V)$$

or its shortened variants

$$\mathrm{Cap}_{T,p}(E), \qquad \mathrm{Cap}_T(E).$$

If the set $M_{T,p}(E, V)$ is empty, then it is natural to set $\mathrm{Cap}_{T,p}(E, V) = \infty$.

Examples of Capacity.

Model example. The Bessel kernel of the order $l > 0$ in the space $\mathbf{R}^n$ is the function of the form

$$G_l(x) = G_l^{(n)}(x) = \frac{(2\pi)^{n/2} 2^{(2-l)/2}}{\Gamma(l/2)} K_{\frac{(n-l)}{2}}(|x|)|x|^{\frac{l-n}{2}},$$

where $K_\nu(r)$, $r > 0$, is the modified Bessel function of the third genus (see, for instance, [71]). For further presentation, we can do without the concrete form of the function $K_\nu(r)$.

The integral operator $f \xrightarrow{I_l} G_l * f$ is called the Bessel potential. The Bessel potential is a strongly positive C-operator.

The capacity induced by the Bessel potential is called $\mathrm{Cap}_{l,p}(\dots)$.

Example 2. Let us assign an arbitrary nonnegative function $K : U \times U \to (R, \infty)$ satisfying the conditions:

K.1) If $x \neq y$, then $K(x,y) < \infty$, and the function K is continuous at the point (x,y). Here

$$\operatorname*{Lim}_{(x,y)\to(y,y)} K(x,y) = \infty$$

for all $y \in U$.

K.2) There exists a number $L < \infty$ such that for every $x \in U$,

$$\int_U K(x,y)\,dy < L.$$

K.3) If $u \in C_0(U)$, then the function

$$T_K u(x) = \int_U K(x,y)u(y)\,dy$$

is continuous. In this case, if u is nonnegative, then $T_K u(x) > 0$ for all $x \in U$.

It is easy to show that the operator T_K is positive. From the Condition K.3, it follows that T_K is a strongly positive C-operator. The Bessel potential I_l is the operator of the kind T_K. The Riesz operator $J_l(J_l(f) = |x|^{l-n} * f)$ is also the T_K operator, as follows, for instance, from [71].

Example 3. The operator $T : L_p^+(U) \to L_{1,\mathrm{loc}}(U)$ is identical. Then for any set E, the capacity $\mathrm{Cap}_{T,p}(E)$ coincides with the outer measure of this set.

1.2. Capacity as the Outer Measure

In Example 3, the capacity $\mathrm{Cap}_{T,p}(E)$ induced by the identical operator T is the outer Lebesgue measure. Let us show that any capacity $\mathrm{Cap}_{T,p}(E,V)$ is an outer measure. First we give the simplest properties of a capacity.

1. The capacity of an empty set equals zero;

2. (Monotonicity relative to a set being measured). For any two sets $E_1, E_2 \subset V$ from the inclusion $E_1 \subset E_2$, it follows that

$$\mathrm{Cap}_{T,p}(E_1, V) \leqslant \mathrm{Cap}_{T,p}(E_2, V).$$

It obviously follows from the inclusion

$$M_{T,p}(E_2, V) \subset M_{T,p}(E_1, V).$$

Corollary. *If a set* $E \subset V$ *has zero capacity* $\mathrm{Cap}_{T,p}(E, V)$, *then any of its subsets also has zero capacity.*

3. (Countable semiadditivity). Let E_ν, $\nu \in N$, be an arbitrary sequence of sets in V. Then

$$\mathrm{Cap}_{T,p}(\cup E_\nu, V) \leqslant \sum_{\nu=1}^{\infty} \mathrm{Cap}_{T,p}(E_\nu, V).$$

Proof. If the sum on the right-hand side is infinite, then the inequality is obvious. Let it be finite. Let us fix $\varepsilon > 0$. Choose the functions $f_\nu \in M_{T,p}(E_\nu, U)$ so that

$$\|f\|^p_{L_p(U)} \leqslant \mathrm{Cap}_{T,p}(E_\nu, V) + \frac{\varepsilon}{2^\nu}.$$

The function $f = \sup\limits_{\nu} f_\nu$ belongs to $L_p(U)$, since

$$\|f\|^p_{L_p(U)} \leqslant \sum_{\nu=1}^{\infty} \|f_\nu\|^p_{L_p(U)} \leqslant \sum_{\nu=1}^{\infty} \mathrm{Cap}_{T,p}(E_\nu, V) + \varepsilon < \infty.$$

On the other hand, the function f is admissible for the set E, since

$$(Tf)(x) \geqslant (Tf_\nu)(x) \quad \text{for } x \in V.$$

So, $(Tf)(x) \geqslant 1$ for all $x \in E = \cup E_\nu$.

From the definition of capacity and from the inequality

$$\mathrm{Cap}_{T,p}(E, V) \leqslant \|f\|^p_{L_p(U)},$$

due to arbitrariness in the choice of ε, we immediately obtain

$$\mathrm{Cap}_{T,p}(E, V) \leqslant \sum_{\nu=1}^{\infty} \mathrm{Cap}_{T,p}(E_\nu, V).$$

This completes the proof.

Corollary. *The union of at the most countable number of sets with zero capacity has zero capacity.*

4. Let there be a set $E \subset V$ and a function $f \in L_p^+(U)$. Then, if for all $x \in V$ the inequality $(Tf)(x) \geqslant \alpha \geqslant 0$ is valid, then

$$\mathrm{Cap}_{T,p}(E, V) \leqslant \tfrac{1}{\alpha} p \|f\|^p_{L_p(U)}.$$

To prove Property 4, it suffices to realize that the function $g = \frac{1}{\alpha} f$ is admissible for the set E; therefore,

$$\mathrm{Cap}_{T,p}(E,V) \leqslant \|g(x)\|^p_{L_p(U)} = \tfrac{1}{\alpha}p\|f\|^p_{L_p(U)}.$$

5. Let V_1 be a measurable subset of V. The operator $T_1 = I \circ T$, where $I : L_{1,\mathrm{loc}}(V) \to L_{1,\mathrm{loc}}(V_1)$ is the restriction operator, $(Iu = u/V_1)$ is a positive operator, a C-operator, or the strongly positive one, respectively, if the operator T is one, too. For an arbitrary set $E \subset V_1$,

$$\mathrm{Cap}_{T_{1},p}(E,V_1) \leqslant \mathrm{Cap}_{T,p}(E,V).$$

The proof is obvious.

We have proved that the capacity $\mathrm{Cap}_{T,p}(\cdot, V)$ is a monotone (Property 2), countably semiadditive (Property 3) function of the set defined on all subsets of the domain V, i.e., $\mathrm{Cap}_{T,p}(\cdot, V)$ is the outer measure.

1.3. Sets of Zero Capacity

The two properties of sets with zero capacity we obtained in the previous subsection are: any subset of a set with zero capacity is a set of zero capacity; the union of a countable number of sets with zero capacity is a set of zero capacity.

Proposition 1.1. *The capacity of the set $E \subset V$ is equal to zero iff there exists a nonnegative function $f \in L_p(U)$ such that*

$$(Tf)(x) = \infty$$

for all x.

Proof. Let $\mathrm{Cap}_{T,p}(E,V) = 0$. Then for every $\nu \in N$, there exists a nonnegative function $f_\nu : U \to \mathbf{R}$ such that $(Tf)(x) \geqslant 1$ for all $x \in E$ and $\|f\|^p_{L_p(U)}, \leqslant 2^{-2\nu}$. Let us define a nonnegative function $f \in L^+_p(U)$ by setting $(f(x))^p = \sum_{\nu=1}^{\infty}(f_\nu(x))^p 2^\nu$. For every ν, we have $f(x) \geqslant 2^{\nu/p} f_\nu(x)$, and due to the positiveness of the operator T,

$$(Tf)(x) \geqslant 2^{\nu/p}(Tf_\nu)(x)$$

for all $x \in V$. By the construction of the functions f_ν, it follows that $(Tf)(x) \geqslant 2^{\nu/p}$ for all $x \in E$, and the inequality holds for all ν, i.e., $(Tf)(x) = \infty$ for all $x \in E$.

It only remains to verify that $f \in L^+_p(U)$.

Indeed,

$$\|f\|^p_{L_p(U)} \leqslant \sum_{\nu=1}^{\infty} 2^\nu \|f\|^p_{L_p(U)} \leqslant \sum_{\nu=1}^{\infty} 2^{-\nu}.$$

The necessity of conditions of the proposition is proved.

Suppose now that for the set $E \subset V$ there exists a nonnegative function $f \in L_p(U)$ such that $(Tf)(x) = \infty$ for all $x \in E$. Due to Property 4 of the capacity, $\mathrm{Cap}_{T,p}(E, V) \leqslant \varepsilon \|f\|^p_{L_p(U)}$. Since $\varepsilon > 0$ was chosen arbitrarily, it follows that $\mathrm{Cap}_{T,p}(E, V) = 0$. The proposition is proved.

Corollary. *For every function $f \in L^+_p(U)$, the set of $x \in V$ for which $(Tf)(x) = \infty$ is a set of zero capacity.*

Some condition A is said to be valid (T,p)-quasieverywhere on the set $E \subset V$ if the capacity of the set of $x \in E$, which does not satisfy this condition, equals zero.

According to Proposition 1.1, for any function $f \in L^+_p(U)$, the capacity of the set $E_{f,\infty} = \{x : (Tf)(x) = \infty\}$ equals zero. An arbitrary function $f \in L_p(U)$ is representable in the form $f = f^+ - f^-$, where $f^+(x) = \max(0, f(x))$, $f^- = \max(0, -f(x))$. The functions f^+ and f^- belong to $L^+_p(U)$. Let us extend the operator $T : L^+_p(U) \to L_{1,\mathrm{loc}}(V)$ as far as the operator acting from $L_p(V)$ into the space $F_{T,p}(V)$ of functions which are finite (T,p)-quasieverywhere on V. We denote the extension of the operator by the same symbol, $T : L_p(U) \to F_{T,p}(V)$. The extension formula is $Tf = Tf^+ - Tf^-$, $Tf \in L_p(U)$. The function Tf is finite (T,p)-quasieverywhere on V, since the functions Tf^+ and Tf^- are finite (T,p)-quasieverywhere on V.

Theorem 1.2. *Let $\sum_{\nu=1}^{\infty} f_\nu$ be an arbitrary series of functions $L_p(U)$ such that*

$$\sum_{\nu=1}^{\infty} \|f_\nu\|_{L_p(U)} < \infty.$$

Then the series $\sum_{\nu=1}^{\infty} Tf$ converges (T,p)-quasieverywhere in V. Besides, if f is the sum of the series $\sum_{\nu=1}^{\infty} f_\nu$, then the series $\sum_{\nu=1}^{\infty} Tf_\nu$ converges (T,p)-quasieverywhere to the function Tf.

Proof. Let us represent each of the functions f_ν in the form $f_\nu = f^+_\nu - f^-_\nu$. From the convergence of the series $\sum_{\nu=1}^{\infty} f_\nu$ to $L_p(U)$, it follows that

$$\sum_{\nu=1}^{\infty} \|f^+_\nu\|_{L_p(U)} < \infty, \qquad \sum_{\nu=1}^{\infty} \|f^-_\nu\|_{L_p(U)} < \infty.$$

The functions $h = \sum_{\nu=1}^{\infty} f^+$ and $g = \sum_{\nu=1}^{\infty} f^-$ belong to the class $L^+_p(U)$. Their nonnegativity and Condition b from the definition of a positive operator allow one to conclude that the equalities

$$(Th)(x) = \sum_{\nu=1}^{\infty} (Tf^+_\nu)(x), \qquad (Tg)(x) = \sum_{\nu=1}^{\infty} (Tf^-_\nu)(x)$$

are valid for all $x \in V$.

Let E_1 be a set of x for which $(Th)(x) = \infty$, and let E_2 be a set of x for which $(Tg)(x) = \infty$. Then E_1 and E_2 are the sets of zero (T,p)-capacity.

The set $E = E_1 \cup E_2$ also has zero (T,p)-capacity. For every $x \in E$, the series $\sum_{\nu=1}^{\infty} Tf_\nu$ obviously converges, the sum being equal to $\sum_{\nu=1}^{\infty}(Tf_\nu^+)(x) - \sum_{\nu=1}^{\infty}(Tf_\nu^-)(x) = (Th)(x) - (Tg)(x) = (Tf)(x)$.

Corollary. *Let $\{f_\nu\}$, $\nu = 1, 2, \ldots$, be an arbitrary sequence of functions from $L_p(U)$ such that $\|f_\nu - f\|_{L_p(U)} \to 0$ for $\nu \to \infty$.*

Then there exists a sequence of indices ν_k such that $\nu_k \to \infty$ for $k \to \infty$ and $(T(|f_k - f|))(x) \to 0$ for $k \to \infty$ (T,p)-quasieverywhere in V.

1.4. Extension of the Set of Admissible Functions

Let us denote by $\overline{M}_{T,p}(E)$ the closure of the set $M_{T,p}(E)$ in the space $L_p(U)$. It is obvious that the set $M_{T,p}(E)$ is convex, and consequently, the set $\overline{M}_{T,p}(E)$ is convex as well.

Theorem 1.3. *The function $f \in L_p(U)$ belongs to the set $\overline{M}_{T,p}(E)$ iff f is nonnegative, and (T,p)-quasieverywhere on the set E the inequality $(Tf)(x) \geqslant 1$ is valid.*

Proof. Suppose that $f \in \overline{M}_{T,p}(E)$. Then there exists a sequence $\{f_\nu\}$, $\nu = 1, 2, \ldots$, of functions from $M_{T,p}(E)$ such that $\|f_\nu - f\|_{L_p(U)} \to 0$ for $\nu \to \infty$. The corollary of the previous theorem allows one to obtain a sequence of subscripts ν_k, $k = 1, 2, \ldots$, such that $(T(|f_{\nu_k} - f|))(x) \to 0$ (T,p)-quasieverywhere in V. Hence we obtain, due to the positivity of the operator T,

$$\begin{aligned} |(Tf_{\nu_k})(x) - (Tf)(x)| &\leqslant |(T(f_{\nu_k} - f)^+)(x)| + |(T(f_{\nu_k} - f)^-)(x)| \\ &= T(f_{\nu_k} - f)^+(x) + T(f_{\nu_k} - f)^-(x) = T(|f_{\nu_k} - f|)(x) \end{aligned}$$

for all $x \in V$. So, $Tf_{\nu_k} \to Tf$ (T,p)-quasieverywhere in V. In addition, $Tf_{\nu_k}(x) \geqslant 1$ for all $x \in E$ for any ν_k. Taking the limit, we obtain $(Tf)(x) \geqslant 1$ (T,p)-quasieverywhere on the set E. The nonnegativity of the function Tf is obvious.

The necessity is proved.

Let $f \geqslant 0$ be a function of the class $L_p(U)$ such that $(Tf)(x) \geqslant 1$ (T,p)-quasieverywhere on the set E. The capacity of the set S of the points from E, for which the inequality $(Tf)(x) \geqslant 1$ does not hold, equals zero. So, according to Theorem 1.1, for any $\nu \in N$ there exists a function $g_\nu \in M_{T,p}(S)$ such that $(Tg_\nu)(x) \geqslant 1$ for all $x \in S$ and $\|g\|_{L_p(U)}^p \leqslant \frac{1}{\nu}$. Let us put $f_\nu = f + g_\nu$. Then, as can easily be seen, $f_\nu \in M_{T,p}(E)$ for every ν and $f_\nu \to f$ in $L_p(U)$ for $\nu \to \infty$. Thus, $f \in M_{T,p}(E)$.

This completes the proof of the theorem.

1.5. Extremal Function for Capacity

For an arbitrary set $E \subset V$ for $p > 1$, the magnitude $(\mathrm{Cap}_{T,p}(E,V))^{1/p}$ equals the distance from the zero point of the space $L_p(U)$ to the convex closed set $\overline{M}_{T,p}(E)$. From the classical results of functional analysis follows the existence of the unique point (the function from $L_p(U)$) which is the nearest one to zero; this is equivalent to the following theorem.

Theorem 1.4. *Let $p > 1$. For every set $E \subset V$ whose capacity is finite, there exists a unique function $f \in \overline{M}_{T,p}(E,V)$ such that*

$$\mathrm{Cap}_{T,p}(E,V) = \int_U |f(x)|^p dx.$$

1.6. Comparison of Various Capacities

Within this section the domain $U \subset \mathbf{R}^n$ is supposed to be fixed, but the positive operators $T : L_p(U) \to F_{T,p}(V)$ vary. By $F_{T,p}(V)$ let us denote an image $T(L_p(V))$. Therefore, we use the shortened terminology: (T,p)-capacity, zero (T,p)-capacity, etc., without indicating the set V to which this capacity is related.

Let $T : L_p(U) \to F_{T,p}(V)$ and $S : L_p(U) \to F_{T,p}(V)$ be two positive operators.

Proposition 1.5. *If for any function $u \in L_p^+(U)$, the inequality $(Tu)(x) \leqslant (Su)(x)$ is valid for all $x \in V$, then*

$$\mathrm{Cap}_{T,p}(E,V) \geqslant \mathrm{Cap}_{S,p}(E,V)$$

for every set $E \subset V$.

The proof follows from the obvious inclusion

$$M_{T,p}(E) \subset M_{S,p}(E).$$

Proposition 1.6. *If $T = \lambda\tau$, then for any set $E \subset V$,*

$$\lambda^p \,\mathrm{Cap}_{T,p}(E,V) = \mathrm{Cap}_{\tau,p}(E,V).$$

Proof. For every function $f \in M_{T,p}(E)$, the function $(1/\lambda)f \in M_{S,p}(E)$, that is, $\lambda^p \,\mathrm{Cap}_{T,p}(E) = \mathrm{Cap}_{S,p}(E)$. We prove the following inequality similarly.

Proposition 1.7. *Let $T : L_p(U) \to F_{T,p}(V)$ and $S : L_p(U) \to F_{S,p}(V)$ be two positive operators for which $F_{T,p}(V) \subset L_p(V)$ and $F_{S,p}(V) \subset L_p(V)$. Suppose that the operator $(T - S) : L_p(U) \to L_p(V)$ is a bounded operator, and that*

$|(T-S)f(x)| \leqslant \|(T-S)\| \, \|f\|_{L_p(V)}$ for almost all $x \in V$. Then, if for the set E, the inequality $\mathrm{Cap}_{T,p}(E,V) \leqslant \|T-S\|^{-p}$ holds,

$$\mathrm{Cap}_{S,p}(E,V) \leqslant \frac{\mathrm{Cap}_{T,p}(E,V)}{(1-\|T-S\| \, \mathrm{Cap}_{T,p}^{1/p}(E,V))^p}.$$

Proof. Let us arbitrarily assign $\varepsilon > 0$ such that $\|T-S\|(C_1+\varepsilon)^{1/p} \leqslant 1$. (To shorten the recording, denote by C_1 the capacity $\mathrm{Cap}_{T,p}(E,V)$.) Let us choose a function $f \in M_{T,p}(E)$ for which $\|f\|_{L_p(U)} < C_1 + \varepsilon$. Then for all $x \in \varepsilon$, we have

$$\begin{aligned}(Sf)(x) = (Tf)(x) - ((T-S)f)(x) &\geqslant 1 - \|T-S\| \, \|f\|_{L_p(U)} \\ &\geqslant 1 - \|T-S\|(C_1+\varepsilon)^{1/p} > 0.\end{aligned}$$

Hence it follows that the function $g = f/a(a = 1-\|T-S\|(C_1+\varepsilon)^{1/p})$ belongs to the set $M_{S,p}(E)$, thus

$$\mathrm{Cap}_{S,p}(E,V) \leqslant \|g\|_{L_p(U)}^p = (1/a)^p \|f\|_{L_p(U)}^p \leqslant \frac{C_1+\varepsilon}{(1-(C_1+\varepsilon)^{1/p}\|T-S\|)^p}.$$

Since $\varepsilon > 0$ is chosen arbitrarily, the proposition is proved.

Corollary 1. *Let the conditions of the previous proposition hold. Then for every $\varepsilon > 0$, there exists $\delta > 0$ such that*

$$(1+\varepsilon)^{-1} \mathrm{Cap}_{T,p}(E,V) \leqslant \mathrm{Cap}_{S,p}(E,V) \leqslant (1+\varepsilon) \mathrm{Cap}_{T,p}(E,V)$$

for any set $E \subset V$ whose (S,p)-capacity is less than δ.

The proof is obtained by applying Proposition 1.7 first to the operator T, then to the operator S.

Remark. Under the conditions of Proposition 1.7 and Corollary 1, the operators T and S are interchangeable since $T-S = S-T$.

Corollary 2. *Let the conditions of Proposition 1.7 be satisfied.*

The (T,p)-capacity of the set E is equal to zero iff its (S,p)-capacity is equal to zero.

1.7. Let us assign an arbitrary locally compact topological space X and some nonnegative measure m defined on the σ-algebra of all Borel sets of the space X. Assume, in addition, that for every point $x \in X$ and for any neighbourhood $U(x)$, $m(U)$ is distinct from zero and there exists a neighbourhood $W(x)$ for which $m(W) < \infty$, $(W(x) \subset U(x))$.

Let us denote by $L_p^+(X)$ a set of nonnegative functions of the class $L_p(X)$. For the positive operators $T : L_p^+(X) \to L_{1,\mathrm{loc}}(X)$ the results of the subsection are valid.

§2 The Classes $W(T,p,V)$

Let us consider a positive operator $T : L_p(U) \to F_{T,p}(V)$.

2.1. Definition of Classes

Let A_p be a kernel of the operator T. The factorspace $L_p(U)/A_p = L_{p,T}(U)$ is a Banach space with respect to the norm:

$$\|\{x\}\|_{L_{p,T}(U)} = \inf_{x\in\{x\}} \|x\|_{L_p(U)},$$

where $\{x\}$ is the equivalence class which is an element of the factorspace $L_{p,T}(U)$. The operator $T_1 : L_{p,T}(U) \to F_{T,p}(V)$ obtained by factorizing the operator T is a one-to-one operator.

In the linear space $F_{T,p}(V)$, let us introduce the norm by means of the operator T_1, by putting

$$\|u\|_{W(T,p,V)} = \|T_1^{-1}u\|_{L_{p,T}(U)}.$$

The space $W(T,p,V)$ is the Banach one, $\{W(T,p,U) = F_{T,p}(U), \|\cdot\|\}$.

Note that $W(I_l,p,\mathbf{R}^n)$ are the Liouville spaces. For the integral l, $W(I_l,p,\mathbf{R}^n)$ coincides with the Sobolev space $W_p^l(\mathbf{R}^n)$, and the norm $\|\cdot\|_{I_l,p,\mathbf{R}^n} \sim \|\cdot\|_{W_p^l(\mathbf{R}^n)}$ (see, for instance, [71]).

2.2. Theorems of Egorov and Luzin for Capacity

Lemma 2.1. *Let the series $\sum_{m=1}^{\infty} v_m$, $v_m \in W(T,p,V)\cap C(V)$ converge absolutely in the space $W(T,p,V)$, with v being its sum.*

Then for every $\varepsilon > 0$, one may find an open set $H \subset V$ such that $\mathrm{Cap}_{T,p}(H,V) < \varepsilon$, and the series $\sum_{m=1}^{\infty} v_m(x)$ uniformly and absolutely converges on $V\backslash H$ to the function v.

Proof. Let m_k be a sequence of natural numbers such that $1 < m_1 < m_2 \ldots$, and for any k the inequality

$$\|v_{m_k+1}\|_{W(T,p,V)} + \|v_{m_k+2}\|_{W(T,p,V)} + \cdots + \|v_{m_{k+1}}\|_{W(T,p,V)} < 2^{-k-(k+2)/p}$$

is valid. By definition of the space $W(T,p,V)$ for every function v_m, there exists a function $u_m \in L_p(U)$ such that

$$Tu_m = v_m, \|u_m\|_{L_p(U)} \leqslant \|v_m\|_{W(T,p,V)} + 1/2^m.$$

First let us assume that all the functions u_m are nonnegative.

Let us put $w_0 = u_1 + u_2 + \cdots + u_{m_1}$, and for $k > 0$, $w_k = u_{m_k+1} + u_{m_k+2} + \cdots + u_{m_{k+1}}$. Then let $a = \|Tw_0\|_{W(T,p,V)}$. Let us arbitrarily assign $\varepsilon > 0$.

Denote by H_0 a set of all $x \in V$ for which $(Tw_0)(x) \geqslant (4/\varepsilon)^{1/pa}$. For $k > 1$, let H_k be the totality of all $x \in V$ for which

$$(Tw_k)(x) > \varepsilon^{-1/p}2^{-k/p}.$$

According to the condition of the lemma, the functions Tw_k are continuous. Consequently, H_k sets are open. Property 4 of (T,p)-capacity allows one to conclude that

$$\mathrm{Cap}_{T,p}(H_k, V) < \varepsilon/2^{k+2}.$$

Let us put $H' = U_{k=0}^{\infty} H_k$. The set H' is open and

$$\mathrm{Cap}_{T,p}(H', V) \leqslant \sum_{k=0}^{\infty} \mathrm{Cap}_{T,p}(H_k, V) \leqslant \sum_{k=0}^{\infty} (\varepsilon/2^{k+2}) = \frac{\varepsilon}{2}.$$

If $x \notin H'$, then $x \in H_k$ for any k, and

$$(Tw_k)(x) \leqslant 2^{-k/p}\varepsilon^{-1/p}.$$

Let us consider the function $F_{m,k} = v_{m+1} + \cdots + v_{m+k}$. If $m_k < m$, and $m + k < m_l$, then for all $x \in V \backslash H$, the inequality

$$\begin{aligned} |F_{m,k}(x)| &= |v_{m_k+1}(x)| + |v_{m_k+2}(x)| + \cdots + |v_{m_l}(x)| \\ &= |Tw_k(x)| + |Tw_{k+1}(x)| + \cdots + |Tw_l(x)| \leqslant \frac{2^{1-k/p}}{\varepsilon^{1/p}} + \frac{1}{2^{m_k}} - 1 \end{aligned}$$

is valid. Thus, for any $m \geqslant m_k$ and for any natural s,

$$|v_{m+1}(x)| + |v_{m+2}(x)| + \cdots + |v_{m+s}(x)| < \varepsilon^{-1/p}2^{1-k/p} + 2^{1-m_k}$$

for all $x \in V \backslash H'$, that is, the series $\sum_{m=1}^{\infty} v_m(x)$ converges uniformly on the set $V \backslash H'$ as well.

Let us look back and consider the functions u_m. The series $\sum_{m=1}^{\infty} u_m$ consisting of nonnegative functions converges everywhere to some function u. Due to the inequality

$$\|u_m\|_{L_p(U)} \leqslant \|v_m\|_{W(T,p,V)} + 1/2^m$$

and to the convergence of the series $\sum_{m=1}^{\infty} v_m$ in $W(T,p,V)$, the function $u \in L_p(U)$. Consequently, by the definition of a positive operator, the series $\sum_{m=1}^{\infty} Tu_m = \sum_{m=1}^{\infty} v_m$ converges everywhere to the function $Tu = v$ which is its sum in the space $W(T,p,V)$, that is, the series $\sum_{m=1}^{\infty} v_m$ converges uniformly on $V \backslash H'$ to the function v.

For nonnegative functions u_m, the lemma is proved.

Now let us consider the case when the functions u_m are arbitrary. Write the function u_m in the form $u_m = u_m^+ - u_m^-$. Due to the above, there exist open sets H^1 and H^2 whose (T,p)-capacities are less than $\varepsilon/2$; outside of them, the series

$$\sum_{m=1}^{\infty}(Tu_m^+)(x) \quad \text{and} \quad \sum_{m=1}^{\infty}(Tu_m^-)(x)$$

converge uniformly. Consequently, the series $\sum_{m=1}^{\infty}(Tu_m)(x)$ converges uniformly and absolutely on the set $V\backslash H(H = H^1 U H^2)$ to the function v, which is the sum with respect to the convergence in $W(T,p,V)$. The capacity of the set $H(H = H^1 \cup H^2)$ is less than ε.

The lemma is proved.

Theorem 2.2 (Luzin's theorem for capacity). *Suppose that T is a C-operator. Let $v \in W(T,p,V)$. Then for every $\varepsilon > 0$, there exists an open set $H \subset V$ such that $\mathrm{Cap}_{T,p}(H\backslash V) < \varepsilon$, and the function is continuous on the set $V\backslash H$.*

Proof. By the definition of the class $W(T,p,V)$, there exists a function $u \in L_p(U)$ such that $Tu = v$. Let us choose a sequence u_m of functions of the class $C_0(U)$ such that $\|u - u_m\| < 2^{-m}$. Let us put $w_m = u_m - u_{m-1}$ for $m > 1$. Then for $m > 1$, $\|w_m\|_{L_p(U)} < 2^{-m+1}$. Hence it follows that

$$\sum_{m=1}^{\infty} \|Tw_m\|_{W(T,p,V)} = \sum_{m=1}^{\infty} \|w_m\| L_p(U).$$

Since T is a C-operator, then each of the functions Tw_m is continuous. According to Lemma 2.1, for any preassigned $\varepsilon > 0$, there exists an open set H of the capacity less than ε, outside of which the series

$$\sum_{m=1}^{\infty}(Tw_m)(x)$$

uniformly converges to the function v. Consequently, the function v is continuous. The theorem is proved.

Theorem 2.3 (Egorov's theorem for capacity). *Suppose that $T : L_p(U) \to W(T,p,V)$ is a C-operator. Let the series $\sum_{m=1}^{\infty} v_m$ of the function $v_m \in W(T,p,V)$ absolutely converge in the space $W(T,p,V)$ and let v be its sum.*

Then for every $\varepsilon > 0$, one may show an open set $H \subset V$ such that $\mathrm{Cap}_{T,p}(H,V) < \varepsilon$, and the series $\sum_{m=1}^{\infty} v_m(x)$ uniformly and absolutely converges on $V\backslash H$ to the function v.

Proof. According to the previous theorem, for each of the functions v_m, there exists an open set H_m such that $\mathrm{Cap}_{T,p}(H_m,V) < \varepsilon/2^m$, and on the set $V\backslash H_m$, the function v_m is continuous. On the set $V_1 = V\backslash H^1(H^1 = \cup_m H_m)$ all the functions v_m are continuous. Due to countable semiadditivity of the capacity,

$$\mathrm{Cap}_{T,p}(H^1,V) < \varepsilon/2.$$

Applying Lemma 2.1 to the series $\sum_{m=1}^{\infty} v_m$, let us construct an open set H^2 such that $\mathrm{Cap}_{T_1,p}(H^2, V_1) < \frac{\varepsilon}{2}$, and on the set $V_2 = V_1 \backslash H^2$, the series $\sum_{m=1}^{\infty} v_m|_{V_1}$ converges uniformly and absolutely to the function v. Here T_1 is a composition of T and of the restriction operator $I : L_{1,\mathrm{loc}}(V) \to L_{1,\mathrm{loc}}(V_1)$. Due to Property 5 of the capacity,

$$\mathrm{Cap}_{T_1,p}(H^2, V_1) \leqslant \mathrm{Cap}_{T,p}(H^2, V).$$

Let us take, as H, the set $H^1 \cup H^2$. It is obvious that the set H is open and $\mathrm{Cap}_{T,p}(H, V) = \mathrm{Cap}_{T,p}(H^1 \cup H^2, V) < \varepsilon$. Outside of the set H, the series $\sum_{m=1}^{\infty} v_m$ converges uniformly and absolutely.

The theorem is proved.

2.3. Dual (T,p)-capacity, $p > 1$. Definition and Basic Properties

Let us consider a set $E \subset V$ and a set $\overline{M}_{T,p}(E)$ of functions admissible for (T,p)-capacity. By the definition of the norm of the functional $u^* \in L_p^*(U)$,

$$\langle u^*, u\rangle \leqslant \|u\|_{L_p(U)} \|u^*\|_{L_p^*(U)} \inf_{u_1 \in \overline{M}_{T,p}(E)} \frac{\langle u^*, u\rangle}{\|u^*\|_{L_p^*(U)}} \leqslant \|u\|_{L_p}(U)$$

for all $u \in M_{T,p}(E)$. If we substitute the extremal function u_0 (extremal for the (T,p) capacity of the set E), instead of an arbitrary function u, in this expression, then we obtain

$$\left(\inf_{u \in \overline{M}_{T,P}(E)} \frac{\langle u^*, u\rangle}{\|u^*\|_{L_p^*(U)}}\right)^p \leqslant \left(\frac{\langle u^*, u_0\rangle}{\|u^*\|_{L_p^*(U)}}\right) \leqslant \|u_0\|^p_{L_p(U)} = \mathrm{Cap}_{T,p}(E, V). \tag{2.1}$$

Hence it follows that

$$\sup_{u^* \in L_p^*(U)} \left[\frac{\inf_{u \in \overline{M}_{T,p}(E)} \langle u^*, u\rangle}{\|u^*\|_{L_p^*(U)}}\right] \leqslant \mathrm{Cap}_{T,p}(E, V). \tag{2.2}$$

Since the set $M_{T,p}(E)$ is dense in $\overline{M}_{T,p}(E)$, then in the expression on the left-hand side, one can only take inf by $M_{T,p}(E)$. The expression on the left-hand side of (2.2) may naturally be called the dual (T,p)-capacity if $\overline{M}_{T,p}(E)$ is substituted for $M_{T,p}(E)$.

The dual (T,p)-capacity is the number

$$\overline{\mathrm{Cap}}_{T,p}(E, V) = \sup_{u^* \in L_p^*(U)} \left[\frac{\inf_{u \in M_{T,p}(E)} \langle u^*, u\rangle}{\|u^*\|_{L_p^*(U)}}\right]^p.$$

For $p > 1$, this expression acquires a more traditional form:

$$\overline{\mathrm{Cap}}_{T,p}(E,V) = \sup_{u^* \in L_q(U)} \left[\frac{\inf_{u \in M_{T,p}(E)} \langle u^*, u \rangle}{\|u^*\|_{L_q(U)}} \right]^p .$$

Let us give an equivalent definition of the dual (T,p)-capacity using a more restricted class of functionals. A functional u^* is said to be admissible for the set E if

$$\inf_{u \in M_{T,p}(E)} \langle u^*, u \rangle \leqslant 1.$$

Then $\overline{\mathrm{Cap}}_{T,p}(E,V) = \sup(\|u^*\|^p_{L^*_p(U)})^{-1}$ where the lowest upper bound is taken by the set of admissible functionals.

From inequality (2.2) follows

Proposition 2.4. *For any set $E \subset V$, the inequality $\overline{\mathrm{Cap}}_{T,p}(E < V) \leqslant \mathrm{Cap}_{T,p}(E,V)$ is valid.*

Proposition 2.5. *For $p > 1$, for any set $E \subset V$ such that $\mathrm{Cap}_{T,p}(E,V) < \infty$, the equality*

$$\overline{\mathrm{Cap}}_{T,p}(E,V) = \mathrm{Cap}_{T,p}(E,V)$$

is valid.

Proof. Let us fix the set $E \subset V$ and consider the extremal function u_0 for the (T,p)-capacity of the set E. By the definition of the extremal function,

$$\int_U u_0^{p-1} u_0 dx = \|u_0\|^p_{L_p(U)} = \mathrm{Cap}_{T,p}(E,V).$$

The function u_0^{p-1} assigns some element from $L_p^*(U)$. Let us use the second variant of the definition of dual (T,p)-capacity. Consider the function $v_0 = u_0^{p-1}/\|u_0\|^p_{L_p(U)}$. It is obvious that

$$\int_U v_0 u_0 dx = 1,$$

therefore, the functional assigned by the function v_0 is admissible for dual (T,p)-capacity.

Let us calculate its norm:

$$\|v_0\|_{L_q(U)} = (\|u_0\|^p_{L_p(U)})^{-1} \|u_0^{p-1}\|_{L_q(U)} = 1/\|u_0\|_{L_p(U)}.$$

Then

$$\overline{\mathrm{Cap}}_{T,p}(E,V) \geqslant \|v_0\|^{-p}_{L_q(U)} = \|u_0\|^p_{L_p(U)} = \mathrm{Cap}_{T,p}(E,V).$$

Comparing it to Proposition 2.4, we have

$$\overline{\mathrm{Cap}}_{T,p}(E,V) = \mathrm{Cap}_{T,p}(E,V).$$

This completes the proof.

Proposition 2.6. *If the sets $E_1 \subset E_2 \subset V$, then*

$$\overline{\mathrm{Cap}}_{T,p}(E_1, V) \leqslant \overline{\mathrm{Cap}}_{T,p}(E_2, V).$$

Proof. Since $M_{T,p}(E_2) \subset M_{T,p}(E_1)$, the set of admissible functionals for E_1 is less than for E_2. Therefore, the lowest upper bound used in the definition of dual (T,p)-capacity will give the greater value for E_2 than for E_1.

The definition of dual (T,p)-capacity is the more convenient the narrower the class of functionals under consideration. Under additional restrictions upon the operator T, this class may be narrowed. The basic tool used to narrow down the class is the investigation of the functional on $W(T,p,V)$ whose image is the functional generated by the function v_0 considered in the proof of Proposition 2.5.

2.4. Calculation of Dual (T,p)-Capacity

We may calculate dual (T,p)-capacity by means of a special class of measures only under additional assumptions upon the operator T. Except for the calculation of dual (T,p)-capacity, we do not use these assumptions elsewhere in the following.

A positive operator $T : L_p(U) \to W(T,p,V)$ is called a C_0-operator if it transforms nonnegative compactly supported in U functions into continuous functions nowhere vanishing.

A positive operator $T : L_p(U) \to W(T,p,V)$ is called lower p-semicontinuous if it transforms the class L_p^* into the class of lower semicontinuous functions.

Proposition 2.7. *If the function $K : U \times U \to \overline{\mathbf{R}}$ satisfies the conditions of Example 2 from Section 1.1, then the positive operator $T_{K,p} : L_p(U) \to L_1(U)$ generated by this function according to the rule*

$$T_{K,p}(u) = \int_U K(x,y)u(y)dy$$

is a C_0-operator lower p-semicontinuous.

Proof. From the Property K.3 of Example 2, it immediately follows that the operator $T_{K,p}$ is a C_0-operator.

Let us prove that the operator $T_{K,p}$ is a p-semicontinuous operator. Let us assign an arbitrary point $x_0 \in U$, and let $x_m, m = 1, 2, \ldots,$ be an arbitrary sequence of points in U, converging to x. Then for every point $y \neq x_m$,

$$K(x_m, y)f(y) \to K(x_0, y)f(y),$$

i.e.,

$$K(x_m, y)f(y) \to K(x_0)f(y)$$

for almost all y. Due to nonnegativity of the functions K and f, on the basis of the Fatou theorem, we obtain

$$\int_U K(x_0,y)f(y)dy \leqslant \lim_{m\to\infty} \int_U K(x,y)f(y)dy,$$

that is,

$$(T_{K,p}f)(x_0) \leqslant \lim_{m\to\infty}(T_{K,p}f)(x_m).$$

The proposition is proved.

Corollary. *If the function $u \in L_p^*(U)$, then its Bessel potential is a lower p-semicontinuous function.*

As was shown in Section 1.1, the Bessel kernel is a special case of the kernel from Example 2.

Lemma 2.8. *Let E be an arbitrary compact subset in V, and let u_0 be an extremal function for the (T,p)-capacity of E. Then the set $A = \{x \in E : (Tu_0)(x) \leqslant 1\}$ is nonempty.*

Proof by Contradiction. If A is empty, then $1 < (Tu_0)(x)$ for all $x \in E$. Since the function Tu_0 is lower semicontinuous and the set E is compact, the function Tu_0 acquires on E the smallest value at some point $x_0 \in E$, i.e., there exists $\delta > 0$ such that $(Tu_0)(x_0) = 1 + \delta$. The function $u_1 = u_0/1+\delta)$ is admissible for the set E. At the same time, $\|u_0\|_{L_p} > \|u_1\|_{L_p}$, which contradicts the extremality of the function u_o. The obtained contradiction proves that the set A is nonempty.

Theorem 2.9. *Let E be a compact subset of V, and let u_0 be an extremal function for the (T,p)-capacity of the set $E(p>1)$.*

Then for any continuous function $\varphi \in L_p(U)$ such that $T\varphi \equiv 0$ on E, the equality

$$\int_U u_0^{p-1}\varphi dx = 0$$

is valid.

Proof. Let $A = \{x \in E : Tu_0(x) = 1\}$. Since the operator T is lower p-semicontinuous, then the function u_0 is lower semicontinuous and the set A is compact. According to Lemma 2.8, this set is nonempty. Denote by $S(A)$ the totality of all functions $g \in L_p(U)$ such that the function Tg is continuous and $(Tg)(x) = 0$ for all $x \in A$. Let us prove that $0 \leqslant \int_U v_0(x)g(x)dx$ for every function $g \in S(A)$. Here $v_0(x) = u_0^{p-1}(x)$. First suppose that the function $g \in S(A)$ is such that $(Tg)(x) > 0$ for all $x \in A$. Then there exists an open set $U \subset A$ such that $(Tg)(x) > 0$ for all x from U. Let us put $H = E \backslash U$. The set H is compact, and $(Tu_0)(x) \geqslant 1$ for all $x \in H$. Due to the fact that the operator T is lower p-semicontinuous, the function Tu_0

is also lower semicontinuous, i.e., there exists $\delta > 0$ such that for all $x \in H$, $(Tu_0)(x) \geqslant 1+\delta$. Let us prove that there exists $t_0 > 0$ such that for $0 < t < t_0$,

$$(Tu_0)(x) + t(Tg)(x) \geqslant 1$$

(T,p)-quasieverywhere on $E \cap U$. Indeed, for $x \in E \cap U$ for all $t > 0$,

$$(Tu_0)(x) + (t(Tg)(x)) \geqslant (Tu_0)(x)$$

(T,p)-quasieverywhere on $E \cap U$. Let $x \in H = A \backslash U$. Let us put $M = \max |(Tg)(x)|$. Then for $0 < t < \delta/M$,

$$(T(u_0 + tg))(x) > 1$$

for all $x \in H$. Hence it follows that for $0 < t < t_0 = \delta/M$,

$$(T(u_0 + tg))(x) \geqslant 1$$

(T,p)-quasieverywhere on E. Let $0 < t < t_0$. Due to the positiveness of the operator T, we have $(T(u_0 + tg))(x) = (T(u_0 + tg))(x)$ for all x. Hence it follows that $|u_0 + tg| \in M_{T,p}(E)$ $(0 < t < t_0)$. Since the function u_0 is extremal for (T,p)-capacity of the set E, then for every $t \in (0, \delta/M)$,

$$\int_U |u_0(x) + tg(x)|^p dx \geqslant \int_U |u_0(x)|^p dx.$$

Hence we have

$$\int_U |u_0(x)|^{p-1} g(x) dx = \lim_{\substack{t \to 0 \\ t > 0}} \int_U \frac{|u_0(x) + tg(x)|^p - |u_0(x)|^p}{t} dx \geqslant 0,$$

and consequently,

$$\int_U v_0(x) g(x) dx \geqslant 0. \tag{2.3}$$

Inequality (2.3) is proved under the assumption that $(Tg)(x) > 0$ for all $x \in A$. Let $g \in S(A)$ be such that $(tg)(x) > 0$ for all $x \in A$. Let $g \in S(A)$ be such that $(Tg)(x) \geqslant 0$ on A. Let us assign an arbitrary nonnegative continuous compactly supported function α which does not identically equal zero. Since T is a C_0-operator, the function $(T\alpha)(x) > 0$ for all $x \in U$ and $T\alpha$ is continuous. The function $g + \tau\alpha \in S(A)$ for any $\tau > 0$ and $(T(g + \tau\alpha))(x) > 0$ for all $x \in E$. Consequently, due to what was proved above,

$$\int_U v_0(x)[g(x) + \tau\alpha(x)]\, dx = 0$$

for any $\tau > 0$. Directing τ to zero, we obtain in the limit:

$$\int_U v_0(x)g(x)dx \geqslant 0. \tag{2.4}$$

Q. E. D.

Now let the function $\varphi \in L_p(U)$ be such that $T\varphi$ is continuous and $(T\varphi)(x) = 0$ for all $x \in E$. Then

$$\int_U v_0(x)\varphi(x)dx = 0.$$

Indeed, the functions φ, $-\varphi$ belong to the class $S(E)$; consequently,

$$\int_U v_0(x)\varphi(x)dx \geqslant 0, \qquad \int_U v_0(x)(-\varphi(x))dx \geqslant 0, \tag{2.5}$$

whence it follows that the integral (2.5) vanishes. The theorem is proved.

Corollary 1. *Let E be a compact set in U, and let u be the extremal function for the (T,p)-capacity of the set E $(p > 1)$.*

If the two functions $g, g_1 \in L_p(U)$ are such that Tg and Tg_1 are continuous, and $Tg \equiv Tg_1$ on E, then

$$\int_U v_0(x)g(x)dx = \int_U v_0(x)g_1(x)dx.$$

Hence it follows that the functional $\int_U v_0(x)g(x)dx$ does not depend on continuous extension of the functions Tg from the compact set E. It is natural that such functionals are said to be concentrated on E.

Corollary 2. *Under the assumptions for the positive operator made at the beginning of this subsection, the equality*

$$\overline{\mathrm{Cap}}_{T,p}(E,U) = \sup \frac{1}{\|u^*\|^p_{L_p(U)}}$$

is valid, where the supremum is taken by all functionals u^* concentrated on E and satisfying the condition

$$\inf_{u \in M_{T,p}(E)} \langle u^*, u\rangle \leqslant 1.$$

Proof. Since the functional generated by the function v_0 is concentrated, the supremum can be taken just by this class.

Let us recall that to every functional $u^* \in L_q(U)$ there corresponds the functional (not necessarily the unique one) $v^* \in W^*(T,p)$ such that $T^*v^* = u^*$. Then $\langle u^*, u\rangle = \langle T^*v^*, u\rangle = \langle v^*, Tu\rangle$, and for dual capacity of compact sets in the case of a one-to-one operator, there is another description

$$\overline{\mathrm{Cap}}_{T,p}(E,V) = \sup(\|v^*\|_{W^*(T,p,V)})^{-1},$$

where the lowest upper bound is taken by all functionals v^* such that

1) $\inf_{u\in M_{T,p}(E)} \langle v^*, Tu\rangle \leqslant 1$;

2) for every function u such that Tu is continuous, the value of the functional $\langle v^*, Tu\rangle$ does not depend on the continuous extension of the function Tu/E from the compact set E.

Under the assumption that all the functionals from $W^*(T,p,E)$ are measures, one may try to simplify the calculation of dual capacity.

Let us show that in the situation under consideration, the functional extremal for the dual capacity of the compact set E generated by the function $|u_0|^{p-1}/\|u_0\|^p_{L_p(U)}$ (where u_0 is the extremal function for $\mathrm{Cap}_{T,p}(E,V)$) is the image of some measure at the mapping T^*.

Theorem 2.10. *Let E be a compact subset of U, and let u_0 be the extremal function for the (T,p)-capacity of the set $E, (p>1)$.*

Then there exists a measure α concentrated on the set E such that

$$\frac{1}{\|u_0\|^p_{L_p(U)}} \int_U |u_0|^{p-1} u\,dx = \int_U (Tu)(x)\,d\alpha(x)$$

for all $u \in L_p(U)$. Here

$$\int_U d\alpha(x) = \|u_0\|^p_{L_p(U)}.$$

Remark. If we use the shortened notation, denoting a function and the functional generated by it by the same symbol, then the formulation of the theorem is as follows:

$$\langle \frac{u_0^{p-1}}{\|u_0\|^p_{L_p(U)}} u, \mu\rangle = \langle \alpha, Tu\rangle,$$

where μ is the Lebesgue measure, and

$$|u_0|^{p-1}/\|u_0\|^p_{L_p(U)} = T^*\alpha.$$

Proof. If $\mathrm{Cap}_{T,p}(E,V) = 0$, then $Tu_0 = 0$ quasieverywhere, and the measure $\alpha(E) = 0$ for any compact $E \subset V$ (that is, identically equal to zero) is suitable for the equality given in the formulation.

Let $\mathrm{Cap}_{T,p}(E, V) > 0$. Let us consider a Banach space of continuous real functions $C(E)$ defined on E, $C^+(E)$ is the totality of all functions of $C(E)$, with each of them being positive everywhere on E. The set $C^+(E)$ is convex and open in $C(E)$.

Let us denote by H' the totality of all functions $v \in C(E)$ admitting the representation $v = Tu$, where the function $u \in L_p$ is such that

$$\int_U u(x)(u_0(x))^{p-1}dx = 0,$$

and denote by H the closure of H' in $C(E)$. The set H is a linear subspace of $C(E)$. Let us prove that H does not contain elements of the set $C^+(E)$. Suppose that there exists a function $v_1 \in C^+(E) \cap H$. The set $C^+(E)$ is open, so there exists the function $v_2 \subset C^+(E) \cap H'$, i.e., $v_2 = Tu_2$ and $v_2(x) > 0$ for all $x \in E$. Let us take an arbitrary continuous compactly supported function $\varphi : U \to \mathbf{R}$. Since T is a C_0-operator, $T\varphi$ is continuous and nonnegative. Thus, there exists $\delta > 0$ such that the function $T(u_2 - \delta\varphi) = v_2 - \delta T\varphi$ is nonnegative on the set E. So, due to (2.5),

$$\int_U (u_2(x) - \delta\varphi(x))(u_0(x))^{p-1}dx \geqslant 0,$$

but, due to the fact that $v_2 \subset H'$,

$$\int_U u_2(x)(u_0(x))^{p-1}dx = 0.$$

Consequently, for any nonnegative continuous compactly supported function,

$$\int \varphi(x)(u_0(x))^{p-1}dx \leqslant 0.$$

This allows us to conclude that $u_0(x) = 0$ almost everywhere, i.e., $0 = \mathrm{Cap}_{T,p}(E, V)$, contrary to the assumption that $\mathrm{Cap}_{T<p}(E, V) > 0$. We reached the contradiction showing that H and $C^+(E)$ do not intersect.

According to the Hahn–Banach theorem, there exists a continuous linear functional $l : C(E) \to \mathbf{R}$ such that $l(v) = 0$ for all $v \in H$ and $l(u) > 0$ for some function $u \in C^+(E)$. According to the Riesz theorem about the representation of a linear functional in $C(E)$, there exists a p ve measure concentrated on E for which the equality

$$l(v) = \int_E v(x)d\alpha(x)$$

is valid for all $v \in C(E)$. The measure α is considered to be extended to the entire space U, setting $\alpha(P) = \alpha(P \cap E)$ for an arbitrary Borel set P.

Let us consider on $C_0(U)$ linear functionals,

$$l_1(u) = \int_U u(x)(u_0(x))^{p-1} dx,$$

$$l_2(u) = \int_U (Tu)(x) d\alpha(x).$$

Let K be an arbitrary compact set in U, $g \in C_0(K)$. Then

$$|l_1(g)| \leqslant \|g\|_{C(U)} \qquad \int_K g(x)|u_0(x)|^{p-1}\, dx \leqslant M\|g\|_{C(U)},$$

i.e., the functional l_1 is continuous on $C_0(K)$, which is considered as a subspace of $C(U)$. Since T is a C_0-operator, then for any function $u \in C_0(K)$, the function Tu is continuous on U. Therefore,

$$|l_2(u)| = |\int_U (Tu)(x) d\alpha(x)| = |l(Tu)| \leqslant \|l\|_{C^*(E)} \|Tu\|_{C(E)},$$

i.e., the functional l_2 is continuous on $C_0(K)$.

Let us prove that the functionals $l_1 : C_0(K) \to \mathbf{R}$, $l_2 : C_0(K) \to \mathbf{R}$ have the same kernel. If $l_1(u) = 0$, then the function Tu is continuous and, by the definition of the set H, the function $Tu \in H'$. By the construction of the functional l_2, we have $l_2(u) = l(Tu) = 0$. Thus, the kernels of the functionals l_1 and l_2 coincide, that is, there exists a constant ζ such that $l_2 = \zeta l_1$, which, in its turn, results in the relations

$$l_2(u) = \zeta l_1(u) = \int_U (Tu)(x) d\alpha(x).$$

We obtain the necessary representation of the integral

$$\int_U (u_0(x))^{p-1} u(x) dx.$$

Let us calculate the integral of the measure $\zeta\alpha$ by the space U. Remember that

$$\mathrm{Cap}_{T,p}(E, U) = \int_U (u_0(x))^{p-1} u_0(x) dx = \int_U (Tu_0)(x) d\zeta\alpha(x).$$

The function $Tu_0(x) \geqslant 1$ (T,p)-quasieverywhere on E. The set $A = \{x : Tu_0(x) = 1\}$ has zero (T,p)-capacity. According to Theorem 1.1, for the entire $m \in N$, there exists a function $g_m \in M_{T,p}(A)$, $\|g_m\|_{L_p(U)} \leqslant m^{-1}$ such that $(Tg_m)(x) \geqslant 1$ for all $x \in A$. Let us consider the functions $u_m = u_0 + g_m$. It is obvious that $\|u_m - u_0\|_{L_p(U)} \to 0$. Due to continuity of the functional l_1 on $L_p(U)$, we obtain

$$\mathrm{Cap}_{T,p}(E,V) = \lim_{m\to\infty} \int\limits_U (u_0(x))^{p-1} u_m(x)dx = \lim_{m\to\infty} \int\limits_U (Tu_m)(x)d\zeta\alpha(x).$$

Since the function $(Tu_m)(x) \geqslant 1$ for all $x \in A$,

$$\int\limits_U (Tu_m)(x)d\alpha(x) = \int\limits_U d\alpha(x)$$

for all $m \in N$. Hence, we directly obtain

$$\mathrm{Cap}_{T,p}(E,V) = \|u_0\|^p_{L_p(U)} = \int\limits_U d\zeta\alpha(x).$$

This completes the proof of the theorem.

Corollary. *Let E be a compact subset of V. Then the relation*

$$\overline{\mathrm{Cap}}_{T,p}(E,V) = \sup(\|T^*\beta\|^p_{L_p(U)})^{-1}$$

is valid, where the lowest upper bound is taken by all possible measures $\beta \in W^*(T,p,V)$ concentrated on E and satisfying the condition $\beta(U) = 1$.

Proof. If β is the measure of the class under consideration, then for any function $u \in M_{T,p}(E)$, the inequalities

$$\|u\|_{L_p(U)} \geqslant (\|T^*\beta\|_{L_q(U)})^{-1} \int u(x)T^*\beta dx = (\|T^*\beta\|_{L_q(U)})^{-1} \int (Tu)(x)d\beta$$

are valid. Since the measure β is concentrated on E, and $(Tu)(x) \geqslant 1$ for all $x \in E$, so we obtain

$$\|u\|_{L_p(U)} = (\|T^*\beta\|_{L_q(U)})^{-1} \int d\alpha(x) = (\|T^*\beta\|_{L_q(U)})^{-1}.$$

Hence, due to Proposition 2.5, there follows

$$\overline{Cap}_{T,p}(E,V) = \mathrm{Cap}_{T,p}(E,V) = \sup(\|T^*\beta\|^p_{L_p(U)})^{-1},$$

where the lowest upper bound is taken by the class of measures in question. On the other hand, from the previous theorem, there follows the existence of the measure α for which $\mathrm{Cap}_{T,p}(E,V) = \|T^*\alpha\|^{-p}_{L_q(U)}$. This completes the proof.

Remark. The requirement $\beta \in W^*(T,p,V)$ may be removed, assuming that the norm $\|T^*\beta\|_{L_p(U)}$ of the measure assigning the discontinuous functional on $W(T,p,V)$ is equal to ∞. This convention is justified, since for any measure β, the norm $\|T^*\beta\|_{L_q(U)}$ may be defined by only using the cone of positive functions in $W(T,p,V)$.

Corollary 2 of Theorem 2.9 and Corollary of Theorem 2.10 are only proved for compact subsets of V. Since T is a C_0-operator, then, as follows from the results of Section 3 (this chapter), a capacity $\mathrm{Cap}_{T,p}$ is a generalized capacity. Remember that the set E is called measurable with respect to the generalized capacity $\mathrm{Cap}_{T,p}$ if $\mathrm{Cap}_{T,p}(E,V) = \sup \mathrm{Cap}_{T,p}(K,V)$, where the lowest upper bound is taken by all possible compact sets $K \subset E$.

Proposition 2.11. *For every set E measurable with respect to (T,p)-capacity $(p > 1)$, the equality*

$$\overline{\mathrm{Cap}}_{T,p}(E,V) = \sup(\|u^*\|^p_{L_q(U)})^{-1}$$

is valid, where the lowest upper bound is taken by all functionals concentrated on E which satisfy the condition

$$\inf_{u \in M(T,p,E)} \langle u^*, u\rangle = 1.$$

This proposition is the immediate corollary of Theorem 2.9 and of the following proposition.

Proposition 2.12. *For any set E measurable with respect to (T,p)-capacity $(p > 1)$, the equality*

$$\overline{\mathrm{Cap}}_{T,p}(E,V) = \sup(\|u^*\|^p_{L_q(U)})^{-1}$$

is valid, where the lowest upper bound is taken by all possible measures $\beta \in W^*(T,p,V)$ satisfying the conditions: the measure β is concentrated on E, and $\beta(V) \leqslant 1$.

Proof. Let us denote the family of measures under consideration by $N(T,p,E)$. If $E_1 \subset E_2$, then it is obvious that $N(T,p,E_1) \subset N(T,p,E_2)$. Hence the inequality

$$\sup_{N(T,p,E_1)} \|T^*\beta\|^{-p}_{L_q(U)} \geqslant \sup_{N(T,p,E_2)} \|T^*\beta\|^{-p}_{L_q(U)}$$

follows.

Let E be a positive set measurable with respect to (T,p)-capacity. Then for any compact set $K \subset E$ from Proposition 2.6 and from the previous remark, we obtain

$$\overline{\mathrm{Cap}}_{T,p}(E,V) = \mathrm{Cap}_{T,p}(E,V) \geqslant \sup_{N(T,p,E)} \|T^*\beta\|_{L_q(U)}^{-p} \geqslant \sup_{N(T,p,K)} \|T^*\beta\|_{L_q(U)}^{-p} = \overline{\mathrm{Cap}}_{T,p}(K,V).$$

Since the set E is measurable with respect to the (T,p)-capacity, then

$$\mathrm{Cap}_{T,p}(E,V) = \sup_{K\subset E} \mathrm{Cap}_{T,p}(K,V).$$

Hence, and from the previous chain of sets, it follows that

$$\mathrm{Cap}_{T,p}(E,V) = \sup_{N(T,p,E)} \|T^*\beta\|_{L_q(U)}^{-p} \geqslant \overline{\mathrm{Cap}}_{T,p}(E,V).$$

This completes the proof.

Remark. As it is seen from the proof of Theorem 2.9, the class $N(T,p,E)$ may still be narrowed if one considers the measures β only, concentrated on the set $A \subset E$ on which the extremal function does not exceed 1. Denote this class by $N_0(T,p,E)$. Every measure $\beta \in N_0(T,p,E)$ satisfies the condition

$$\inf_{u\in M_{T,p}(E)} \langle T^*\beta, u\rangle \leqslant 1.$$

Indeed, $\beta(A) = \beta(V) \leqslant 1$, the measure β is concentrated on A, and the function $u_0 \in M_{T,p(E)}$ does not exceed 1 on A. Thus,

$$\langle T^*\beta, u_0\rangle = \langle \beta, Tu_0\rangle = \int_U (Tu_0)(x)d_\beta(x) \leqslant \beta(V) \leqslant 1.$$

Theorem 2.10 remains valid for the class $N_0(T,p,E)$ as well. Since the class $N_0(T,p,E)$ belongs to the class of functionals admissible in Proposition 2.11, whence the statement of this proposition follows.

In conclusion, let us give more variants of calculating dual (T,p)-capacity which directly follow from Proposition 2.12.

Let $L(T,p,E)$ be a family of all measures β concentrated on the set E and belonging to the class $W^*(T,p,E)$ and such that $\|T^*\beta\|_{L_q(V)} \leqslant 1$, and $L_0(T,p,E)$ is a family of all measures $\beta \in W^*(T,p,E)$ concentrated on the set E.

Proposition 2.13. *For any set E measurable with respect to (T,p)-capacity $(p>1)$, the equalities*

$$\overline{\mathrm{Cap}}_{T,p}(E,V) = \sup_{L_0(T,p,E)} \frac{|\beta(V)|^p}{\|T^*_\beta\|_{L_q(U)}^p} = \sup_{L(T,p,E)} |\beta(V)|^p$$

are valid.

§3 Sets Measurable with Respect to Capacity

For a wide class of spaces under additional assumptions easily performed on the operator T, the (T,p)-capacity in question is a special case of the notion of generalized capacity [12].

3.1. Definition and the Simplest Properties of Generalized Capacity

Let X be a locally compact topological space, and let φ be a nonnegative real function with values in $\overline{\mathbf{R}}$, defined on the class of all subsets of the set X. The function is called a generalized capacity if it satisfies the following conditions:

1. Monotonicity. If $A \subset A'$, then $\varphi(A) \leqslant \varphi(A')$.
2. For every increasing sequence of sets $\{A_m \subset X\}, m = 1, 2, \ldots,$

$$\varphi(\cup_m A_m) = \sup_m \varphi(A_m).$$

3. For every decreasing sequence of compact sets (A_m), $m = 1, 2, \ldots,$

$$\varphi(\cap A_m) = \inf_m \varphi(A_m).$$

The set $A \subset X$ is called measurable with respect to the generalized capacity φ if the number $\varphi(A)$ is the lowest upper bound of values of φ on all possible compact subsets K of the set A.

Let X be a separable topological space. The set $A \subset X$ is called a set of the type K_σ if A is a countable union of compact sets. The set A is a set of the type $K_{\sigma,\delta}$ if A is a countable intersection of the sets of the K type. The set $A \subset X$ is called a K-analytical set if A is a continuous image of sets of the type $K_{\sigma,\delta}$ lying in some compact set in X.

Theorem 3.1 [12]. *Let φ be a generalized capacity on a locally compact topological space X. Then every K-analytical set in X is measurable with respect to generalized capacity.*

Remark. Any Borel set is measurable with respect to generalized capacity since it is K-analytical.

3.2. (T,p)-Capacity as Generalized Capacity

Proposition 3.2. *Let $T : L_p(U) \to W(T,p,V)$ be a C-operator. Let us consider a set $E \subset V$ having finite (T,p)-capacity. For every $\varepsilon > 0$, there exists an open set $H \supset E$ such that*

$$\mathrm{Cap}_{T,p}(H,V) < \mathrm{Cap}_{T,p}(E,V) + \varepsilon.$$

Proof. Let us choose an admissible function $f \in M_{T,p}(E)$ such that $\mathrm{Cap}_{T,p}(E,V) > \|f\|^p_{L_p(U)} - \varepsilon/3$. According to the Egorov-type theorem (Theorem 2.3) for capacity, there exists an open set $H_1 \subset V$ such that $\mathrm{Cap}_{T,p}(H_1, V) < \varepsilon/3$, and on the set $V \backslash H_1$ the function Tf is continuous. The set $G_h = \{x \in V\backslash H_1 : Tf(x) > 1 - h\}, h \in (0,1)\}$, is open. Let us put $g = (1-h)^{-1} f$. Then $Tg(x) \geqslant 1$ for all $x \in G_h$, therefore,

$$\begin{aligned}\mathrm{Cap}_{T,p}(G_h, V) \leqslant \|g\|^p_{L_p(U)} &= (1-h)^{-p}\|f\|^p_{L_p(U)} \\ &\leqslant (\mathrm{Cap}_{T,p}(E,V) + \varepsilon/3)(1-h)^{-p}.\end{aligned}$$

Let us choose $h \in (0,1)$ so that the right-hand side of the inequality is less than $\mathrm{Cap}_{T,p}(E,V) + (2/3)\varepsilon$.

Let us take, as H, an open set $G_h \cup H_1$. It is obvious that $H \supset E$. From the properties of the (T,p)-capacity and from the above, it follows that

$$\begin{aligned}\mathrm{Cap}_{T,p}(H,V) = \mathrm{Cap}_{T,p}(G_h \cap H_1, V) &\leqslant \mathrm{Cap}_{T,p}(G_h, V) + \mathrm{Cap}_{T,p}(H_1, V) \\ &\leqslant \mathrm{Cap}_{T,p}(E,V) + \varepsilon.\end{aligned}$$

The proposition is proved.

Remark. As can easily be seen from the proof of Egorov's theorem, it remains valid not only for C-operators, but for C_0-operators also. Therefore, Proposition 3.2 remains valid for C_0-operators.

Theorem 3.3. *If $T : L_p(U) \to W(T,p,V)(p > 1)$ is a C-operator and (T,p)-capacity of any compact set is finite, then (T,p)-capacity is a generalized capacity.*

Proof. Condition 1 from the definition of generalized capacity holds due to monotonicity of (T,p)-capacity. Condition 3 follows from Proposition 3.2. Condition 2 alone needs to be verified.

Let $\{A_m \subset V\}$ be an arbitrary increasing sequence of sets, $A = \cap_{m=1} A_m$. Due to monotonicity of (T,p)-capacity,

$$\mathrm{Cap}_{T,p}(A,V) \geqslant \sup_m \mathrm{Cap}_{T,p}(A_m, V). \tag{3.1}$$

If $S = \sup_m \mathrm{Cap}_{T,p}(A_m, V) = \infty$, then Condition 2 holds. Suppose that supremum is finite. Then each of the sets $M_m = \overline{M}_{T,p}(A_m, V)$, $(M_{m+1} \subset M_m)$ is convex, nonempty and closed.

Let us fix $\varepsilon > 0$. Let us consider in the space $L_p(U)$ a closed ball $B_\varepsilon = B(0, S+\varepsilon)$. The intersections $K_m = M_m \cap B_\varepsilon$ form a monotonically decreasing sequence of bounded closed convex sets in $L_p(U)$. Due to reflectivity of $L_p(U)$, each of these sets is weakly compact. Consequently, their intersection $M = \cap_{m=1}^{\infty} K_m$ is nonempty. Any function $u_0 \in M$ belongs to $K_m \subset M_m$ for all m, i.e., $(Tu_0)(x) \geqslant 1$ (T,p)-quasieverywhere on A_m. So, $(Tu_0)(x) \geqslant 1$ (T,p)-

quasieverywhere on $A = \cup_m A_m$ and $u_0 \in \overline{M}_{T,p}(A_m)$ (due to Theorem 1.3). Therefore,

$$\mathrm{Cap}_{T,p}(A, V) \leqslant \|u_0\|^p_{L_p(U)} \leqslant S + \varepsilon = \sup_m \mathrm{Cap}_{T,p}(A_m, V) + \varepsilon.$$

Since ε was chosen arbitrarily, and due to arbitrariness of inequality (3.1), Condition 2 of the definition of generalized capacity is fulfilled.

The theorem is proved.

Let us formulate the conditions under which $\mathrm{Cap}_{T,p}(E, V) < \infty$ for any compact set E.

A positive operator $T : L_p(U) \to W(T, p, V)$ is said to be separating if for any compact set $E \subset V$, the image $T(M_{T,p}(E))$ of the set of admissible functions for E contains at least one continuous positive function.

Every C_0-operator is a separating one. Indeed, the image Tg of every function $g \in C_0(U)$ compactly supported on U is a continuous positive function according to the definition of a C_0-operator. Consequently, on any compact set $E \subset V$, the function $(Tg)(x) > 0$ for all $x \in E$, i.e., the function $g_1 = g/\min_{x\in E}(Tg)(x)$ is admissible for E and $\mathrm{Cap}_{T,p}(E, V) \leqslant \|g_1\|^p_{L_p(U)}$. Therefore, the operators of Examples 1–3 from Section 1 are separating, since they are C_0-operators.

Proposition 3.4. *If a set $E \subset V$ is compact, and an operator $T : L_p(U) \to W(T, p, V)$ is a separating one, then*

$$\mathrm{Cap}_{T,p}(E, V) < \infty.$$

The proof is obvious.

Taking Proposition 3.4 into account, we obtain the obvious corollary of Theorem 3.3.

Theorem 3.5. *If $T : L_p(U) \to W(T, p, V) (p > 1)$ is a separating C-operator, then the (T, p)-capacity is generalized capacity.*

Taking into account the remark to Proposition 3.2 and Theorem 3.5, we immediately obtain

Theorem 3.6. *If $T : L_p(U) \to W(T, p, V) (p > 1)$ is a C_0-operator, then the (T, p)-capacity is generalized capacity.*

§4 Variational Capacity

The concept of variational capacity connected with positive operator implies that the space under consideration was obtained from L_p by means of a special type of operator. But for the spaces $W^l_p(U)$ (U is a domain in $\mathbf{R}^n$) of such a type, the representations are only known under special restrictions upon the domain. On the other hand, locally the functions of the class $W_p^l(U)$ are the restrictions of functions from $W_p^l(\mathbf{R}^n)$. This means that for the Sobolev classes in domains of Euclidean space, the application of (J_l, p)-capacity is bounded by local properties of functions of these classes. Due to this, let us introduce the notion of variational capacity which is more appropriate for the study of boundary behaviour of spaces of functions with "generalized smoothness." For the spaces $W_p^l(U)$, this notion (in a somewhat different technical variant) was considered in the paper of Reshetnyak [66], and as "conductivity," it was studied in the papers by Maz'ya and Khavin [47, 49].

4.1. Definition of Variational Capacity

We shall consider linear subspaces $F(G)$ of a space of measurable functions $M(G)$ which are defined in the domain $G \subset \mathbf{R}^n$ and at the same time are seminormed spaces, i.e., it is supposed that for every function $u \in F(G)$ some semi-norm $\|u\|_{F(G)}$ is finite. The space $F(G)$ is supposed to be complete. For this class of spaces we shall use the term "a seminormed space of functions."

4.1.1. Definition of Variational F-Capacity in a Domain G

Let $F(G)$ be a seminormed space of functions. Let us consider a pair of sets $(F_0, F_1) \subset G$. The function $u \subset F(G)$ is called F-admissible for the pair (F_0, F_1) in the domain G if $u(x) \geqslant 1$ for all x belonging to some neighbourhood of the set F_1, $u(x) = 0$ for all x belonging to some neighbourhood of the set F_0. The neighbourhoods of sets in G are meant, i.e., such open sets in G whose closure contains F_0 or F_1, respectively.

The number $\mathrm{Cap}_F(F_0, F_1, G)$ equal to $\inf \|u\|_{F(G)}$, where the greatest lower bound is taken by all possible functions admissible in the domain G for the pair (F_0, F_1), is called the variational F-capacity of the pair (F_0, F_1).

If the admissible functions for the pair (F_0, F_1) do not exist, we suppose $\mathrm{Cap}_F(F_0, F_1, G) = \infty$.

We shall often use just "F-capacity" instead of "variational F-capacity of the pair" when ambiguity is impossible.

Let us give some simple properties of F-capacity. The set of F-admissible functions for the pair (F_0, F_1) we denote by $M_F(F_0, F_1, G)$.

Property 1. Monotonicity relative to a pair. For pairs of sets $(F_0, F_1) \subset G$ and $(F_0', F_1') \subset G$ from the inclusions $F_0' \subset F_0$, $F_1' \subset F_1$, the inequality

$$\mathrm{Cap}_F(F_0', F_1', G) \leqslant \mathrm{Cap}_F(F_0, F_1, G)$$

follows.

This inequality is the corollary of the definition of capacity since

$$M_F(F_0, F_1, G) \subset M(F_0', F_1', G).$$

Property 2. Let G, U be domains in $\mathbf{R}^n$, $G \subset U$,

$$F_1(G) \subset F(U)|_G, \|u|_G\|_{F_1(G)} \leqslant K\|u\|_{F(U)}$$

for all $u \in F(U)$.

Then for any pair $(F_0, F_1) \subset G$, the inequality

$$\mathrm{Cap}_{F_1}(F_0, F_1, G) \leqslant K\,\mathrm{Cap}_F(F_0, F_1, U)$$

is valid.

Proof is obvious and follows from the definition of capacity.

Let $H_0 = \{u \in F(G) : \|u\|_{F(G)} = 0\}$, then the space $F(G)/H_0$ is the Banach space relative to the norm $\|\{u\}\|_{F(G)/H_0} = \|u\|_{F(G)}$, where u is any element of the equivalence class $\{u\}$.

Property 3. If the factorspace $F(G)/H_0$ is uniformly convex, then for any pair $(F_0, F_1) \subset G$ for which $\mathrm{Cap}_{F(G)}(F_0, F_1) < \infty$, there exists to within the functions of H_0, the unique extremal function $u_0 \in M_F(F_0, F_1, G)$ such that $\|u_0\|_{F(G)} = \mathrm{Cap}_F(F_0, F_1, G)$.

Proof. The set $M_F(F_0, F_1, G)$ is convex in the space $F(G)$. Let us consider the set $M_F^\circ(F_0, F_1, G) = \{v \in F(G) :$ there exists $u \in M_F(F_0, F_1, G)$ such that $u - v \in H_0\}$. This set is said to be an extended set of admissible functions. It is obvious that

$$\mathrm{Cap}_F(F_0, F_1, G) = \inf_{u \in M_F^\circ(F_0, F_1, G)} \|u\|_{F(G)}.$$

Since $M_F^\circ(F_0, F_1, G)$ contains, together with any function, its equivalence class $\{u\}$, one may consider the factor set $M_{F(G)/H_0}(F_0, F_1, G)$ of the set $M_F^\circ(F_0, F_1, G)$. It is obvious that the extended set of admissible functions and its factor set are convex.

Let $\mathrm{Cap}_F(F_0, F_1, G) > 0$. Then the closure of the set $M_F(F_0, F_1, G)$ contains no elements of H_0. Consequently, $M_{F(G)/H_0}(F_0, F_1, G)$ does not contain zero either. Since the Banach space $F(G)/H_0$ is uniformly convex, then there exists the unique element $\{u_0\} \in F(G)/H_0$ for which

$$\mathrm{Cap}_F(F_0, F_1, G) = \inf_{\{u\} \in M_{F(G)/H_0}(F_0, F_1, G)} \|\{u\}\|_{F(G)/H_0} = \|\{u_0\}\|_{F(G)/H_0}.$$

Let us choose any function $v_0 \in \{u_0\}$. According to the construction of the set $M_{F(G)/H_0}(F_0, F_1, G)$, there exists the function $\overline{u}_0 \in M_F(F_0, F_1, G)$ such that $v_0 - \overline{u}_0 \in H_0$. If u_1 is another extremal function for capacity of the pair (F_0, F_1), then, due to the uniqueness of extremality of the class u_0, the function $u_1 \in \{\overline{u}_0\}$. Thus, $u_1 - \overline{u}_0 \in H_0$. This completes the proof.

Property 4. Let $F(G)$ be a Banach space, $F_0 \subset G$, and let $\{F_{1,m}\}$ be an arbitrary sequence of sets in G. Then

$$\mathrm{Cap}_F(F_0, \cup F_{1,m}, G) \leqslant \sum_{m=1}^{\infty} \mathrm{Cap}_F(F_0, F_{1,m}, G).$$

Proof. Let us fix $\varepsilon > 0$. Choose the functions $f_m \in M_F(F_0, F_{1,m}, G)$ so that

$$\mathrm{Cap}_F(F_0, F_{1,m}, G) + \varepsilon/2^m \geqslant \|f_m\|_{F(G)}.$$

If the sum

$$\sum_{m=1}^{\infty} \mathrm{Cap}_F(F_0, F_{1,m}, G)$$

is infinite, there is nothing to prove. If it is finite, then the function $\sum_{m=1}^{\infty} f_m = f$ belongs to $F(G)$ and is admissible for the pair $(F_0, F_{1,m})$, since $f(x) \geqslant f_m(x) \geqslant 1$ for all $x \in F_{1,m}$ at all m, and

$$\|f\|_{F(G)} \leqslant \sum_{m=1}^{\infty} \mathrm{Cap}_F(F_0, F_{1,m}, G) + \varepsilon.$$

Since $\varepsilon > 0$ was chosen arbitrarily, the proof is completed.

4.1.2. Strong Variational Capacity

Let $F_0, F_1 \subset \overline{G}$. Let us denote by $N(F_0, F_1, G)$ a set of all functions u belonging to the space $F(G)$ such that $u(x) = 0$ in some neighbourhood F_0 in G, $u(x) = 1$ in some neighbourhood F_1 in G, $0 \leqslant u(x) \leqslant 1$ everywhere in G.

$$\overline{\mathrm{Cap}}_F(F_0, F_1, G) = \inf_{u \in N(F_0, F_1, G)} \|u\|_{F(G)}$$

is said to be the strong variational capacity of the pair (F_0, F_1) in G.

Strong variational capacity preserves all the properties of variational capacity except for the property of countable semiadditivity.

4.1.3. Weak Variational Capacity

The definition is the same as that of variational capacity, but instead of the condition $u(x) = 0$ in some neighbourhood F_0, one should require $u(x) \leqslant 0$ in some neighbourhood F_0. Consequently, the set $N^\circ(F_0, F_1, G)$ of functions admissible for weak variational capacity $\mathrm{Cap}_F^\circ(F_0, F_1, G)$ is wider than for variational capacity and $\mathrm{Cap}_F^\circ(F_0, F_1, G) \leqslant \mathrm{Cap}_F(F_0, F_1, G) \leqslant \overline{\mathrm{Cap}}_F(F_0, F_1, G)$. Properties 1–4 for weak variational capacity are valid.

4.2. Comparison of Variational Capacity and (T,p)-Capacity

Let us compare (T,p)-capacity and the variational $W(T,p,G)$-capacity in the domain $G \subset \mathbf{R}^n$. To make notations less cumbersome, we write $V.\,\mathrm{Cap}_{T,p}(\dots)$ instead of $\mathrm{Cap}_{W(T,p,G)}(F_0, F_1, G)$ (we use the same new notation for strong and weak variational capacities).

Proposition 4.1. *Let (F_0, F_1) be an arbitrary pair of subsets of G. Then $\mathrm{Cap}_{T,p}(F_1, G) \leqslant (V.\,\mathrm{Cap}_{T,p}(F_0, F_1, G))^p$.*

Proof. A set $M_{T,p}(F_1)$ of functions admissible for (T,p)-capacity consists of all functions $u(x) \geqslant 1$ for all $x \in F_1$. The set $M(T,p,F_0,F_1,G)$ of functions admissible for variational (T,p)-capacity consists of all functions $v(x) \geqslant 1$ for all x belonging to some neighbourhood of the set F_1, and $v(x) = 0$ for all x belonging to some neighbourhood of the set F_0. It is obvious that

$$M_{T,p}(F_1, G) \supset M(T,p,F_0,F_1,G).$$

Then

$$\mathrm{Cap}_{T,p}(F_1, G) = \inf_{u \in M_{T,p}(F_1,G)} \|u\|^p_{W(T,p,G)} \leqslant \inf_{v \in M(T,p,F_0,F_1,G)} \|v\|^p_{W(T,p,G)}$$
$$= (\mathrm{V.\,Cap}_{T,p}(F_0, F_1, G))^p.$$

This completes the proof.

If the space $W(T,p,G)$ is sufficiently poor, then the variational (T,p)-capacity need not be estimated via (T,p)-capacity.

Example. Let us consider a space $L_1(\mathbf{R})$ and a positive operator $T : L_1^+(\mathbf{R}) \to L_{1,\mathrm{loc}}(\mathbf{R})$ which puts the function identically equal to $\|u\|_{L_1(\mathbf{R})}$ into correspondence to the function $u \in L_1^+(\mathbf{R})$. It is obvious that for any set F_1, $\mathrm{Cap}_{T,1}(F_1, \mathbf{R}) = 1$, while $\mathrm{V.\,Cap}_{T,1}(F_0, F_1, \mathbf{R}) = \infty$ for $F_0 \neq 0$, since the functions admissible for variational capacity do not exist.

The space $W(T,p,G)$ is said to be multiplicatively separable if for every pair of nonintersecting compact sets F_0, F_1, there exists the function $g_{F_0,F_1} \in W(T,p,G)$ satisfying the conditions: 1) $g_{F_0,F_1} \geqslant 0$ almost everywhere; 2) $g_{F_0,F_1} \geqslant 1$ for all $x \in F_1$, $g_{F_0,F_1} = 0$ for all x belonging to some neighbourhood of the set F_0; 3) the function g_{F_0,F_1} is the multiplicator in $W(T,p,G)$.

Proposition 4.2. *If an operator $T : L_p(U) \to W(T,p,G)$ is a C-operator, and the space $W(T,p,G)$ is multiplicatively separable, then for every pair of compact sets $F_0, F_1 \in G$, $F_0 \cap F_1 = \emptyset$, the estimate*

$$\mathrm{V.\,Cap}_{T,p}(F_0, F_1, G) \leqslant K(\mathrm{Cap}_{T,p}(F_1, G))^{1/p}$$

is valid. The constant K depends on the choice of the pair F_0, F_1.

Proof. Let us arbitrarily choose $\varepsilon > 0$ and a continuous function u admissible for (T,p)-capacity of the set F_1 whose norm satisfies the inequality

$$\|u\|_{L_p(u)} = \|Tu\|_{W(T,p,G)} \leqslant (\mathrm{Cap}_{T,p}(F_1, G))^{1/p} + \varepsilon.$$

Let us recall that $(Tu)(x) = 1$ for all $x \in F_1$. Since the operator T is a C-operator, the function $(Tu_\varepsilon)(x) = (1+\varepsilon)(Tu)(x)$ is continuous, $(Tu_\varepsilon)(x) \geqslant 1$ for all $x \in F_1$. Consequently, the set $F_{1,\varepsilon} = \{x : u(x) + \varepsilon > 1\}$ is an open neighbourhood of the set F_1.

The condition of multiplicative separability of the space $W(T,p,G)$ allows us to construct a continuous function g_{F_0,F_1} which is equal to zero in some neighbourhood $H(F_0)$ of the set F_0 and is equal to 1 in some neighbourhood $H(F_1)$ of the set F_1. Let us consider the function $v = u_\varepsilon g_{F_0,F_1}$. It is equal to zero in $H(F_0)$, it exceeds 1 in $F_{1,\varepsilon} \cap H(F_1)$, and it belongs to the class $W(T,p,G)$, i.e., the function v_ε is admissible for weak variational capacity of the pair (F_0, F_1).

From the condition of multiplicative separability, due to the choice of the function u_ε, it follows that

$$\begin{aligned}&\mathrm{V.Cap}_{T,p}(F_0,F_1,G)\\ &\quad\leqslant \|v_\varepsilon\|_{W(T,p,G)} \leqslant K(F_0,F_1)\|u_\varepsilon\|_{W(T,p,G)}\\ &\quad\leqslant (1+\varepsilon)K(F_0,F_1)(1+\varepsilon)(\mathrm{Cap}_{T,p}(F_1,G)^{1/p}+\varepsilon).\end{aligned}$$

Since $\varepsilon > 0$ is arbitrary, the statement is proved.

Remark. The constant K in Proposition 1.2 depends on the norm of the multiplication operator by the function g_{F_0,F_1}.

As an example, let us show that for the Sobolev classes $W_p^l(\mathbf{R}^n)$, the condition of multiplicative separability is satisfied and the constant in the inequality of Proposition 4.2 only depends on the distance between the compact sets F_0 and F_1.

Lemma 4.3. *For any pair of compact sets* $(F_0, F_1 \subset \mathbf{R}^n)$, *there exists a function* $g \in C^\infty$ *such that* $g(x) = 0$ *for all* $x \in F_0$, $g(x) = 1$ *for all* $x \in F_1$ *and* $|D^\alpha g(x)| \leqslant K_\alpha/[\rho(F_0,F_1)]^{|\alpha|}$ *for all* $x \in \mathbf{R}^n$, *where* K_α *does not depend on the choice of the sets* F_0 *and* F_1.

Proof. Let us put $h(x) = \exp(\frac{1}{x^2-n})$ for $|x| < \sqrt{n}$, $h(x) = 0$ for $|x| \geqslant \sqrt{n}$.
Let us put

$$Q(x) = \sum_m h(x-m),$$

where m runs the set of all vectors with integral coordinates in $\mathbf{R}^n$. It is easy to see that $Q(x) > 0$ for all x. Here, for every point x, there exists a neighbourhood V in which only the finite number of terms on the right-hand side is distinct from zero. Hence it follows that $Q \subset C^\infty(\mathbf{R}^n)$.

We set $s(x) = g(x)/Q(x)$. Then

$$s(x) \subset C^\infty(\mathbf{R}^n), \qquad s(x) = 0 \quad \text{for} \quad x \geqslant \sqrt{n},$$

and for all $x \in \mathbf{R}^n$,

$$\sum_m s(x-m) = 1.$$

The functions $s(x-m)$ form the partition of unit in the space $\mathbf{R}^n$. In this case, the ball with the center m and radius n is the support of the function $s(x-m)$.

Let now $d = (1/2\sqrt{n})\rho(F_0, F_1)$. Let us consider a system of functions $s(x-ad)$, where a is a vector with integral coordinates. Let $a_1, a_2, \ldots, a_i$ be all vectors a for which the support of the function $s(\frac{x-ad}{d})$ intersects the set F_1. Let

$$g_{F_0,F_1}(x) = \sum_{i=1}^{t}((x-a_id)/d).$$

It is obvious that $g_{F_0,F_1} \subset C^\infty(\mathbf{R}^n)$ and $g_{F_0,F_1}(x) = 1$ for all $x \in F_1$. Further, due to the choice of d, each of $s((x-a_id)/d)$ functions equals zero for $x \in F_0$. Consequently, $g_{F_0,F_1}(x) = 0$ for $x \in F_0$. Note that every point x has the neighbourhood V in which no more than k functions $s((x-a_id)/d)$ are distinct from zero. It is easy to see that

$$|D^\alpha s((x-a_id)/d| \leqslant \frac{M_\alpha}{dh^{|\alpha|}},$$

where $M_\alpha = \max|D^\alpha s(x)|$. Hence it follows that

$$|D^\alpha g_{F_0,F_1}(x)| \leqslant kM_\alpha/d^{|\alpha|} = K_\alpha(\rho(F_0, F_1)).$$

The lemma is proved.

Corollary. *The space $W_p^l(\mathbf{R}^n)$ possesses the multiplicative separability property. The norm of the operator of multiplication by the separating function g_{F_0,F_1} constructed in the lemma only depends on the numbers n, l, p and on the distance between the compact sets F_0 and F_1.*

Proof. According to the previous lemma, for every pair of compact sets $F_0, F_1 \subset \mathbf{R}^n$, there exists a function $g_{F_0,F_1} \subset C^\infty(\mathbf{R}^n) \cap W_p^l(\mathbf{R}^n)$ equal to zero on F_0 and equal to 1 on F_1; for its derivatives, the inequality

$$|D^\alpha g_{F_0,F_1}(x)| \leqslant K_\alpha/(\rho(F_0, F_1))^{|\alpha|}$$

holds. Let us take any function $u \in W_p^l(\mathbf{R}^n)$. Let us estimate the norm of the operator L,

$$L = L_{g_{F_0,F_1}} : W_p^l(\mathbf{R}^n) \to W_p^l(\mathbf{R}^n),$$

$$\begin{aligned} L_{g_{F_0,F_1}}(u) &= u g_{F_0,F_1}, \|g_{F_0,F_1} u\|_{W_p^l(\mathbf{R}^n)} \\ &= \sum_{l \geqslant |\alpha| \geqslant 0} \|D^\alpha (g_{F_0,F_1} u)\|_{L_p(\mathbf{R}^n)} \\ &= \sum_{l \geqslant |\alpha| \geqslant 0} \Big\| \sum_{\alpha \geqslant \beta \geqslant 0} C_\alpha^\beta D^\beta g_{F_0,F_1} D^{\alpha-\beta} u \Big\|_{L_p(\mathbf{R}^n)} \\ &\leqslant \sum_{l \geqslant |\alpha| \geqslant 0} \sum_{\alpha \geqslant \beta \geqslant 0} \left(C_\alpha^\beta K_\beta \left(\rho(F_0, F_1)\right)^{|\beta|} \right\| D^{\alpha-\beta} u\|_{L_p(\mathbf{R}^n)} \\ &\leqslant \left[\sum_{l \geqslant |\beta| \geqslant 0} (l K_\beta l!)/(\rho(F_0, F_1))^{|\beta|} \right] \sum_{l \geqslant |\alpha| \geqslant 0} \|D^\alpha u\|_{L_p(\mathbf{R}^n)}. \end{aligned}$$

Remember that for the integer l, the space $W(J_l, p, \mathbf{R}^n) = W_p^l(\mathbf{R}^n)$. The coincidence is understood as follows: for every function $u \in W_p^l(\mathbf{R}^n)$, there exists a function $u' \in W(J_l, p, \mathbf{R}^n)$ coinciding with it almost everywhere, and

$$C_1 \|u\|_{W_p^l(\mathbf{R}^n)} \leqslant \|u'\|_{W(J_l,p,\mathbf{R}^n)} \leqslant C_2 \|u\|_{W_p^l(\mathbf{R}^n)}, \tag{4.1}$$

where the constants C_1 and C_2 do not depend on the choice of the function u.

From Proposition 4.1, 4.2, and from the corollary of Lemma 4.3, we obtain for the space $W_p^l(\mathbf{R}^n)$:

Proposition 4.4. *Let F_0, F_1 be an arbitrary pair of compact sets in $\mathbf{R}^n$. Then the inequalities*

$$K_1 \operatorname{Cap}_{J_l,p}(F_1, \mathbf{R}^n) \leqslant \operatorname{Cap}_{W_p^l(\mathbf{R}^n)}(F_0, F_1, \mathbf{R}^n) \leqslant K_2 \operatorname{Cap}_{J_l,p}(F_1, \mathbf{R}^n)$$

are valid, where the constants K_1 and K_2 only depend on the numbers n, l, p, and on the distance between the compact sets F_0 and F_1.

Proof. The left-hand equality follows from inequality (4.1). From Proposition 4.2 and inequality (4.2), the right-hand inequality follows.

4.3. Sets of Zero Variational Capacity

Let G be a domain in $\mathbf{R}^n$. A set $A \subset G$ is a set of zero variational F-capacity, if for every compact set $F_0 \subset G$, $F_0 \cap A = \emptyset$, $\operatorname{Cap}_F(F_0, A, G) = 0$.

From the properties of variational F-capacity, there follows:

1. Any subset of the set of zero variational capacity is a set of zero variational capacity.

2. The union of no more than a countable number of sets of zero variational capacity is a set of zero variational capacity.

Theorem 4.5. *If an operator $T : L_p(U) \to W(T,p,G)$ is a C-operator and the space $W(T,p,G)$ is multiplicatively separable, then any closed set of zero (T,p)- capacity is a set of zero variational (T,p)-capacity. Conversely, any set of zero variational (T,p)-capacity is a set of zero capacity.*

Proof. Due to Proposition 4.1, the (T,p)-capacity of the set A is less than the variational (T,p)-capacity of any pair (A, F_1). Therefore, from the vanishing of variational (T,p)-capacity, the vanishing of the (T,p)-capacity of A follows.

Since the space $W(T,p,G)$ is multiplicatively separable and T is a C-operator, then, by Proposition 4.2, from vanishing of the (T,p)-capacity of the compact set A, vanishing of the variational (T,p)-capacity of any pair of compact sets F_0, A follows. Due to multiplicative separability of $W(T,p,G)$, the vanishing of variational capacity of the pair (F_0, A) follows. Using countable semiadditivity of zero variational capacity, we obtain the statement of the theorem.

4.4. Examples of Variational Capacity

1. Let G be a domain in $\mathbf{R}^n$, $F(G) = L_p(G)$. Then $\mathrm{Cap}_{L_p}(F_0, F_1, G) = (m(F_1))^{1/p}$ if F_1 is a measurable set. The capacity is independent of the set F_0.

2. Let G be a domain in $\mathbf{R}^n$, $F(G) = C(G)$. Then for any pair of sets $(F_0, F_1) \subset G$ such that $F_0 \cap F_1 = \emptyset$, $\mathrm{Cap}_{C(G)}(F_0, F_1, G) = 1$. But if $F_0 \cap F_1 \neq \emptyset$, then $\mathrm{Cap}_{C(G)}(F_0, F_1, G) = \infty$.

3. Let G be a domain in $\mathbf{R}^n$, and $L_p^l(G)$ be a seminormed space of locally summable functions in the domain G, having generalized derivatives up to order l inclusive,

$$\|u\|_{L_p^l(G)} = \sum_{|\alpha|=l} \|D^\alpha u\|_{L_p(G)}.$$

Below, we calculate or estimate variational L_p^l-capacity for several simple pairs of sets. In the case where no ambiguity might arise, we call it (l,p)-capacity.

For bounded domains with smooth boundary,

$$\mathrm{Cap}_{L_p^l(G)}(F_0, F_1, G) \leqslant \mathrm{Cap}_{W_p^l(G)}(F_0, F_1, G) \leqslant K\, \mathrm{Cap}_{L_p^l(G)}(F_0, F_1, G).$$

4.5. Refined Functions

Let us consider a domain $G \subset \mathbf{R}^n$ and a seminormed space of functions $F(G)$. Suppose that there exists a C-operator $T : L_p(U) \to W(T,p,G)$ satisfying the conditions: 1) $W(T,p,G) \subset F(G)$; 2) if $u \in W(T,p,G)$, then $\|u\|_{W(T,p,G)} \sim \|u\|_{F(G)}$; 3) for any function $v \in F(G)$ there exists $u \in W(T,p,G)$ such that

$\|u-v\|_{F(G)} = 0$. This function is called (T,p)-refined. According to Theorem 1.6, for every $\varepsilon > 0$, there exists an open set H with (T,p)-capacity smaller than ε, outside of which the function u is continuous. In the papers [47,50], Theorem 1.6 is used as the definition of the refined function of the class $W_p^l(G)$.

The simplest properties of refined functions: 1) if u is a refined function, then it is finite (T,p)-quasieverywhere; 2) every sequence u_m of refined functions which converges in $F(G)$ to a refined function u contains a subsequence converging to u quasieverywhere.

Property 1 is obvious. Property 2 follows from Theorem 1.7 and from the equivalence of the norms $\|\cdot\|_{W(T,p,G)}$, $\|\cdot\|_{F(G)}$.

4.5.1. *Examples of Functional Classes Admitting Refinement of Functions*

Example 1. The spaces $W_p^l(\mathbf{R}^n)$. The Bessel potential J_l maps $L_p(\mathbf{R}^n)$ onto $W_p^l(\mathbf{R}^n)$ in the following sense: in every equivalence class, which is an element of the space $W_p^l(\mathbf{R}^n)$, there exists the function $u \in W(J_l, p, \mathbf{R}^n)$. The norm $\|u\|_{W(J_l,p,\mathbf{R}^n)} \sim \|u\|_{W_p^l(\mathbf{R}^n)}$ [83]. Consequently, every function from $W_p^l(\mathbf{R}^n)$ may be refined. If two refined functions coincide almost everywhere, they coincide (l,p)-quasieverywhere [66]. Moreover, if in the equivalence class, which is an element of $W_p^l(\mathbf{R}^n)$, one chooses the function taking its natural values in the sense of Lebesgue everywhere where these values exist, then the function is a refined one [83]. Therefore, the refinement property for the class W_p^l proves to be local. The function $u \in W_p^l(G)$ (G is the domain in $\mathbf{R}^n$) is said to be refined if it is refined in every ball $B \subset G$.

(Indeed, every function $u \in W_p^l(B)$ may be extended to the function $u' \in W_p^l(\mathbf{R}^n)$ and, having refined the function u', one may return to the ball B.)

All the above remarks apply to the classes $L_p^l(G)$ and $B_{p,p}^l(G)$.

Example 2. The spaces $L_p^l(G)$. Since every function $u \in L_p^l(G)$ locally belongs to $W_p^l(G)$, the notion of refinement for $L_p^l(G)$ coincides with the notion of refinement for $W_p^l(G)$.

Example 3. The spaces $B_{p,p}^l(G)$. Let us begin with the case $B_{p,p}^l(\mathbf{R}^n)$. This space is the space of traces for $W(J_\alpha, p, \mathbf{R}^{n+1})$, where $\alpha = l + 1/p$. The positive operator A_l equal to the composition J_α and to the trace operator $Sp_\alpha : W(J_\alpha, p, \mathbf{R}^{n+1}) \to B_{p,p}^l(\mathbf{R}^n)$ [44] is, obviously, a C-operator. Let us consider the space $W(A_l, p, \mathbf{R}^n)$. Note that

$$\|u\|_{W(A_l,p,\mathbf{R}^n)} = \inf_{\substack{v \in L_p(\mathbf{R}^{n+1}) \\ A_l v = u}} \|v\|_{L_p(\mathbf{R}^{n+1})} = \inf_{\substack{w \in W(J_\alpha,p,\mathbf{R}^{n+1}) \\ Sp_\alpha w = u}} \|w\|_{W(J_\alpha,p,\mathbf{R}^{n+1})}.$$

From the existence of the bounded extension operator $T_l : B_{p,p}^l(\mathbf{R}^n) \to W_p(\mathbf{R}^n)$ [8], from the coincidence of the spaces $W_p(A_l, p, \mathbf{R}^n)$ and $W(J_\alpha, p,$

$\mathbf{R}^{n+1}$), and from the boundedness of the operator Sp_α, $\|u\|_{W(A_l,p,\mathbf{R}^n)} \sim \|u\|_{B^l_{p,p}}(\mathbf{R}^n)$ follows. Consequently, the functions from $B^l_{p,p}(\mathbf{R}^n)$ may be (A_l, p)-refined. Localization of the refinement notion and refinement in $B^l_{p,p}(G)$ are performed similarly to the spaces $W^l_p(G)$.

Example 4. The space $L_p(G)$. Any function from $L_p(G)$ is refined with respect to the identical operator $T : L_p(G) \to L_p(G)$.

4.6. Theorems of Imbedding into the Space of Continuous Functions

In terms of capacity, Maz'ya obtained the necessary and sufficient conditions for the existence of the imbedding operator of spaces W^l_p into a space of continuous functions. We propose here an abstract variant of the Maz'ya method to study imbedding operators of Banach spaces of functions into a space of continuous functions.

In a domain G, let us consider two seminormed spaces of functions $H_0(G)$ and $H_1(G)$. A bounded imbedding operator $I : H_0(G) \to H_1(G)$ is said to exist if $I(u) = u$ for all the functions $u \in H_0(G)$ and $\|I(u)\| \leqslant \|I\|\,\|u\|$.

Theorem 4.6. *If $I : H_0(G) \to H_1(G)$ is a bounded imbedding operator, then $\|I\| \operatorname{Cap}_{H_0}(F_0, F_1, G) \geqslant \operatorname{Cap}_{H_1}(F_0, F_1, G)$ for any pair of sets $(F_0, F_1) \subset G$.*

Proof. Let us take any function $u \subset M_{H_2}(F_0, F_1, G)$ for some fixed pair of sets $(F_0, F_1) \subset G$. (If such a function does not exist, then $\operatorname{Cap}_{H_0}(F_0, F_1, G) = \infty$, and the inequality is obvious.) The function u belongs to the set $M_{H_1}(F_0, F_1, G)$, since $u(x) \geqslant 1$ for all x lying in some neighbourhood of the set F_1, and $u(x) = 0$ for all x lying in some neighborhood of the set F_0 and $u \in H_1(G)$, due to the existence of the imbedding operator. From the boundedness of the imbedding operator, it follows that $\|u\|_{H_0(G)} \leqslant \|u\|_{H_1(G)}\|I\|$. Hence,

$$\begin{aligned}\operatorname{Cap}_{H_1}(F_0, F_1, G) &= \inf_{u \in M_{H_1}(F_0,F_1,G)} \|u\|_{H_1(G)} \leqslant \|I\| \inf_{u \in M_{H_0}(F_0,F_1,G)} \|u\|_{H_0(G)} \\ &= \operatorname{Cap}_{H_0}(F_0, F_1, G).\end{aligned}$$

The theorem is proved.

Corollary. *If $I : H(G) \to C(G)$ is a bounded imbedding operator, then for any pair of points $x, y \in G$, the inequality*

$$\operatorname{Cap}_H(\{x\}, \{y\}, G) > \|I\|^{-1} > 0$$

is valid.

Proof. Remember that for any pair of points, $\operatorname{Cap}_{C(G)}(\{x\}, \{y\}, G) = 1$. According to the previous theorem, we obtain

$$\|I\| \operatorname{Cap}_H(\{x\}, \{y\}, G) \geqslant \operatorname{Cap}_{C(G)}(\{x\}, \{y\}, G) = 1,$$

i.e., $\mathrm{Cap}_H(\{x\},\{y\},G) = I^{-1} > 0$.

Remark. The corollary may be formulated in a formally stronger variant: $\mathrm{Cap}_H(F_0,F_1,G) \geqslant \|I\|^{-1}$ for any pair of sets $(F_0,F_1) \in G$. However, due to monotonicity of variational capacity, the statement for any pair of sets easily follows from the statement for a pair of points.

Theorem 4.6. *Let G be a bounded domain in $\mathbf{R}^n$, $H(G)$ is a Banach space of functions.*

If the imbedding operator $I : H(G) \to C(G)$ is compact, then

$$\lim_{\delta\to 0} \inf_{\substack{x,y\in G\\ |x-y|<\delta}} \mathrm{Cap}^{\circ}_H(\{x\},\{y\},G) = \infty.$$

Proof. Suppose the inverse. Then there exist sequences of numbers $\{\delta_m\}$ and of points $\{x_m\}$, $\{y_m\} \in G$, $|x_m - y_m| < \delta_m$ for which

$$\mathrm{Cap}^{\circ}_H(\{x_m\},\{y_m\},G) < K < \infty.$$

Every function of the space $H(G)$ is continuous due to the existence of the imbedding operator I. Therefore, the inequality for the capacity $\mathrm{Cap}^{\circ}_H(.,.,.)$ implies the existence of a sequence of continuous functions $\{u_m\}$ belonging to the space $H(G)$, $u_m(x_m) = 0$ for all m, $\|u_m\|_{H(G)} < 2K$, $u_m(y_m) = 1$ for all m.

From the compactness of the imbedding operator I, the equicontinuity of the sequence of functions $\{u_m\}$, bounded in $H(G)$, follows. Equicontinuity of functions u_m contradicts their property $u_m(x_m) - u_m(y_m) = 1$ under the condition that the distance $|x_m - y_m| \to 0$.

This completes the proof.

The Banach space of functions $H(G)$ satisfies the approximativity condition if $C(G) \cap H(G)$ is dense in $H(G)$.

Theorem 4.8. *Let a Banach space $H(G)$ (G is a bounded domain in $\mathbf{R}^n$) satisfy the approximativity and H-separability conditions.*

Suppose that in the domain G there exist two non-intersecting compact subdomains W and W_1 such that

$$a = \inf_{x\in G\setminus\overline{W}} \mathrm{Cap}^{\circ}_H(\overline{W},x,G) > 0,$$

$$a_1 = \inf_{x\in G\setminus\overline{W}_1} (\mathrm{Cap}^{\circ}_H(\overline{W},x,G)) > 0.$$

Then there exists a bounded imbedding operator $I : H(G) \to C(G)$.

Proof. Let us choose any compact subdomain $W' \supset \overline{W}$, $W' \cap W_1 = \emptyset$, $W' \subset G$ of the domain G. Denote by g the function of the class $H(G)$ which is equal to zero on W, equal to 1 on $G\backslash W'$, belongs to the class $H(G)$, and has the following property: for any function $u \in H(G)$, the function $ug \in H(G)$ and the inequality

$$\|ug\|_{H(G)} \leqslant K(G)\|u\|_{H(G)}$$

is valid.

The existence of such a function g follows from the separability condition.

Let y be an arbitrary point from $G\backslash W'$ at which the function $u \in H(G)$ does not vanish. Due to the above, the function $v(x) = (u(x)/u(y))g(x)$ belongs to the class $H(G)$. It equals zero on W and is equal to 1 at the point y. So, the function v is admissible for the variational capacity of the pair $(\overline{W}, y)$,

$$\mathrm{Cap}^{\circ}_{H}(\overline{W}, y, G) \leqslant \|v\|_{H(G)}.$$

By the construction of the function v,

$$\|v\|_{H(G)} \leqslant \frac{1}{|u(y)|} K(g)\|u\|_{H(G)}.$$

Comparing this inequality with that in the formulation of the theorem, we obtain

$$|u(y)| \leqslant \frac{K(g)\|u\|_{H(G)}}{\mathrm{Cap}^{\circ}_{H}(\overline{W}, y, G)} = a^{-1}K(g)\|u\|_{H(G)}$$

for all $y \in G\backslash W'$.

Let us choose any compact subdomain $W_1' \supset \overline{W}_1$, $\overline{W}_1' \cap W' = \emptyset$, $\overline{W}_1' \subset G$. The condition of H-separability allows us to construct a function g_1 of the class $H(G)$ which equals zero on W_1, equals 1 on $G\backslash\overline{W}_1'$, belongs to the class $H(G)$, and satisfies the multiplicativity condition

$$\|g_1 u\|_{H(G)} = K(g_1)\|u\|_{H(G)}.$$

Just as in the first part of the proof, we obtain the inequalities

$$|u(y)| \leqslant \frac{K(g_1)\|u\|_{H(G)}}{\mathrm{Cap}^{\circ}_{H}(\overline{W}_1, y, G)} = a_1^{-1}K(g_1)\|u\|_{H(G)}$$

for all $y \in G\backslash\overline{W}'$.

Thus, for all $y \in G$, the inequality

$$|u(y)| \leqslant (\max(a^{-1}K(g), a_1^{-1}K(g_1)))\|u\|_{H(G)}$$

is valid.

From the approximativity condition, the statement of the theorem follows directly.

Theorem 4.9. *Let G be a domain in $\mathbf{R}^n$, and let $H(G)$ be a seminormed space of functions; the kernel of the seminorm contains the function which is identically equal to* 1.

If $H(G)$ satisfies the condition of weak approximativity, then the fulfillment of the relation

$$\lim_{\delta\to 0} \inf_{\substack{x,y\in G\\ |x-y|<\delta}} \mathrm{Cap}_H^{\circ}(x,y,G)=\infty$$

implies the existence of the imbedding operator $I: H(G)\to C(G)$. Here, the operator I transfers every bounded set from $H(G)$ to the equicontinuous set of continuous functions.

Condition of weak approximativity. A seminormed space of functions $H(G)$ satisfies the weak approximativity condition if for every function $u\in H(G)$, one can construct a sequence of continuous functions $\{u_m\}\in H(G)$ converging to u almost everywhere. Here the inequality

$$\varlimsup_{m\to\infty} \|u_m\|_{H(G)} \leqslant K\|u\|_{H(G)}$$

is valid, where the constant K is independent of the choice of the function u.

Proof. Let us choose an arbitrary continuous function $f\in H(G)$ which is not identically equal to the constant. By the definition of weak variational H-capacity, the function

$$u_\varepsilon(z)=\left(\frac{f(x)-f(z)}{f(x)-f(y)}\right)(1+2\varepsilon)-\varepsilon \qquad (\varepsilon>0)$$

is admissible for weak variational H-capacity of a pair of points $x,y\in G$, $x\neq y$, for all $\varepsilon>0$. Indeed, u_ε is continuous, $u_\varepsilon(x)=0, u_\varepsilon(y)=1$, and $u_\varepsilon\in H(G)$, since $u_\varepsilon(z)=v(z)+c$, where $v(z)\in H(G)$, and the kernel of the seminorm $\|\cdot\|_{H(G)}$ contains constants.

Then, according to the definition of weak variational capacity,

$$(1+2\varepsilon)\left\|\frac{f(z)-f(x)}{f(x)-f(y)}\right\|_{H(G)} \geqslant \mathrm{Cap}_H^{\circ}(\{x\},\{y\},G),$$

that is,

$$|f(x)-f(y)|\,\mathrm{Cap}_H^{\circ}(\{x\},\{y\},G)\leqslant \|f\|_{H(G)}. \tag{4.2}$$

From the assumption of the theorem, we directly obtain that any set of continuous functions bounded in $H(G)$ is equicontinuous.

Now, let a function $u\in H(G)$ be arbitrary. The weak approximation condition allows us to construct a sequence $\{u_m\}$ of continuous functions which is bounded in $H(G)$ and converges to u almost everywhere. Since the sequence $\{u_m\}$ is equicontinuous according to the abovesaid, the function u is also continuous, and the sequence $\{u_m\}$ may be considered to be converging to u uniformly on every compact set.

Remember that due to the weak approximativity condition,

$$\overline{\lim_{m\to\infty}} \|u_m\|_{H(G)} \leqslant K\|u\|_{H(G)}.$$

From this inequality, from the uniform convergence of $\{u_m\}$ on every compact set and from inequality (4.2), it follows that

$$|u(x) - u(y)| \operatorname{Cap}^{\circ}_H(\{x\}, \{y\}, G) \leqslant 2K\|u\|_{H(G)}, \tag{4.3}$$

i.e., the equicontinuity of every bounded in $H(G)$ set of functions.

This completes the proof.

Theorem 4.10. *Let G be a domain in $\mathbf{R}^n$, let $m(G) < \infty$, and let $H(G)$ be a Banach space of functions, with the norm in it being representable as a sum of two seminorms $\|\cdot\|_{H(G)} = \|\cdot\|_{1,H(G)} + \|\cdot\|_{2,H(G)}$. Suppose that for the seminorms the following conditions are satisfied: 1) if $u \in H(G)$ and $u(x) \geqslant 1$ for all $x \in G$, then $\|u\|_{1,H(G)} \geqslant K_0 m(G)$, where the constant K_0 does not depend on the choice of the function u; 2) the kernel of the seminorm $\|\cdot\|_{2,H(G)}$ contains the function identically equal to 1.*

If $H(G)$ satisfies the weak approximativity condition, then, with the relations

$$\inf_{x,y\in G} \operatorname{Cap}^{\circ}_H(\{x\}, \{y\}, G) \geqslant a^2 > 0,$$

$$\lim_{\delta\to 0} \inf_{\substack{x,y\in G \\ |x-y|<\delta}} \operatorname{Cap}^{\circ}_{2,H}(\{x\}, \{y\}, G) = \infty$$

being fulfilled, the existence of the compact imbedding operator $I : H(G) \to C(G)$ follows.

Remark. $\operatorname{Cap}^{\circ}_{2,H}(\cdot,\cdot,\cdot)$ is the weak variational capacity connected with the second seminorm $\|\cdot\|_{2,H(G)}$.

Proof. Equicontinuity of the image of every bounded set from $H(G)$ under the action of imbedding operator, just as the existence of the imbedding operator, follows from Theorem 4.9. It remains to prove boundedness in $C(G)$ of the image of every ball $B(0,r) \subset H(G)$. It suffices to consider $B(0,1)$. From Condition 1 it follows that every continuous function $u \in B(0,1)$ has, at least at one point, the value $|u(x)| < K_0^{-1}(m(G))^{-1} + 1$. But, due to (4.3),

$$|u(y) - u(x)| \leqslant \frac{2K\|u\|_{H(G)}}{\operatorname{Cap}^{\circ}_H(\{x\}, \{y\}, G)} \leqslant \frac{2K}{\operatorname{Cap}^{\circ}_H(\{x\}, \{y\}, G)} = 2a^2K,$$

that is,

$$|u(z)| \leqslant 2Ka^2 + (K_0 m(G))^{-1} + 1 = K_1$$

for all $z \in G$. So, for any function $u \in B(0,1)$, $\|u\|_{C(G)} \leqslant K_1$. Hence, it follows that the imbedding operator is completely continuous.

The theorem is proved.

§5 Capacity in Sobolev Spaces

5.1. Three Types of Capacity

Let G be a domain in $\mathbf{R}^n$. We will consider three types of capacity induced by the spaces $W_p^l(G)$ and $L_p^l(G)$. The space $W_p^l(\mathbf{R}^n)$ coincides with the space $W(G_l, p, \mathbf{R}^n)$. Let us recall that there exist two-sided estimates

$$K\|u\|_{W_p^l(\mathbf{R}^n)} \leqslant \|u\|_{W(G_l,p,\mathbf{R}^n)} \leqslant K_1\|u\|_{W_p^l(\mathbf{R}^N)}.$$

The first type of capacity is the (l,p)-capacity,

$$C_p^l(E) = C_p^l(E, \mathbf{R}^n) = \mathrm{Cap}_{(G_l,p)}(E, \mathbf{R}^n) = \inf_{u \in M_{(G_l,p)}(E,\mathbf{R}^n)} \|G_l u\|_{W(G_l,E,\mathbf{R}^n)}.$$

Unfortunately, the representation of $W_p^l(G)$ as a space of the type $W(T,p, G)$ is only known for $G = \mathbf{R}^n$, and this capacity will only be used to study problems of local behaviour for the functions from $W_p^l(G)$.

The second type of capacity is the weak variational $L_p^l(G)$- capacity:

$$\widetilde{C}_p^l(F_0, F_1, G) = \inf_{u \in \widetilde{M}_{L_p^l(G)}(F_0,F_1)} \|u\|_{L_p^l(G)}.$$

Remember that $\widetilde{M}_{L_p^l(G)}(F_0, F_1)$ consists of all functions of the class $L_p^l(G)$ which are equal to zero on F_0 and are greater than 1 on F_1.

This capacity is more easily calculated and is more convenient for estimates than the $L_p^l(G)$-capacity. But it has an essential shortcoming: at $l \geqslant 2$, the kernel of the seminorm $\|\cdot\|_{L_p^l(G)}$ contains linear functions. Consequently, $C_p^l(\{x\}, \{y\}, G) = 0$ for all pairs of points x, y belonging to the domain G of Euclidean space. This example shows that weak l,p-capacity cannot be used to study necessary conditions of the imbedding of $W_p^l(l \geqslant 2)$ into $C(G)$.

The third type of capacity is $[l,p]$-capacity

$$C_p^l(F_0, F_1, G) = \inf_{u \in M_{L_p^l(G)}(F_0,F_1)} \|u\|_{L_p^l(G)}.$$

Let us recall that $M_{L_p^l(G)}(F_0, F_1)$ consists of all functions of the class $L_p^l(G)$ which are equal to zero in some neighbourhood of the set F_0 and are greater than 1 in some neighbourhood of the set F_1.

It is obvious that $\widetilde{C}_p^l(F_0, F_1, G) \leqslant C_p^l(F_0, F_1, G)$.

From the results of the previous section, it follows that the supply of sets of zero capacity is the same for all three types of capacity.

Let us show that the sets of admissible functions

$$M_{(l,p)}(E) = M_{(G_l,p)}(E, \mathbf{R}^n) \quad \text{and} \quad M_{[l,p]}(F_0, F_1, G) = M_{L_p^l(G)}(F_0, F_1, G)$$

for the compact sets E, F_0, F_1 may be reduced to smooth functions, i.e., to intersections of $C^\infty(\mathbf{R}^n) \cap M_{(l,p)}(\mathbf{R}^n)$ and $C^\infty(G) \cap M_{[l,p]}(F_0, F_1, G)$.

Proposition 5.1. *For every compact set $E \subset \mathbf{R}^n$, the (l,p)-capacity of E equals the greatest lower boundary of the magnitude $(\|u\|_{L_p^l})^p$ taken on the set of all functions $u \in C^\infty(\mathbf{R}^n)$ that belong to $(L_p^l)^+$ and such that $u(x) \geqslant 1$ on the set E.*

Proof. Let $u = G_l v$, where $v \in M_{(G_l,p)}(E, \mathbf{R}^n)$, $v \geqslant 0$. Then $u(x) \geqslant 1$ for all $x \in E$. Let us assign an arbitrary number $h > 0$, and let $E_h = \{x \in \mathbf{R}^n; \rho(x, E) \leqslant h\}$. Let us put $1 - \delta(h) = \inf_{x \in E_h} u(x)$. The function $u(x)$ is lower semicontinuous. Hence, one can easily conclude that $\delta(h) \to 0$ for $h \to 0$. Let the function u undergo the averaging operation with the parameter h. We obtain the function

$$u_h(x) = M_h u(x) = (G_l M_h v)(x).$$

Note that for all $x \in E_h$,

$$u_h(x) \geqslant 1 - \delta(h).$$

Let

$$\varphi_h(x) = \frac{u_h(x)}{1 - \delta(h)}, \qquad \psi_h(x) = \frac{M_h v(x)}{1 - \delta(h)}.$$

Then $\psi_h \in (E)$, $\psi_h \to v$ in $L_p(\mathbf{R}^n)$, and the function $\varphi_h \in C^\infty$. Thus we see that the set of $v \in M_{(G_l,p)}$, for which $G_l v \in C$, is everywhere dense in $M_{G_l,p}$. Hence, the desired result obviously follows.

Proposition 5.2. *Let F_0, F_1 be compact sets. Then the $[l,p]$-capacity of the pair (F_0, F_1) equals the greatest lower bound of the magnitude $\|u\|_{L_p^l(G)}$ taken on the set of functions $u \in C^\infty(G) \cap L_p^l(G)$ such that $u(x) = 0$ in some neighbourhood of the set F_0, and $u(x) \geqslant 1$ in some neighbourhood of the set F_1.*

Remark. The same result is valid for strong variational capacity in Sobolev spaces.

Proof. Let $u \in M_{l,p}(F_0, F_1, G)$. Thus, there exist the compact neighbourhood U_0 of the set F_0 and the compact neighbourhood U_1 of the set F_1, such that $u(x) = 0$ on U_0, $u(x) \geqslant 1$ on U_1. Let us consider the function $u_h(x) = M_h(x)$, which is the averaging of the function u. As soon as h is less than $\min(\rho(F_0, \partial U_0), \rho(F_1, \partial U_1))$, $u_h(x)$ is an admissible function for the pair (F_0, F_1). Since $\|u_h - u\|_{L_p^l(G)} \to 0$ for $h \to o$, then due to the arbitrariness in the choice of the admissible function, we obtain the desired statement.

For the weak $[l,p]$-capacity, the situation is more complicated. The answer is positive if F_0, F_1, G are compacts with smooth boundary. In the general case, the answer depends on the structure of the boundary. We shall not deal with this problem any further, since we do not need it in the following.

Proposition 5.3. *Let $l = 1$. Then the $[l,p]$-capacity of the pair of compact sets $F_0, F_1 \subset G$ equals the greatest lower bound of the magnitude $\|u\|_{L^1_p(G)}$ taken on the set of all continuous functions of the class $L^1_p(G)$ such that $u(x) \geqslant 1$ for all $x \in F_1$ and $u(x) \leqslant 0$ for all $x \in F_0$ (i.e., in this case, the variational capacity is equivalent to the weak variational capacity).*

Proof. Let us arbitrarily choose $\varepsilon \in (0,1)$ and a continuous function u satisfying the condition of Proposition 5.3. Let us consider the function $u_1(x) = \frac{1}{1-2\varepsilon}(u(x) - \varepsilon)$. It is obvious that $u_1(x) < 0$ for all $x \in F_0$ and $u_1(x) > 1$ for all $x \in F_1$. Due to the continuity of the function u_1, there exist a neighbourhood U_0 of the set F_0 in which $u_1(x) < 0$, and the neighbourhood U_1 of the set F_1 in which $u_1(x) > 1$. The function $u_2(x) = \max(0, \min(1, u_1(x)))$ belongs to the class $L^1_p(G)$ and equals 1 on U_1, it equals 0 on U_0. Consequently, it is admissible for the pair (F_0, F_1) in G. Since

$$\|u_1(x)\|_{L^1_p(G)} \leqslant \frac{1}{1-2\varepsilon}\|u(x)\|_{L^1_p(G)},$$

we obtain the inequality

$$C^1_p(F_0, F_1, G) \leqslant \inf \|u(x)\|_{L^1_p(G)},$$

where the greatest lower bound is taken by the class of functions mentioned in the formulation of the proposition.

The converse inequality is obvious. The proposition is proved.

We have proved that our definition of $[1,p]$-capacity is equivalent to that traditionally used for $p = n$ (see, for instance, [2]).

The proof of Proposition 5.4 is similar to that of Proposition 5.3.

Proposition 5.4. *Let $l = 1$. Then the $[1,p]$-capacity of the pair of compact sets $F_0, F_1 \in \overline{G}$ equals the greatest lower boundary of the magnitude $\|u\|_{L^1_p(G)}$ taken on the set of all continuous functions of the class $L^1_p(G)$ such that $u(x) \geqslant 1$ in some neighbourhood of the set F_1 and $u(x) \leqslant 0$ in some neighbourhood of the set F_0.*

Remark. Using the same method as in the proof of Proposition 5.2, one can show that in Propositions 5.3 and 5.4, instead of continuous functions, it suffices to consider smooth functions.

Theorem 5.5. *Let $\{F_{1,m} \subset \overline{G}\}$ be a monotonically decreasing sequence of compact sets, $F_1 = \bigcap_{m=1}^{\infty} F_{1,m}$.*

Then for any compact set $F_0 \subset G$, $F_0 \cap F_{1,m} = \emptyset$, the equality

$$\lim_{m\to\infty} C^l_p(F_{1,m} F_0, G) = C^l_p(F_1, F_0, G)$$

is valid.

Proof. Let us choose an arbitrary function u admissible for the pair F_1, F_0 and belonging to the class $C^\infty(G)$. Since the function u that vanishes in some neighbourhood U_0 of the set F_0 is greater than 1 in some neighbourhood U_1 of the set F_1, then beginning with some m, it will be admissible for the pair $(F_{1,m}, F_0)$ as well, i.e.,

$$C_p^l(F_{1,m}, F_0, G) \leqslant \|u\|_{L_p^l(G)}$$

for sufficiently large m. Since the admissible function u was chosen arbitrarily, by taking the limit, we obtain

$$\lim_{m\to\infty} C_p^l(F_{1,m} F_0, G) \leqslant C_p^l(F_1, F_0, G).$$

From the monotonicity property of variational capacity with respect to a pair it follows that $C_p^l(F_1, F_0, G) \leqslant C_p^l(F_{1,m}, F_0, G)$ for all m.

The theorem is proved.

Theorem 5.6. *Let E_1, E_2 be compact sets in the closure of a domain G. Then*

$$C_p^l(F_0, E_1, G) + C_p^1(F_0, E_2, G) \geqslant C_p^1(F_0, E_1, \cup E_2, G)$$

for all compact sets $F_0 \subset G$.

Proof. According to Proposition 5.4, as a set of admissible functions one may consider continuous functions. Let $u_1(x)$ be a continuous function of the class L_p^1, exceeding 1 in some neighbourhood E_1 and equal to zero in some neighbourhood of the set F_0, and let a function u_2 have the same properties for the pair (F_2, F_0). Then, according to Proposition 5.4, $u_1(x) + u_2(x)$,[1] will do to calculate the capacity of the pair $(F_0, E_1 \cup E_2)$. From the inequality

$$\|u_1 + u_2\|_{L_p^1(G)} \leqslant \|u_1\|_{L_p^1(G)} + \|u_2\|_{L_p^1(G)}$$

and from the arbitrariness in the choice of admissible functions u_1 and u_2, the statement of the theorem follows.

5.2. Extremal Functions for Capacity

As was shown in the previous subsection, for every pair of sets F_0, F_1 belonging to the closure of the domain G, there exists an extremal function for (l,p)-capacity. This function is defined to within an element from the kernel of the seminorm $\|\cdot\|_{L_p^l}$. The kernel K_p^l of the seminorm consists of polynomials of the order not exceeding $l-1$.

First let us note that the term "(l,p)-quasieverywhere" is applicable to $[l,p]$-capacity as well, since the notion of the set of zero capacity is of local character, and the supply of sets of zero-(l,p)-capacity and zero-$[l,p]$-capacity is the same.

[1] The functions u_1 and u_2 may be considered to be nonnegative.

Theorem 5.7. *For every pair of sets $F_0, F_1 \subset G$, $\operatorname{Int} F_0 \neq \emptyset$ from the inequality $0 < C_p^l(F_0, F_1, G) < \infty$, $p > 1$, there follows the existence of the unique extremal function u_0 for $C_p^l(F_0, F_1, G)$, which is equal to zero (l,p)-quasieverywhere on F_0 and is equal to 1 (l,p)-quasieverywhere on F_1.*

Proof. The set of the admissible function $M_{l,p}(F_0, F_1, G)$ is convex in $L_p^l(G)$. Let us consider the factor space $L_p^l(G)/K_p^l(G)$. Since the functions from $M_{l,p}(F_0, F_1, G)$ are constant on the open set, and the kernel K_p^l consists of polynomials, then in every equivalence class, which is an element of the factor space, there may appear at most one function from $M_{l,p}(F_0, F_1, G)$. If $\{\overline{u}_0\}$ is the extremal class for the capacity of $C_p^l(F_0, F_1, G)$, then, taking this into account, one may obtain a sequence of functions $\{u_m \in M_{l,p}(F_0, F_1, G)\}$, $m = 1, 2, \ldots$, converging to $\{u_0\}$ in the factor space. Let us fix the bounded domain $V \subset G$.

From Theorem 4.2 of Chapter 2 follows the existence of the sequence of polynomials $\{\Pi_m\} \subset K_p^l$, $m = 1, 2, \ldots$, such that $(u_m - \Pi_m) \to \overline{u}_0 - \Pi_0$ in $W_p^l(V)$. Then $(u_m - \Pi_m) \to \overline{u}_0 - \Pi_0$ $(l,p,)$-quasieverywhere in V. On the set $\operatorname{Int} F_0 \cap V$, all the function $u_m \equiv 0$ and $(u_m - \Pi_m) \to (\overline{u}_0 - \Pi_0)$ almost everywhere, i.e., $\Pi_m \to (u_0 - \Pi_0)$ almost everywhere. Thus, on $\operatorname{Int} F_0$, the function $\overline{u}_0 - \Pi_0$ is a polynomial (on $\operatorname{Int} F_0 \cap V$) which is denoted by Π. According to what was proved above, $\Pi_m \to \Pi$ on $\operatorname{Int} F_0 \cap V$, therefore, $\Pi_m \to \Pi$ uniformly on V, i.e., $(\Pi_m - \Pi) \to 0$ uniformly on V. The sequence of functions $\{[u_m - (\Pi_m - \Pi)]\}$ converges in $W_p^l(V)$ to the function $u_0 - (\Pi_0 - \Pi) = (\overline{u}_0 - \Pi_0) + \Pi$. Since $(\Pi_m - \Pi) \to 0$ is uniform, it follows that $u_m \to (\overline{u}_0 - \Pi_0) + \Pi$ in $W_p^l(V)$ and (l,p)-quasieverywhere in V. Let $(\overline{u}_0 - \Pi_0) + \Pi = u_V$.

It is obvious that from the inclusion $V \subset V_1$, where V_1 is the bounded domain in G, it follows that $u_{V_1}|_V = u_V$.

Let us denote by u_0 the function coinciding with u_V for every bounded subdomain V. Since $u_V - \overline{u}_0 \subset K_p^l(V)$, $\|u_m - u_V\|_{L_p^l(V)} \leqslant \|u_m - u_0\|_{L_p^l(G)}$, and consequently, $\|u_m - u_0\|_{L_p^l(G)} \leqslant \|u_m - \overline{u}_0\|_{L_p^l(G)}$. Hence it follows that $u_0 \in \{\overline{u}_0\}$. On every subdomain V $u_m \to u_0$ (l,p)-quasieverywhere. Since the union of countable numbers of sets of zero capacity has zero capacity, $u_m \to u_0$ (l,p)-quasieverywhere in G. Hence it follows that $u_0(x) = 0$ (l,p)-quasieverywhere on F_0 and $u_0(x) \geqslant 1$ (l,p)-quasieverywhere on F_1.

The uniqueness of the function u_0, which equals zero quasi everywhere on F_0 and equals 1 (l,p)-quasieverywhere on F_1, follows from the fact that the polynomial, equal to zero almost everywhere on $\operatorname{Int} F_0$, is equal to zero.

Remark. For $l = 1$, the condition $\operatorname{Int} F_0 \neq \emptyset$ is superfluous, because in this case, the polynomials Π_m are just constants, and in order for them to converge in the domain, it suffices that they converge at least at one point. The existence of such a point follows from the condition $C_p^l(F_0, F_1, G) > 0$ and from the convergence (l,p)-quasieverywhere of the sequence $\{u_m - \Pi_m\}$ constructed in the proof of the theorem.

For the rest of l, the condition $\operatorname{Int} F_0 \neq \emptyset$ may also be weakened, but we do not need it below.

We cannot give up this condition entirely. To illustrate our remark, let us consider an example. If $x, y \in \mathbf{R}^n$ $(n > 1)$, $x \neq y$, then for the $(2,p)$-capacity $(p > 1)$ of this pair let us take any extremal function u_0. The kernel of the seminorm $\|\cdot\|_{2,p}$ consists of linear functions $az + b$. One can always select a_0 and b_0 so that the function $u_0 = a_0 z + b$ should vanish at the point x and should turn to 1 at the point y. This can obviously be done in more than one way, i.e., there cannot be uniqueness for this case in Theorem 5.7.

Theorem 5.8. *For every pair $F_0, F_1 \subset G$ from the inequality $C_p^l(F_0, F_1, G) < \infty$, there follows the existence of the extremal function u_0 which equals zero (l,p)-quasieverywhere on F_0 and is greater than 1 (l,p)-quasieverywhere on F_1.*

With minor variations, the proof is the same as that for the previous theorem.

5.3. Capacity and the Hausdorff h-Measure

Let us recall the notion of the Hausdorff h-measure. Let $h(r)$, $0 \leqslant r < \infty$ be a nondecreasing function, and $h(0) = 0$ and $h(r) \to \infty$ for $r \to \infty$.

Let A be an arbitrary set in $\mathbf{R}^n$. Let us assign $\varepsilon > 0$, and let $B_1, B_2, \dots,$ $B, \dots$ be an arbitrary sequence of open balls, such that $A \subset \bigcup_\nu B_\nu$ and their radii $r_1, r_2, \dots, r_\nu, \dots$ do not exceed ε. The greatest lower boundary of the sum

$$\sum_\nu h(r_\nu),$$

taken by the set of all sequences of balls satisfying the above conditions, is denoted by $\mu_h(A, \varepsilon)$. The magnitude $\mu_h(A, \varepsilon)$ is the nonincreasing function of ε. The limit $\lim_{\varepsilon \to 0} \mu_h(A, \varepsilon) = \mu_h(A)$ is called the Hausdorff h-measure of the set A. In the case $h(r) = r^\alpha$, $\alpha > 0$, $\mu_h(A)$ is called the α-dimensional Hausdorff measure and is denoted by $\mu_\alpha(A)$. The measure $\mu_1(A)$ is also called the linear Hausdorff measure.

If the functions $h_1(r)$ and $h_2(r)$ are such that $h_1(r) = h_2(r)$ for $0 \leqslant r \leqslant r_0$, then the Hausdorff measures μ_{h_1} and μ_{h_2} corresponding to them coincide. Due to this, in the definition of the Hausdorff h-measure one may consider the function to be initially determined only in some interval $[0, r_0]$ where $r_0 > 0$, which arbitrarily extends it outside of this interval. The final result does not depend on the way this extension is realized.

Besides the Hausdorff measure, we shall need one more characteristic of the set. Let $h(r)$ be a monotone nondecreasing function defined for all $r \geqslant 0$; in addition, $h(0) = 0$ and $h(r) \to \infty$ for $r \to \infty$. We consider all possible sequences of open balls $\{B_\nu\}$, $\nu = 1, 2, \dots,$ which cover the given set A. The greatest lower boundary of the sums

$$\sum_{\nu=1}^{\infty} h(r_\nu),$$

where r_ν is the radius of the ball B_ν, $\nu = 1, 2, \ldots$, taken by all such sequences of balls is called the h-imbeddability of the set. We denote it by the symbol $\gamma_h(A)$. In the case $h(r) = r^\alpha$, we write $\gamma_\alpha(A)$ instead of $\gamma_h(A)$.

Note that if A and B are arbitrary sets in $\mathbf{R}^n$, then from the inclusion $A \subset B$, it follows that $\gamma_h(A) \leqslant \gamma_h(B)$.

The h-imbeddability is a simpler characteristic of the set. Due to this, further estimates are more convenient to obtain for this characteristic, than for the Hausdorff measure. At the same time, h-imbeddability, in some sense, proves to be equivalent to the Hausdorff h-measure, as follows from the next lemma:

Lemma 5.9. *The Hausdorff h-measure of a set A is equal to zero iff its h-imbeddability equals zero.*

We leave it to the reader to carry out the proof of the lemma.

Lemma 5.10. *Let μ be an arbitrary measure in the space $\mathbf{R}^n$, such that $\mu(\mathbf{R}^n) < \infty$, and $h(r)$, $0 \leqslant r \leqslant \infty$ is a nondecreasing function; besides, $h(0) = 0$ and $h(r) \to \infty$ for $r \to \infty$. Denote by A_λ, $\lambda > 0$ a set of all $x \in \mathbf{R}^n$ for which at any $r > 0$, the inequality $\mu[B(x,r)] \leqslant h(r)/\lambda$ holds. Then the following estimate is valid:*

$$\gamma_h(R^n \setminus A_\lambda) \leqslant C_n \lambda \mu(R^n).$$

This is the well-known Cartan lemma. For its proof for the two-dimensional case, see, for instance, [58]. For the case of arbitrary n, the proof is contained in [43].

Let us establish some formula of transformation of multiple integrals.

Lemma 5.11. *Let $F(r)$, $0 \leqslant r < \infty$, be a nonnegative decreasing function such that $F(r) \to 0$ for $r \to \infty$. Suppose that $F(r)$ has a continuous derivative $F'(r)$ for all $r > 0$. Then for every nonnegative measurable function $u(x)$, $x \in \mathbf{R}^n$, the equality*

$$\int_{R^n} F(|x-y|)\, u(y)\, dy = -\int_0^\infty \left(\int_{B(x,r)} u(y)\, dy \right) F'(r)\, dr$$

is valid.

Proof. Let $\chi_r(x,y) = 1$ for $|x-y| < r$, and $\chi_r(x,y) = 0$ for $|x-y| \geqslant r$. Then

$$\int_{B(x,r)} u(y)\, dy = \int_{R^n} \chi_r(x,y)\, u(y)\, dy.$$

Hence,

$$\int_0^\infty \left(\int_{B(x,r)} u(y)\,dy \right) F'(r)\,dr = \int_0^\infty \left(\int_{R^n} \chi_r(x, y)\,u(y)\,dy \right) F'(r)\,dr.$$

We apply the Fubini theorem to the right-hand integral. As a result,

$$\int_0^\infty \left(\int_{B(x,r)} u(y)\,dy \right) F'(r)\,dr = \int_{R^n} \left(\int_0^\infty \chi_r(x, y)\,F'(y)\,dr \right) u(y)\,dy.$$

It is easy to see that for any x, y,

$$\int_0^\infty \chi_r(x, y)\,F'(r)\,dr = -F(|x-y|).$$

The lemma is proved.

Let $\beta_l(|x|) = G_l(x)$, where G_1 is the Bessel kernel.

Theorem 5.12. *Let $h(r)$, $0 \leqslant r_m < \infty$, be a nondecreasing function such that $h(0) = 0$ and $h(r) \to \infty$ for $r \to \infty$. Suppose that*

$$\int_0^\infty [h(r)]^{1/p}\, r^{n-n/p}\, |\beta_l'(r)|\,dr = h_0 < \infty.$$

Then for every set $E \subset \mathbf{R}^n$, the following inequality

$$\gamma_k(E) \leqslant \sigma_n^{p-1} C_n h_0^p \operatorname{Cap}_{l,p}(E) \tag{5.1}$$

holds, where σ_n is the volume of the unit ball in $\mathbf{R}^n$, and C_n is the constant of Lemma 5.10.

Proof. Let $E \subset \mathbf{R}^n$ be an arbitrary set. We set $\operatorname{Cap}_{l,p} E < \infty$, since otherwise inequality (5.1) is obvious.

Let us introduce the following notation. For an arbitrary nonnegative measurable function $f(x)$ in the space $\mathbf{R}^n$, we put

$$\theta(x, r, f) = \int_{B(x,r)} f(y)\,dy.$$

Now let $u \in L_p(\mathbf{R}^n)$ be an arbitrary nonnegative function such that $G_l u \geqslant 1$ for all $x \in E$.

By transforming the integral according to Lemma 5.11, we obtain

$$(G_l u)(x) = \int_0^\infty \theta(x, r, u)\,|\beta_l'(r)|\,dr. \tag{5.2}$$

The magnitude $\theta(x, r, u)$ is estimated by means of the Hölder inequality. This yields the following result:

$$\theta(x, r, u) \leqslant \sigma_n^{1-1/p} r^{n-n/p} [\theta(x, u^p, r)]^{1/p}.$$

Hence,

$$(G_l u)(x) \leqslant \sigma_n^{1-1/p} \int_0^\infty [\theta(x, r, u^p)]^{1/p} r^{n-n/p} |\beta_l'(r)| \, dr.$$

Let us arbitrarily assign $\lambda > 0$ and denote by A_λ a set of all $x \in \mathbf{R}^n$ for which

$$\theta(x, r, u^p) \leqslant h(r)/\lambda.$$

By applying Lemma 5.10 to the measure $\mu(E) = \int_E u^p dx$, we obtain

$$\gamma_h(R^n \setminus A_\lambda) \leqslant C_n \lambda \int_{R_n} [u(x)]^p \, dx.$$

For $x \in A_\lambda$, we have

$$(G_l u)(x) \leqslant \sigma_n^{1-1/p} \int_0^\infty [\theta(x, r, u^p)]^{1/p} r^{n-n/p} |\beta_l'(r)| \, dr$$

$$\leqslant \frac{\sigma_n^{1-1/p}}{\lambda^{1/p}} \int_0^\infty [h(r)]^{1/p} r^{n-n/p} |\beta_l'(r)| \, dr = \frac{\sigma_n^{1-1/p} h_0}{\lambda^{1/p}}.$$

Now let λ be such that $\lambda^{1/p} > \sigma_n^{1-1/p} h_0$. Then for all $x \in A_\lambda$, we have $(G_l u)(x) < 1$. Since for all $x \in E$, $(G_l u)(x) \geqslant 1$, $E \subset \mathbf{R}^n \setminus A_\lambda$. Thus, we obtain the estimate

$$\gamma_h(E) \leqslant \gamma_h(R^n \setminus A_\lambda) \leqslant C_n \lambda \int_{R^n} [u(x)]^p \, dx. \tag{5.3}$$

Since here λ is an arbitrary number exceeding $\sigma_n^{p-1} h_0$, it follows from inequality (5.3) that

$$\gamma_h(E) \leqslant C_n \sigma_n^{p-1} h_0^p \int_{R^n} [u(x)]^p \, dx. \tag{5.4}$$

Due to arbitrariness of $u \in M_{l,p}(E, \mathbf{R}^n)$ from inequality (5.4), it follows that

$$\gamma_h(E) \leqslant C_n \sigma_n^{p-1} h_0^p \operatorname{Cap}_{l,p} E,$$

and this proves the theorem.

Corollary 1. *If the (l,p)-capacity of the set $E \subset \mathbf{R}^n$, $0 < l < n$, equals zero, then for every nondecreasing function $h(r)$, $0 \leqslant r \leqslant \infty$, such that $h(0) = 0$,*

$$\int_0^1 \frac{[h(r)]^{1/p}}{r^{n/p-l+1}} \, dr < \infty, \tag{5.5}$$

the Hausdorff h-measure equals zero.

Proof. Let us redefine the function $h(r)$ on the segment $[1, \infty)$ and let us assume it to be equal to K where $K = \text{const}$. The Hausdorff h-measure A does not change here. For $r \to 0$,

$$|\beta_l'(r)| = \frac{c}{r^{n-l+1}} [1 + o(r)]$$

and $|\beta_l'(r)| = 0(l^{-r})$ for $r \to \infty$. Hence it follows that if inequality (5.5) holds for the function h, then

$$h_0 = \int_0^\infty [h(r)]^{1/p} r^{n-n/p} |\beta_l'(r)| \, dr < \infty,$$

and so, if $\mathrm{Cap}_{l,p} E = 0$, then, due to Lemma 5.9, $\gamma_h E = 0$; consequently, $\mu_h(E) = 0$, due to Lemma 5.11.

Corollary 2. *Let $E \subset \mathbf{R}^n$ be such that $\mathrm{Cap}_{l,p} E = 0$ where $lp \leqslant n$. Then for every $\alpha > n - lp$, $\mu_\alpha(E) = 0$.*

To prove this, it suffices to take $h(r) = r^\alpha$ in Corollary 1.

5.4. Sufficient Conditions for the Vanishing of (l,p)-Capacity

Lemma 5.13. *Let $\psi_{l,p}(r)$ be (l,p)-capacity of the ball with radius r, where $lp \leqslant n$. Then for $r \to 0$,*

$$\psi_{l,p}(r) = O(r^{n-lp})$$

in the case $n > lp$, and

$$\psi_{l,p}(r) = O\left[\left(\ln \frac{1}{r}\right)^{1-p}\right]$$

in the case $lp = n$.

Proof. Let $B_r = B(0,r)$ be a ball with radius r and with the centre in the origin of coordinates. We set $r < 1$. Let $u(x) = 1/|x|^k$, where $k \geqslant n/p$ for $r \leqslant |x| \leqslant 1$, and $u(x) = 0$ for the remaining x. We have

$$\int_{R^n} [u(x)]^p \, dx = \omega_{n-1} \int_r^1 \frac{\rho^{n-1} d\rho}{\rho^{pk}} = \frac{\omega_{n-1}}{pk-n} \left(\frac{1}{r^{pk-n}} - 1 \right)$$

in the case $k > n/p$, and

$$\int_{R^n} [u(x)]^p \, dx = \omega_{n-1} \ln \frac{1}{r}$$

in the case $k = n/p$.

Now let us establish estimates from below for the potential $G_l u$ on the ball B_r. For $x \in B_r$, we have

$$(G_l u)(x) = \int_{r \leqslant |y| \leqslant 1} G_l(x-y) \frac{dy}{|y|^k} .$$

From the properties of G_l, it follows that there exists a constant $K_l > 0$ such that for $|x| \leqslant 2$,

$$G_l(x) \geqslant K_l/|x|^{n-l}. \tag{5.6}$$

From inequality (5.6), it follows that

$$(G_l u)(x) > K_l \int_{r \leqslant |y| \leqslant 1} \frac{dy}{|x-y|^{n-l}|y|^k}$$

for all $x \in B_r$. For $x \in B_r$ and $y \neq B_r$, we obviously have $|x-y| \leqslant |x|+|y| \leqslant 2|y|$, since in this case, $|x| \leqslant |y|$. Hence we obtain that for all $x \in B_r$,

$$(G_l u)(x) \geqslant \frac{K_l}{2^{n-l}} \int_{r \leqslant |y| \leqslant 1} \frac{dy}{|y|^{n-l+k}}.$$

The latter integral equals $K'(1/r^{k-l}-1)$ for $k \neq l$ and $K' ln(1/r)$ for $k = l$. For the (l,p)-capacity of the ball B_r, we have the following estimate:

$$\mathrm{Cap}_{(l,p)} B_r \leqslant \int_{R^n} [u(y)]^p \, dy / [\min_{x \in B_r} (G_l u)(x)]^p.$$

Let $n > lp$. Let us put $K > n/p > l$. Inequality (5.7) results in the following estimate:

$$\mathrm{Cap}_{(l,p)} B_r \leqslant K'' \frac{r^{kp-lp}(r^{n-pk}-1)}{(1-r^{k-l})^p} = K'' \frac{r^{n-lp}(1-r^{kp-n})}{(1-r^{k-l})^p}.$$

This proves that $\mathrm{Cap}_{(l,p)}(B_r) = 0(r^{n-lp})$ for $r \to 0$.

Now let $n = lp$. Let us put $k = n/p = l$. Inequality (5.7) yields

$$\mathrm{Cap}_{(l,p)} B_r \leqslant K'' \left(\ln \frac{1}{r}\right)^{1-p}.$$

Theorem 5.14. *Let* $h(r) = r^{n-lp}$ *for* $n > lp$ *and* $h(r) = (\ln(1/r))^{1-p}$ *for* $0 < r \leqslant 1/2$ *in the case* $lp = n(p > 1)$. *If the Hausdorff h-measure of the set* $E \subset \mathbf{R}^n$ *equals zero, then its* (l,p)*-capacity equals zero.*

Proof. Let $B_1, B_2 \ldots$ be an arbitrary sequence of balls covering the set E, such that their radii $r_\nu \leqslant 1/2$, $\nu = 1, 2, \ldots$. Then, due to Lemma 5.13, we have

$$\mathrm{Cap}_{(l,p)} E \leqslant \sum_\nu \mathrm{Cap}_{(l,p)} B_\nu \leqslant K \sum_\nu h(r_\nu),$$

where K is a constant. Due to the arbitrariness in choice of the sequence $\{B_\nu\}$, we have $\mathrm{Cap}_{(l,p)} E \leqslant K\mu_h(E,\varepsilon)$. Taking the limit for $\varepsilon \to 0$, we finally obtain

$$\mathrm{Cap}_{(l,p)} E \leqslant K\mu_h(E) = 0, \quad \mathrm{Cap}_{(l,p)} E = 0.$$

The theorem is proved.

§6 Estimates of $[l,p]$-Capacity for Some Pairs of Sets

A pair of nonintersecting compact sets $(F_0, F_1) \subset G$ is called a condenser in a domain G, or just a condenser when the definition domain is fixed beforehand.

6.1. Estimates of Capacity for Spherical Domains

Let ω be a domain on a sphere $S(0,1)$. The domain $D_{R,r}(\omega) = \{(\rho,\theta)|r < \rho < R, \theta \in \omega\}$ is said to be spherical (θ are current coordinates on the unit sphere). If $\omega = S(0,1)$, then $D_{R,r}(\omega)$ coincides with the ring $\{x/r < |x| < R\}$. In the domain $D_{R,r}(\omega)$ let us consider the condenser $(\overline{R\omega}, \overline{r\omega})$, where $\lambda\omega = \{x \in \mathbf{R}^n | x = \lambda\theta, \theta \in \omega\}$. We are interested in the (l,p)-capacity of the condenser $(\overline{R\omega}, \overline{r\omega})$ with respect to the domain $D_{R,r}(\omega)$.

Let us consider an arbitrary function $u \in M_{(l,p)}(\overline{R\omega}, \overline{r\omega}; D_{R,r})$. From the definition of an admissible function, it follows that the derivatives of the function u up to the order l inclusive vanish in some neighbourhood of the set $\overline{r\omega} \cup \overline{R\omega}$. For almost all $\theta \in \omega$, the function $u_\theta(\rho) = u(\rho,\theta)$ belongs to $L_p^l((r,R))$. For such θ, let us apply the Taylor formula with the residual term in the integral form to the function $u_\theta(\rho)$:

$$u_\theta(R) - u_\theta(r) = \frac{(-1)^l}{(l-1)!}\int_r^R \rho^{l-1} u_\theta^{(l)}(\rho)\, d\rho.$$

According to the above, $u_\theta^{(k)}(\rho) = 0$ for $\rho = R$, $\rho = r$, $k = 1, 2, \ldots, l-1$. Recalling that for any admissible function u, $u_\theta(R) - u_\theta(r) \geqslant 1$, we obtain the estimate

$$1 \leqslant \frac{1}{(l-1)!}\int_r^R \rho^{l-1} |\nabla_l u|(\rho,\theta)\, d\rho.$$

Integrating by θ, we obtain the inequality

$$|\omega| < \frac{1}{(l-1)!}\int_r^R\int_\omega \rho^{l-1} |\nabla_l u|(\rho,\theta)\, d\rho\, d\theta, \tag{6.1}$$

where $|\omega|$ is the Lebesgue measure of ω on the sphere.

If $p = 1$, then by setting $\rho^{l-1} = \rho^{l-n}\rho^{n-1}$ in (6.1), we obtain

$$|\omega| \leqslant \frac{1}{(l-1)!}\int_r^R\int_\omega \rho^{l-n} |\nabla_l u|(\rho,\theta)\rho^{n-1}\, d\rho\, d\theta$$

$$\leqslant \begin{cases} \dfrac{1}{(l-1)!} R^{l-n} \displaystyle\int_{D_{R,r}} |\nabla_l u|\, dx & \text{for } l \geqslant n, \\ \dfrac{1}{(l-1)!} r^{l-n} \displaystyle\int_{D_{R,r}} |\nabla_l u|\, dx & \text{for } l < n. \end{cases} \tag{6.2}$$

For $p > 1$, let us represent ρ^{l-1} in the form $\rho^{\frac{n-1}{p}} \cdot \rho^{-\frac{n-1}{p}+l-1}$ and let us estimate the integral in (6.1) by the Hölder inequality:

$$|\omega| \leqslant \frac{1}{(l-1)!} \int_r^R \int_\omega |\nabla_l u|(\rho, \theta) \rho^{\frac{n-1}{p}} \rho^{\frac{lp-n-p+1}{p}} d\rho \, d\theta$$
$$\leqslant \frac{1}{(l-1)!} \left(\int_{D_{R,r}(\omega)} |\nabla_l u|^p \, dx \right)^{\frac{1}{p}} \left(|\omega| \int \rho^{\frac{lp-n-p+1}{p-1}} d\rho \right)^{\frac{p-1}{p}}. \quad (6.3)$$

From (6.3), we obtain

$$\int_{D_{R,r}} |\nabla_l u|^p \, dx \geqslant \begin{cases} |\omega| [(l-1)!]^p \left[\dfrac{(R^{\frac{lp-n}{p-1}} - r^{\frac{lp-n}{p-1}})(p-1)}{lp-n} \right]^{1-p}, & lp \neq n, \\ |\omega| [(l-1)!]^p \left[\ln \dfrac{R}{r} \right]^{1-p}, & lp = n, \end{cases}$$
$$(6.4)$$

Proposition 6.1. *If $lp > n$ for $l \geqslant n$, $p = 1$, then for $R > 2r$,*[2]

$$C_p^l(\overline{R\omega}, \overline{r\omega}; D_{R,r}\omega) \sim R^{n-lp} |\omega|^1.$$

Proof. Let us consider a monotone function $\alpha \in C^\infty(1/2, 1)$ which equals zero on the interval $(1 - 1/8, 1)$ and equals 1 on the interval $(1/2, 5/8)$. The estimate from above,

$$C_1^l(\overline{R\omega}; \overline{r\omega}; D_{R,r}(\omega)) \leqslant |\omega| R^{n-l} \int_{1/2}^1 \alpha(\rho) \, d\rho.$$

is obvious.

For $p = 1$, from inequality (6.2), due to the arbitrariness in the choice of the admissible function u, the estimate from below follows:

$$(l-1)! \, |\omega| \, R^{n-l} \leqslant C_1^l(\overline{R\omega}, \overline{r\omega}; D_{R,r}(\omega)).$$

If $p > 1$, then the desired estimate of the (l, p)-capacity $C_p^l(\overline{R\omega}, \overline{r\omega}; D_{R,R}(\omega))$ is obtained from the inequalities (6.3).

Proposition 6.2. *If $lp = n$ and $p > 1$, then*

$$C_p^l(R\omega, r\omega; D_{R,r}(\omega)) \sim |\omega| \left(\ln \frac{R}{r} \right)^{1-p}.$$

Proof. The estimate from below follows directly from (6.4):

---

[2] The sign $\sim$ hereafter denotes the existence (for capacity) of estimates from above and below via R^{n-lp}, where the constants only depend on l, p, and on the dimension of the space.

$$[(l-1)!]^p\,|\omega|\left[\ln\frac{R}{r}\right]^{1-p}\leqslant C_p^l(\overline{R\omega},\overline{r\omega};D_{R,r}(\omega)).$$

To estimate the capacity from above, let us consider on the domain the function

$$v(x)=\left[\ln\frac{R}{r}\right]^{-1}\ln\frac{R}{|x|}.$$

Let $u(x)=\alpha[2/3v(x)+1/2)]$, where α is the same function as in the proof of Proposition 6.1. It is clear that $u\in M_{l,p}(\overline{R\omega},\overline{r\omega};D_{R,r}(\omega))$. Differentiating the function u, we obtain that $|\nabla_l u(x)|\leqslant C[\ln(R/r)]^{-1}|x|^{-l}$. Therefore,

$$C_p^l(\overline{R\omega},\overline{r\omega};D_{R,r}(\omega))\leqslant\int_{D_{R,r}(\omega)}|\nabla_l u|^p\,dx$$

$$\leqslant C\left[\ln\frac{R}{r}\right]^{-p}\int_{D_{R,r}}|x|^{-lp}dx=C\left[\ln\frac{R}{r}\right]^{1-p}.$$

Proposition 6.3. *If $lp<n,\ p>1$ or $l<n,\ p=1$, then for $R>2r$,*

$$C_p^l(\overline{R\omega},\overline{r\omega};D_{R,r}(\omega))\sim r^{n-lp}|\omega|.$$

Proof is similar to that of Proposition 6.1.

Proposition 6.4. *If $l=1$, then*

$$C_p^l(\overline{R\omega},\overline{r\omega};D_{R,r}(\omega))=|\omega|\left\{\frac{R^{\frac{p-n}{p-1}}-r^{\frac{p-n}{p-1}}}{p-\dfrac{n}{p-1}}\right\}^{1-p}\quad\text{for}\quad p\neq n;$$

$$C_p^l(\overline{R\omega},\overline{r\omega};D_{R,r}(\omega))=|\omega|\left(\ln\frac{R}{r}\right)^{1-n}\quad\text{for}\quad p=n,$$

$$C_p^l(\overline{R\omega},\overline{r\omega};D_{R,r}(\omega))=|\omega|\,r^{n-1}\quad\text{for}\quad p=1.$$

Proof. For $p\neq 1$ it suffices to verify that the lower estimates for capacity in Propositions 6.2, 6.3 for $l=1$ are obtained for the function

$$u(x)=\left(R^{\frac{p-n}{p-1}}-|x|^{\frac{p-n}{p-1}}\right)\left(R^{\frac{p-n}{p-1}}-r^{\frac{p-n}{p-1}}\right)^{-1}$$

if $p\neq n$, and for the function

$$\left(\ln\frac{R}{r}\right)^{-1}\ln\frac{R}{|x|}$$

if $p=n$.

If $p=1$, then from (6.2), we obtain the estimate from below:

$$|\omega|\,r^{n-1}\leqslant C_1^1(\overline{R\omega},\overline{r\omega};D_{R,r}(\omega)).$$

Let us consider the function $\alpha_\varepsilon(\rho)\in C^\infty\cup(\mathbf{R})$ which equals 1 for $\rho=r$ and equals zero for $\rho\geqslant r+\varepsilon$. For this function, we have a chain of inequalities,

$$C_1^1(\overline{R\omega}, \overline{r\omega}; D_{R,r}(\omega)) \leqslant |\omega| \int_r^{r+\varepsilon} \rho^{n-1} |\nabla \alpha_\varepsilon(\rho)| d\rho = \frac{|\omega|}{\varepsilon} \int_r^{r+\varepsilon} \rho^{n-1} d\rho.$$

By taking the limit for $\varepsilon \to 0$, we obtain the inequality

$$C_1^1(\overline{R\omega}, \overline{r\omega}; D_{R,r}(\omega)) \leqslant \omega_{n-1} r^{n-1}.$$

Proposition 6.5. *If $p = \infty$ and $R = 2r$, then*

$$C_\infty^l(\overline{R\omega}, \overline{r\omega}; D_{R,r}(\omega)) \sim R^{-l}.$$

Proof. Let $u \in M_{l,\infty}(\overline{R},\omega), \overline{r,\omega}; (D_{R,r}(\omega)))$. Then from (6.4) for $lp > n$, we obtain the inequality

$$|\omega|^{1/p} \operatorname*{ess\,sup}_{x \in D_{R,r}(\omega)} |\nabla_l u| \geqslant \left[\int_{D_{R,r}(\omega)} |\nabla_l u|^p dx \right]^{\frac{1}{p}}$$

$$\geqslant |\omega|^{1/p} [(l-1)!] \left[\frac{R^{\frac{l-n/p}{1-1/p}} - r^{\frac{l-n/p}{1-1/p}}}{(l-n/p)/(1-1/p)} \right]^{\frac{1}{p-1}}.$$

By taking the limit for $p \to \infty$, we obtain the estimate from below:

$$C_\infty^l(\overline{R\omega}, \overline{r\omega}; D_{R,r}(\omega)) \geqslant (l-1)!\, R^{-l}.$$

Since for $R = 1$, the $[l,\infty]$-capacity is bounded, by taking the transformation of similarity into account, we obtain the desired estimate from above.

6.2. Estimates of Capacity for Pairs of Continuums Connecting Concentric Spheres

Let us consider in the ring $D_{R,r} = \{x \in \mathbf{R}^2 | r < x < R\}$ two continuums F_0, F_1 connecting the spheres S_r and S_R, i.e., two connected closed sets $F_0, F_1 \subset \overline{D}_{R,r}$ such that $F_i \cap \zeta_r \neq \emptyset$, $F_i \cap S_R \neq \emptyset$, $i = 0, 1$.

Proposition 6.6. *For the (l,p)-capacity of the condenser (F_0, F_1) with respect to the domain $D_{R,r}$, the estimates*

$$C_p^l(F_0, F_1; D_{R,r}) \geqslant \begin{cases} C \ln \frac{R}{r} & \text{for } lp = 2, \\ \frac{C}{lp-2}(r^{2-lp} - R^{2-lp}) & \text{for } lp > 2, \\ \frac{C}{2-lp}(r^{2-lp} - R^{2-lp}) & \text{for } lp < 2. \end{cases}$$

are valid.

Proof. Let us consider an arbitrary function $u \in M_{l,p}(F_0, F_1; D_{R,r})$. The integral $\int_{D_{R,r}} |\nabla_l u|^p dx$ is denoted by polar coordinates,

$$\int_{D_{R,r}} |\nabla_l u|^p \, dx = \int_r^R \rho \, d\rho \int_0^{2\pi} |\nabla_l u| (\rho, \theta) \, d\theta.$$

For almost all $\rho \in (r, R)$, the function $u_\rho(\theta) = u(\rho, \theta) \in L_p^l(S(\rho))$, where $S(\rho)$ is a circle with the centre at zero, with radius ρ. It is obvious that the function $u_\rho(\theta) \in M_{l,p}(F_0 \cap S(\rho), F_1 \cap S(\rho), S(\rho)$. From Proposition 6.1 applied to the circle $S(\rho)$, we obtain for $p > 1$,

$$\int_{S_\rho} |\nabla_l u|^p (\rho, \theta) \, d\theta = \int_0^{2\pi} |\nabla_l u|^p (\rho, \theta) \rho \, d\theta$$

$$\geqslant C_p^l (F_0 \cap S(\rho), \; F_1 \cap S(\rho), \; S(\rho)) \geqslant C \rho^{1-lp}.$$

By integrating by ρ, we obtain

$$\int_{D_{R,r}} |\nabla_l u|^p \, dx \geqslant C \int_r^R \rho^{1-lp} d\rho = \frac{C}{lp-2} [r^{2-lp} - R^{2-lp}]$$

for $lp \neq 2$, and

$$\int_{D_{R,r}} |\nabla_l u|^p \, dx \geqslant C \ln \frac{R}{r}$$

for $lp = 2$.

The case $p = 1$ is obtained by limit transition by p; the case $p = \infty$ is proved in the way similar to the proof of Proposition 6.5.

Proposition 6.7. *Let $U \subset \mathbf{R}^n$ be a domain with smooth boundary,*[3] *with (F_0, F_1) being a pair of nonintersecting nonempty compact sets belonging to U.*

Then for $lp > n$, the (l,p)-capacity $C_p^l(F_0, F_1; U)$ exceeds zero.

Proof. Due to monotonicity of the (l,p)-capacity with respect to the pair of sets (F_0, F_1), it suffices to prove the theorem for a pair of points only.

Suppose that $C_p^l(\{a\}, \{b\}, U) = 0$. According to the definition of $[l,p]$-capacity, there exists a sequence $\{u_m\}$ of $[l,p]$-admissible functions for the pair (F_0, F_1) in the domain U which converge to zero in $L_p^l(U)$. Since $u_m(a) \leqslant 0$, $u_m(b) \geqslant 1$, and since the domain U has smooth boundary, one may consider the sequence $\{u_m\}$ to converge to zero in the norm of $W_p^l(U)$.

Due to the imbedding theorem for the spaces $W_p^l(U)$ for $lp > n$, it follows that the sequence $\{u_m\}$ converges to zero uniformly in the domain U. But this contradicts the inequalities $u_m(a) \leqslant 0$ and $u_m(b) \geqslant 1$.

This completes the proof.

For bounded domains, Proposition 6.7 may be strengthened.

[3] It is sufficient that the boundary belongs to the class C^1. This is exactly what the term "smooth boundary" means in all the remaining cases.

Corollary. *Under the conditions of Proposition 6.7, we impose additional boundedness upon the domain U. Then for $lp > n$, there exists a constant $\alpha^2(l,p,U) > 0$ such that*

$$C_p^l(F_0, F_1; U) \geqslant \alpha^2(l, p, U).$$

Proof almost literally repeats that of Proposition 6.7. It is only necessary to assume the pair of the points $(\{a\},\{b\})$ to be mobile.

Proposition 6.8. *Let $lp > n$. There exists a constant $\alpha^2(l,p) > 0$ such that*

$$C_p^l(F_0, F_1; B(0, r)) \geqslant [\rho(F_0, F_1)]^{n-lp}\alpha^2(l, p)$$

for any pair of nonintersecting compact sets $F_0, F_1 \subset B(0,r)$.

Proof. The monotonicity of (l,p)-capacity with respect to the pairs (F_0, F_1) allows us to consider the pairs of points $(\{a\},\{b\})$ instead of arbitrary pairs of compact sets (F_0, F_1).

Let us first consider the case for the ball $B(0,1)$. For any l, p, there exists an extension operator θ_p^l from $L_p^l(B(0,1))$ to $L_p^l(\mathbf{R}^n)$, $\theta_p^l u|_{B(0,1)} = u$ for any function $u \in L_p^l(B(0,1))$. Recalling the definition of $[l,p]$-capacity, we directly obtain inequalities of $[l,p]$-capacities of the pairs of points from the existence of the extension operator:

$$C_{a,b} = C_p^l(\{a\}, \{b\}, R^n) \leqslant \|\theta_p^l\|^p C_p^l(\{a\}, \{b\}, B(0, 1)).$$

Since the $[l,p]$-capacity in $\mathbf{R}^n$ is invariant in orthogonal transformations, it suffices to estimate the $[l,p]$-capacity of the pairs of points in $\mathbf{R}^n$:

$$(\{a_t = (t/2, 0, \ldots, 0)\}, \{b_t = (-t/2, 0, \ldots, 0)\}).$$

For $t = 1$, from the corollary of Proposition 6.6 and from the monotonicity of $[l,p]$-capacity with respect to the domain, the inequality

$$C_t^{l,p} = C_p^l(\{a_t\}, \{b_t\}, R^n) \geqslant \alpha^2(l, p, B(0, 1))$$

immediately follows. Taking into account the obtained inequalities and the behaviour of the $[l,p]$-capacity for similarities, we obtain the estimate from below:

$$C_p^l(\{a\}, \{b\}, B(0, 1)) \geqslant \frac{C_t^{l,p}}{\|\theta_p^l\|^p} \geqslant \frac{[\rho(\{a\}, \{b\})]^{n-lp}}{\|\theta_p^l\|^p}\alpha^2(l, p).$$

The theorem is proved for $r = 1$.

Again using the behaviour of the $[l,p]$-capacity for similarities, let us compute the $[l,p]$-capacity of the pair of points in $B(0,r)$ via $[l,p]$-capacity, like the transformed pair in $B(0,1)$:

$$C_p^l\left(\left\{\frac{a}{r}\right\}, \left\{\frac{b}{r}\right\}, B(0, 1)\right) = r^{n-lp} C_p^l(\{a\}, \{b\}; B(0, r)).$$

This allows us to complete the proof of the theorem with the simple estimate

$$C_p^l(\{a\}, \{b\}; B(0, r)) \geqslant r^{n-lp}\left[o\left(\left\{\frac{a}{r}\right\}\left\{\frac{b}{r}\right\}\right)\right]^{n-lp} \frac{\alpha^2(l, p)}{\|\theta_p^l\|^p}.$$

Corollary. *Let $lp > n$. Then for any pair of points $\{a\}, \{b\} \subset B(0,r)$,*

$$C_p^l(\{a\}, \{b\}; B(0, r)) \sim |a-b|^{n-lp}.$$

Proof. Just as in Proposition 6.8, we reduce the problem to the case $\mathbf{R}^n$ and use the inequality

$$C_p^l(\{a\}, \{b\}; B(0, r)) \leqslant C_p^l(\{a\}, \{b\}; R^n).$$

The invariance of $[l,p]$-capacity with respect to translations and orthogonal transformations allows us to regard the points a and b as symmetric with respect to zero of the space $\mathbf{R}^n$ and as lying on the axis $0x_1$.

Let us apply the similarity $\varphi(x) = \frac{2x}{a-b}$. Now $\varphi(a)$ and $\varphi(b)$ are symmetric points on the unit sphere. Let u be any function from $L_p^l(\mathbf{R}^n)$ admissible for the pair $\{a\}, \{b\})$. As a result of the same calculations as those at the end of the proof of the previous proposition, we obtain

$$C_p^l(\{a\}, \{b\}, R^n) \leqslant |a-b|^{n-lp} \frac{\|u\|_{L_p^l(R^n)}}{2^{n-lp}}.$$

This completes the proof.

§7 Capacity in Besov–Nickolsky Spaces

7.1. Preliminary Information

Let us recall the definition of Nickolsky–Besov spaces in domains of Euclidean space. When studying necessary extension conditions, it makes no difference which variant of definitions we use. Therefore, we give here the one which is most familiar to the authors from a technical point of view. In the definition and in the properties of these classes of spaces, we follow the book of Besov, Ilyin, and Nickolsky [8].

We use the notations of finite differences,

$$\begin{aligned}\Delta^m(z)f(x) &= \Delta(z)[\Delta^{m-1}(z)f(x)] \\ &= \sum_{j=0}^{m}(-1)^{m-j}C_m^j f(x+jz)\end{aligned}$$

for $x \in \mathbf{R}^n$, $z \in \mathbf{R}^n$, for the natural $m > 1$. Here $\Delta(z)f(x) = f(x+z) - f(x)$. In the domain $G \subset \mathbf{R}^n$,

$$\Delta^m(z,G)f(x) = \begin{cases} \Delta^m f(x) & \text{for } [x, x+mz] \subset G \\ 0 & \text{for } [x, x+mz] \not\subset G. \end{cases}$$

Definition. A linear normed space of functions $f(x)$ which are defined on G with the norm

$$\|f\|_{B^l_{p,\theta}(G,h)} = \|f\|_{L_p(G)} + \|f\|_{(l,p,\theta,G,h)},$$

where

$$\|f\|_{(l,p,\theta,G,h)} = \left\{ \int\limits_{|t|<h} \frac{\|\Delta^k(t,G)f\|_{L_p(G)}}{t^{\theta l-n}} \right\}^{1/\theta}, \quad k = [l]$$

is called the space $b^l_{p,\theta}(G,h)$.

A seminormed space of functions f which are defined on G with the seminorm $\|f\|_{l,p,\theta,G,h}$ is called the space $b^l_{p,\theta}(G,h)$.

7.2. Capacities in $b^l_{l,p,\theta,G,h}$. Simplest Properties

The capacity associated with the seminorm $\|\cdot\|_{l,p,\theta,G,h}$ is denoted by $C^l_{p,\theta}(F_0, F_1, G, h)$.

Let us give the properties specific for this type of capacities:

1) *Monotonicity relative to h.* If $0 < h_1 < h < \infty$, then

$$\tfrac{1}{K} C^l_{p,\theta}(F_0, F_1, G, h) \leqslant C^l_{p,\theta}(F_0, F_1, G, h_1) \leqslant K C^l_{p,\theta}(F_0, F_1, G, h).$$

The constant K does not depend on F_0, F_1.

The proof is based on the equivalence of seminorms $\|\cdot\|_{l,p,\theta,G,h}$ for different finite h [10].

2) *Positiveness.* Let $G \subset \mathbf{R}^n$ be a domain with smooth boundary, and (F_0, F_1) are pairs of nonintersecting compact sets belonging to G.

Then for $lp > n$, the capacity $C^l_{p,\theta}(F_0, F_1, G, h)$ exceeds zero.

Proof. For $lp > n$, the imbedding $B^l_{p,\theta}(G,h) \to C(G)$ is valid [10]. Consequently, by the corollary of Theorem 4.6 for any of equivalent seminorms $\|\cdot\|_{l,p,\theta,G,h}$, our statement holds.

3) Let $\varphi_{a,x_0}(x) = a(x - x_0)$. Then

$$C^l_{p,\theta}(F_0, F_1, \mathbf{R}^n, \infty) = |a|^{\frac{n}{p}-l} C^l_{p,\theta}(\varphi_{a,x_0}(F_0), \varphi_{a,x_0}(F_1), \mathbf{R}^n, \infty)$$

Proof. In the transformation of the similarity $\varphi_{a,x}(x) = a(x - x_0)$, the function f admissible for the pair of compact sets (F_0, F_1) is transformed into the function $f \circ \varphi_{a,x_0}$ which is admissible for the pair of compact sets $(\varphi_{a,x_0}(F_0))$, $\varphi_{a,x_0}(F_1))$. From the definition of the seminorm $\|\cdot\|_{l,p,\theta,G,h}$, we have

$$\|f\|_{l,p,\theta,\mathbf{R}^n,\infty} = |a|^{\frac{n}{p}-l} \|f \circ \varphi_{a,x_0}\|_{l,p,\theta,\mathbf{R}^n,\infty}.$$

Hence, due to the definition of the capacity $C^l_{p,\theta}$, we obtain the desired equality for capacities.

7.3. Comparison of Capacity of a Pair of Points to Capacity of a Point Relative to a Complement of a Ball

Let x, y be points from $\mathbf{R}^n$. Let us consider a capacity $C^l_{p,\theta}(\{x\},\{y\},\mathbf{R}^n,\infty)$. Due to Property 3 for the capacity of $C^l_{p,\theta}$, the function $C^l_{p,\theta}(\{x\},\{y\},\mathbf{R}^n,\infty)$ is invariant with respect to isometries. Any pair of points (x,y) by means of isometry may be transformed to the pair $(0,\dots,0)$, $(0,\dots,|x-y|)$. Consequently, $C^l_{p,\theta}(\{x\},\{y\},\mathbf{R}^n,\infty)$ is the function of the distance between the points $|x-y|$. Let us denote it by $\gamma^l_{p,\theta}(x-y)$.

Similarly, we denote the function $C^l_{p,\theta}(\{x\},\mathbf{R}^n\backslash B(x,r),\mathbf{R}^n,\infty)$ by $\delta^l_{p,\theta}(r)$.

Lemma 7.1. *For the space $B^l_{p,\theta}(\mathbf{R}^n)$ for $lp>n$, the relation*

$$\gamma^l_{p,\theta}(t) = Ac^{l-\frac{n}{p}}\delta^l_{p,\theta}(ct)$$

is valid for some $A>0$, which only depends on n, l, p, θ.

Proof. Let us consider the capacities $\gamma(a)$ and $\delta(a)$, $a>0$. Let us prove that the relation does not depend on the choice of the number a.

Due to Property 2 for the capacities $C^l_{p,\theta}$, $\gamma(a)>0$ and $\delta(a)>0$ for any a. In the transformation of the similarity $\varphi_a:\mathbf{R}^n\to\mathbf{R}^n$, $\varphi_a(x)=ax$, due to Property 3 for capacities, we have:

$$C^l_{p,\theta}(F_0,F_1,\mathbf{R}^n,\infty) = a^{\frac{n}{p}-l}C^l_{p,\theta}(\varphi(F_0),\varphi(F_1),\mathbf{R}^n,\infty).$$

Consequently, for the functions $\gamma(a)$ and $\delta(a)$, the relations

$$\gamma(a) = a^{\frac{n}{p}-l}\gamma(1), \qquad \delta(a) = a^{\frac{n}{p}-l}\delta(1)$$

hold, i.e.,

$$\gamma(a)/\delta(a) = \gamma(1)/\delta(1) = A,$$

where $A>0$ only depends on n, l, p, θ.

Further,

$$\frac{\gamma(t)}{\delta(ct)} = \frac{A\gamma(t)}{\gamma(ct)} = \frac{At^{\frac{n}{p}-l}\gamma(1)}{(ct)^{\frac{n}{p}-l}\gamma(1)} = Ac^{l-\frac{n}{p}},$$

whence our statement follows.

Corollary. *For $lp<n$, the functions $\delta(t)$ and $\gamma(t)$ are identically equal to zero.*

Proof. Due to monotonicity of the capacity $\delta(a)\geqslant\delta(1)$ if $a<1$. On the other hand, $\delta(a)=a^{\frac{n}{p}-l}\delta(1)$. Therefore, $a^{(n/p)-l}\delta(1)>\delta(1)$, i.e., $\delta(1)=0$ and $\delta(a)=0$ for all a.

From the lemma. we obtain that $\gamma(t)=0$.

This completes the proof.

7.4. Capacity of the Spherical Layer

Proposition 7.2. *Let us consider in* $\mathbf{R}^n$ *two concentric balls* $B(0,r)$, $B(0,R)$, $R > r$. *There exists a function* $\psi^l_{p,\theta} : \mathbf{R} \to \mathbf{R}$, *which monotonically decreases at* $t \to \infty$, $\lim \psi^l_{p,\theta}(t) = 0$ *for* $t \to \infty$, *for which the inequality*

$$C^l_{p,\theta}(\mathbf{R}^n \backslash B(0,R), B(0,r), \mathbf{R}^n, n) \leqslant \psi^l_{p,\theta}(R/r)$$

is valid for all h, R, r. *Here* $lp = n$, $2 \leqslant \theta$, *if* $1 \leqslant p \leqslant 2$; $p \leqslant \theta$ *if* $2 \leqslant p \leqslant \theta$.

Proof. The capacity under consideration for $h = \infty$ is invariant for similarities, therefore, it only depends on the relation R/r. As $\psi(R/r)$ let us take this capacity $C^l_{p,\theta}(\mathbf{R}^n \backslash B(0,R), \overline{B(0,r)}, \mathbf{R}^n, \infty)$. Without loss of generality, one may set $R = 1$. From monotone capacity, it follows that the function $\psi(t)$ does not decrease for $t \to \infty$.

For $\psi(t)$, there are only two possibilities: $\lim\limits_{t\to\infty} \psi(t) = a^2$ or $\lim\limits_{t\to\infty} \psi(t) = 0$. Taking into account the continuity of capacity (Property 3), the realization of the former possibility means that $C^l_{p,\theta}(\mathbf{R}^n \backslash B(0,1), \{0\}, \mathbf{R}^n, \infty) = a^2 > 0$, and from the invariance for similarities, we directly obtain that

$$A_{l,p,\theta} = C^l_{p,\theta}(\mathbf{R}^n \backslash B(0,R), \{0\}, \mathbf{R}^n, \infty) = a^2 \geqslant 0$$

for all R. Under the constraints upon θ mentioned in the formulation of the proposition, the Liouville space $\mathcal{L}^l_p(\mathbf{R}^n)$ is imbedded into the space $B^l_{p,\theta}(\mathbf{R}^n, \infty)$. Consequently,

$$\begin{aligned} A^l_{p,\theta} &\leqslant C^l_{p,\theta}(\mathbf{R}^n \backslash B(0,R), \{o\}, \mathbf{R}^n, \infty) \\ &\leqslant K \operatorname{Cap}_{(l,p)}(\mathbf{R}^n \backslash B(0,R), \{0\}, \mathbf{R}^n) = 0, \end{aligned}$$

where K is the norm of the imbedding operator. Hence it follows that $\lim\limits_{t\to\infty} \psi(t) = 0$.

Corollary. *Let* G *be an arbitrary domain in* $\mathbf{R}^n$, $x_0 \in G$. *Then for any closed set* $F_1 \subset G$, $F_1 \cap \{x_0\} = \emptyset$, *it holds that* $C^l_{p,\theta}(\{x_0\}, F_1, G, h) = 0$ *for* $lp = n$; $2 < \theta$ *if* $1 \leqslant p \leqslant 2$, $p \leqslant \theta$ *if* $2 \leqslant p \leqslant \infty$.

Remark. In proving the proposition, it was shown that for $lp = n$ and for any θ, the capacity $C^l_{p,\theta}(\mathbf{R}^n \backslash B(0,R), \{0\}, \mathbf{R}^n, \infty) < \infty$, and its value is the constant of $A^l_{p,\theta}$ independent of R, i.e., $\lim\limits_{t\to\infty} \psi^l_{p,\theta}(t) = A^l_{p,\theta} \geqslant 0$.

For further presentation, we need two imbedding theorems.

Theorem 7.3 [8, 71]. *Let* $G \subset \mathbf{R}^n$ *be a domain with smooth boundary. Then, for* $l_1 < l$, *the imbedding* $B^l_{p,\theta_1} \to B^{l_1}_{p,\theta}$ *for any* θ_1, θ.

Theorem 7.4 [8]. *Let* $G \subset \mathbf{R}^n$ *be a domain with smooth boundary. Then for* $lp > n$, *the imbedding* $B^l_{p,1} \to C(G)$ *is valid.*

The arrow $\to$ denotes the existence of the linear bounded imbedding operator.

Lemma 7.5. *Let $lp > n-1$, and let a sequence of functions $\{f_m \in B^l_{p,\theta}(B(0,1))\}$ converge in $B^l_{p,\theta}(B(0,1))$ to the function f. Then for almost all r, the functions f_m are continuous on the spheres $S(0,r)$ and converge on these spheres uniformly to the function f.*

Proof. Let us imbed the space $B^l_{p,\theta}$ into $B^{l-\delta}_{p,1}$ according to Theorem 7.4. The number δ is chosen to be so small that $(l-\delta)p > n-1$. Let us imbed the space $B^{l-\delta}_{p,1}$ into $\mathcal{L}^{l-\delta}_p$ [8].

Any function u from $\mathcal{L}^{l-\delta}_p$ is continuous on almost all spheres due to the fact that $(l-\delta)p > n-1$. Indeed, due to Theorem 1.7 for every $\varepsilon > 0$ outside of some set H whose $[l-\delta, p]$-capacity is less than ε, the function u is continuous. Due to the fact that the set having small $[l-\delta, p]$- capacity because the inequality $[l-\delta)p > n-1$ has a small linear Hausdorff measure (Theorem 5.10), we obtain the continuity of the function u on almost all spheres $S(0,r)$. Due to the imbeddings considered above, we obtain continuity on almost all spheres for any function from $B^l_{p,\theta}$.

Similarly, by using the same imbedding in $\mathcal{L}^{l-\delta}_p$ and the analogy of Egorov's theorem (Theorem 1.6, Chapter 3), we obtain uniform convergence $f_m \to f$ on almost all spheres.

This completes the proof.

Proposition 7.6. (Estimate for the Teichmüller capacity). *Let the continuums F_0, F_1 connect two concentric spheres $S_1 = S(0,1)$ and $S_2 = S(0,2)$. Then for $lp > n-1$, $p \geqslant 1$ for any θ, the inequality*

$$C^l_{p,\theta}(F_0, F_1, B(0,2), h) \geqslant \gamma(n, l, p, h) > 0$$

is valid.

Suppose the contrary. Then there exists a sequence of continuous functions $u_m \in B^l_{p,\theta}$ converging to zero in the seminorm $b^l_{p,\theta,h}$ and equal to 1 on F_1. This sequence must uniformly converge to some function u on almost all spheres $S(0,r)$ which follows from Proposition 7.5. But then on every one of these spheres, the oscillation of the limit function should be equal to 1, since all the spheres $S(0,r)$ intersect the sets F_0 and F_1, i.e., $u_m \not\to u$ on almost all spheres $S(0,r)$. The obtained contradiction proves the statement of the above proposition.

Proposition 7.7. *Let $\{F^m_0\}$, $\{F^m_1\}$ be two monotone increasing sequences of continuums, $F^m_0 \cap S(0,1) \neq \emptyset$, $F^m_1 \cap S(0,1) \neq \emptyset$ for all m,*

$$\operatorname{dist}(F^m_0, \{0\}) \to 0, \qquad \operatorname{dist}(F^m_1, \{0\}) \to 0$$

for $m \to \infty$.

Then for $lp = n$, $1 \leqslant p < \infty$, $1 \leqslant \theta < \infty$,

$$C^l_{p,\theta}(F^m_0, F^m_1, \mathbf{R}^n, \infty) \to \infty$$

for $m \to \infty$.

Proof. Due to the existence of the imbedding $b^l_{p,\theta}(\mathbf{R}^n)$ into $b^l_{p,\theta_1}(\mathbf{R}^n)$ for $\theta_1 > \theta$ for any pair of the sets (F_0, F_1), the inequality

$$C^l_{p,\theta}(F_0, F_1, \mathbf{R}^n, \infty) \geqslant C^l_{p,\theta}(F_0, F_1, \mathbf{R}^n \infty)$$

is valid. Therefore, it suffices to prove the statement for the case $\theta \geqslant p$.

From the monotonicity of capacity,

$$C^l_{p,\theta}(F_0^m, F_1^m, \mathbf{R}^n, \infty) \geqslant C^l_{p,\theta}(F_0^m, F_1^m, B(0,1), \infty).$$

follows.

Let us prove that the limit of capacities $C^l_{p,\theta}(F_0^m, F_1^m, B(0,1)\infty)$ equals ∞. Let us consider a sequence of balls $\{B_s = B(0, 1/2^s)\}$. According to Proposition 7.6 and Lemma 7.1,

$$\alpha_s = \lim_m C^l_{p,\theta}(F_0^m, F_1^m, B(0,1/2^s)\backslash B(0,1/2^{s+1}), \infty) > \gamma(n,l,p) > 0$$

for all s.

Since $\theta \geqslant p$, then for any function $u \in b^l_{p,\theta}(B(0,1))$, the inequality

$$\|\Delta^m(t, B_1)u\|^\theta_{L_p(B_1)} \geqslant \sum_{\zeta=1}^{\infty} \|\Delta^m(t, B_s\backslash B_{s+1})u\|_{L_p(B_s B_{s+1})}$$

is valid for all t, that is,

$$\|u\|^\theta_{b^l_{p,\theta}(B_1,\infty)} \geqslant \sum_{\zeta=1}^{\infty} \|u\|^\theta_{b^l_{p,\theta}(B_\zeta\backslash B_{\zeta+1},\infty)}.$$

Therefore,

$$\lim_{m\to\infty} [C^l_{p,\theta}(F_0^m, F_1^m, B(0,1), \infty]^\theta \geqslant \sum_{\zeta=1}^{\infty} \alpha_s^\theta = \infty.$$

The proposition is proved.

CHAPTER 4

Density of Extremal Functions in Sobolev Spaces With First Generalized Derivatives

Every function of a class $L_p(G)$ (G is a domain in $\mathbf{R}^n$) may be represented as a sum of the series of step-functions having nonintersecting supports. This representation is ambiguous, but it proves helpful in the technical sense, since step-functions are the "simplest" of summable functions. On the other hand, step-functions are extremal for (T,p)-capacity associated with the identical operator $T : L_p(G) \to L_p(G)$.

It is natural to consider extremal functions for capacity of a pair of compact sets to be "the simplest ones" in space that generates this capacity. We do not know how far this analogy may be carried in the case of arbitrary types of spaces. We give the result showing that decomposition into "simplest" functions is true for Sobolev spaces with first generalized derivatives. Here, the role of "simplest" functions is performed by extremal functions for variational capacity.

This decomposition is used in the capacitance description of sets which are removable singularity sets for the $L_p^1(W_p^1)$ spaces. Analogy with the L_p scale is absent here, since the space L_p does not react to the change of the definition domain for a set with zero measure. The zero measure for (T,p)-capacity associated with the identical operator $T : L_p(G) \to L_p(G)$ is the zero (T,p)-capacity. For the spaces $W_p^1(\mathbf{R}^n)$ for p exceeding the dimension of the domain, the sets of zero (T,p)-capacity do not exist. This follows from Proposition 5.10 of Chapter 3 which shows that (l,p)-capacity of a point exceeds zero. At the same time, any function of the space $W_p^1(B(0,1)\setminus\{0\})$ is extended to a function of the class $W_p^1(B(0,1))$, i.e., the point is the removable singularity for any of the W_p^1 classes. The possibility of extension obviously follows from the theorem about the coincidence of the classes W_p^1 and ACL_p (Theorem 5.5, Chapter 2).

Capacitance description of sets of removable singularities will be used in the next chapter, where removable singularities for quasi-conformal and quasi-isometric homeomorphisms are studied.

§1 Extremal Functions for $(1,p)$-Capacity

In $L^1_p(G)$, we would rather use the seminorm

$$\|u\|_{L^1_p(G)} = \left(\int_G |\nabla u|^p\,dx\right)^{1/p},$$

where $\nabla u = \left(\frac{\partial u}{\partial x_1}, \frac{\partial u}{\partial x_2}, \ldots, \frac{\partial u}{\partial x_n}\right)$. Variational $[1,p]$-capacity is considered to be connected with just this seminorm. Since the seminorm in question is equivalent to the seminorm $\sum_{|\alpha|=1} \|D^\alpha u\|_{L_p(G)}$, then the properties of $[1,p]$-capacity remain the same. Note that the estimates obtained at the end of the previous chapter in the case of L^1_p are adjusted to this very norm.

1.1. Simplest Properties of Extremal Functions

A specific property of the spaces $L^1_p(G)$ is the possibility to use the section of functions. In a more general form, this implies that for $f,g \in L^1_p(G)$ $\max(f,g) \in L^1_p(G)$ and $\min(f,g) \in L^1_p(G)$. Also, $\|\max(f,g)\|_{L^1_p(G)} \leqslant \|f\|_{L^1_p(G)} + \|g\|_{L^1_p(G)}$. Therefore, the section $f_{a^2} = \max(f,a^2) \in L^1_p(G)$ and $\|f_{a^2}\|_{L^1_p(G)} \leqslant \|f\|_{L^1_p(G)}$. From these simple remarks, we extract a few useful properties of extremal functions.

Proposition 1.1. *Let* $F_0, F_1 \subset \overline{G}$, $F_0 \cap G \neq \emptyset$, $F_1 \cap G = \emptyset$, *and* $0 < C^1_p(F_0,F_1;G) < \infty$ $(p > 1)$. *Then there exists a function* u_0, *extremal for* $[1,p]$*-capacity of the pair* F_0, F_1, *which equals zero* $(1,p)$*-quasi-everywhere on* $F_0 \cap G$ *and equals 1* $(1,p)$*-quasi-everywhere on* $F_1 \cap G$ *and* $0 \leqslant u_0(x) \leqslant 1$ *for all* $x \in G$.

Proof. Every continuous function admissible for the pair (F_0,F_1) equals 1 on $F_1 \cap G$ and equals zero on $F_0 \cap G$. The extremal function is the limit of admissible functions in $W^1_{p,\mathrm{loc}}(G)$ (see the proof of Theorem 5.7, Chapter 3). Then, from the general theorems about convergence (Theorem 1.3, Chapter 3), we obtain that u_0 equals zero $(1,p)$-quasi-everywhere on F_0 and equals 1 $(1,p)$-quasi-everywhere on F_1. Let us consider the function

$$\tilde{u}_0(x) = \min(1, \max(u_0(x), 0)).$$

According to the above, $\|\overline{u}_0\|_{L^1_p(G)} \leqslant \|u_0\|_{L^1_p(G)}$, $0 \leqslant \overline{u}_0(x) \leqslant 1$ for all $x \in G$. If the sequence of admissible functions $\{u_m\}$ converges in $L^1_p(G)$ to u_0, i.e., $\|u_m - u_0\|_{L^1_p(G)} \to 0$, then the sequence

$$\tilde{u}_m(x) = \min(1, \max(u_m(x), 0))$$

obviously converges to the function $\overline{u}_0$ in $L^1_p(G)$. Consequently, $\overline{u}_0 \in \overline{M}_{1,p}$ $(F_0,F_1;G)$ and $\|\overline{u}_0\| = \|u_0\|$. Due to uniqueness of $(1,p)$-extremal function $\overline{u}_0 = u_0$. Proposition is proved.

Below, we assume that among extremal functions we have chosen the function which is equal to zero $(1,p)$-quasi-everywhere on F_0 and equal to 1 $(1,p)$-quasi-everywhere on F_1. Such an extremal function is unique. The measurable function u is called monotone is the sense of Lebesgue in the domain G if for every compact subdomain V of the domain G, the equalities

$$\operatorname*{ess\,sup}_{x\in V} u(x) = \operatorname*{ess\,sup}_{x\in\partial V} u(x),$$
$$\operatorname*{ess\,inf}_{x\in V} u(x) = \operatorname*{ess\,inf}_{x\in\partial V} u(x) \tag{1.1}$$

are valid.[1]

Proposition 1.2. *Let F_0, F_1 be closed subsets of G and $0 < C_p^l(F_0, F_1; G) < \infty$ $(p > 1)$. Then the extremal function u_0 of the pair F_0, F_1 is monotone in the domain $G_1 = G\backslash\{F_0 \cup F_1\}$.*

Proof by contradiction. Let u_0 not be monotone. Then there exists a compact subdomain V of the domain G for which none of equalities (1.1) is valid. Suppose that the first one does not hold, i.e., $b^2 = \operatorname*{ess\,sup}_{x\in V} u_0(x) > \operatorname*{ess\,sup}_{x\in\partial V} u_0(x) = a^2$. According to the above, $a^2 \in (0, 1)$. Let us consider the function $v(x)$ which equals $u_0(x)$ outside of V and equals $\min(u_0(x), a^2 + (b^2 - a^2)/2)$ in the domain V. As in the proof of the previous proposition, it is proved that $\|v\|_{L_p^1(G)} \leqslant \|u_0\|_{L_p^1(G)}$ and $v \in \overline{M}_{1,p}(F_0, F_1; G)$. Consequently, due to uniqueness of the extremal function, $v(x) = u_0(x)$ almost everywhere. This contradicts the assumption about the nonmonotonicity of the function $u_0(x)$, since there should be $u_0(x) \geqslant v(x)$ on the set of positive measure. The obtained contradiction proves that the former relation of (1.1) is valid. To prove the latter, it suffices to apply the previous argument to the function $1 - u_0$.

The proposition is proved.

Corollary. *Under the conditions of Proposition 1.2, the extremal function is continuous in the domain G_1 (for $p > n - 1$).*

Proof. The monotone function of the class $L_p^1(G)$ is continuous according to Proposition 3.3 of Chapter 5 (for $p > n - 1$).

A set F is said to have smooth boundary in G if $\operatorname{Int} F \neq \emptyset$ and $\partial F \cap G$ is a manifold of the class C^∞.

Proposition 1.3. *Let $F_0, F_1 \subset \overline{G}$ be closed sets with smooth boundary in G and $0 < C_p^1(F_0, F_1; G) < \infty$ $(p > n - 1)$. Then the extremal function u_0 of the pair F_0, F_1 is continuous in the domain G.*

Proof. Taking into account Proposition 1.2, it suffices to prove that the function u_0 is continuous at the points of the boundary ∂F_0 of the set F_0 and of the boundary ∂F_1 of the set F_1. Let us take any point $x \in \partial F_1$, and a ball

[1] According to Theorem 5.11 of Chapter 3, the function u is defined on ∂V almost everywhere in the sense of the $(n-1)$-dimensional Hausdorff measure.

$B(x,r)$ so small that it does not entirely contain any connected component of the set F_1. It is easy to choose such a ball. Since ∂F_1 is a manifold, the point x has a neighbourhood $U(x)$ which is diffeomorphic to the unit ball in $\mathbf{R}^n$ for some diffeomorphism $\varphi : U(x) \to B(0,1) \cap L$. This diffeomorphism maps the intersection $U \cap \partial F_1$ to the set $B(0,1) \cap L$, where L is a coordinate plane in $\mathbf{R}^n$. Any ball $B(x,r) \subset U(x)$ will do. According to Proposition 1.1, $u_0(x) = 1$ $(1,p)$-quasi-everywhere on $B(x,r) \cap F_1$, i.e., $\operatorname*{ess\,sup}_{x \in B(x,r)} u_0(x) = \operatorname*{ess\,sup}_{x \in S(x,r)} u_0(x) = 1$. To prove the monotonicity of the function u_0 on the ball $B(x,r)$, it suffices to verify the coincidence of essential infimums on the ball and on its boundary. This is done just as in the proof of Proposition 1.2. Since the ball $B(x,r)$ belonging to $U(x)$ was chosen arbitrarily, the monotonicity follows and, as a result, continuity of $u_0(x)$ on the set $U(x)$. The arbitrariness in the choice of the point $x \in \partial F_1$ completes the proof.

1.2. The Dirichlet Problem and Extremal Functions

In a domain $G \in \mathbf{R}^n$ let us consider two closed sets F_0 and $F_1 \subset G$, $F_0 \neq \emptyset$ and $F_1 \cap F_0 = \emptyset$ relative to the domain G. In addition, suppose there exists a continuous function $f \in L^1_p(G)$ which is equal to zero on F_0 and is equal to 1 on F_1.

The space $\overset{\circ}{L}{}^1_p(F_0, F_1; G)(p > 1)$ is a set of functions $u \in L^1_p(G \setminus (F_0 \cup F_1))$ such that the function $\overline{u} : G \to \mathbf{R}$, which is equal to u on $G \setminus (F_0 \cup F_1)$ and is equal to zero at the rest of the points of G, belongs to the space $L^1_p(G)$.

Let us minimize the integral

$$\int_{G \setminus (F_0 \cup F_1)} |\nabla (f+u)|^p \, dx,$$

where $u \in \overset{\circ}{L}{}^1_p(F_0, F_1; G)$.

The same arguments that were used in the investigation of extremal functions for capacity, being literally repeated, result in the following.

Proposition 1.4. *There exists the unique function* $v \in f + \overset{\circ}{L}{}^1_p(F_0, F_1; G)$ *for which*

$$\int_{G \setminus (F_0 \cup F_1)} |\nabla v|^p \, dx = \inf_{u \in \overset{\circ}{L}{}^1_p(F_0, F_1; G)} \int_{G \setminus (F_0 \cup F_1)} |\nabla (f+u)|^p \, dx.$$

Let us consider the equivalent v $(1,p)$-refined function. Not to change notations, we denote it by v. Valid for this function is

Proposition 1.5. *Let* $F_0, F_1 \subset G$ *be closed sets with respect to* G *with smooth boundary in* G. *If there exists a continuous function* $f \in L^p_1(G)$, $f = 0$ *on* F_0, $f = 1$ *on* F_1, *then the extremal function for the Dirichlet problem is continuous in the domain* G *[42].*

Corollary. *Under the conditions of Proposition 1.5 the extremal function for the Dirichlet problem is the extremal function for $[1,p]$-capacity.*

Proof. The proof is obvious, because due to Proposition 1.5 the extremal function for the Dirichlet problem is continuous, and consequently it is admissible for the pair F_0, F_1.

Remark. If the inequality $C_p^1(F_0, F_1; G) < \infty$ is valid for the pair of closed sets $F_0, F_1 \subset G$ with respect to G, then the Dirichlet problem for this pair is correct.

Indeed, due to subsection 4.1 of Chapter 3, the inequality $C_p^1(F_0, F_1; G) < \infty$ implies that there exists a continuous function $f \in L_p^1(G)$ which is equal to zero in some neighbourhood of the set F_0 and is equal to 1 in some neighbourhood of the set F_1.

1.3. Extremal Functions for Pairs of Smooth Compacts

In the present subsection we study the construction of extremal functions for pairs of closed sets with smooth boundary in G, each of them consisting of a finite number of connected components. Note that a compact set $K \subset G$ having smooth boundary consists of a finite number of connected components.

Let us denote by $F_p(G)$ a set of extremal functions of pairs (F_0, F_1) of sets closed with respect to the domain G with smooth boundaries in G. Each of these sets has a finite number of connected components.

Let $E_p(\overline{G})$ be a subclass of a class of extremal functions for pairs (F_0, F_1) of sets closed with respect to G selected according to two conditions:

a) for every function $u \in E_p(\overline{G})$ and for any number $0 < a < 1$ the set $u^{-1}(0, a)$ is connected.

b) for every function $u \in E_p(\overline{G})$ and for any $0 < a < 1$ the set $u^{-1}(a, 1)$ is connected.

Recall that, according to the above, the functions $u \in F_p(G)$ or $u \in E_p(\overline{G})$ are continuous. The extremal functions of pairs of connected compact sets $F_0, F_1 \subset G$ lie in the class $E_p(\overline{G})$.

Theorem 1.6. *Any function u of a class $F_p(G)$ is representable as the sum $u = c_0 + \sum_{i=1}^{l} c_i v_i$, where $c_0, c_1, \ldots, c_l$ are constants, and $v_1, v_2, \ldots, v_l$ are functions of the class $E_p(\overline{G})$. Here $\|u\|^p_{L_p^1(G)} = \sum_{i=1}^{l} |c_i|^p \cdot \|v_i\|^p_{L_p^1(G)}$.*

Proof. Let us consider any function $u \in F_p(G)$. We associate two sets: $V_a = u^{-1}(-1, a)$ and $W_a = u^{-1}(a, 2)$ to a real number $0 \leqslant a \leqslant 1$. The function $\tau_{0,u}(a)$, which is equal to the number of connected components of the set V_a, is said to be the lower connectivity function of the function u; the function $\tau_{1,u}(a)$, which is equal to the number of connected components of the set W_a, is called the upper connectivity function of the function u. Since $u \in F_p(G)$, $F_{0,u} = u^{-1}(0)$ and $F_{1,u} = u^{-1}(1)$ consist of a finite number of connected components. Let this number be equal to K_0 for $F_{0,u}$; let it equal K_1 for $F_{1,u}$. We consecutively study the properties of the functions $\tau_{0,u}, \tau_{1,u}$.

Property 1. For all $a \in (0,1)$, the inequalities $0 \leqslant \tau_{0,u}(a) \leqslant K_0$, and $0 \leqslant \tau_{1,u}(a) \leqslant K_1$ are valid.

Let us assume the converse: there exists a number a for which the set V_a consists of $l > K_0$ connected components. Since on each connected component of the set $F_{0,u}$ the function equals zero (this follows from the positivity of $(1,p)$-capacity of any connected component of this set), then $V_a \supset F_{0,u}$. From our assumption it follows that there exists a connected component $\overline{V}_a$ of the set V_a which does not intersect $F_{0,u}$. For the function $\overline{u} = u$ outside of V_a and $\overline{u} \equiv a$ on $\overline{V}$, we have $\|\overline{u}\|_{L_p^1(G)} \leqslant \|u\|_{L_p^1(G)}$. Due to uniqueness of the extremal function $u = \overline{u}$. The resulting contradiction proves that $0 \leqslant \tau_{0,u}(a) \leqslant K_0$. The inequality for the function $\tau_{1,u}$ is proved in a similar way.

Property 2. The function $\tau_{0,u}$ is non-increasing on the interval $(0,1)$; the function $\tau_{1,u}$ is non-decreasing on the interval $(0,1)$.

Let us prove this property for the function $\tau_{0,u}$ only, because for the function $\tau_{1,u}$ the argument is the same. Let us assume the converse.

Let $0 < a < a_1 < 1$ and $\tau_{0,u}(a_1) > \tau_{0,u}(a)$. Then the set V_{a_1} consists of a larger number of connected components than V_a. Therefore, from the inclusion $V_{a_1} \supset V_a$, there follows the existence of a connected component $\overline{V}$ of the set V_{a_1}. This component does not intersect the set V_a. Having assumed the functions u to be equal to u outside of $\overline{V}$ and to be equal to a_1 on $\overline{V}$, we reach a contradiction, just as in the proof of Property 1.

Property 3. If for the functions $\tau_{0,u}$ and $\tau_{1,u}$ the inequalities $0 \leqslant \tau_{0,u} \leqslant 1$ and $0 \leqslant \tau_{1,u} \leqslant 1$ are valid, then $u \in E_p(\overline{G})$.

This property directly follows from the definition of the class $E_p(\overline{G})$.

Property 4. $\tau_{0,u} = \tau_{1,u}$. Let us turn back to the function u. Suppose that the function $\tau_{0,u}$ is not constant on the interval $(0,1)$. Let $0 = a_0 < a_1 < \cdots < a_s < a_{s+1} = 1$ be its discontinuity points. Let us consider the sets $F_{0,k} = \overline{V}_{a_k}$ and $F_{1,k} = \overline{W}_{a_{k+1}}$. Due to the continuity of the function u, the intersections $F_{0,k} \cap F_{1,k} \cap G$ are empty. The function

$$u_k = (a_k - a_{k-1})^{-1} \min(\max(u, a_{k-1}), a_k) - a_{k-1}(a_k - a_{k-1})^{-1}$$

is continuous, belongs to the class $L_p^1(G)$, equals zero on $F_{0,k}$, and equals 1 on $F_{1,k}$. Consequently, the function u_k is admissible to the pair $(F_{0,k}, F_{1,k})$. According to the construction of the functions u_k, we have $u = \sum_{k=1}^{s+1}(a_k - a_{k-1})u_k$.

Now let us prove that the function u_k is extremal for $(1,p)$-capacity of the pair $(F_{0,k}F_{1,k})$. Assume the converse. Let there exist an admissible function $\overline{u}_k$ for the pair $(F_{0,k}, F_{1,k})$ and $\|\overline{u}_k\|_{L_p^1(G)} \leqslant \|u_k\|_{L_p^1(G)}$. The function $\overline{u} = \sum_{l=1}^{k-1}(a_l - a_{l-1})u_l + (a_k - a_{k-1})\overline{u}_k + \sum_{l=k+1}^{s+1}(a_l - a_{l-1})u_l$ is admissible for the pair $u^{-1}(0)$ and $u^{-1}(1)$. By the construction of the functions $\overline{u}_l$ and of the function $\overline{u}_k$, we obtain

$$\|u\|^p_{L^1_p(G)} = \int_G |\nabla u|^p\,dx = \sum_{l=1}^{s+1} (a_l - a_{l-1})^{-1} \int_G |\nabla u_l|^p\,dx$$
$$> \sum_{l=1}^{k-1} (a_l - a_{l-1})^{-1} \int_G |\nabla u_l|^p\,dx + (a_k - a_{k-1})^{-1} \int_G |\nabla \tilde{u}|^p\,dx$$
$$+ \sum_{l=k+1}^{s-1} (a_l - a_{l-1})^{-1} \int_G |\nabla u_l|^p\,dx = \|\tilde{u}\|^p_{L^1_p(G)}.$$

This inequality contradicts the extremality of the function u.

We have proved that u_k are extremal functions for which the lower connectivity function τ_{0,u_k} is constant on the interval $(0,1)$. Applying the above argument to the function $1-u_k$ one may construct the representation of this function in the form of a linear combination of constants and functions $w_{k,l}$ for which the lower connectivity function is constant on the interval $(0,1)$. Then, according to Property 4, for the functions $1-w_{k,l}$ in the decomposition of the function u_k, the upper connectivity function and the lower connectivity function are constant simultaneously on the interval $(0,1)$. Finally we obtain that there exist extremal functions v_i, for which τ_{0,v_i} and τ_{1,v_j} are constant, and there exist the numbers $c_0, c_1, \ldots, c_i$ such that $u = c_0 + \sum_i c_i v_i$ and $\|u\|^p_{L^1_p(G)} = \sum |c_i|^p \|v_i\|^p_{L^1_p(G)}$.

Our theorem is proved to within the following statement.

Proposition 1.7. *Every continuous extremal function u for which the upper and lower connectivity functions are constant on $(0,1)$ is representable in the form $u = \sum_{l=1}^{s} u_l$ where $u_l \subset E_p(\overline{G})$ for all l.*

Proof. Let us consider a set $W = \cup_{0<a<1} W_a$. The function $\tau_{1,u}$ is constant. Therefore, for every a the number of connectivity components of the set W_a is the same. Then W has the same number of connectivity components as each of the sets W_a. Let us denote these components by $W_1, W_2, \ldots, W_l$. Consider the functions u_i $i = 1, 2, \ldots, l$ such that $u_i = u$ on W_i and $u_i = 0$ at all the rest of the points of the domain G. An argument similar to the one used in the first part of the proof of the theorem shows that u_i is the extremal function for the pair $(G \backslash W_i, W_i \cap u^{-1}(1))$. Relative closure in G of the set W is obvious from the construction of the set W as of the union of W_a. Besides, $\|u\|_{L^1_p(G)} = \sum_{k=1}^{l} \|u_i\|_{L^1_p(G)}$ and $\tau_{1,u_i} = 1$. (Since $u_j(x) = 0$ for all $j \neq i$ follows from the inequality $u_i(x) > 0$.) This representation of the function u and Property 3 stating that $\tau_{0,u}$ does not increase allow us to remark that the functions τ_{0,u_i} are constant for any i. Now it remains to apply the same argument to each of the functions $1 - u_i$ as a sum of functions with the upper and lower connectivity functions constant on $(0,1)$, the lower connectivity function being equal to 1. From Property 4 we obtain the representation of u_i in the form of the sum $u_i = \sum_{j=1}^{l_i} u_{i,j}$, where the functions $u_{i,j}$ have the upper and lower connectivity functions equal to 1, and $\|u_i\|^p_{L^1_p(G)} = \sum_j \|u_{i,j}\|^p_{L^1_p(G)}$.

Finally we obtain the representation $u = \sum_{i,j} u_{i,j}$, where $u_{i,j} \subset E_p(\overline{G})$ (according to Property 4) and $\|u\|^p_{L^1_p(G)} = \sum_{i,j} \|u_{i,j}\|^p_{L^1_p(G)}$.

This completes the proof of Proposition 1.7 and Theorem 1.6.

§2 Theorem on the Approximation of Functions from L_p^1 by Extremal Functions

First let us give some necessary information from the functional analysis.

2.1. Auxiliary Statements

Lemma 2.1. *Let a sequence of functions $\{v_m \in L_p^1(G)\}$ weakly converge to a function $v \in L_p^1(G)$. Then*

$$\|v\|_{L_p^1(G)} \leqslant \varliminf_{m\to\infty} \|v_m\|_{L_p^1(G)}.$$

Proof. The factor-space of the space $L_p^1(G)$ by the identically constant functions is the Banach one, and the seminorm $\|\cdot\|_{L^1_p(G)}$ is the norm in it.

The inequality is the well-known property of the norm of a Banach space.

Corollary. *Let a sequence $\{v_m \in L_p^1(G)\}$ be bounded in $L_p^1(G)$ and let it converge almost everywhere to the function $v \in L_p^1(G)(p > 1)$. Then there exists a subsequence $\{v_{mk}\}$ of the sequence $\{v_m\}$ which weakly converges in $L_p^1(G)$ to the function v. In this case the inequality*

$$\|v\|_{L_p^1(G)} \leqslant \varliminf_{m\to\infty} \|v_{m_k}\|_{L_p^1(G)}$$

is valid.

Proof. Since the sequence $\{v_m\}$ is bounded, one may single out from it the weakly converging subsequence $\{v_{m_k}\}$. Due to the imbedding theorem, the subsequence $\{v_{m_k}\}$ converges to its weak limit u in L_p on any ball $B(B \subset G)$. Therefore, one may assume that $v_{m_k} \to u$ almost everywhere, i.e., $u = v$. The inequality follows from the lemma about semicontinuity.

Lemma 2.2. *If a sequence $\{v_m\}$, $v_m \in L_p^1(G)$, converges weakly to a function v_0 and $\|v_m\|_{L^1_p(G)} \to \|v_0\|_{L^1_p(G)}$, then v_m converges to v_0 in $L_p^1(G)$, $p > 1$.*

Proof. Let $v \in L_p^1(G)$. Then $\nabla v \in L_p(G)$, and the set of all ∇v such that $v \in L_p^1(G)$ is a closed subspace of the space $L_p(G)$. Consequently, the weak convergence of the sequence $\{v_m\}$ to v_0 in $L_p^1(G)$ results in the weak convergence of ∇v_m to ∇v_0 in $L_p(G)$. Now, from the well-known theorem [35] about weak convergence in $L_p(G)$, we obtain the statement of the lemma.

2.2. The Class $\mathrm{Ext}_p(G)$

Let us denote by $\mathrm{Ext}_p(G)$ the totality of extremal functions for $[1,p]$-capacity of all possible pairs of connected compact sets $(F_0, F_1) \subset G$ having smooth boundaries.

Lemma 2.3. *A set* $\mathrm{Ext}_p(G)$ *is dense in* $E_p(\overline{G})$ *in the sense of convergence in* $W^1_{p,\mathrm{loc}}(G)$ *and in* $L^1_p(G)$.

Proof. Let $u \subset E_p(\overline{G})$. Let us fix $1/2 > \varepsilon > 0$. Since the lower connectivity function for u is identically equal to 1, the set $F_{0,\varepsilon} = \overline{V}_\varepsilon$ is connected. Using the upper connectivity function, we obtain that $F_{1,\varepsilon} = \overline{W}_{1-\varepsilon}$ is also connected. The function $v_\varepsilon(x) = \{\min[1-\varepsilon, \max(u(x)]\}/(1-2\varepsilon)$ is extremal for the pair $(F_{0,\varepsilon}, F_{1,\varepsilon})$. Let us prove that $v_\varepsilon \to v$ in $L^1_p(G)$. Indeed,

$$\|(1-2\varepsilon)\, v_\varepsilon\|^p_{L^1_p(G)} = (1-2\varepsilon)\int_G |\nabla v_\varepsilon|^p\, dx = \int_{G\setminus(F_{0,\varepsilon}\cup F_{1,\varepsilon})} |\nabla u|^p dx.$$

Thus,

$$\|(1-2\varepsilon)\, v_\varepsilon - u\|^p_{L^1_p(G)} = \int_{(u^{-1}(0)\cup u^{-1}(1))\setminus(F_{0,\varepsilon}\cup F_{1,\varepsilon})} |\nabla u|^p dx.$$

Since for $\varepsilon \to 0$ the sequence of sets $\{A_\varepsilon = [u^{-1}(0) \cup u^{-1}(1)]\setminus(F_{0,\varepsilon} \cup F_{1,\varepsilon})\}$ decreases and tends to the empty set, it is easy to see that the integral in its right-hand part tends to zero. Therefore, $v_\varepsilon \to v$ in $L^1_p(G)$. Hence it follows that for every compact subdomain V of the domain G, there exist constants $c_{\varepsilon,v}$ such that $v_\varepsilon - c_{\varepsilon,v} \to v$ in $W^1_p(V)$. Since the functions v_ε and v coincide on an open set, then $v_\varepsilon \to v$ in $W^1_p(V)$.

Let us construct two sequences of connected compact sets $\{F^m_{0,\varepsilon}\}$ and $\{F^m_{1,\varepsilon}\}$ having smooth boundaries and "exhausting" $F_{0,\varepsilon}$ and $F_{1,\varepsilon}$ in the following sense:

1) $F^m_{i,\varepsilon} \subset F_i$ for all $m \geqslant 1$, $i = 0,1$;
2) $F^{m_1}_{i,\varepsilon} \subset F^{m_2}_{i,\varepsilon}$ for $m_1 \leqslant m_2$;
3) $\cup F^m_{i,\varepsilon} = \mathrm{Int}\, F_{i,\varepsilon}$, $i = 0,1$.

The function v_ε is admissible for any of the pairs $F^m_{0,\varepsilon}, F^m_{1,\varepsilon}$ for all m. Therefore, $C^1_p(F^m_{0,\varepsilon}, F^m_{1,\varepsilon}; G) < \infty$. Let v_m be extremal functions for these pairs of compact sets. It is obvious that $v_m \subset \mathrm{Ext}_p\, G$. Due to the extremality of v_m, we have $\|v_m\|^p_{L^1_p(G)} \leqslant \|v_\varepsilon\|^p_{L^1_p(G)}$. The sequence $\{v_m\}$ is bounded in $L^1_p(G)$. It may be considered to be weakly converging to some function $v \in L^1_p(G)$. The limit functions equals 1 on $\mathrm{Int}\, F_1$, and equals zero on $\mathrm{Int}\, F_0$. According to Lemma 2.1, we have $\|v\|_{L^1_p(G)} \leqslant \lim_{m\to\infty} \|v_m\|_{L^1_p(G)}$, consequently, $\|v\|_{L^1_p(G)} \leqslant \|v_\varepsilon\|_{L^1_p(G)}$. Due to the uniqueness of solution of the Dirichlet problem $v_\varepsilon = v$. The sequence v_m weakly converges to v_ε, and choosing the subsequence v_{m_k} one may obtain from $\|v_\varepsilon\|_{L^1_p(G)} = \|v\|_{L^1_p(G)}$ the equality $\lim \|v_{m_k}\|_{L^1_p(G)} = \|v_\varepsilon\|_{L^1_p(G)}$. Consequently, v_m converges to v_ε in $L^1_p(G)$

(according to Lemma 2.2). Since $v_\varepsilon \to v$ in $L_p^1(G)$ for $\varepsilon \to 0$, by using the diagonal process, one may construct a sequence of functions $\{u_m \in \mathrm{Ext}_p G\}$, which converges to u in $L_p^1(G)$.

Now let us note that the functions v_m and v coincide on an open set. Therefore, $v_m \to v$ in $W_{p,\mathrm{loc}}^1(G)$, and the same diagonal process allows us to see that the sequence of functions u_m converges to u in $W_{p,\mathrm{loc}}^1(G)$. The lemma is proved.

Proposition 2.4 (On the continuity of $(1,p)$- capacity). *Let F_0, F_1 be a pair of compact sets having smooth boundary, and let $\{F_{0,m}\}, \{F_{1,m}\}$ be two sequences of compact sets such that* $\sup\limits_{x\in\partial F_{0,m}} \rho(x,\partial F_0) \to 0$ *and* $\sup\limits_{y\in\partial F_{1,m}} (y,\partial F_1) \to 0$ *for* $m \to \infty$. *Then*

$$\lim_{m\to\infty} C_p^1(F_{0,m}, F_{1,m}; G) = C_p^1(F_0, F_1; G).$$

Proof. Let u be an admissible function of the pair (F_0, F_1). Then $u(x) \geqslant 1$ in some neighbourhood $U(F_1) \subset G$ of the compact set F_1, and $u(x) \equiv 0$ in some neighbourhood $U(F_0)$ of the compact set F_0. According to the condition of the lemma, for sufficiently large m,

$$C_p^1(F_{0,m}, F_{1,m}; G) \leqslant \|u\|_{L_p^1(G)}.$$

Since the admissible function u is chosen arbitrarily,

$$\lim_{m\to\infty} C_p^1(F_{0,m}, F_{1,m}; G) \leqslant C_p^1(F_0, F_1; G).$$

Consequently, the sequence $\{u_m\}$ of extremal functions for $(1,p)$-capacity of the pairs $(F_{0,m}, F_{1,m})$ is bounded in $L_p^1(G)$. From it one can extract a weakly converging subsequence $\{u_{m_k}\}$. The limit function u of this sequence is admissible for the pair F_0, F_1 since, due to the imbedding theorems of $L_p^1(G)$ into $L_{q,\mathrm{loc}}(G)$ from the weak convergence in $L_p^1(G)$, there follows the convergence in $L_q(G)$; that is, $\overline{u} = 0$ on F_0 and $\overline{u} \geqslant 1$ on F_1. Thus u is admissible for the Dirichlet problem. Due to the smoothness of the boundaries ∂F_0 and ∂F_1, we obtain

$$\begin{aligned} C_p^1(F_0, F_1; G) &\leqslant \|\tilde{u}\|_{L_p^1(G)} \leqslant \lim_{m\to\infty} \|u_{m_k}\|_{L_p^1(G)} \\ &= \lim_{m\to\infty} C_p^1(F_{0,m_k}, F_{1,m_k}, G). \end{aligned}$$

Comparing it to the above equality we obtain the statement of the proposition, if we take into account the fact that from every subsequence of the sequence $\{u_m\}$, according to the proved above, one can extract the subsequence for which the proposition is valid.

Lemma 2.5. *There exists a countable set of functions $v_i \in \mathrm{Ext}_p G$ which is dense in $E_p(G)$.*

Proof. If there exists in $\mathrm{Ext}_p G$ a countable dense set, then Lemma 2.5 follows from Lemma 2.3.

Let us consider a totality $\mathcal{P}$ of all polynomials $P : \mathbf{R}^n \to \mathbf{R}$ with rational coefficients. For each polynomial we choose in $\mathbf{R}$ a dense set A countable eveywhere of regular values. The totality $\mathfrak{N}$ of $(n-1)$-dimensional manifolds is countable if the manifolds are connected components of the pre-images $P^{-1}(t)$, $t \subset A$.

Let us point out in $\operatorname{Ext}_p G$ all the functions v extremal for all possible pairs (F_0, F_1) of sets whose boundaries belong to $\mathfrak{N}$. The set U of such functions is countable.

If $u \in \operatorname{Ext}_p G$ is the extremal function of the pair (F_0, F_1), then $\partial F_0, \partial F_1$ are smooth manifolds. They are surfaces of the level of smooth compactly supported functions $\varphi_0, \varphi_1 \in C_0^\infty(G)$. According to the Weierstrass theorem, the functions φ_0 and φ_1 may be approximated by polynomials of $\mathcal{P}$. Using the Sard theorem one can construct a sequence of surfaces of the level of these polynomials in such a way that these surfaces are smooth manifolds approximating ∂F_0 and ∂F_1 in the sense of Proposition 2.4. After this, the result follows from Proposition 2.4.

Lemma 2.6. *Let $\{F_{0,m}\}$ be a sequence of sets closed with respect to a domain G,*

$$F_0 = \bigcup_{m=1}^{\infty} F_{0,m}.$$

Then for any set $F_1, F_{0,m} \cap F_1 = \emptyset$ closed with respect to G for $p > 1$, the relation

$$\lim_{l\to\infty} C_p^1\left(\bigcup_{m=1}^{l} F_{0,m}, F_1; G\right) = C_p^1(F_0, F_1; G)$$

is valid.

Proof. Let u be the extremal function of the pair (F_0, F_1). Then it is admissible for any of the pairs $(\cup_{m=1}^{l} F_{0,m}, F_1)$, i.e., $C_p^1(\cup_{m=1}^{l} F_{0,m}, F_1; G)$. For the extremal functions u_l of these pairs, the inequalities

$$\|u_l\|_{L_p^1(G)} \leqslant \|u\|_{L_p^1(G)}, \quad \|u_l\|_{L_p^1(G)} \leqslant \|u_{l_1}\|_{L_p^1(G)}$$

are valid for $l_1 > l$. Let us choose from u_l a weakly converging subsequence $\{u_{l_k}\}$. Then, according to the lemma about semicontinuity, $\lim_{k\to\infty} \|u_{l_k}\|_{L_p^1(G)} \geqslant \|u\|_{L_p^1(G)}$. Since the sequence of norms $\|u_m\|$ is monotone, we finally obtain

$$\|u\|_{L_p^1(G)} \geqslant \lim_{l\to\infty} \|u_l\|_{L_p^1(G)} \geqslant \|u\|_{L_p^1(G)}.$$

The lemma is proved.

Corollary. *If, under the conditions of the lemma, all the above sets have smooth boundaries, then the extremal functions u_l of the pairs $(\cup_{m=1}^{l} F_{0,m}, F_1)$ converge in $L_p^1(G)$ to the extremal function u of the pair (F_0, F_1).*

Proof literally repeats the end of the proof of Proposition 2.4.

2.3. Proof of the Theorem on Approximation

Theorem 2.7. *For every function $u \in L_p^1(G)$ and for every $\varepsilon > 0$ there exist an open set A whose $(1,p)$- capacity is smaller than ε, and a linear combination $c_0 + \sum_{k=1}^{l} c_k v_k$ of functions $v_k \notin E_p(\overline{G})$ which satisfy the conditions:*

$$\left| u(x) - c_0 - \sum_{k=1}^{l} c_k v_k(x) \right| < \varepsilon$$

for all $x \notin A$;

$$\left\| u - \sum_{k=1}^{l} c_k v_k \right\|_{L_p^1(G)} < \varepsilon; \tag{2.1}$$

$$\left\| \sum_{k=1}^{l} c_k v_k \right\|_{L_p^1(G)}^p = \sum_{k=1}^{l} |c_k|^p \| v_k \|_{L_p^1(G)}^p.$$

Proof. Let us divide the proof into several parts.

A. According to Theorem 4.2 of Chapter 2, the function $u \in W_{p,\mathrm{loc}}^1(G)$. The functions of the class $C^\infty(G) \cap L_p^1(G)$ are dense in $L_p^1(G)$ and according to the Poincaré–Sobolev inequality, using subsection 2.1, one may choose a sequence $\{u_m\}$ of functions of the class $C^\infty(G) \cap L_p^1(G)$ which converges to u in $W_{p,\mathrm{loc}}^1(G)$ and in $L_p^1(G)$ simultaneously. Then $u_m \to u$ $(1,p)$-quasieverywhere. According to Egorov's theorem for capacity, there exists an open set A_1 of capacity smaller than ε. Outside of this set A_1, $u_m \to u$ is uniform, i.e., for fixed $\varepsilon > 0$ one can find a smooth function of the class $C^\infty(G) \cap L_p^1(G)$ such that outside of A $|u(x) - f(x)| < \varepsilon/4$. It is obvious that the approximating smooth functions may be considered to be bounded.

B. Approximation of a smooth function by piecewise-extremal ones.

Definition of a Piecewise-Extremal Function. In a domain G let us consider a monotone sequence $V_0 \subset V_1 \subset \cdots \subset V_l$ of sets closed with respect to G and having smooth boundaries. Suppose that $\partial V_i \cap \partial V_j \cap G = \emptyset$ for all $i \neq j$ and that for every pair $\{V_{k-1}, (G \backslash V_k) \cup \partial V_k\}$ the $(1,p)$-capacity is finite.

A function v is called a piecewise-extremal function associated with the sets $V_0, V_1, \ldots, V_l$ and with the real numbers $a_0, \ldots, a_l$ if

$$v = a_0 + \sum_{k=1}^{l} (a_k - a_{k-1}) v_k,$$

where v_k is the extremal function of the pair $(\overline{V}_{k-1}, G \backslash V_k)$.

To assign a piecewise-extremal function, it suffices to know a collection of sets $\{V_k\}$ and a collection of numbers $\{a_k\}$. The piecewise-extremal function $v \in L_p^1(G)$

$$\| v \|_{L_p^1(G)}^p = \sum_{k=1}^{l} |a_k - a_{k-1}|^p \| v_k \|_{L_p^1(G)}^p.$$

Let us turn back to the smooth bounded function $f \in L_p^1(G)$ obtained in A. Let us consider a partition τ of a segment $[\min_{x\in G} f(x), \max_{x\in G} f(x)]$ by means of real numbers $\min_{x\in G} f(x) < a_0 < a_1 \cdots < a_l < \max_{x\in G} f(x)$ for which $f^{-1}(a)$ are smooth manifolds. According to Sard's theorem, almost all the values of the function f satisfy this condition.

The open sets $V_k = f^{-1}(-\infty, a_k)$ form a monotone chain. Consequently, for the collection of sets $\{V_k\}$ and for the collection of numbers $\{a_k\}$ the piecewise-extremal function

$$v_\tau = a_0 + \sum_{k=1}^{l} (a_k - a_{k-1}) v_k$$

is defined.

In each of the open sets $\overline{V}_k \setminus V_{k-1}$ the inequality

$$\int_{\overline{V}_k \setminus V_{k-1}} |\nabla v_k|^p dx \leqslant (a_k - a_{k-1})^{-p} \int_{\overline{V}_k \setminus V_{k-1}} |\nabla f|^p dx$$

is valid. Therefore, in the entire domain G the inequality

$$\int_G |\nabla v_\tau|^p dx \leqslant \int_G |\nabla f|^p dx \tag{2.2}$$

is valid.

A sequence $v_m = v_{\tau_m}$ of piecewise-extremal functions is put into correspondence to the sequence τ of partitions of the segment $[\min f(x), \max f(x)]$.

By the construction of piecewise-extremal functions it is obvious that

$$|f(x) - v_m(x)| \leqslant \operatorname{diam} \tau_m \operatorname{diam} \tau_m = \max_{i=1,l} |a_i - a_{i-1}|.$$

Suppose, in addition, that $\operatorname{diam} \tau_m \to 0$ for $m \to \infty$. In this case the sequence v_m uniformly converges to the function f.

The sequence $\{v_m\}$ is bounded in $L_p^1(G)$. Let us choose from it a weakly converging in $L_p^1(G)$ subsequence $\{v_{m_k}\}$. According to the lemma about semicontinuity,

$$\underline{\lim}_{k\to\infty} \|v_{m_k}\|_{L_p^1(G)} \geqslant \|f\|_{L_p^1(G)}.$$

From inequality (2.2) and from Lemma 2.2 it follows that $v_{m_k} \to f$ in $L_p^1(G)$. For sufficiently large k, due to uniform convergence of v_{m_k} to f we obtain $|v_{m_k}(x) - f(x)| \leqslant \varepsilon/4$ for all $x \in G$.

Then

$$|v_{m_k}(x) - f(x)| + |f(x) - u(x)| \leqslant \varepsilon/4 + \varepsilon/4 \leqslant \varepsilon/2$$

for all $x \in A_1$.

Now it only remains to prove the theorem for piecewise-extremal functions.

C. A piecewise-extremal function $v = a_0 + \sum_{k=1}^{l} (a_k - a_{k-1}) v_k$ is assigned by a chain of sets $\{V_k\}$, each of which may have a countable number of connected components.

We are going to prove that for every $\varepsilon > 0$ there exists a piecewise-extremal function v_ε satisfying the conditions:

a) $\|v - v_\varepsilon\|_{L^1_p(G)} < \varepsilon$;

b) the chain of sets V_{k-1}, which assigns v_ε, only consists of sets having a finite number of connected components, here the sets $G \backslash V_k$ consist of a finite number of connected components $k = 1, 2, \ldots, l$.

The proof is held by induction by successive reconfiguration of the sets $V_{k-1}, G \backslash V_k$, $k = 1, 2, \ldots, l$.

Let us fix $\varepsilon > 0$.

The Induction Assumption. Suppose that there exists a piecewise-extremal function $u = a_0 + \sum_{k=1}^{l} (a_k - a_{k-1}) u_k$ satisfying the conditions:

a) for $l \geqslant q \geqslant r > 0$ the sets V^u_{q-1} and $G \backslash V^u_q$ consist of a finite number of connected components;[2]

b) $\|v - u\|_{L^1_p(G)} < \varepsilon(l - r + 1)/l$;

c) there exists a set A_3 of $(1,p)$-capacity smaller than $\varepsilon/4$, such that $|v(x) - u(x)| < \varepsilon/4$ for $x \notin A_3$.

The base of induction is $r = l$. Suppose that the set $G \backslash V_l$ consists of a finite number of connected components U_i, $i = 1, 2, \ldots$.

According to the corollary of Lemma 2.6, the extremal functions $v_{l,j}$ of the pairs $(V_{l-1}, \cup U_i)$ converge to the function v_i in the space $L^1_p(G)$.

Since $v_{l,j}$ coincide with v_l on an open set, then according to Theorem 4.2 of Chapter 2, $v_{l,j} \to v_l$ in $W^1_{p,\mathrm{loc}}$. Then, due to Egorov's theorem, there exists a set A_2 such that $(1,p)$-capacity of A_2 is smaller than $\varepsilon/4$ and

$$|v_{l,j}(x) - v_l(x)| \leqslant \varepsilon/4$$

for all $x \in G \backslash A_2$.

Not to encumber the proof by the standard technical experiment, let us note that every time the corollary of Lemma 2.6 is used, the above argument allows us to construct a set of small $[1,p]$-capacity. Outside of this set the functions differ for a magnitude comparable to ε. Therefore, we do not repeat this argument and only give the result of it at the end of the proof of this part of the theorem.

Let us suppose V^w_l to be equal to $G \backslash \cup_{i=1}^{j_0} \overline{U}_i$ where j_0 is sufficiently large for $\|v_{l,j_0} - v_l\| \leqslant \varepsilon \backslash 2l(a_l - a_{l-1})$.

A new piecewise-extremal function w is constructed by the sets $V_0, V_1, \ldots, V_{l-1}, V^w_l$ and by the numbers $a_0, a_1, \ldots, a_l$.

It is obvious that $\|v - w\|_{L^1_p(G)} < \varepsilon/2l$.

Now let V_{l-1} consist of a countable number of connected components U_i, $i = 1, 2, \ldots$. For every pair $(\cup_{i=1}^{j} \overline{U}_i,\ G \backslash V^w_l)$ the Dirichlet problem is solvable. According to Lemma 2.6, the extremal functions $u_{l,j}$ of the corresponding

---

[2]The index u of the set V^u_k implies that V^u_k is a set from the chain assigning the function u.

pairs converge to the function w_l in the space $L_p^1(G)$.

Suppose that $V_{l-1,j} = \cup_{i=1}^{j} U_i$ and $V_{k,j} = V_k^v \cap V_{l-1,j}$ for all $0 \leqslant k \leqslant l-1$. Let us consider a sequence of piecewise-extremal functions

$$u_j = a_0 + \sum_{k=1}^{l} (a_k - a_{k-1})\, u_{k,j}, \quad j = 1, 2, \ldots$$

The function $u_{k,j}$ is the extremal function of the pair $[V_{k-1,j}, (G \setminus V_{k,j}) \cup \partial V_{k,j}]$ for all $k < l$ and for all $j \geqslant 1$. Let us prove that $\|w - u_j\|_{L_p^1(G)} \to 0$ for $j \to \infty$. Indeed,

$$\|w - u_j\|^p_{L_p^1(G)} = \sum_{k=1}^{l-1} (a_k - a_{k-1}) \int_G |\nabla u_k - \nabla u_{k,j}|^p\, dx$$
$$+ \|w_l - u_{l,j}\|^p_{L_p^1(G)}.$$

According to Lemma 2.6 the latter summand tends to zero. For $k \geqslant l$ from the equalities

$$\int_G |\nabla u_k - \nabla u_{k,j}|^p\, dx = \int_{\overset{v}{V_k} \setminus V_{k,j}} |\nabla u_k - \nabla u_{k,j}|^p\, dx$$

$$= \int_{\overset{v}{V_k} \setminus\ k,j} |\nabla u_k|^p dx = \sum_{i=j+1}^{\infty} \int_{U_i} |\nabla u_k|^p dx$$

the convergence to zero of $\|u_k - u_{k,j}\|_{L_p^1(G)}$ follows for $j \to \infty$.

Let us choose j_0 to be such that

$$\|\bar{w} - u_{j_0}\|_{L_p^1(G)} \leqslant \varepsilon/2l.$$

We have constructed the piecewise-extremal function $u = u_{j_0}$ for which the induction assumption for $r = l$ holds: the sets $\overline{V_{l-1}^u}, (G \setminus \overline{V_l^u}) \cup \partial \overline{V_l^u}$ consist of a finite number of connected components, and the estimate

$$\|v - \hat{u}\|_{L_p^1(G)} \leqslant \|v - \bar{w}\|_{L_p^1(G)} + \|\bar{w} - u_{j_0}\|_{L_p^1(G)} \leqslant \varepsilon/l$$

is valid.

The Induction Step. Suppose that the induction assumption is valid for $r = s < l$, $s > 1$. Let us construct a function u for which the induction assumption is valid for $r = s - 1$.

Let V_{s-2}^u consist of a finite number of connected components U_i. For every pair $(\cup_{i=1}^{j} \overline{U}_i, G \setminus V_{s-1}^u)$, $(1,p)$-capacity is finite. According to Lemma 2.6, the extremal functions $u_{s-1,j}$ of the corresponding pairs converge to the function u_{s-1} in the space $L_p^1(G)$.

Let us put $V_{s-2,j} = \cup_{i=1}^{j} U_i$ and $V_{k,j} = V_k^u \cap V_{s-2,j}$ for all $0 \leqslant k < s-2$. Let us consider a sequence of piecewise-extremal functions

$$u_j = a_0 + \sum_{k=1}^{s-1} (a_k - a_{k-1})\, u_{k,j} + \sum_{k=s}^{l} (a_k - a_{k-1})\, u_k, \; j = 1, 2 \ldots$$

The functions $u_{k,j}$ are the extremal function of the pair $(V_{k-1,j}, V_{k,j} \cup \partial V_{k,j})$ for all $k < s - 1$ and for all $j \geqslant 1$.

Let us prove that

$$\|u - u_j\|_{L_p^1(G)} \to 0.$$

Indeed,

$$\|u - u_j\|^p_{L_p^1(G)} = \sum_{k=1}^{s-2} (a_k - a_{k-1})^p \int_G |\nabla u_k - \nabla u_{k,j}|^p dx$$
$$+ \|u_{s-1} - u_{s-1,j}\|^p_{L_p^1(G)}.$$

According to Lemma 2.6, the latter summand tends to zero. For $k < s - 1$ from the equalities

$$\int_G |\nabla u_k - \nabla u_{k,j}|^p dx = \int_{V_k^u \setminus V_{k,j}} |\nabla u_k - \nabla u_{k,j}|^p dx$$
$$= \int_{V_k^u \setminus V_{k,j}} |\nabla u_k|^p dx = \sum_{i=j+1}^{\infty} \int_{U_i} |\nabla u_k|^p dx$$

there follows the convergence to zero of $\|u_k - u_{k,j}\|_{L_p^1(G)}$ for $j \to \infty$.

Let us choose j_0 so that

$$\|u - u_{j_0}\|_{L_p^1(G)} < \varepsilon/2l. \tag{2.3}$$

Let us construct a piecewise-extremal function $\overline{w} = u_{j_0}$ for which the induction assumption holds for $r = s$, the set $V^{\overline{w}}_{s-2}$ consists of a finite number of connected components, and estimate (2.3) is valid.

The complement of the set $V^{\overline{w}}_{s-1}$ may consist of a countable number of components W_j, $j = 1, 2, \ldots$. In this case the set $V^{\overline{w}}_{s-2}$ undergoes further reconstruction.

Due to the induction assumption, every set $G \setminus \overline{V^{\overline{w}}_k}$ for $k \geqslant s$ consists of a finite number of connected components. The inclusion $G \setminus \overline{V^{\overline{w}}_k} \subset G \setminus V^{\overline{w}}_{s-1}$, $k \geqslant s$, yields that for j larger than some j_0, $W_j \cap (G \setminus V^u_k) = \emptyset$ for all $k \geqslant s$. Hence it follows that $W_j \subset V^{\overline{w}}_s$ for $j > j_0$.

Let us consider the sets $W_{s-1,j}$ which are the interior of the closure of the sets $V^{\overline{w}}_{s-1} \cup (\cup_{k=j} W_k)$ for $j > j_0$ with respect to the domain G. The number of connected components of the sets $W_{s-1,j}$ does not exceed the number of connected components of the set $V^{\overline{w}}_{s-1}$ for all $j > j_0$.

From the inclusion $W_j \subset V^u_s = V^{\overline{w}}_{s-1}$ for $j > j_0$ if follows that the extremal function of the pair $\overline{V^{\overline{w}}_{s-1}}, G \setminus V^{\overline{w}}_s)$ vanishes on W_j. Therefore, it is the extremal function for any of the pairs $(\overline{W}_{s-1,j}, G \setminus V^{\overline{w}}_s)$ for $j > j_0$.

For every pair $(\overline{V^{\overline{w}}_{s-2}}, G\backslash W_{s-1,j})$ the Dirichlet problem is solvable. The extremal functions $w_{s-1,j}$ of these problems converge to the extremal function w_{s-1} of the pair $(\overline{V^{\overline{w}}_{s-2}}, G\backslash V^{\overline{w}}_{s-1})$ in the space $L^1_p(G)$. This allows us to choose $j_1 > 0$ such that the piecewise-extremal function w assigned by the partition $V^{\overline{w}}_0 \subset \cdots \subset V^{\overline{w}}_{s-2} \subset W_{s-1,j} \subset V^{\overline{w}}_s \cdots \subset V^{\overline{w}}_l$ and by the numbers $a_0, a_1, \dots, a_l$ satisfies the inequality

$$\| \tilde{w} - w \|_{L^1_p(G)} \leqslant \varepsilon/2l.$$

Hence and from (2.3),

$$\| u - w \|_{L^1_p(G)} \leqslant \varepsilon/l.$$

The induction assumption is proved.

D. The above three items proved the possibility to approximate an arbitrary function $u \in L^1_p(G)$ by the piecewise-extremal function $w = a_0 + \sum_{k=1}^{l}(a_k - a_{k-1})w_k$. Each of the functions w_k is extremal for a pair of closed with respect to the domain G sets $(F_{0,k}, F_{1,k})$ whose interiors consist of a finite number of connected components. The functions w_k are continuous, $\|w\|^p_{L^1_p(G)} = \sum_k |a_k - a_{k-1}|^p \, \|w\|^p_{L^1_p(G)}$, $|u(x) - w(x)| < \varepsilon$ outside of the set of capacity smaller than ε.

To complete the proof of the theorem it suffices to show the possibility to represent any of the functions w_k in the form $w_k = c_0^{(k)} + \sum_{i=1}^{l_k} c_i^{(k)} w_i^{(k)}$ where the functions $w_i^{(k)} \in \operatorname{Ext}_p G$ and $\|w_k\|^p_{L^1_p(G)} = \sum_{i=1}^{l_k} |c_i^{(k)}|^p \|w_i^{(k)}\|^p_{L^1_p(G)}$. This follows from Proposition 1.7.

This completes the proof of the theorem.

Remark 1. From the proof of the theorem it is clear that for the positive function $u \in L^1_p(G)$ the number $0 < c_0 < 2 \operatorname{ess\,inf} u(x) + \varepsilon$. The number 2 in front of ess inf u appears if u is being approximated by smooth functions.

Remark 2. If the function $u \in L^1_p(G)$ is bounded, then the function $w = c_0 + \sum_k c_k v_k$, constructed in the theorem, is also bounded and $|w(x)| < 2|u(x)|$ for all $x \in G$.

Corollary 1. *The set* $\operatorname{Ext}_p(G)$ *is dense in* $L^1_p(G)$.

This follows from Lemma 2.3.

2.4. Representation in Form of a Series

Theorem 2.8. *Every function $u \in L_p^1(G)$ is representable in the form $u = c_0 + \sum_i c_i v_i$ where c_i are real numbers, the functions v_i belong to the class $\mathrm{Ext}_p(G)$ for all $i \geqslant 1$. Here $\|u - \sum_{i=1}^{l} c_i v_i\|_{L_p^1(G)} \to 0$ for $l \to \infty$.*

Proof. Let us fix $\varepsilon > 0$. According to Theorem 2.7, there exists an open set A_i, whose $(1,p)$-capacity is smaller than $\varepsilon/2$, the functions $\overline{v}_1, \overline{v}_2, \ldots, \overline{v}_{l_0} \in \mathrm{Ext}_p(G)$, and the numbers $c_0, c_1, \ldots, c_{l_0}$ for which

$$\left| u(x) - \sum_{i=1}^{l_0} c_i \widetilde{v}_i(x) - c_{0,1} \right| < \varepsilon/2$$

for all $x \in G \backslash A_1$ and

$$\left\| u - \sum c_i \widetilde{v}_i \right\|_{L_p^1(G)} < \varepsilon/2.$$

Let us put $\overline{u}_1 = u - \sum_{i=1}^{l_0} c_i \overline{v}_i - c_{0,1}$. It is obvious that $\|\overline{u}_1\|_{L_p^1(G)} < \varepsilon/2$. Let us consider the functions $u_1^+(x)$ and $u_1^-(x)$. It is obvious that $\operatorname*{ess\,sup}_{x \in G\backslash A_1} u_1^+(x) < \varepsilon/2$ and $\operatorname*{ess\,sup}_{x \in G\backslash A_1} u_1^-(x) < \varepsilon/2$. Let us apply Theorem 2.7 and Remark 1 to Theorem 2.7 to the functions u_1^+ and u_1^-. Then there exist open sets A_2, A_3 such that the $(1,p)$-capacity of A_2 is smaller than $\varepsilon/4$, the $(1,p)$-capacity of A_3 is smaller than $\varepsilon/8$; there exist the functions $\overline{v}_{l_1+1}, \overline{v}_{l_2}, \overline{v}_{l_2+1}, \ldots, \overline{v}_{l_3} \in \mathrm{Ext}_p(G)$ and the numbers $0 < c_{0,2} < \varepsilon + \varepsilon/4, 0 < c_{0,3} < \varepsilon + \varepsilon/8$ such that the inequality

$$\left| u_1^+(x) - \sum_{i=l_1+1}^{l_2} c_i \widetilde{v}_i(x) - c_{0,2} \right| < \varepsilon/4$$

is valid for all $x \notin A_2 \cup A_1$, and the inequality

$$\left| u_1^-(x) - \sum_{i=l_2+1}^{l_3} c_i \widetilde{v}_i(x) - c_{0,3} \right| < \varepsilon/8$$

is valid for all $x \in A_3 \cup A_1$,

$$\left\| u_1^+ - \sum_{i=l_1+1}^{l_2} c_i \widetilde{v}_i \right\|_{L_p^1(G)} < \varepsilon/4, \qquad \left\| u_1^- - \sum_{i=l_2+1}^{l_3} c_i \widetilde{v}_i \right\|_{L_p^1(G)} < \varepsilon/8.$$

Hence it follows that

$$\left| u(x) - \sum_{i=1}^{l_3} c_i \widetilde{v}_i(x) - c_{0,1} - c_{0,2} - c_{0,3} \right| < \varepsilon/2 + \varepsilon/4 + \varepsilon/8$$

for all $x \notin A_1 \cup A_2 \cup A_3$, $\|\sum_{i=l_1+1}^{l_2} c_i v_i\|_{L_p^1(G)} \leqslant \varepsilon/2 + \varepsilon/4$, $\|\sum_{i=l_2+1}^{l_3} c_i v_i\| \leqslant \varepsilon/4 + \varepsilon/2$, $\|u - \sum_{i=1}^{l_3} c_i \bar{v}_i\| \leqslant \varepsilon/2 + \varepsilon/4 + \varepsilon/8$ and $|c_{0,1}| + |c_{0,2}| + |c_{0,3}| < |c_{0,1}| + 2\varepsilon + \varepsilon/4 + \varepsilon/8$.

Let us also note that $Cap_{(1,p)}(A_1 \cup A_2 \cup A_3, \mathbf{R}^n) \leqslant \varepsilon/2 + \varepsilon/4 + \varepsilon/8$.

Continuing this proof by induction, we construct on the kth step the open sets A_{2k} and $_{2k+1}$ such that $(1,p)$-capacity of A_{2k}, A_{2k+1} is smaller than $\varepsilon/2^{2k}, \varepsilon/2^{2k+1}$, the functions $v_{l_{2k-1}+1}, v_{l_{2k}}, v_{l_{2k}+1}, \dots, v_{l_{2k+1}}$, and the numbers $0 < c_{0,2k} < 2\left(\frac{\varepsilon}{2^{2k-2}} + \frac{\varepsilon}{2^{2k-1}}\right) + \frac{\varepsilon}{2^{2k}}, 0 < C_{0,2k+1} < 2\left(\frac{\varepsilon}{2^{2k-2}} + \frac{\varepsilon}{2^{2k-1}}\right) + \frac{\varepsilon}{2^{2k+1}}, c_{l_{2k-1}+1}, \dots, c_{l_{2k}}, c_{l_{2k+1}+1}, \dots, c_{l_{2k+1}}$ such that

$$\left| u(x) - \sum_{i=1}^{l_{2k+1}} c_i \tilde{v}_i(x) - \sum_{i=1}^{l_{2k+1}} c_{0,i} \right| < \frac{\varepsilon}{2^{k+1}}$$

for all $x \notin \bigcup_{i=1}^{l_{2k+1}} A_i$, and

$$\left\| \sum_{i=l_{2k-1}+1}^{l_{2k}} c_i \tilde{v}_i \right\|_{L_p^1(G)} < \frac{\varepsilon}{2^{2k-1}} + \frac{\varepsilon}{2^{2k}},$$

$$\left\| \sum_{i=l_{2k}+1}^{l_{2k+1}} c_i \tilde{v}_i \right\|_{L_p^1(G)} < \frac{\varepsilon}{2^{2k-1}} + \frac{\varepsilon}{2^{2k+1}},$$

$$\left\| u - \sum_{i=1}^{l_{2k+1}} c_i \tilde{v}_i \right\| \leqslant \frac{\varepsilon}{2^{2k+1}}.$$

In this case,

$$\sum_{i=1}^{2k+1} |c_{0,i}| < |c_{0,1}| + \sum_{i=1}^{2k+1} \frac{\varepsilon}{2^i} + \sum_{i=1}^{2k+1} \frac{\varepsilon}{2^i} = |c_{0,1}| + 6\varepsilon,$$

$$\mathrm{Cap}_{(1,p)} \left(\bigcup_{i=1}^{2k+1} A_i \right) < \frac{\varepsilon}{2^i}.$$

Let us denote the sum $\sum_{i=l_j-1}^{l_j} c_i \bar{v}_i$ by w_j. The series $\sum w_j$ absolutely converges in $L_p^1(G)$, since, due to the above,

$$\sum_{j=m}^{N} \|w_j\|_{L_p^1(G)} < \left(\sum_{j=m}^{N} \frac{\varepsilon}{2^{j-1}} \right) \frac{5}{4}.$$

Consequently, there exist the constants $\{b_j\}$ such that the sequence $\sum_{i=1}^{j} w_i + b_j$ converges almost everywhere to some function $\bar{u} \in L_p^1(G)$. Since by construction $\sum_{i=1}^{j} w_i + \sum_{i=1}^{j} c_{0,i}$ converges to u on the set of positive measure, then $(\sum_{i=1}^{j} c_{0,i} - b_j)$ tends to some constant b_0, i.e., on this set of positive measure $\bar{u} - b_0 = u$, and the sequence $\sum_{i=1}^{j} w_i + \sum_{i=1}^{j} c_0, i$ converges to $\bar{u} - b_0$ in $L_p^1(G)$. Consequently, $\bar{u} - b_0$ and u coincide. Then for every compact subdomain V of the domain G, the series $\sum_{j=1} w_j + c_0 (c_0 = \sum_{j=1} c_{0,j})$ converges to u in $W_p^1(V)$; i.e., $(1,p)$-quasieverywhere. Since zero capacity is countably additive, then $\sum_{j=1} w_j(x) + c_0 = u(x)$ $(1,p)$-quasi-everywhere.

The proof of the theorem is completed if we prove that, besides the series $\sum_j w_j + c_0$, the series $\sum_i c_i \overline{v}_i + c_0$ also converges in $L_p^1(G)$ and $(1,p)$-quasieverywhere to the function u, and if we manage to successfully approximate v_i by the functions from $\text{Ext}_p(G)$ having preserved the properties of the series.

The first goal is reached by a simple estimate which follows from the previous ones and from the form of the function w_j. Let $m < l_j$,

$$\left\| w_j - \sum_{i=l_{j-1}+1} c_i v_i \right\|^p_{L_p^1(G)} = \left\| \sum_{i=m}^{l_j} c_i v_i \right\|^p_{L_p^1(G)} \leqslant \| w_j \|^p \leqslant \left(\frac{\varepsilon}{2^{j-1}} + \frac{\varepsilon}{2^{j-1}} \right)^p.$$

To reach the second goal we use Lemma 2.3 about the density of $\text{Ext}_p(G)$ in the sense of $L_p^1(G)$ and $W_{p,\text{loc}}^1(G)$. Let us fix a compact subdomain V of the domain G, and let us for every function $\overline{v}_i$ choose a function v_i from $\text{Ext}_p(G)$ so that $\|\overline{v}_i - v_i\|_{L_p^1(G)} \leqslant \varepsilon/|c_i|2^i$ and $|\overline{v}_i(x) - v_i(x)| < \varepsilon/|c_i|2^i$ for all x lying in V outside of some set B whose measure is smaller than $\varepsilon/2^i$. Then in all the above arguments $\overline{v}_i$ may be substituted by v_i.

Theorem 2.9. *There exists a countable totality of functions $v_i \in \text{Ext}_p(G)$, $i = 1, 2, \ldots$, such that for any function $u \in L_p^1(G)$ and for any $\varepsilon > 0$ there exists a representation in the form $u = c_0 + \sum_{i=1}^{\infty} c_i v_i$ for which*

$$\|u\|_{L_p^1(G)} \leqslant \sum_{i=1}^{\infty} \|c_i v_i\|_{L_p^1(G)} \leqslant \|u\|_{L_p^1(G)} + \varepsilon.$$

Proof. If at the end of the proof of the previous theorem we choose the functions v_i from the countable totality of functions of the class $\text{Ext}_p(G)$ constructed in Lemma 2.4, and if we use Theorems 2.6 and 2.8, we obtain the representation $u = c_0 + \sum_{i=1}^{\infty} c_i v_i$ with the required properties.

For the bounded domains G, Corollary 1 of Theorem 2.7 is transferred to the space $W_p^1(G)$ of the functions which are summable in the degree p by the domain G and which have in G generalized derivatives summable in the degree p. In the space W_p^1 we consider the norm

$$\|u\|_{W_p^1(G)} = \|u\|_{L_p(G)} + \|u\|_{L_p^1(G)}.$$

Theorem 2.10. *A linear hull of the set* $\text{Ext}_p(G)$ *is dense in the set* $W_p^1(G)$.

Proof. Let us consider a bounded function $u \in W_p^1(G)$. According to Theorem 2.7 and to Remark 1 of Theorem 2.7, there exists a sequence of functions $\{u_k\}$ which are bounded in total. Each of these functions is a linear combination of elements of the set $\text{Ext}_p(G)$, which converges to u in $L_p^1(G)$.

Then one can find a bounded sequence of real numbers $\{c_k\}$ such that the sequence $\{u_k + c_k\}$ converges to u almost everywhere. Due to the Lebesgue theorem $\{u_k + c_k\} \to u$ in $L_p^1(G)$.

This completes the proof of the theorem.

§3 Removable Singularities for the Spaces $L_p^1(G)$

3.1. Two Ways of Describing Removable Singularities

Domains G_1 and G_2 $(G_1 \subset G_2)$ are said to be $(1,p)$-equivalent if the restriction operator $\theta : L_p^1(G_1) \to L_p^1(G_2)(\theta u = u|_{G_2})$ is the isomrophism of the vector spaces $L_p^1(G_2)$ and $L_p^1(G_1)$.

The fact that the set $E = G_1 \backslash G_2$ belongs to the class NC_p in the domain G_1 (Theorem 3.1) is the criterion of $(1,p)$-equivalence of the domains G_1 and G_2.

The definition of the class NC_p. A set E closed with respect to the domain G is called the NC_p-set if for any pair of continuums $F_0, F_1 \subset G\backslash E$ $C_p^1(F_0, F_1, G) = C_p^1(F_0, F_1, G\backslash E)$.

The basic properties of NC_p-sets are direct corollaries of Theorem 3.1. Among them we distinguish the localization principle: the set $E \subset G$ is the NC_p-set in the domain G iff it is the NC_p-set in any ball $B \subset G$.

On a plane the class NC_2 exactly coincides with NED, i.e., removable sets for AD classes, quasiconformal mappings and for the space L_2^1 are the same. In the case of dimension $n = 2$ for a class similar to AD, removable singularities are less in number than for the space L_n^1 or for the class of conformal mappings.

The above criterion of $(1,p)$-equivalence of domains is the corollary of the theorem for the possibility of approximating to any accuracy an arbitrary function $v \in L_p^1$ $(p > 1)$ by a linear combination $c_0 + \sum_{i=1}^{l} c_i v_i$ of the extremal functions for $(1,p)$-capacity. The gradient supports of these functions do not pairwise intersect (Theorem 2.7).

Theorem 3.1. *Domains G_1 and G_2 $(G_1 \supset G_2)$ are $(1,p)$-equivalent iff the set $G_1 \backslash G_2$ is an NC_p-set in G_1.*

Proof. *Necessity.* Let the spaces $L_p^1(G_1)$ and $L_p^1(G_2)$ be isomorphic as linear spaces for the restriction isomorphism $\theta u = u|_{G_2}$ and $u \in L_p^1(G_1)$. Passing to the factor-spaces $L_p^1(G_1)$ and $L_p^1(G_2)$ and using the Banach theorem, we obtain the boundedness of the operator θ^{-1}.

Let us prove that $|(G_1 \backslash G_2)| = 0$. Assume the converse. Then the set $G_1 \backslash G_2$ has at least one density point x_0. Let us consider a sequence of open cubes $Q_m = Q(x_0, 1/m)$ with the centre at the point x_0, with an edge of length $1/m$ and with sides parallel to coordinate planes.

Let us consider a function u_m which equals zero outside of the cube Q_m, it equals $1/2m$ at the point x_0 and it is linear on every segment connecting the point x_0 with an arbitrary point of the cube Q_m boundary. It is obvious that $|\nabla u_m(x)| = 1$ almost everywhere in Q_m.

From the boundedness of the operator θ^{-1} we have

$$|Q_m| = \int_{G_1} |\nabla u_m|^p dx \leqslant \|\theta^{-1}\| \int_{G_2} |\nabla u_m|^p dx = \|\theta^{-1}\| \, |(Q_m \setminus (G_1 \setminus G_2))|.$$

If the point x_0 is the density point, then the inequality is not valid for m. The resulting contradiction proves that $|(G_1 \backslash G_2)| = 0$.

Consequently, θ is an isometric operator and $G_1 \backslash G_2$ is an NC_p-set.

Sufficiency. Let $E = G_1 \backslash G_2$ be an NC_p-set in G_1. For a pair of connected sets with smooth boundaries $F_0, F_1 \subset G_2$, $F_0 \cap F_1 = \emptyset$, let us consider the extremal function u_1 in the domain G_1 and the extremal function u_2 in the domain G_2. Due to the definition of NC_p-sets

$$\int_{G_1} |\nabla u_1|^p dx = \int_{G_2} |\nabla u_2|^p dx.$$

The function u_1 is equal to zero on F_0, to 1 on F_1, and

$$\int_{G_2} |\nabla u_1|^p dx \leqslant \int_{G_1} |\nabla u_1|^p dx = \int_{G_2} |\nabla \tilde{u}_2|^p dx.$$

Due to the uniqueness of the extremal function $u_1 \equiv u_2$ on G_2. Consequently, every function $u \subset \mathrm{Ext}_p(G_2)$ may be extended to the domain G_1, and the norm is preserved. Lemma 2.1 and the semicontinuity lemma allow us to transfer the same conclusion to the class $E_p(\overline{G})$.

Let us choose an arbitrary function $v \in L_p^1(G_2)$.

According to Theorem 2.7, for every $\varepsilon > 0$ there exists a function $v_\varepsilon = \sum_{k=1}^{l} c_{k,\varepsilon} v_{k,\varepsilon}$ satisfying the conditions: a) $\|v - v_\varepsilon\|_{L_p^1(G)} < \varepsilon$; b) $\|v_\varepsilon\|^p_{L_p^1(G_1)} = \sum_{k=1}^{l} |c_{k,\varepsilon}|^p \|v_{k,\varepsilon}\|^p$; c) $v_{k,\varepsilon} \in E_p(\overline{G}_2)$ for all k. According to the above, each of the functions v_ε is extended to G_1, the norm being preserved. For the extensions $\overline{v}_\varepsilon$ of the functions v_ε the inequalities

$$\|\tilde{v}_\varepsilon\|^p_{L_p^1(G_1)} = \left\| \sum_{k=1}^{l} c_{k,\varepsilon} \tilde{v}_{k,\varepsilon} \right\|^p_{L_p^1(G_1)} \leqslant \sum_{k=1}^{l} |c_{k,\varepsilon}|^p \|\tilde{v}_{k,\varepsilon}\|^p_{L_p^1(G_1)}$$

$$= \sum_{k=1}^{l} |c_{k,\varepsilon}|^p \|v_{k,\varepsilon}\|^p_{L_p^1(G_2)} = \|v_\varepsilon\|^p_{L_p^1(G_2)} \leqslant \|v\|^p_{L_p^1(G_2)} + \varepsilon$$

are valid.

Choosing $\varepsilon_n \to 0$ for $n \to \infty$ let us construct a sequence of functions $v_{\varepsilon_n} \to v$ in $L_p^1(G_2)$ such that the sequence $\sum_k c_{k,\varepsilon_n} \overline{v}_{k,\varepsilon_n}$ weakly converges in $L_p^1(G_1)$ to some function $\tilde{v}$. Here $\tilde{v} - v = T$ on G_2. Putting $\overline{v} = \tilde{v} - T$ we obtain the extension $\overline{v}$ of the function v on G_1. Due to the lemma about semicontinuity,

$$\|\tilde{v}\|_{L^1_p(G_1)} = \|v\|_{L^1_p(G_2)}.$$

We have proved that every function $v \in L^1_p(G_2)$ is extended onto G_1, the norm being preserved.

To complete the proof of $(1,p)$-equivalence of the domains G_1 and G_2, the one-to-one correspondence of the extension operator remains to be proved. It is sufficient to prove that the measure of the set $G_1 \backslash G_2$ equals zero.

Let $x \in \partial(G_1 \backslash G_2) \cap G_1$. Let us take a spherical ring $D = \{y \in \mathbf{R}^n : 0 < a < |x-y| < b\}$ lying in G_1. The set $F_1 = \{y \in \mathbf{R}^n : |y-x| \leqslant a\}$ has non-empty intersection with the domain G_2. For sufficiently small b the same property has the set $F_0 = \{y \in \mathbf{R}^n : |y-x| \geqslant b\}$ as well.

The gradient of the extremal function u of the pair (F_0, F_1) is distinct from zero on D. According to the above, the function $u|_{G_2}$ may be extended to the function $\overline{u} \in L^1_p(G_1)$, besides $\|\overline{u}\|_{L^1_p(G_1)} = \|u\|_{L^1_p(G_2)}$. Hence,

$$\int_{G_1 \backslash G_2} |\nabla \tilde{u}|^p \, dx = 0. \tag{3.1}$$

If $|(G_1 \backslash G_2)| \neq 0$, then $|\nabla \overline{u}| \equiv 0$ almost everywhere on $G_1 \backslash G_2$; i.e., $|\nabla \overline{u}| = 0$ almost everywhere on $F_0 \cup F_1$, $\overline{u} = 0$ almost everywhere on F_0, and $\overline{u} = 1$ almost everywhere on F_1 (since $F_0 \cap G_2 \neq \emptyset$, $F_1 \cap G_2 \neq \emptyset$).

The fact that the function $\overline{u}$ is equal to zero on F_0 and equal to 1 on F_1, and the equality $\|\overline{u}\|_{L^1_p(G_1)} = \|u\|_{L^1_p(G_2)}$ result in the inequality

$$\int_D |\nabla \tilde{u}|^p \, dx \leqslant \int_D |\nabla u|^p \, dx.$$

Due to the uniqueness of the extremal function, $\overline{u} \equiv u$. From inequalities (3.1) and $|\nabla u| > 0$ on D, it follows that $|(D \cap (G_1 \backslash G_2))| = 0$. Countable additivity of measure and arbitrariness in the choice of the ring D allow us to conclude that the set $G_1 \backslash G_2$ does not contain interior points and, moreover, it has zero measure.

This completes the proof.

Corollary. *The restriction operator in the definition of* $(1,p)$*-equivalent domains is the isometry of the spaces* L^1_p.

This follows from the fact that the difference measure of $(1,p)$-equivalent domains is equal to zero.

3.2. Properties of NC_p-Sets. Localization Principle

$E \subset G$ is an NC_p-set in G iff for any open ball $B(x,r) \subset G$ the set $E \cap B(x,r)$ is the NC_p-set in the ball.

Sufficiency. Let us take a countable covering of the domain G by the balls $B_i \subset G$ and let us consider the partition of unit $\{\varphi_i\}$ subordinate to this covering. According to Theorem 3.1, the domains B_i and $B_i \backslash E$ are $(1,p)$-equivalent. Let us take any function $u \in L^1_p(G \backslash E)$. Due to $(1,p)$-equivalence of the domains B_i and $B_i\backslash E$ there exists the extension of the function $\varphi_i u$ to the function $\overline{u}_i$ from $L^1_p(B_i)$. Since such a function $\overline{u}_i$ is unique, the function $\overline{u}_i$ is compactly supported (its natural values are equal to zero near ∂B_i because the measure of the set $E\cap B_i$ equals zero). Then $u = \sum_i \overline{u}_i$ is the extension of the function u. Due to the fact that $|E| = 0$, we obtain $\|u\|_{L^1_p(G)} = \|\tilde{u}\|_{L^1_p(G)}$. Thus the extension operator is isometry, and the domains $G\backslash E$ and G are $(1,p)$-equivalent; i.e., according to Theorem 3.1, E is an NC_p-set.

Necessity. Let us choose an arbitrary ball $B_1 = B(x, r_1)$, $0 < r_1 < r$, and a function $v \in L^1_p(B\backslash E)$. Multiplying v by a smooth compactly supported function ψ which equals 1 on the ball B_1 and equals 0 outside of the ball $B(x,r)$, and putting $\overline{v} = v\psi$ in the ball $B(x,r)$ and $\overline{v} = 0$ outside of the ball $B(x,r)$, we obtain the function belonging to $L^1_p(G\backslash E)$. According to Theorem 3.1, $\overline{v}$ is extended (the class is preserved and the norm is not increased) to some function $\overline{w}$ defined on the domain G. Thus, we obtain the unique extension of the function $v \in L^1_p(B(x,r_1)\backslash E)$ to some function $\omega \in L^1_p(B(x,r_1))$. Since r_1 is arbitrary, the domains $B(x,r)\backslash E$ and $B(x,r)$ are $(1,p)$-equivalent, i.e., E is an NC_p-set in$B(x,r)$.

Let us fix the domain G and the set $E \subset G$. With the exception of Property 3.2, the properties of NC_p-sets below are the corollaries of the localization principle.

Property 3.2. *If there exists a sequence of balls $\{B_m\}$ which covers the set $E \subset G$, and a set $E_m = E \cap B_m$ is an NC_p-set in B_m, then E is an NC_p-set in G.*

This fact was proved in the proof of sufficiency in the localization principle.

Property 3.3. *Any closed subset of an NC_p-set is an NC_p-set.*

Proof. Let E_1 be a closed subset of the set E. Let us choose an arbitrary function $v \in L^1_p(B\backslash E_1)$. Then $v \in L^1_p(G\backslash E)$, and according to Theorem 3.1, it is extended uniquely to the function $\overline{v} \in L^1_p(G)$. Consequently, the domains G and $G\backslash E_1$ are $(1,p)$-equivalent. According to the same theorem, E_1 is an NC_p-set.

Property 3.4. *The intersection of any number of NC_p-sets is an NC_p-set.*

Property 3.5. *Let G be a domain in $\mathbf{R}^n$, and let E_m, $m = 1, 2, \ldots, M$, be NC_p-sets in G. Then their union $E = \cup_{m=1}^M E_m$ is an NC_p-set.*

Proof. It is sufficient to consider two NC_p-sets, E_1 and E_2. The intersection $E_1 \cap E_2$ is an NC_p-set according to Property 3.3. Let us consider a domain $G_1 = G\backslash(E_1 \cap E_2)$. The domain G and the set $(E_1 \cup E_2)\backslash(E_1 \cap E_2)$ satisfy the condition of Property 3.2. Let us choose an arbitrary function $v \in L^1_p(G\backslash(E_1 \cup E_2))$. In the domain G_1, applying Property 3.2 and Theorem 3.1, let us extend v up to some function $\overline{v} \in L^1_p(G_1)$. Since $E_1 \cap E_2$ is an NC_p-set in G_1, then $\overline{v}$ is extended up to the function $w \in L^1_p(G)$.

This completes the proof.

Property 3.6. *Let G_1 be a subdomain of G and let E be an NC_p-set in G. Then $E_1 = G_1 \cap E$ is an NC_p-set in G.*

This follows from the localization principle.

Property 3.7. *Every compact subset of an NC_p-set E in a domain G is an NC_p-set in any domain.*

Proof. Due to Property 3.6, it is sufficient to prove that every compact subset of the set E is an NC_p-set in $\mathbf{R}^n$. This follows easily from Theorem 3.1.

Property 3.8. *Every NC_q-set E in a domain G is an NC_p-set in G for all $p > q$.*

Proof. Due to the localization principle, it suffices to verify the statement of the theorem for a ball $B \subset G$. If the function $v \in L_p^1(B\backslash E)$, then $v \in L_q^1(B\backslash E)$, $q < p$. According to the condition, E is an NC_p-set in B. Therefore, v has generalized derivatives in the ball B. Since $|E| = 0$, $v \in L_p^1(B)$.

Property 3.9. *Let E be a closed set in a domain G. Then*

a) if E is an NC_p-set, then $|E| = 0$;

b) if E is an NC_p-set, then $\dim(E) \leqslant n-2$;

c) if the $(n-1)$-dimensional Hausdorff measure $\Lambda_{n-1}(E) = 0$, then E is an NC_p-set.

Proof. The property a) is proved in Theorem 3.1.

From the localization principle it follows that the intersection of a set with any ball $B \subset G$ is an NC_p-set in B. Suppose that there exists a ball B which is partitioned by a set into two non-empty open sets B_0 and B_1 $(B\backslash E = B_0 \cup B_1)$. Let us choose in each set a closed ball $F_0 \subset B_0$, $F_1 \subset B_1$. A function v which equals zero on F_0 and equals 1 on F_1 is admissible for the capacity $C_p^1(F_0, F_1; B\backslash E)$. Consequently, $C_p^1(F_0, F_1; B\backslash E) = 0$. At the same time, it is well known that $C_p^1(F_0, F_1; B) > 0$. We obtain the contradiction with the fact that $E \cap B$ is an NC_p-set in B.

By contradiction it was proved that for any ball $B \subset G$ the set $B\backslash E$ is connected. The statement b) is proved.

If the set E satisfies the property c), then every function having generalized derivatives in the domain $G\backslash E$ is extended to some function $\overline{v}$ having generalized derivatives in the domain G. Since $|E| = 0$, then G and $G\backslash E$ are $(1,p)$-equivalent. Consequently, E is an NC_p-set.

Remark. Property 3.9 for $p = n$ and for $G = \mathbf{R}^n$ was proved in paper [113].

Remark. There exists an example of an NC_p-set having non-zero $(n-1)$-dimensional Hausdorff measure [3].

CHAPTER 5

Change of Variables

This chapter deals with the problems concerning mappings that preserve the Sobolev classes. First, one has to construct the analogy of mapping degree for discontinuous mappings of the class L_n^1, which, in our opinion, is of special interest. The introductory section presents the properties of mapping degree and of the notion of multiplicity of mapping, close to mapping degree.

§1 Multiplicity of Mapping, Degree of Mapping, and Their Analogies

1.1. The Multiplicity Function of Mapping

For every set E belonging to a definition domain U of a continuous mapping $f : U \to \mathbf{R}^n$ let us denote by $N_f(y, E)$ the number of elements of a set $f^{-1}(y) \cap E$. If this set is infinite, we put $N_f(y, E) = \infty$. The function $y \to N_f(y, E)$ is called the function of multiplicity of mapping f on the set E. Let us show some simple properties of the multiplicity function:

1) If $E_1 \subset E_2 \subset U$, then for any $y \in \mathbf{R}^n$, $N_f(y, E_1) \leqslant N_f(y, E_2)$.

2) Let $\{E_m \subset U\}$ be an arbitrary increasing sequence of sets. Then

$$N_f(y, E) = \lim_{m\to\infty} N_f(y, E_m), \quad \text{where } E = \bigcup_{m=1}^{\infty} E_m.$$

Proof. If $N_f(y, E)$ is finite, then $f^{-1}(y) \in E$ consists of a finite number of points: $x_1, x_2, \ldots, x_N$. Since the sequence $\{E_m\}$ increases, there exists E_m containing $x_1, x_2, \ldots, x_N$. Thus,

$$\lim_{m\to\infty} N_f(y, E) = N_f(y, E_m) = N_f\left(y, \bigcup_{m=1}^{\infty} E_m\right).$$

But if $N_f(y, E) = \infty$, then $f^{-1}(y) \cap E$ contains a sequence of points $x_1, x_2, \ldots, x_N, \ldots$. For any N fixed so far, there exists E_{m_0} containing the points $x_1, x_2, \ldots, x_N$, i.e., $\lim\limits_{m\to\infty} N_f(y, E_m) = \infty$.

3) Let $E_1, E_2 \subset U$ and $E_1 \cap E_2 = \emptyset$. Then $N_f(y, E_1, E_2) = N_f(y, E_1) + N_f(y, E_2)$ for any $y \in E_1 \cup E_2$.
4) Let $\{E_m \subset U\}_{m=1,2,\dots}$ be an arbitrary sequence of pairwise nonintersecting sets. Then for any y,

$$N_f\left(y, \bigcup_{m=1}^{\infty} E_m\right) = \sum_{m=1}^{\infty} N_f(y, E_m).$$

Let us investigate the properties of the multiplicity function for mappings satisfying additional requirements. The continuous mapping $f : U \to \mathbf{R}^n$ has the N-property if the image of every set of zero measure is a set of zero measure.

Proposition 1.1. *If $f : U \to \mathbf{R}^n$ satisfies the N-condition, then the image of every measurable set is measurable.*

Proof. If the set $A \subset U$ is compact, then $f(A)$ is compact and, therefore, measurable. Let $E \subset U$ be a set of the type K_σ, i.e., a countable union of compact sets. Then $f(E)$ is measurable as the union of a countable number of measurable sets. Let us consider an arbitrary measurable set E. There exists a set F of the type K_σ such that $F \subset E$ and $m(E \backslash F) = 0$.

From the equality $f(E) = f(E) \cup f(E \backslash F)$ the measurability of $f(E)$ follows. Since $f(E)$ is measurable, $f(E \backslash F)$ due to the N-condition has zero measure, i.e., it is also measurable.

Theorem 1.2. *If the mapping $f : U \to \mathbf{R}^n$ satisfies the N-condition, then for any measurable set $E \subset U$, the multiplicity function $y \to N_f(y, E)$ is measurable in $\mathbf{R}^n$.*

Proof. Let the set E be bounded. For every natural number m construct the partition of the set E into pairwise non-intersecting measurable sets $E_1^{(m)}, E_2^{(m)}, \dots, E_{k_m}^{(m)}$. The diameter of each of these sets does not exceed $1/m$. Here $E = \cup_{i=m}^{k_m} E_i^{(m)}$. Let us denote by $N_m(y)$ the sum of characteristic functions of the sets $f(E_i^{(m)})$. Each of the sets $f(E_i^{(m)})$ is measurable. Consequently, the function N_m is measurable. It is obvious that $N_m(y) \leqslant N_f(y, E)$ for $y \in \mathbf{R}^n$. Let us prove that $N_m(y) \leqslant N_f(y, E)$ for $y \in \mathbf{R}^n$. First let us consider the case $N_f(y, E) < \infty$, i.e., $f^{-1}(y) \cap E = (a_1, a_2, \dots, a_{N_f(y,E)})$. If $m_0 > 1/\min_{i,j} |a_i - a_j|$, then none of the sets $\{E_i^{(m)}\}$ contains two different points a_i and a_j. The sets $E_i^{(m)}$ do not pairwise intersect. We only consider the sets $E_{i_k}^{(m)}$ which contain the point y. It is obvious that $N_m(y) \geqslant N_f(y, E)$ for all $m > m_0$, i.e., $N_m(y) \to N_f(y, E)$ for $m \to \infty$. But if $N_f(y, E) = \infty$, then the same argument shows that $\lim N_m(y) = \infty$.

Each of the functions $N_m(y)$ is measurable; therefore, their limit $N_f(y, E)$ is also measurable.

Theorem 1.3. *If a mapping $f : u \to \mathbf{R}^n$ is continuous and open, then for* *every open set $V \subset U$, the function $y \to N_f(y, V)$ is lower semicontinuous in* $\mathbf{R}^n$.

Proof. First let us consider the case $N_f(y_0, V) < \infty$. Let us choose $N = N_f(y_0, V)$ of pairwise nonintersecting open balls $\{B_1 \subset V, \dots, B_N \subset V\}$ with centres at the points $a_1, \dots, a_N$ of the set $f^{-1}(y_0) \cap V$. Let us find $\varepsilon > 0$ such that the ball $B(y_0, \varepsilon) \subset f(B_i)$ for all $i = 1, \dots, N$. Since the mapping f is open, such ε does exist. For every point $y \in B(y_0, \varepsilon)$ for all $i = 1, 2, \dots, N$, $N_f(y, B_i) \geqslant 1$. We have proved that for $N_f(y_0, V) < \infty$ (due to Property 4),

$$\underline{\lim}_{y \to y_0} N_f(y, V) \geqslant N_f(y_0, V).$$

If $N_f(y_0, V) = \infty$, then the same argument proves that $\lim_{y \to y_0} N_f(y, V) = \infty$.

We finish this subsection by showing that the multiplicity function introduced here is sometimes called "the rough multiplicity function" (see [108]). The notion of multiplicity function admits various modifications often used in the theory of functions.

1.2. The Approximate Differential

Let U be an open set in $\mathbf{R}^n$, and let $f : U \to \mathbf{R}^n$ be continuous. For $x \in U$, we set

$$L_h(X) = \frac{f(x + hX) - f(x)}{h},$$

where $x \in B(0, 1)$. For sufficiently small h, $L_h(X)$ as the function of h is defined in the ball $B(0, 1)$. The linear mapping $L : \mathbf{R}^n \to \mathbf{R}^n$ is called the approximate differential of the function f at the point x if L_h converges to L by measure on the ball $B(0, 1)$ for $h \to 0$. We denote the approximate differential of the function f at a point x by means of the symbol $(\mathrm{app})df_x$. The determinant of the matrix which assigns the approximate differential is called the approximate Jacobian. We denote the approximate Jacobian by $J(x, f)$, like the usual one.

Theorem 1.4. *Let $U \subset \mathbf{R}^n$ be an open set; let $f : U \to \mathbf{R}^n$ be a continuous mapping. Suppose that f possesses the N-property, has almost everywhere in U the approximate differential, and that the Jacobian $J(x, f)$ is locally summable in U.*

Then for every measurable set $E \subset U$, the inequality

$$m(f(E)) \leqslant \int_E |J(x, f)|\, dx.$$

holds.

Proof. Let E_1 be a set of the points in U in which the approximate differential of the function f does not exist, and let E_2 be the set of the points where $J(x, f)$ has no natural value in the sense of Lebesgue. According to the assumption of the theorem, $m(E_1 \cup E_2) = 0$. Since f possesses the N-property, $m(f(E_1 \cup E_2)) = 0$. Therefore, one may assume that the set E under consideration does not intersect $E_1 \cup E_2$. We also assume that $\overline{E} \subset U$. It is clear that the general case is reduced to this.

Let us assign arbitrarily $\varepsilon > 0$ and let us construct an open set $G_1 \subset E$, $\overline{G}_1 \subset U$, and $m(G_1) < m(E)+\varepsilon$. Consider a measure $\eta(A) = \int_A |J(x, f)|dx$ in U. This measure is absolutely continuous and bounded in G_1. Consequently, there exists $\delta > 0$ such that for every set $A \subset G_1$, measurable in the sense of Lebesgue, and such that in $m(A) < \delta$, the inequality $\eta(A) < \varepsilon$ holds.

Let us consider an open set $E \subset G \subset G_1$ such that $m(G) < m(E) + \delta$. Let $x \in E$, $L = (\text{app})df_x$ and $L_h(X) = (f(x + hX) - f(x))/h$. For $h \to 0$, the function L_h converges to L by the measure in the ball $B(0,1)$. Let us put $H_h = L_h(B_1)$. Find $1 > h_0 > 0$ such that

$$m(H_h) < m(L(B_1)) + \varepsilon m(B(0, 1))/(m(E) + \delta)$$

for $h < h_0$.

Let us denote by $Q(h)$ a set of all $x \in B(0,1)$ such that $|L_h(x) - L(x)| > h_0$. The set $Q(h)$ is measurable, and $m(Q(h)) \to 0$ for $h \to 0$, since at the point $x \in E$, the function f has the approximate differential. Let us put

$$\tau = \min\left(\frac{\delta}{m(E) + \varepsilon}, \frac{\varepsilon}{m(E) + \varepsilon}\right)$$

Let us find $h_1(x) < 1$ such that for $0 < h < \min(h_0, h_1(x))$, $m(Q(h)) < \tau m(B(0,1))$. Let us set $P(h) = B(0,1)\backslash Q(h)$. Then, since $h < h_0$,

$$m(L_h(P(h))) \leqslant m(L(B(0, 1))) + \frac{\varepsilon m(B(0, 1))}{m(E) + \delta}.$$

Since x is the Lebesgue point of the function $J(x, f)$, there exists $1 > h_2(x) > 0$ such that for $0 < h < h_2(x)$,

$$|J(x, f)|\, m(B(0, 1)) < \int_{B(0,1)} |J(x + hX, f)|\, dX + \frac{\varepsilon m(B(0, 1))}{m(E) + \delta}. \quad (1.1)$$

The mapping $x \to x + hX$ transforms the ball $B(0,1)$ into the ball $B(0,h)$, the set $Q(h)$ is transformed into the set which we denote by $Q(x,h)$, and the set $P(h)$ into the set $P(x,h) = B(x,h)\backslash Q(x,h)$. Obvious are the inclusion $Q(x,h) \subset B(x,h)$ and the inequality $m(Q(x,h)) < \tau h^n m(B(x,h))$. The set $f(P(x,h))$ is obtained from the set $L_h(P(h))$ by the transformation $y \to f(x) + hy$. Hence it follows that

$$m(f(P(x, h)) = h^n m(L_h(P(h))).$$

Finally for $0 < h < h_1$, we obtain the inequality

$$m(f(P(x, h))) \leqslant m(L(B(x, h))) + \frac{\varepsilon m(B(x, h))}{m(E) + \delta}. \quad (1.2)$$

Performing the change of variables of $x + hX = y$ in each of the integrals in inequality (1.1), we see that for all $h < h_2(X)$, the inequality

$$|J(x,f)|\, m(B(x,h)) < \int_{B(x,h)} |J(y,f)| + \frac{em(B(x,h))}{m(E)+\delta} \tag{1.3}$$

holds.

By comparing inequalities (1.2) and (1.3), we obtain that for all $h < \min(h_1, h_2)$ the inequality

$$m(f(P(x,h))) < \eta(B(x,h)) + \frac{\alpha em(B(x,h))}{m(G)+\delta} \tag{1.4}$$

is valid. Let us denote by $h_3(x)$ the distance from the point $x \in E$ to the set ∂G. Since $E \subset G$, then $h_3(x) > 0$. Let us put $h(x) = \min(h_1(x), h_2(x), h_3(x))$. For every $x \in E$, let us denote by $\mathcal{F}_x$ the set of all balls with centre x and radius $h < h(x)$. Let us put $\mathcal{F} = \cup_{x\in E}\mathcal{F}_x$. The set forms the covering of E in the sense of Vitali. Therefore, one can single out in it a countable subset of pairwise nonintersecting balls $B_1 = B(x_1, h_1), \ldots, B_m(x_m, h_m)$ such that $m(E \setminus \cup_{m=1}^{\infty} B_m) = 0$. We set $T = \cup_{m=1}^{\infty} B_m$. Each of the balls $B_m \subset G$. Therefore,

$$m(T) = \sum_m m(B_m) \leqslant m(G) < m(E) + \delta.$$

Let $S = E\setminus T$. According to the abovesaid, $m(S) = 0$. Since the mapping f has the N-property, then $m(f(S)) = 0$

$$m(f(A)) \leqslant m(f(T)) + m(f(S)) = m(f(T)).$$

Let us put $Q = \cup_m Q(x_m, h_m)$, $P = \cup_m P(x_m, h_m)$. The, according to the abovesaid, $m(Q) \leqslant \tau \sum_m m(B(x_m, h_m)) = \tau m(T) \leqslant \tau(m(E) + \delta)$.

Due to the definition of the number τ, it therefore follows that $m(Q) < \min(\varepsilon, \delta)$. Further we have from (1.4),

$$m(f(E \cap P)) \leqslant \sum_m m(f(B(x_m, h_m)) \leqslant \sum_m \eta(B(x_m, h_m))$$

$$+ \sum_m \frac{2em(B(x_m, h_m))}{m(E)+\delta} \leqslant \sum h(B(x_m, h_m)) + 2e = \eta(T) + 2e.$$

Besides, $\eta(T) < \eta(G) < \eta(E) + \varepsilon = \eta(E \cup P) + \eta(E \cap Q) + \varepsilon \leqslant \eta(E \cap P) + \eta(Q) + \varepsilon < \eta(E \cap P) + 2\varepsilon$ since $m(Q) < \delta$; therefore, due to the choice of δ, $\eta(Q) < \varepsilon$.

Finally, we obtain

$$m(f(E \cap P)) \leqslant \eta(E \cap P) + 4\varepsilon.$$

Now, from the equality $E \cap P = E\setminus Q$, it follows that for every $\varepsilon > 0$, there exists a set Q such that $m(Q) < \varepsilon$ and $m(f(E\setminus Q)) < \eta(E\setminus Q) + 4\varepsilon$.

To complete the proof of the theorem, we construct a sequence of sets $\{E_k\}_{k=0,1,\ldots} E_k \subset E_{k+1}$ for all k. Suppose that $E_0 = E\setminus Q$. Let E_k be constructed. According to what we proved above, there exists a set $E'_k \subset E_k$ such that $m(E'_k) < \varepsilon/2^k$ and $m(f(E_k \backslash E'_k)) < \eta(E_k \backslash E'_k) + 4\varepsilon/2^k$. We set $E_{k+1} + E'_k$. Let $\widetilde{E} = \cup_{k=0}^{\infty}(E_k \backslash E_{k+1})$. It is obvious that $\widetilde{E} \subset E$ and $m(\widetilde{E}) = m(E)$. By the construction of the sequence $\{E_k\}$, we therefore obtain

$$\mu(f(E)) = \mu(f(\bar{E})) \leqslant \sum_{k=0}^{\infty} \mu(f(E_k \backslash E_{k+1}))$$

$$< \sum_{k=0}^{\infty} \eta(E_k \backslash E_{k+1}) + 4 \sum_{k=0}^{\infty} \varepsilon/2^k = \eta(E) + 8\varepsilon.$$

Since $\varepsilon > 0$ was chosen arbitrarily, the theorem is proved.

Corollary 1. *Under the conditions of the previous theorem, let $E_0 = \{x : J(x, f) = 0\}$. Then $m(E_0) = 0$.*

Corollary 2. *Under the conditions of the previous theorem, for every measurable set E, the inequality*

$$\int_{R^n} N_f(y, E)\, dy \leqslant \int_E |J(x, f)|\, dx$$

is valid.

Proof. For every natural m let us construct a partition of the set E into pairwise nonintersecting measurable sets $E_i^{(m)}$, $i = 1, 2, \ldots, k_m$, $E = \cup_{i=1}^{k_m} E_i^{(m)}$:

$$d(E_i^{(m)}) < 1/\text{m}.$$

Due to Theorem 1.4,

$$\sum_{i=1}^{k_m} m(f(E_i^{(m)})) \leqslant \sum_{i=1}^{k_m} \int_{E_i^{(m)}} |J(x, f)|\, dx = \int_E |J(x, f)|\, dx.$$

Let us denote by $N_m(y)$ the sum of characteristic functions $f(E_i^{(m)})$. Then

$$\sum_{i=1}^{k_m} m(f(E_i^{(m)})) = \int_{R^n} N_m(y)\, dy.$$

In the proof of Theorem 1.2, it was shown that $N_m(y) \to N_f(y, E)$ for $m \to \infty$. The inequality $N_m(y) \leqslant N(y, f, E)$ is obvious. Taking the limit in the inequality

$$\int_{R^n} N_m(y)\, dy \leqslant \int_{R^n} |J(x, f)|\, dx,$$

we obtain the desired result.

1.3. The K-Differential

Let $f : U \to \mathbf{R}^n$ be a continuous mapping of domain $U \subset \mathbf{R}^n$. A linear mapping $L : \mathbf{R}^n \to \mathbf{R}^n$ is called the K-differential of the mapping f at a point $x \in U$ if $L = (\text{app})df_x$, and if there exists a sequence $h_m \to 0$ for $m \to \infty$ such that the functions $L_{h_m}(X) = (f(x + h_m X) - f(x))/h_m$ for $m \to \infty$ converge to the function L uniformly on the sphere $S(0,1)$.

Lemma 1.5. *Let $f : U \to \mathbf{R}^n$ be continuous. Then, if f has the K-differential at a point $a \in U$, there exists a sequence of balls $B_m = B(a, r_m)$ where $r_m \to 0$ for $m \to \infty$ such that*

$$\lim_{m\to\infty} \frac{m(f(B_m))}{m(B_m)} \geqslant |J(a, f)|.$$

Proof. The lemma is obvious if $J(a, f) = 0$. By the definition of the K-differential, f has the approximate differential L at the point a and there exists a sequence of numbers $\{h_m\}_{m=1,2,\dots}$, such that $h_m \to 0$ for $m \to \infty$, and the functions $L_{h_m} \to L$ uniformly on $S(0,1)$ for $m \to \infty$. Let us set $\varepsilon_m = \max_{|x|=1} |L_{h_m}(x) - L(x)|$. Then $\varepsilon_m \to 0$ for $m \to \infty$. Let $B_m = B(a, h_m)$, $S_m = S(a, h_m)$. For every $x \in S_m$, we obviously have

$$|f(x) - f(a) - L(x-a)| < \varepsilon_m h_m.$$

Let us assign an arbitrary number $z \in (0,1)$. Together with the ball B_m, let us consider the ball $B_m(z) = B(a, zh_m)$. Then $m(B_m(z)) = z^n(B_m)$. The affine mapping $A = f(a) + L(x-a)$ transforms the ball $B_m(z)$ into an ellipsoid whose volume is equal to

$$m(B_m(z))\,|\det L| = m(B_m(z))\,|J(a, f)|.$$

Let us prove that for sufficiently large m, $f(B_m) \supset A[B_m(z)]$. Indeed, let $\delta = \min_{|x|=1} |L(x)| > 0$. Let us consider $A(S_m)$. Let us choose arbitrarily the points $y_1 \in A(S_m)$, $y_2 \in A(B_m(z))$, i.e., $y_1 = f(a) + L(x_1 - a)$ and $y_2 = f(a) + L(x_2 - a)$. Since $|x_1 - a| = h_m$, and $|x_2 - a| < zh_m$, then $|x_1 - x_2| \geqslant |x_1 - a| - |x_2 - a| > (1 - z)h_m$; therefore, $|A(y_1) - A(y_2)| = |L(x_1 - x_2)| \geqslant \delta(1 - z)h_m$. Thus, every point $y \in A(S_m)$ is away from the set $A(B_m(z))$ at a distance not less than $\delta(1 - z)h_m$.

Now let us find m_0 such that for $m > m_0$, $\varepsilon_m < \delta(1 - z)$. Let us take an arbitrary value of $m > m_0$ and construct a homotopy of the mappings f and A by setting $f(x,t) = (1 - t)f(x) + tA(x)$ where $0 \leqslant t \leqslant 1$. For fixed x, the point $f(x,t)$ for $t \in [0,1]$ circumscribes in $\mathbf{R}^n$ some segment with the ends $f(x)$ and $A(x)$. For $x \in S_m$, the length of this segment does not obviously exceed $\varepsilon_m < \delta(1-z)$. In the case $x \in S_m$, one end of the given segment belongs to the set; consequently, this segment does not contain points of the set $A(B_m(S))$. We obtain that $f|_{S_m}$ and $A|_{S_m}$ map S_m in $V_m = \mathbf{R}^n \setminus A(B_m(z))$ and $f(x,t)$ is the homotopy of the mappings $f|_{S_m}$ and $A|_{S_m}$ on the set V_m. Hence it follows that every point $y \in A(B_m(z))$ for the given value of z belongs to $f(B_m)$, thus we have proved that $f(B_m) \supset A(B_m(z))$. Consequently, for every $m > m_0$, we have

$$m(f(B_m)) \geqslant m(A(B_m(z)) = z^n m(B_m)\,|J(a, f)|.$$

Since $z \in (0,1)$ is arbitrary, the lemma is proved.

Remark. When proving the lemma, it was shown that for sufficiently large m the mappings $f|_{B_m}$ and $A|_{B_m} (A = f(a) + L(x-a))$ are homotopic.

Theorem 1.6. *Let a continuous mapping f of a domain U in $\mathbf{R}^n$ possess the N-property, have K-differential almost everywhere, and let the Jacobian $J(x,f)$ be locally summable in U.*

Then for every measurable set $A \subset U$, the equality

$$\int_A |J(x,f)|\,dx = \int_{\mathbf{R}^n} N_f(y,A)\,dy$$

is valid.

Proof. Due to Corollary 2 of Theorem 1.4, we have the inequality

$$\int_E |J(x,f)|\,dx \geqslant \int_{\mathbf{R}^n} N_f(y,E)\,dy$$

for every measurable set $E \subset U$.

Let us establish the inverse inequality.

Let $\varphi(A) = \int_{\mathbf{R}^n} N_f(y,A)\,dy$. Let us prove the absolute continuity of the function of the sets $\varphi(A)$. Let A be an arbitrary set of zero measure. From the N-property it follows that $m(f(A)) = 0$; therefore, $N_f(y,A) = 0$ almost everywhere in $\mathbf{R}^n$. Hence it follows that $\varphi(A) = 0$. Due to absolute continuity, the representation (by the Radon–Nikodym theorem) $\varphi(A) = \int_A \Phi(x)\,dx$ exists, where $\Phi(x) \geqslant 0$ is a measurable function. Let us choose a point $x \in U$ at which f has K-differential. Due to Lemma 1.5, a sequence of balls $\{B_m = B(x,h_m)\}_{m=1,2,\ldots}$, $h_m > 0$, $h_m \to 0$ for $m \to \infty$ exists for which the inequality

$$\varliminf_{m\to\infty} \frac{m(f(B_m))}{m(B_m)} \geqslant |J(x,f)|,$$

is valid; but, on the other hand, it is obvious that

$$\varphi(B_m) = \int_{\mathbf{R}^n} N_f(y,B_m)\,dy \geqslant m(f(B_m)),$$

and therefore,

$$\varliminf_{m\to\infty} \frac{\varphi(B_m)}{\mu(B_m)} \geqslant |J(x,f)|.$$

Due to the theorem about differentiation of functions of a set, it follows that $\Phi(x) \geqslant |J(x,f)|$ almost everywhere in $\mathbf{R}^n$. So,

$$\varphi(A) \geqslant \int_A |J(x,f)|\,dx.$$

This completes the proof of the theorem.

1.4. The Change of Variable Theorem for the Multiplicity Function

Let us consider a continuous mapping $f : u \to \mathbf{R}^n$ of a domain $U \subset \mathbf{R}^n$ and a measurable function $u : \mathbf{R}^n \to \mathbf{R}$. Suppose that the function u is only defined at the points where it acquires natural values in the sense of Lebesgue. At the remaining points, the function u is not defined. Then the composition $v = u \cdot f$ may not be defined on a set of positive measure. In this case let us set $v(x) = 0$.

Lemma 1.7. *Let a continuous mapping $f : U \to \mathbf{R}^n$ of a domain $U \subset \mathbf{R}^n$ possess the N-property, have K-differential almost everywhere, and let the Jacobian $J(x, f)$ be locally summable.*

Then the function $v(x) = (u \cdot f)(x)$ is measurable.

Proof. Let $E_0 = \{x \in U : J(x, f) = 0\}$. Let us prove that for every measurable $A \subset \mathbf{R}^n$, the set $E_0 \cap f^{-1}(A)$ is measurable. Let us consider Borel sets P, Q, $P \subset A \subset Q$, $m(Q \backslash P) = 0$. The set $f^{-1}(P)$ is the Borel one; consequently, it is measurable. The set $f^{-1}(M)$, $M = Q \backslash P$ is also measurable, and according to Theorem 1.6, $J(x, f) = 0$ almost everywhere on $f^{-1}(M)$, since

$$\int_{f^{-1}(M)} |J(x, f)| \, dx = \int_{R^n} N_f(y, f^{-1}(M)) = \int_M N_f(y, f^{-1}(M)) = 0.$$

This means that $m(f^{-1}(M) \backslash E_0) = 0$.

The sets $f^{-1}(P) \backslash E_0$ and $f^{-1}(Q) \backslash E_0$ are measurable and their difference $f^{-1}(M) \backslash E_0$ has zero measure. From the inclusion

$$f^{-1}(P) \backslash E_0 \subset f^{-1}(A) \backslash E_0 \subset f^{-1}(Q) \backslash E_0,$$

it now follows that $f^{-1}(A) \backslash E_0$ is measurable. Let $t \in \mathbf{R}$, $A_t = \{y \in \mathbf{R}^n : u(y) \leqslant t\}$, $P_t = \{x \in U \backslash E_0 : v(x) < t\}$. Then $P_t = f^{-1}(A_t) \backslash E_0$, i.e., the set P_t is measurable for all t; consequently, the function $v(x)$ is measurable.

Corollary. *Under the conditions of Lemma 1.7, the function $w(x) = v(x) |J(x, f)|$ is measurable.*

Indeed, according to Lemma 1.5, the function $w(x)$ is measurable on $U \backslash E_0$. On the set E_0, $w(x) = 0$; consequently, w is measurable on the entire set U.

Theorem 1.8. *Let a continuous mapping f of a domain U in $\mathbf{R}^n$ possess the N-property, have K-differential almost everywhere, and let the Jacobian $J(x, f)$ be locally summable in U.*

Then for every nonnegative measurable function $v : \mathbf{R}^n \to \mathbf{R}$ and for any measurable set E such that $E \subset U$, the equality

$$\int_E (v \circ f)(x) \, |J(x, f)| \, dx = \int_{R^n} N_f(y, E) v(y) \, dy$$

is valid.

Proof. Let $u : U \to \mathbf{R}^n$ be a measurable nonnegative function. Let us construct a new function $y \to N_f(y, u, E)$. If $N_f(y, E) < \infty$, then let $N_f(y, u, E)$ be equal to the sum of all values of the function $u(x)$ at the points of the set $f^{-1}(y) \cap E$. But if $N_f(y, E) = \infty$, then $N_f(y, u, E)$ is equal to the limit of the sums of values of the function $u(x)$ by all possible finite subsets of the set $f^{-1}(y) \cap E$. In the case when this limit does not exist, the function $N_f(y, u, E)$ is considered to be undefined.

Since $\overline{E} \subset U$, then $N_f(y, u, E)$ is defined and finite almost everywhere.

If the function $u(x)$ is bounded, then

$$N_f(y, u, E) \leqslant \sup_{x \in U} u(x) N_f(y, E)$$

for almost all $y \in \mathbf{R}^n$.

Let us consider a characteristic function χ_A of a measurable set A, $\bar{A} \subset U$. Then

$$N_f(y, \chi_A, A) = N_f(y, A \cap E)$$

and

$$\int_E \chi_A |J(x, f)| \, dx = \int_{A \cap E} |J(x, f)| \, dx.$$

Hence, due to Theorem 1.6, we obtain

$$\int_E \chi_A |J(x, f)| \, dx = \int_{R^n} N_f(y, \chi_A, E) \, dy.$$

If the function $u(x)$ is simple, i.e., if it equals the linear combination of a finite number of characteristic functions of the measurable sets A, $\bar{A} \subset U$, then we obtain

$$\int_E u(x) |J(x, f)| \, dx = \int_{R^n} N_f(y, u, E) \, dy. \tag{1.5}$$

For an arbitrary nonnegative measurable function $u : U \to \mathbf{R}$, let us construct an increasing sequence $\{u_m\}$ of simple functions which converges to u for almost all x. Then the sequence of functions $N_m(y) = N_f(y, u_m, E)$ is also nondecreasing; and for $m \to \infty$, $N_m(y) \to N_f(y, u, E)$ for almost all y. Hence follows the measurability of the function $y \to N_f(y, u, E)$. For every function u_m, equality (1.5) is valid. According to the Beppo–Levi theorem, we obtain

$$\int_E u(x) |J(x, f)| \, dx = \int_{R^n} N_f(y, u_m, E) \, dy. \tag{1.6}$$

Now let $u(x) = (v \circ f)(x)$, where $v : \mathbf{R}^n \to \mathbf{R}$ is a nonnegative measurable function. Let us establish that $N_f(y, u, E) = v(y) N_f(y, E)$ almost everywhere. Indeed, let $y \in \mathbf{R}^n$ and let $N_f(y, E) < \infty$, and let the value $v(y)$ be defined (i.e., let a natural value for $u(y)$ exist at the point y. Then the set $f^{-1}(y) \cap E$ is finite, and at every point x of this set, $u(x) = (v \circ f)(x) = v(y)$. Therefore, $N_f(y, u, E) = v(y) N_f(y, E)$ at these points. Now the statement of the theorem follows from (1.6) and from the measurability of the function $u(x)$ proved in Lemma 1.7.

Corollary. *Let $f : U \to \mathbf{R}^n$ satisfy the condition of Theorem 1.8, let E be measurable, and let $u(y) : \mathbf{R}^n \to \mathbf{R}$ be a measurable function such that the function $u(y)N_f(y,E)$ is summable.*

Then the function $v(x) = (u \circ f)|J(x,f)|$ is summable on E, and

$$\int_E (u \circ f)(x)\,|J(x,f)|\,dx = \int_{R^n} u(y)\,N_f(y,E)\,dy.$$

To prove the corollary, it suffices to apply Theorem 1.8 to the functions u^+ and u^-.

1.5. The Degree of Mapping

Let us give a short summary of the properties of the degree of mapping. We will need this information later.

Let us define the degree of mapping for a mapping of the class C^1. Let U be a domain in $\mathbf{R}^n$ whose closure is compact. Let us consider a continuous mapping $\varphi : \overline{U} \to \mathbf{R}^n$ of the class $C^1(U)$. Let $y_0 \in \mathbf{R}^n \backslash \varphi(\partial U)$ be a regular value of the mapping φ. First we give the definition of the degree of the mapping φ at the point y_0. Since y_0 is a regular value of this mapping, then, as follows from the implicit function theorem, the set $\varphi^{-1}(y_0)$ is finite. In this case, the degree of mapping is defined by the formula $\mu(y_0, \varphi, U) = \sum_{x \in \varphi^{-1}(y_0)} \operatorname{sign} J(x, \varphi)$.

Since, due to Sard's theorem, almost all values of the mapping φ are regular, the degree is defined almost everywhere in $\mathbf{R}^n \backslash \varphi(\partial U)$.

Properties of the degree:

1) The degree of the mapping $\varphi : U \to \mathbf{R}^n$ is constant on every connected component of the set $\mathbf{R}^n \backslash \varphi(\partial U)$.

   This property allows us to define the degree by limit transition at the points $y \in \mathbf{R}^n \backslash \varphi(\partial U)$, which are not regular values of the mapping φ.

2) The degree is equal to zero on an unbounded connected component of the set $\mathbf{R}^n \backslash \varphi(\partial U)$.

3) Let $\varphi_m : \overline{U} \to \mathbf{R}^n$ be a sequence of mappings of the class $C^1(U)$, uniformly converging to the continuous mapping $\varphi : \overline{U} \to \mathbf{R}^n$ for $m \to \infty$. For any point $y_0 \in \mathbf{R}^n \backslash \varphi(\partial U)$, there exists a number $m_0(y_0) = m_0$ such that for all $m_1, m_2 > m_0$, $\mu(y_0, \varphi_{m_1}, U) = \mu(y_0, \varphi_{m_2}, U)$.

This property allows us to construct the degree $\mu(y_0, \varphi)$ for the limit continuous mapping φ. It is easy to see that $\mu(y_0, \varphi) = \lim \mu(y_0, \varphi_m)$ does not depend on the choice of the approximating sequence $\{\varphi_m \in C^1(U)\}$. The existence of such a sequence is obvious. For instance, this is the averaging of the function φ (Subsection 2.1 of Chapter 1).

Properties 1 and 2 remain valid for the continuous mapping $\varphi : \bar{U} \to \mathbf{R}^n$ as well.

4) Let $\varphi_m : \overline{U} \to \mathbf{R}^n$ be a sequence of continuous mappings which uniformly converge to φ for $m \to \infty$. For any point $y_0 \in \mathbf{R}^n \setminus \varphi(\partial U)$, there exists m_0 such that for any $m > m_0$, $\mu(y_0, \varphi_m, U) = \mu(y_0, \varphi_0, U)$.
5) *(Homotopic invariance).* Let us consider a one-parameter family of mappings $\varphi_t(x) : \overline{U} \times [0,1] \to \mathbf{R}^n$ which are continuous on $\overline{U} \times [0,1]$. Suppose that for every t, the condition $y_0 \notin \varphi_t(\partial U)$ is satisfied. Then the degree $\mu(y_0, \varphi_t, U)$ is independent of t.

Theorem 1.9. *Let $f : U \to \mathbf{R}^n$ be a continuous mapping of a domain $U \subset \mathbf{R}^n$ and let it possess the N-property, have K-differential, and let the Jacobian $J(x, f)$ be locally summable.*

Then for every compact domain $G \subset U$, such that $\overline{G} \subset U$ and $m(\partial G) = 0$, the function $y \to \mu(y, f, G)$ is summable in $\mathbf{R}^n$, and the equality

$$\int_G J(x, f)\,dx = \int_{R^n} \mu(y, f, G)\,dy$$

is valid.

Proof. The function $u(x) = \operatorname{sign} J(x, f)$ is measurable. By proving Theorem 1.8, equality (1.5) was obtained for positive measurable functions v

$$\int_G v(x)\,|J(x, f)|\,dx = \int_{R^n} N_f(y, v, G)\,dy.$$

Since the function $u(x)$ is summable on G, it follows that

$$\int_G J(x, f)\,dx = \int_G u(x)\,|J(x, f)|\,dx = \int_{R^n} N_f(y, u, G)\,dy.$$

It remained to obtain $N_f(y, u, G) = \mu(y, f, G)$ for almost all $y \in \mathbf{R}^n$. Indeed, let E_1 be a set of $x \in U$, where f has no K-differential, $E_2 = \{x \in U | J(x, f) = 0\}$, $S = \{y \in \mathbf{R}^n | N_f(y, G) = \infty\}$. Since $\overline{G} \subset U$ is compact, the function $N_f(y, G)$ is summable, and therefore, $m(S) = 0$. Let $A = S \cup f(E_0) \cup f(E_1) \cup f(\partial G)$. Then $m(A) = 0$.

Let us arbitrarily choose $y \notin A$. The set $f^{-1}(y) \cap G$ is finite. At each of its points the function f has K-differential $J(x, f) \neq 0$ for all $x \in f^{-1}(y) \cap G$. All the points of this set are the inner points of G. Let $a_1, a_2, \dots, a_N$ be all points of the set $f^{-1}(y) \cap G$. Due to Lemma 1.5, its Remark, and due to the homotopic invariance of degree, there exists a sequence of closed balls $B_m^{(i)} = B(a_i, r_m)$, $m = 1, 2, \dots$, $i = 1, 2, \dots, k$, whose radii tend to zero for sufficiently large m, such that

$$\mu(y, f, B_m^{(i)}) = \operatorname{sign} J(a_i, f).$$

For sufficiently large m, the balls $B_m^{(i)}$ and $B_m^{(j)}$, $i \neq j$, do not intersect. Then

$$\mu(y, f, G) = \sum_{i=1}^{N} \mu(y, f, B_m^{(i)}) = \sum_{i=1}^{N} \operatorname{sign} J(a_i, f) = N_f(y, U, G).$$

This proves the theorem.

Remark. Incidentally, it was established that for mappings of the class under consideration, the formula

$$\mu(y, f, G) = \sum_{x \in \varphi^{-1}(y) \cap G} \operatorname{sign} J(x, f)$$

is valid almost everywhere in G.

Theorem 1.10. *Let f be a continuous mapping of a domain U in $\mathbf{R}^n$ which possesses the N-property, and has K-differential almost everywhere, such that the Jacobian $J(x, f)$ is locally summable.*

Suppose that $G \subset U$ is a compact domain, $\overline{G} \subset U$, $m(\partial G) = 0$.

If a measurable real function u is such that the function $y \to u(y)$ $\mu(y, f, G)$ is summable in $\mathbf{R}^n$, then the function $(u \circ f)(x)J(x, f)$ is summable on G, and the equality

$$\int_G (u \circ f)(x) J(x, f) = \int_{R^n} u(y) \mu(y, f, G)\, dy$$

is valid.

We leave it to the reader to prove the theorem.

§2 The Change of Variable in the Integral for Mappings of Sobolev Spaces

2.1. The Change of Variable Theorem for Continuous Mappings of the Class L_n^1

First let us give an important special case of the change of variable theorems that was formulated in the previous section.

A mapping $f : U \to \mathbf{R}^n$ of a domain $U \subset \mathbf{R}^n$ is called the Lipschitz mapping if for every pair of points $x, y \in U$, the inequality $|f(x) - f(y)| \leqslant C|x - y|$ is valid, where the constant C does not depend on the choice of the pair of points x, y.

The mapping $f : U \to \mathbf{R}^n$ is called locally Lipschitz if it is Lipschitz in every ball $B \subset U$.

Every mapping f of the class $C^1(U)$ is locally Lipschitz.

Proposition 2.1. *Every locally-Lipschitz mapping $f : U \to \mathbf{R}^n$ possesses the N-property.*

Proof. It suffices to prove that every bounded set E, $\bar{E} \subset U$, $m(E) = 0$ is transferred by the mapping f to the set of zero measure.

The set $E_h = \{x \in \mathbf{R}^n | \rho(x, E) \leqslant h\}$ is compact, and for sufficiently small h_0, $E_{h_0} \subset U$. Let us arbitrarily assign $\varepsilon > 0$. Let us construct a sequence of balls $B_m = B(x_m, r_m)$ contained in the set E_h, such that $E \subset \cup_m B_m$ and $\sum_m m(B_m) < \varepsilon$; since the mapping f is locally Lipschitz, it is Lipschitz on E_{h_0}, i.e., there exists a constant C such that for all $x, y \in E_{h_0}$, $|f(x) - f(y)| \leqslant C|x-y|$. Thus, the set $f(B_m)$ lies in the ball with radius Cr_m and $m(f(B_m)) \leqslant C^n m(B_m)$. It is obvious that $f(E) \subset \cup_m f(B_m)$. Hence it follows that $m(f(E)) \leqslant C^n m(E) = 0$.

Proposition 2.2. *Every locally-Lipschitz mapping $f : U \to \mathbf{R}^n$ belongs to the class $W^1_{\infty,\mathrm{loc}}(U)$.*

Proof. It is obvious that coordinate functions of the mapping f are absolutely continuous on all segments lying in the domain U, i.e., $f \in ACL(U)$. Now, from the differentiability almost everywhere of mappings of the class ACL, it follows that partial derivatives of coordinate functions are bounded on every compact set $E \subset U$. Hence and from Theorem 5.7 of Chapter 2, our statement follows.

Corollary. *A locally-Lipschitz mapping has K-differential almost everywhere.*

It follows from Theorem 5.1 of Chapter 2.

Now it is clear that all the theorems of the previous section remain valid for locally-Lipschitz mappings as well.

Let us recall that every mapping of the class $L^1_n(U)$ has K-differential almost everywhere (Theorem 5.2 of Chapter 2).

Let us formulate Theorems 1.6, 1.8 and its corollary of the previous section for the mappings of the class $L^1_n(G)$.

Theorem 2.3. *Let a continuous mapping $\varphi : U \to \mathbf{R}^n$ of the class $L^1_n(U)$ possess the N-property. Then for every compact subdomain $G \subset U$, the equality*

$$\int_G |J(x, \varphi)|\, dx = \int_{R^n} N_\varphi(y, G)\, dy$$

is valid.

If $m(\partial G) = 0$, then for all $y \in \mathbf{R}^n$, $\mu(y, \varphi, G) \leqslant N_\varphi(y, G)$, and the equality

$$\int_G J(x, \varphi)\, dx = \int_{R^n} \mu(y, \varphi, G)\, dy$$

is valid.

Theorem 2.4. *Let a continuous mapping $\varphi : U \to \mathbf{R}^n$ of the class $L^1_n(U)$ possess the N-property; $G \subset U$ is a compact domain, and for the measurable function $u : \mathbf{R}^n \to \mathbf{R}$, the product $N_\varphi(y, U)u(y)$ is summable.*

Then

$$\int_G (u \circ \varphi)(x) J(x, \varphi)\, dx = \int_{R^n} N_\varphi(y, G) u(y)\, dy.$$

If here $m(\partial G) = 0$, then the equality

$$\int_G (u\circ\varphi)(x)\, J(x,\varphi)\, dx = \int_{R^n} u(y)\,\mu(y,\varphi,G)\, dy$$

is valid.

2.2. The Linking Index

Throughout this chapter the functions from the space $L_n^1(G)$ are considered to be $(1,n)$-refined.

Let G be a compact domain in $\mathbf{R}^n$. The image ∂G for the continuous mapping $f : \partial G \to \mathbf{R}^n$ is said to be a cycle. For an arbitrary continuous extension $F : G \to \mathbf{R}^n$ of the mapping f, the degree $\mu(y, F, G)$ of the mapping F at the point $y \in \mathbf{R}^n \backslash f(\partial G)$ is defined. Let us assume the linking index $\chi(y, f(\partial G))$ to be equal to $\mu(y, F, G)$ for all $y \notin f(\partial G)$. The magnitude of $\chi(y, f(\partial G))$ does not depend on the continuous extension F of the mapping f.

Let us formulate a few simple properties of the linking index which are obvious corollaries of the properties of the degree of mapping:

1) The linking index $\chi(y, f(\partial G))$ is constant on every connected component of the set $\mathbf{R}^n \backslash f(\partial G)$.
2) The linking index equals zero on an unbounded component of the set $\mathbf{R}^n \backslash f(\partial G)$.
3) Let $f_m : \partial G \to \mathbf{R}^n$ $(m = 1, 2, \dots)$ be a sequence of continuous mappings which uniformly converges to f for $m \to \infty$. For any point $y_0 \in \mathbf{R}^n \backslash f(\partial G)$, there exists m_0 such that for any $m > m_0$, $\chi(y_0, f_m(\partial G) = \chi(y_0, f(\partial G))$. From this and from Property 1), it follows that for every compact domain Δ not intersecting $f(\partial G)$, there exists m_0 such that for $m > m_0$, the functions $\chi(y, f_m(\partial G))$ are constant and coincide with $\chi(y, f(\partial G)$. The choice of the number m_0 depends on $\rho(\Delta, f(\partial G))$.
4) (Homotopic invariance.) Let us consider a one-parameter family of mappings $f(x,t) : \partial G \times [0,1] \to \mathbf{R}$ which are continuous on $\partial G \times [0,1]$. Suppose that for every t, the condition $y_0 \notin f_t(\partial G)(f_t(x) = f(x,t))$ is satisfied. Then the linking index $\chi(y_0, f_t(\partial G))$ is independent of t.

From this and from Property 1), it follows that in every compact domain Δ which intersects none of the cycles $f_t(\partial G)$, the linking index is constant and is independent of t.

The boundary of the domain $U \subset \mathbf{R}^n$ is called a locally-quasiconformal boundary if for every point $x \in \partial U$, there exists a neighbourhood $V \ni x$ and a quasiconformal homeomorphism $\varphi : V \to Q^n(0,1)$ (Q^n is a cube) which transfers $\partial U \cap V$ into the set

$$Q^{n-1} = Q^n \cap \{x_n = 0\}, \qquad \varphi(U \cap V) = Q^n(0,1) \cap \{x_n < 0\}.$$

Theorem 2.5. *Let a mapping $\varphi : G \to \mathbf{R}^n$ of a domain $G \subset \mathbf{R}^n$ belong to the class $L_n^1(G)$. Then for every compact domain $U \subset G$, which has locally-*

quasiconformal boundary, on which the mapping $\varphi|_{\partial U}$ is continuous, the equality

$$v(U, \varphi) = \int_U J(x, \varphi)\, dx = \sum_i \chi_i m(V_i) \tag{2.1}$$

is valid.

Here V_i, $i = 1, 2, \ldots$, are bounded components of the set $\mathbf{R}^n \setminus \varphi(\partial U)$, X_i is the linking index corresponding to the component V_i. The series in formula (2.1) absolutely converges.

Then, if u is a Borel function in $\mathbf{R}^n \setminus \varphi(\partial U)$, for which the product $u(y)\chi(y, \varphi(\partial U))$ is summable in $\mathbf{R}^n \setminus \varphi(\partial U)$, the formula

$$\int_U u(\varphi(x)) J(x, \varphi)\, dx = \int_{R^n \setminus \varphi(\partial U)} u(y) \chi(y, \varphi(\partial U))\, dy \tag{2.2}$$

is valid. (If $\varphi(x) \in \varphi(\partial U)$, $x \in U$, then we put $u(\varphi(x)) = 0$ as well.)

Proof. For smooth mappings, the formulae (2.1)–(2.2) follow from Theorem 2.4. In the nonsmooth case a sequence $\{\varphi_m : U \to \mathbf{R}^n\}$ is constructed of once differentiable mappings which are continuous up to the boundary of the domain U. These mappings coincide with φ on ∂U and approximate the mapping φ in the norm of L_n^1.

Approximation. The coordinate functions $\varphi_1, \varphi_2, \ldots \varphi_n$ of the mapping φ are continuous on ∂U. The boundary of the domain U is locally quasiconformal, therefore, there exist continuous functions $\overline{\varphi}_1, \overline{\varphi}_2, \ldots, \overline{\varphi}_n$ of the class $L_n^1(V)$ which coincide with $\varphi_1, \varphi_2, \ldots, \varphi_n$ on ∂U. For each of the functions $\overline{\varphi}_i$, we solve the variational problem to find the function minimizing the integral $\int_U |\nabla v|^n\, dx$ among all the functions v_i which coincide with $\overline{\varphi}_i$ on ∂U.

The coincidence of the functions $\overline{\varphi}_i$ and v_i on the boundary is understood as follows: $\overline{\varphi}_i - v_i \in \overset{\circ}{L}{}_n^1(U)$, where $L_n^1(U)$ is the closure of the set of compactly supported, infinitely differentiable functions in the domain U by the norm

$$\|\psi\| = \left(\int_U |\nabla\psi|^n\, dx\right)^{1/n}.$$

The variational problem is solvable [42], the extremal function $v_{0,i} \in L_n^1(U)$ which is continuous up to the boundary, coincides with $\overline{\varphi}_i$ on ∂U and is at least continuously differentiable [42].

Since $\varphi_i - \overline{\varphi}_i \in \overset{\circ}{L}{}_n^1(U)$, then $\varphi_i - v_{0,i} \in \overset{\circ}{L}{}_n^1(U)$. To construct the approximation, let us take a sequence $\{h_m^{(i)}\}$ of differentiable functions from $\overset{\circ}{L}{}_n^1$ which converges to $\varphi_i - v_{0,i}$ in this space. The sequence of mappings

$$\varphi_m = \{v_{0,1} + h_m^{(1)},\ v_{0,2} + h_m^{(2)},\ \ldots,\ v_{0,n} + h_m^{(n)}\}$$

is the desired one.

For all the mappings φ_m, the right-hand side of formula (2.1) acquires the same value. The sequence converges to φ in $L_n^1(U)$. Hence follows the possibility of the limit transition on the left-hand side of (2.1).

The estimate

$$\sum_{i\geqslant 1} |\chi_i| \, m(V_i) \leqslant \int_{R^n\setminus\varphi(\partial U)} |N(y, \varphi_m, U)| \, dy = \int_U |J(x, \varphi_m)| \, dx$$

proves absolute convergence of the series on the right-hand side.

Formula (2.1) is proved.

For the smooth mappings φ_m, $m = 1, 2, \ldots$, and for the continuous bounded function u, the formula

$$\int_U u(\varphi_m(x)) \, J(x, \varphi_m) \, dx = \int_{R^n\setminus\varphi(\partial U)} u(y) \, \chi(y, \varphi_m(\partial U)) \, dy$$

follows from Theorem 2.4. In this formula, the limit transition is possible for $m \to \infty$.

The validity of formula (2.1) for arbitrary Borel functions follows from the Lebesgue–Hausdorff analytical representation theorem, i.e., from the possibility to obtain any Borel function by a number of pointwise limit transitions, beginning with the class of continuous functions [40].

2.3. The Change of Variable Theorem for Discontinuous Mappings of the Class L_n^1

Let us formulate a few additional conditions for a mapping of the class L_n^1 under which formula (2.1) is valid for functions measurable in the sense of Lebesgue.

A mapping $\varphi \in L_n^1(G)$ (G is a domain in $\mathbf{R}^n$) has the N^{-1}-property if the pre-image of a set of zero measure is the set of zero measure. It is obvious that this property is equivalent to the following one: the image of a set of nonzero measure is a set of nonzero measure.

Theorem 2.6. *Let a mapping $\varphi : G \to \mathbf{R}^n$ of a domain $G \subset \mathbf{R}^n$ belong to the class $L_n^1(G)$ and let one of the following conditions hold for it: 1) the Jacobian $J(x,\varphi) \geqslant 0$ almost everywhere; 2) φ possesses the property N^{-1}. Suppose that the compact domain $U \subset G$ has locally-quasiconformal boundary, and on ∂U the mapping φ is continuous.*

If for a measurable function $u : \mathbf{R}^n \to \mathbf{R}$ the product $u(y)\chi(y, \varphi(\partial U))$ is summable, then the formula

$$\int_U (u \circ \varphi)(x) \, J(x, \varphi) \, dx = \int_{R^n\setminus\varphi(\partial U)} u(y) \, \chi(y, \varphi(\partial U)) \, dy$$

is valid.

Proof. First, let the function u be bounded, i.e., $|u(y)| \leqslant C$ for all $x \in U$. There exists a sequence $\{u_m\}$ of Borel functions which converges to $u(x)$ almost everywhere, and the functions satisfy the inequality $|u_m(x)| \leqslant 2C$. According to the previous theorem,

$$\int_U (u_m \circ \varphi)(x) J(x, \varphi)\, dx = \int_{R^n \setminus \varphi(\partial U)} u_m(y) \chi(y, \varphi(\partial U))\, dy.$$

For the mappings φ possessing the N^{-1}-property, the sequence $(u_m \circ \varphi)(x)$ converges to the function $(u \circ \varphi)(x)$ almost everywhere; in this case, the theorem is proved by means of limit transition in both parts of the equality.

Let $J(x, \varphi) \geqslant 0$ almost everywhere. The characteristic function χ_E of any set of the G_δ type belongs to the class of Borel functions. If $B \cap \varphi(\partial U) = \emptyset$, then according to the previous theorem, we obtain

$$\int_{U_\Delta} \chi_B(\varphi(x)) J(x, \varphi)\, dx = \int_{\varphi^{-1}(B)} J(x, \varphi)\, dx$$
$$= \int_{R^n \setminus \varphi(\partial U)} \chi_B(y) \chi(y, \varphi(\partial U))\, dy.$$

For the sets B having zero measure, it is clear from this formula that $J(x, \varphi) = 0$ almost everywhere on $\varphi^{-1}(B)$. Let us take an arbitrary set A having zero measure. Let $A \cap \varphi(\partial U) = \emptyset$. There exists a set B of the type G_δ, $B \supset A$, and $m(B \setminus A) = 0$, i.e., $m(B) = 0$. According to the above, $J(x, \varphi) = 0$ almost everywhere on $\varphi^{-1}(B)$; consequently, $J(x, \varphi) = 0$ almost everywhere on $\varphi^{-1}(A)$.

The set A_0 on which the sequence $u_m(x)$ does not converge to $u(x)$ has zero measure. The Jacobian $J(x, \varphi) = 0$ almost everywhere on $\varphi^{-1}(A_0)$. Therefore, the sequence $(u_m \circ \varphi)(x) J(x, \varphi)$ converges to the function $(u \circ \varphi)(x) J(x, \varphi)$ almost everywhere. In this case, the theorem is proved by means of limit transition.

For an unbounded function u, let us consider the section $u_{m,n}(y) = \max\{-m, \min\{n, u(y)\}\}$. The sequence increases monotonically for $n \to \infty$ and $\lim_{n\to\infty} u_{m,n}(y) = \max\{-m, u(y)\} = u_m(y)$. By the property of the mapping φ proved above, the sequences

$$(u_{m,n} \circ \varphi)(x) J(x, \varphi) \to (u_m \circ \varphi)(x) J(x, \varphi)$$

almost everywhere for $n \to \infty$.

For the functions $u_{m,n}$, the theorem is proved. Therefore,

$$\left| \int_U (u_{m,n} \circ \varphi)(x) J(x, \varphi)\, dx \right| = \left| \int_{R^n \setminus \varphi(\partial U)} u_{m,n}(y) \chi(y, \varphi(\partial U))\, dy \right|$$
$$\leqslant \int_{R^n \setminus \varphi(\partial U)} |u_{m,n}(y)| \chi(y, \varphi(\partial U))\, dy$$
$$\leqslant \int_{R^n \setminus \varphi(\partial U)} |u(y)| \chi(y, \varphi(\partial U))\, dy.$$

By consecutively applying the limit transition first by n, then by m, we obtain the statement of the theorem.

In conclusion, it should be noted that the class of subdomains of the domain G which have locally-quasiconformal boundary is rather wide. Some fixed mapping φ of the class L_n^1 is continuous on this boundary. Recall that a mapping φ of the class L_n^1 is $(1,n)$-refined. Let us consider a family of spheres $\{S(x_0,r)\}$, $x_0 \in G$, $r < \rho(x_0, \partial G)$. According to Egorov's theorem for $(1,n)$-capacity, the mapping is continuous on almost all spheres $S(x,r)$ except, perhaps, for the set of zero $(1,n)$-capacity. It is easy to show that a family of spheres may be replaced in this argument by any family of locally-quasiconformal surfaces which are boundaries of families of domains.

Let us consider two spheres: $S_0 = S(x_0,r_0)$, $S_1 = S(x_0,r_1)$ lying in G, $r_0 < r_1$. Let us emit all possible rays from the point x_0. On each ray, there exists a segment connecting the spheres S_0 and S_1. The segment l_x of this family is uniquely defined by the point x through which it passes on the sphere S_0. From Egorov's theorem for $(1,n)$-capacity, it follows that the mapping φ is continuous on almost all segments of the family l_x.

§3 Sufficient Conditions of Monotonicity and Continuity for the Functions of the Class L_n^1

A mapping $\varphi : G \to \mathbf{R}$ of the class L_n^1 is continuous outside of some set V of small capacity. The set of small capacity has small linear Hausdorff's measure. The comparison of these facts yields that φ is continuous on the spheres $S(x_0,r)$ ($x_0 \in G$ is supposed to be fixed) for almost all r smaller than $d(x_0,\partial G)$. For all such r, one can define the linking index $\chi(y, \varphi(S(x_0,r)))$.

The mapping $\varphi : G \to \mathbf{R}^n$ of the class $L_n^1(G)$ is topologically nondegenerate at the point $x_0 \in G$ if there exists a positive number $r(x_0) < d(x_0,\partial G)$ such that for almost all $r \in (0, r(x_0))$, the set $\mathbf{R}^n \backslash (S(x_0,r))$ contains a non-empty bounded component V_i whose linking index χ_i is distinct from zero.

In all remaining cases, the mapping φ is said to be topologically degenerate at the point x_0.

Let us investigate the connection of topological nondegeneracy of mapping with "functional nondegeneracy."

If the mapping $\varphi : G \to \mathbf{R}^n$ belongs to the class $L_n^1(G)$, then for almost all $x_0 \in G$, due to the Lebesgue theorem, the Jacobian $J(x_0,\varphi)$ coincides with the limit

$$\lim_{r\to 0} [m(B(x_0, r))]^{-1} \int_{B(x_0,r)} J(x, \varphi)\, dx = J(x_0, \varphi). \tag{3.1}$$

We are interested in the points at which equality (3.1) holds and the Jacobian is nondegenerate.

Lemma 3.1. *If at a point x_0, equality (3.1) holds and the Jacobian $J(x_0,\varphi) \neq 0$, then at this point the mapping φ is topologically nondegenerate.*

Proof. Let us choose r_0 so that for all $r < r_0$,

$$\int_{B(x_0,r)} J(x,\varphi)\,dx \neq 0.$$

For almost all $r < r_0$, the restriction of the mapping φ upon the sphere $S(x_0, r)$ is continuous. Consequently, for these spheres the conditions of Theorem 2.4 are satisfied and the equality

$$\int_{B(x_0,r)} J(x,\varphi)\,dx = \sum \chi_i \mu(V_i)$$

is valid, where V_i are connected bounded components of the set $\mathbf{R}^n \setminus \varphi(S(x_0, r))$, and χ_i is the corresponding linking index of the mapping $\varphi(S(x_0,r))$. The left-hand part of the equality is nonzero; therefore, the right-hand side is nonzero and at least one of the numbers of χ_i is distinct from zero.

The lemma is proved.

If the degree $\mu(y,\varphi,U)$ of the continuous mapping φ is distinct from zero, then the equation $\varphi(x) = y$ has a solution, or in other words, the component of the set $\mathbf{R}^n \setminus \varphi(\partial U)$ which contains the point y is covered by the mapping φ. This property of degree admits generalization to mappings of the class L^1_n.

Lemma 3.2. *Let $U \subset G$ be a compact domain with locally-quasiconformal boundary on which the mapping $\varphi|_{\partial U}(\varphi \in L^1_n(U))$ is continuous. If on the connected component V_i of the set $\mathbf{R}^n \setminus \varphi(\partial U)$ the linking index $\chi_i \neq 0$, then for almost all $y \in V_i$, there exists a point $x \in U$ whose image coincides with y.*

Proof. In the set $A \subset V_i$ of points which are not covered by the mapping $\varphi|_U$, let us take any compact subset. The characteristic function χ_F of any compact set $F \subset A$ is the Borel function. Let us apply Theorem 2.4 to it:

$$\int_U \chi_F(\varphi(x))\,J(x,\varphi)\,dx = \int_{R^n \setminus \varphi(\partial U)} \chi_F(y)\,\chi(y,\varphi(\partial U))\,dy$$
$$= \int_F \chi(y,\varphi(\partial U))\,dy = \chi_i(y,\varphi(\partial U))\,m(F).$$

The function $\chi_F(\varphi(x)) \equiv 0$, and $\chi_i \neq 0$, therefore, $m(F) = 0$. The compact set $F \subset A$ was chosen arbitrarily; consequently, $m(A) = 0$.

Corollary. *If at the point $x_0 \in G$,*

$$\lim_{r\to 0} [\mu(B(x_0,r))]^{-1} \int_{B(x_0,1)} J(x,\varphi)\,dx = J(x_0,\varphi)$$

and $J(x_0,\varphi) \neq 0$, then for every domain $U \subset G$, it follows from the condition $x_0 \in U$, that $m(\varphi(U)) \neq 0$.

The mapping $\varphi : G \to \mathbf{R}^n$ of the class $L^1_{p,\mathrm{loc}}$ $(p > n-1)$ is said to be monotone at the point $x \in G \subset \mathbf{R}^n$ if for almost all $r \in (0, r(x))$, $0 < r(x) < \rho(x, \partial G)$, which is the pre-image of the intersection of the set $\varphi(B(x,r))$ with unbounded component of the set $\mathbf{R}^n \setminus \varphi(S(x,r))$, has zero measure in $B(x,r)$. The mapping is monotone in the domain G if it is monotone at each point of this domain. Every coordinate function of the monotone mapping is monotone.

Proposition 3.3. *A monotone function of the class $L^1_{n,\mathrm{loc}}(G)$ is continuous.*

Proof. Let us take any point $x_0 \in G$ and a family of balls $\{B(x_0,r)\}$, $r < \rho(x_0,\partial G)$, $r < 1$. For any of the balls $B(x_0,r_0)$ of this family, due to the Fubini theorem, the equality

$$\int_0^{r_0} \int_{S(x_0,r)} |\nabla u|^n \, d\sigma_r = \int_{B(x_0,r_0)} |\nabla u|^n \, dx$$

is valid. Here σ_r is the Lebesgue measure on the sphere $S(x_0,r)$. Therefore, there exists a set $T(r) \subset (0,r)$ such that for every $r > 0$, the intersection $T(r) \cap (0,r)$ has nonzero measure, $u|_{s(x_0,r)} \in L^1_n(S(x_0,r))$ for all $r \in T(r)$, and the inequality

$$\int_{S(x_0,r)} |\nabla u|^n \, d\sigma_n \leqslant \frac{2}{r} \int_{B(x_0,r)} |\nabla u|^n \, dx$$

is valid.

From the imbedding theorem for the spaces $L^1_n(S(x_0,r))$, it follows that $u|_{s(x_0,r)}$ is continuous and the inequality

$$r^{n-1} \int_{S(x_0,r)} |\nabla u|^n \, dx \geqslant C \operatorname*{osc}_{x \in S(x_0,r)} u(x)$$

holds, where the constant C only depends on the dimensionality of the space.

Finally, we obtain that

$$\operatorname*{osc}_{x \in S(x_0,r)} u(x) \leqslant \frac{2r^{n-2}}{C} \int_{B(x_0,r)} |\nabla u|^n \, dx.$$

Since the function u is monotone, we have for almost all $r \in T(r)$,

$$\operatorname*{ess\,sup}_{B(x_0,r)} u(x) - \operatorname*{ess\,inf}_{B(x_0,r)} u(x) < \operatorname*{osc}_{S(x_0,r)} u(x).$$

It remains to note that for $r \to 0$

$$\int_{B(x_0,r)} |\nabla u|^n \, dx \to 0.$$

Hence, due to monotonicity of the function u, it follows that

$$\lim_{r \to 0} |\operatorname*{ess\,sup}_{x \in B(x_0,r)} u(x) - \operatorname*{ess\,inf}_{x \in B(x_0,r)} u(x)| = 0.$$

Consequently, the function $u(x)$ is continuous.

The proposition is proved.

Thus, the sufficient conditions of monotonicity of mappings of the class $L^1_{n,\mathrm{loc}}$ are at the same time the sufficient conditions of continuity.

To prove monotonicity of mapping, we study the components V_i of the set $\mathbf{R}^n \setminus \varphi(S(x,r))$ on which the linking index χ_i vanishes. Namely, in particular, the exterior components to the cycle $\varphi(S(x,r))$. If a component with zero linking index is called the mapping φ, then (just as in the smooth case) it is covered at least twice for different signs of the Jacobian. Positivity of the Jacobian or the conditions substituting its conditions make repeated covering impossible and lead to monotonicity of the mapping.

Let us consider the restriction of the mapping φ upon the domain U with locally quasiconformal boundary, assuming that on ∂U the mapping is continuous. Let us choose any component V_i of the set $\mathbf{R}^n \setminus \varphi(\partial U)$ on which the linking index $\chi(y, \varphi(\partial U))$ vanishes. Let us divide the pre-image $W = (\varphi|_U)^{-1}(V_i)$ of the component V_i into two parts:

$$W^+ = \{x \in W : J(x, \varphi) \geqslant 0\}, \quad W^- = W \setminus W^+.$$

Lemma 3.4. *If $m(W) > 0$ and the set of points $x \in W$, in which the mapping φ is topologically nondegenerate, has positive measure, then $m(W^+) > 0$, $m(W^-) > 0$, and $m(\varphi(W^+) \cap \varphi(W^-)) > 0$.*

Proof. From the conditions of Lemma and from the theorem for the weak differential (Theorem 5.1, Chapter 2) follows the existence of a point $x \in W$ such that:

1) There exists a sphere $S(x,r)$ whose image $\varphi(S(x,r)) \subset V_i$, and the mapping $\varphi(S(x,r))$ is continuous.

2) $\mathbf{R}^n \setminus \varphi(S(x,r))$ contains a bounded component $\tilde{V}$ on which the linking index $\chi(y, \varphi(S(x,r))) \neq 0$. By applying Theorem 2.5 to the characteristic function $\chi_{\tilde{V}}$, we obtain

$$\int_{B(x,r)} \chi_{\tilde{V}}(\varphi(x)) J(x, \varphi)\, dx = \chi(y, \varphi(S(x, r)))\, m(\tilde{V}), \quad y \in \tilde{V}. \tag{3.2}$$

The right-hand side of the formula is distinct form zero. The function $\chi_{\tilde{V}}(\varphi(x)) \equiv 0$ on $B(x,r) \setminus \varphi^{-1}(\tilde{V})$. Therefore, $m(W) > m(\varphi^{-1}(\tilde{V}) \cap B(x,r)) > 0$.

According to Lemma 3.2, $m(\varphi(W)) \geqslant m(\tilde{V}) > 0$. Suppose that $m(\varphi(W^+) \cap \tilde{V} > 0$, while $m(\varphi(W^+) \cap \varphi(W^-)) = 0$.

In the set $(\varphi(W^+) \cup \tilde{V}) \setminus \varphi(W^-)$, let us choose a compact set F of nonzero measure. Let us apply Theorem 2.5 to the characteristic function χ_F:

$$\int_{B(x,r)} \chi_F(\varphi(x)) J(x, \varphi)\, dx = \chi(y, \varphi(S(x, r))\, m(F), \quad y \in \tilde{V}.$$

Hence, since the right-hand side is distinct from zero and $\varphi^{-1}(F) \subset W^+$, we obtain $m(\varphi^{-1}(F)) > 0$, $m(W^+) > 0$. Further,

$$\int_U \chi_F(\varphi(x)) J(x,\varphi)\,dx = \int_{R^n \setminus \varphi(\partial U)} \chi_F(y)\chi(y,\varphi(\partial U))\,dy$$

$$= \int_{\tilde V} \chi_F(y)\chi(y,\varphi(\partial U))\,dy = 0.$$

On the other hand,

$$\int_U \chi_F(\varphi(x)) J(x,\varphi)\,dx \geqslant \int_{B(x,r)} \chi_F(\varphi(x)) J(x,\varphi)\,dx > 0.$$

The obtained contradiction proves the inequality $m(\varphi(W^+) \cap \varphi(W^-)) > 0$. From $m(W^+) > 0$ and from

$$\int_W J(x,\varphi)\,dx = \int_{\tilde V} \chi(y,\varphi(\partial U))\,dy = 0$$

it follows that $m(W^-) > 0$.

Corollary. *Let V_i be a connected component of the set $\mathbf{R}^n \setminus \varphi(\partial U)$ for which $\chi(y,\varphi(\partial U)) = 0$, $y \in V_i$. If the Jacobian $J(x,\varphi) \neq 0$ almost everywhere in U, and $m(W) > 0$, then $m(W^+) > 0$, $m(W^-) > 0$, and $m(\varphi(W^+) \cap \varphi(W^-)) > 0$.*

The statement of the corollary follows from Lemmas 3.1 and 3.4.

Theorem 3.5. *Let a mapping $\varphi : G \to \mathbf{R}^n$, $\varphi \in L^1_{n,\mathrm{loc}}(G)$ be given. Let one of the following conditions be satisfied:*

1) *The mapping φ is topologically nondegenerate at almost all points of the domain G (the Jacobian $J(x,\varphi) \neq 0$ almost everywhere), and φ is a one-to-one mapping almost everywhere, i.e., for any two nonintersecting sets $A, B \subset G$ from $m(A) > 0$, $m(B) > 0$, it follows that $m(\varphi(A) \cap \varphi(B)) = 0$;*
2) *The mapping φ is topologically nondegenerate at almost all points of the domain G, and $J(x,\varphi) \geqslant 0$ for almost all $x \in G$ (for almost all $x \in G$, the Jacobian $J(x,\varphi) > 0$);*
3) *The mapping φ has a nonnegative Jacobian and has the property: if $J(x,\varphi) = 0$, then*

$$\sum_{i,j=1} \left[\frac{\partial \varphi_i}{\partial x_j}(x)\right]^2 = 0.$$

Then the mapping φ is monotone in the domain G.

Remark. Condition 3 of Theorem 3.5 is satisfied by mappings with bounded distortion first studied by Reshetnyak [69]. Let us recall that the mapping $\varphi : G \to \mathbf{R}^n$, $\varphi \in L^1_{n,\mathrm{loc}}$ is called the mapping with bounded distortion if there exists a constant $K \geqslant 1$ for which the inequality

$$\left\{\sum\left[\frac{\partial\varphi_i}{\partial x_j}\right]^2\right\}^{n/2}\leqslant n^{n/2}KJ(x,\varphi)$$

holds almost everywhere in G.

Proof of the theorem. Suppose that at a point x_0 the mapping φ is nonmonotone. This means that a set T is a set of positive measure in the set of real numbers if $r \in (0, \rho(x_0, \partial G))$ for which the pre-image W of the intersection of the set $\varphi(B(x_0, r))$ with the unbounded component V to the cycle $\varphi(S(x_0, r))$ has nonzero measure. Reducing the set T for a set of zero measure, we may assume that on all the spheres $S(x_0, t)$, $t \in T$, the mapping φ is continuous. Let us divide the set W into two parts: $W^+ = \{x \in W : J(x, \varphi) \geqslant 0\}$, $W^- = W \backslash W^+$. In cases 1 and 2, by applying Lemma 3.4 or its corollary, we conclude that $m[\varphi(W^+) \cap \varphi(W^-)] > 0$. This contradicts the condition of the theorem.

In case 3 two situations are possible:

a) The measure is positive for the set of the points at which the mapping φ is topologically nondegenerate. In this case, the proof is completed just as in the two previous cases.

b) At almost all points $x \in W$, the mapping φ is topologically degenerate. In this case, the proof is completed differently. Let us fix a point $x_1 \in W$ at which the mapping is topologically degenerate. Let us consider the sphere $S(x_1, t) \subset B(x_0, r)$, $r \in T$, on which the mapping is continuous and the function $\chi(y, \varphi(S(x_1, t)) \equiv 0$. The existence of such a sphere follows from the definition of a topologically degenerate mapping.

By applying Theorem 2.5 to the function $u \equiv 1$ and to the ball $B(x_1, t)$, we obtain

$$\int\limits_{B(x,t)} J(x,\varphi)\,dx = \int\limits_{R^n\setminus\varphi(S(x,t))} \chi(y, \varphi(S(x_1, t)))\,dy = 0.$$

Thus, $J(x_1, \varphi)$ almost everywhere in $B(x_1, t)$, and $\sum\left[\frac{\partial\varphi_i}{\partial x_j}(x)\right]^2 = 0$ almost everywhere in $B(x, r)$, that is, a) $\varphi(y) = \varphi(x)$ for all $y \in B(x_1, t)$; b) $\varphi(B(x_1, t) \subset V$; c) a set $\widetilde{W}$, of the points from W where the mapping φ is topologically degenerate, is an open set; d) on any connected component of the set $\widetilde{W}$, the mapping φ is constant.

Let us fix a point $x_2 \in \widetilde{W}$ and a sphere $S_2 = S(x_2, q)$, $q < \rho(x_2, \partial\widetilde{W})$. Let us emit all the possible rays from the point x_0. On each of the rays, consider a segment l_x with the origin at a point $x \in S_2$. This segment connects the spheres S_2 and $S_0 = S(x_0, r)$. The mapping φ is continuous on the segments l_x for almost all $x \in S_0$. Furthermore, the mapping φ is absolutely continuous on the segments l_x for almost all x.

From the above it follows that the mapping φ is constant on the sphere S_2. According to the assumption $y_0 = \varphi(S_2) \in \varphi(S_0)$. A family $\mathcal{L}$ of all segments on which φ is absolutely continuous is transferred by the mapping φ to a family of curves connecting y_0 and $\varphi(S(x_0, r))$.

The set $\varphi(\widetilde{W})$ is countable. Using this fact and the absolute continuity of φ on any segment $l_x \in \mathcal{L}$, we obtain that on each of them there exists a set of positive measure l_x whose image lies outside of $\varphi(\widetilde{W})$ but still in V. Consequently, $l_x \in W\backslash\widetilde{W}$. From the Fubini theorem, we directly obtain that $m(\cup_{\mathcal{L}} l_x) > 0$, thus $m(W\backslash\widetilde{W}) > 0$, which contradicts the assumption that $m(W\backslash\widetilde{W}) = 0$.

This completes the proof of the theorem.

Corollary 1. *If for the mapping $\varphi \in L^1_{n,\mathrm{loc}}(G)$ the conditions of Theorem 3.5 are satisfied, then the mapping φ is continuous.*

This follows from Theorem 3.5 and Lemma 3.4.

Corollary 2. *If for the mapping $\varphi \in L^1_n(G)$ the conditions of Theorem 3.5 are satisfied, then for every compact domain $U \in G$ such that φ is not constant on U and $y \in \varphi(U)\backslash\varphi(\partial U)$, $\chi(y, \varphi(\partial U)) \neq 0$ is valid. (If the conditions of the theorem are satisfied, then $\chi(y, \varphi(\partial U)) \neq 0$.)*

Proof. The set $U_1 = U\backslash\varphi^{-1}(\varphi(\partial U))$ is open, the set U_2 on which φ is a one-to-one mapping is a set of complete measure. Let us choose the point $x_0 \in U_1 \cap U_2$. Since at the point x_0 the mapping is topologically nondegenerate, the existence of r such that $\varphi(B(x_0, r)) \subset \varphi(U)\backslash\varphi(\partial U)$ follows from the continuity of φ and from the closure of $\varphi(\partial U)$, and the set $(\varphi(U)\backslash\varphi(\partial U))\backslash\varphi(S(x_0, r))$ contains a nonempty connected component V_0 on which the linking index $\chi_0 \neq 0$. Consequently, due to the fact that the set $B(x_0, r)\backslash\varphi^{-1}(\varphi(S(x_0, r)))$ is open, $\varphi^{-1}(V_0)$ is open. Therefore, in $\varphi^{-1}(V_0)$, one can find the point x_1 at which φ is a one-to-one mapping. Let $y_2 = \varphi(x_2)$. Then $\chi(y_2, \varphi(S(x_0, r)) = \chi(y_2, \varphi(\partial U))$ is not equal to zero, since $y_2 \in V_0$.

For Condition 1 of Theorem 3.5, our corollary is proved. For Condition 2, we use the nonnegativity of the differential, due to which the linking index coincides with the degree and multiplicity function. After this observation, the proof is carried out in the same way.

For Condition 3 of Theorem 3.5, we use the part of the proof of Theorem 3.5 which shows that the image of a set of points is no more than countable if at these points the mapping is topologically nondegenerate and $J(x, \varphi) = 0$. After this, the proof is reduced to Case 2.

Corollary 3. *If for the mapping φ, Conditions 2 or 3 of Theorem 3.5 are satisfied and $J(x, \varphi) > 0$ almost everywhere, then the mapping φ has the N-property.*

Proof. Let us take an arbitrary set $A \in \mathbf{R}^n$ of zero measure. According to Lemma 3.2, for every compact domain U with boundary of zero measure, the equality

$$\int_U J(x, \varphi)\, \chi_A(\varphi(x))\, dx = \int_{R^n \setminus \varphi(\partial U)} \chi_A(y)\, \chi(y, \varphi(\partial U))\, dy$$

is valid where $\chi_A(y)$ is the characteristic function of the set A. Then the right-hand side of the equality equals zero. Consequently, the left-hand side also equals zero. Since $J(x,\varphi) > 0$ almost everywhere, the function $\chi_A(\varphi(x))$ must be a characteristic function of a set of zero measure. Hence it follows that $m(\varphi^{-1}(A) \cap U) = 0$, and due to arbitrariness in the choice of the domain U, the corollary is proved.

Corollary 4. *If Condition 1 of Theorem 3.5 is satisfied and if $J(x,\varphi) \neq 0$ almost everywhere, then the mapping φ has the N^{-1}-property.*

Proof. Let us consider an arbitrary compact subdomain U of a domain G whose boundary has zero measure. Due to one-to-one uniqueness almost everywhere, the set $\varphi^{-1}(\varphi(\partial U))$ also has zero measure. Let $V_1, V_2, \dots, V_i, \dots$ be bounded connected components of the set $\mathbf{R}^n \backslash \varphi(\partial U)$. The pre-images $W_i = \varphi^{-1}(V_i) \cap U$ are open sets. Let us choose an arbitrary set $A \in \mathbf{R}^n$, $m(A) = 0$. Consider the mappings $\varphi_i = \varphi|W_i$. Let $B_i = \varphi_i^{-1}(A_i)$. According to Corollary 2, $\chi(y, \varphi, (\partial W_i)) \neq 0$ and is constant. Due to one-to-one uniqueness almost everywhere of φ, $\chi(y, \varphi(\partial W_i))$ equals either 1 or -1. Now, from Corollary 3, we obtain $m(\varphi^{-1}(A)) = 0$. Since $m(\varphi^{-1}(\varphi(\partial U))) = 0$ and $\varphi^{-1}(A) \cap U \subset \cup_{i=1} \varphi_i^{-1}(A) \cup \varphi^{-1}(\varphi(\partial U))$, the corollary is proved.

Corollary 5. *If the conditions of Theorem 3.5 are satisfied, then the mapping φ has the N-property.*

Proof. First let us make a remark. Let U be a compact domain in $\mathbf{R}^n$. Since the mapping φ is continuous, $\varphi(U)$ is a set of finite measure. Let us choose a point $x_0 \in U$ and consider a family of spheres $S(x_0, t)$, $t < \rho(x_0, \partial U)$. Let us prove that for almost all t, $m(\varphi(S(x_0,t))) = 0$. Indeed, if Condition 1 is satisfied, the images of spheres $\varphi(S(x_0,t))$ intersect by the set of zero measure for different t. Then the number of spheres for which $m(\varphi(S(x_0))) > 1/n$ does not exceed $nm(\varphi(U))$. Consequently, the spheres for which $m(\varphi(S(x_0,t))) > 0$ are at most countable.

In the cases of Condition 2 and 3, the proof is a bit more complicated. Recall that the mapping has a K-differential distinct from zero. Therefore, the multiplicity function is finite almost everywhere, i.e., for almost all points from $\varphi(U)$ the pre-image is finite. Let us consider the same family of spheres as in the previous case. Let the sphere $S(x_0,t)$ be such that $m(\varphi(S(x_0,t))) > 0$. First let us prove that the set of spheres $S(x_0,t_0)$ for which $m(\varphi(S(x_0,t_0))) \cap \varphi(S(x_0,t)) > 0$ is at most countable. Let $A_n = \{y \in \varphi(S(x_0,t_0)) : N_\varphi|_U(y,U) < n\}$. Then $m([\cup A_n] \Delta(\varphi(S(x_0,t_0)))) = 0$. The set A_n may belong to at the most n images $\varphi(S(x_0,t)$ of the spheres $S(x_0,t)$. Consequently, the number of spheres $S(x_0,t)$ for which $m(\varphi(S(x_0, t)) \cap \varphi(S(x_0,t_0))) > 0$ is at most countable. Now, the same arguments as those for Condition 1 of Theorem 3.5 prove that the number of spheres for which $m(\varphi(S(x_0,t_0))) > 0$ is at most countable.

Consequently, for every point $x_0 \in U$, there exists a sphere $S(x_0,t)$ such that $m(\varphi(S(x_0,t))) = 0$. The balls $B(x_0,t)$ form the covering of the compact domain U. Let us select from it a finite subcovering $B_1, B_2, \dots, B_k$. Consider an arbitrary set $A \subset B$, $m(A) = 0$. Let $T = \{x : \varphi^{-1}(\varphi(x))\backslash\{x\} \neq \emptyset\}$. If the mapping φ is a one-to-one mapping almost everywhere, then $m(T) = 0$. If in this case φ is topologically nondegenerate almost everywhere, then from Corollary 2 we obtain that the linking index χ_i on any bounded connected component V_i of the set $\varphi(B_i)\backslash\varphi(\partial B_i)$ is distinct from zero. The pre-images $W_i = (\varphi|_{B_i})^{-1}(V_i)$ are open, and due to the fact that φ is a one-to-one mapping almost everywhere, $m(\cup W_i) = \sum m(W_i)$ and $W_i \cap W_j = \emptyset$.

Let us apply Theorem 3.5 to the characteristic function χ_i of the set $\varphi(A \cap W_i) \cap \varphi(B_i)$. Due to the fact that φ is a one-to-one mapping almost everywhere, $\chi_i(\varphi(x)) = 0$ for all $x \notin A \cup T$. We obtain

$$0 = \int_{B_i} \chi_i(\varphi(x))\, J(x,\varphi)\, dx = \int_{R^n\setminus\varphi(B_i)} \chi_i(y)\, \chi(y, \varphi(\partial B_i))\, dy$$

$$= \int_{R^n} \chi_i(y)\, \chi(y, \varphi(\partial B_i))\, dy \qquad (\text{for } \; m(\varphi(\partial B_i)) = 0).$$

Hence it follows that $m(\varphi(A) \cap V_i) = 0$. Thus, $m(\varphi(A) \cap \varphi(B_i)) = 0$. Due to countable additivity of measure, it follows that $m(\varphi(A)) = 0$ for any set $A \subset U$ having zero measure.

Let us turn back to Conditions 2 and 3 of Theorem 3.5. As was proved in Corollary 2, for such mappings and for any compact domain $U \subset G$, the linking index is positive everywhere in $\varphi(U)\backslash\varphi(\partial U)$, i.e., $\chi(y,\varphi(\partial U)) \geqslant 1$ for all $y \in \varphi(U)\backslash\varphi(\partial U)$. Let $A \subset U$ and $m(A) = 0$. According to the above, for every $a > 0$, there exists an at most countable set of balls $B_1, B_2, \dots, B_k, \dots$ such that: a) $m(\varphi(\partial B_i)) = 0$; b) $\sum_i m(B_i) < \varepsilon$; c) $\cup B_i = A$. Let us reconstruct the covering as follows: $\tilde{B}_1 = \bar{B}_1$, $\tilde{B}_2 = \overline{B_1\backslash B_2}, \dots, \tilde{B}_k = \overline{B_1\backslash(\cup_{i=1}^{k-1} B_i)}$. Compact domains of $\tilde{B}_i$ cover A. It is obvious that $m(\varphi(\partial\tilde{B}_i)) = 0$ and $\sum' m(\tilde{B}_i) < \varepsilon$,

$$\sum \chi_i m(V_i) = \int_{\operatorname{Int}\tilde{B}_i} J(x,\varphi)\, dx = \int_{\tilde{B}_i} J(x,\varphi)\, dx,$$

where V_i are bounded connected components of the set $\varphi(B_i)\backslash\varphi(\partial B_i)$, and χ_i are the corresponding linking indices. Hence, due to the fact that $\chi_i > 1$ and due to the choice of the balls B_i, we obtain that

$$m(\varphi(\tilde{B}_i)) \leqslant \sum_i \chi_i m(V_i) = \int_{\tilde{B}_i} J(x,\varphi)\, dx,$$

that is,

$$m(\varphi(A)) \leqslant \sum_i m(\varphi(\tilde{B}_i)) \leqslant \sum_i \int_{\tilde{B}_i} J(x,\varphi)\, dx = \int_{\cup\tilde{B}_i} J(x,\varphi)\, dx.$$

Since $\varphi \in L_n^1(G)$, then for $\varepsilon \to 0$, due to the fact that $m(\cup \tilde{B}_i) \to 0$, we obtain $\|J(x,\varphi\|_{L_1(\cup B_i)} \to 0$ for $\varepsilon \to 0$.

This completes the proof.

§4 Invariance of the Spaces $L_p^1(G)(L_n^1(G))$ for Quasiisometric (Quasiconformal) Homeomorphisms

4.1. Preliminary Information on the Mappings

Let $\varphi : G \to G'$ be a homeomorphism of domains in $\mathbf{R}^n$. We associate with φ the following magnitudes:

$$H_\varphi(p) = \overline{\lim_{r\to 0}} \frac{L_\varphi(p,r)}{l_\varphi(p,r)}, \qquad l_\varphi'(p) = \overline{\lim_{r\to 0}} \frac{L_\varphi(p,r)}{r},$$

$$J_\varphi(p) = \overline{\lim_{r\to 0}} \frac{m(\varphi(B_r(p)))}{m(B_r(p))}.$$

The homeomorphism $\varphi : G \to G'$ is said to be quasiconformal if the magnitude H_φ is bounded on G. The mapping φ is called K-quasiconformal if H_φ is bounded on G and $H_\varphi(x) \leqslant K$ almost everywhere on G. The smallest $q(\varphi)$ of the constants K is called the metric coefficient of quasiconformality.

Theorem 4.1. *A quasiconformal mapping $\varphi : G \to G'$ is differentiable almost everywhere, and φ belongs to the class L_n^1 on every compact subdomain of the domain G.*

Proof. Let $P(x_1, x_2, \ldots, x_n) = (x_1, x_2, \ldots, x_{n-1}, 0)$ be the projection; A is a Borel set in $L = \{x \in \mathbf{R}^n : x_n = 0\}$. The set $P^{-1}(A)$ is the Borel one. Let us fix the cube $Q \subset G$ with sides parallel to coordinate axes. The set $E_A = P^{-1}(A) \cap Q$ is measurable. Let us consider the function of the set $\eta(A) = m(\varphi(E_A))$ on the cube $P(Q)$. According to the Lebesgue theorem, the function η has the finite derivative $\eta'(y)$ for almost all points $y \in P(Q)$. Let us fix such a point y. Let us prove that η is absolutely continuous on the segment E_y.

Let $F \subset E_y$ be the compact and $F \subset \operatorname{Int} Q$. Let us choose H such that $H(x,t) < H$ for all $x \in G$. Let $k > \rho(F, \partial Q)$; from $F_k = \{x \in F : 0 < x < 1/k\}$, it follows that $L_\varphi(x,r) < Hl_\varphi(x,r)$. The compact sets F_k form a monotonically increasing sequence and $\cup F_k = F$. Fix $\delta \in (0, 1/k)$. For every $r \in (0, \delta)$, there exists a finite covering of the compact set F_k by the intervals $\Delta_1, \Delta_2, \ldots$ such that $m_1(\Delta_i) = 2r$ for all r; the centre of Δ_i belongs to F_k, every three intervals have empty intersection; $m_1(\cup_i \Delta_i) < m_1(F_k) + \varepsilon$. Let, in addition, $|\varphi(x) - \varphi(z)| < t$, if $x, z \in Q$ and $|x - z| < 2r$. Let us denote the centres of the segments Δ_i by x_i. Since $x_i \in F_k$, then $L_\varphi(x_i, r) \leqslant H l_\varphi(x_i, r)$. Since $d(\varphi(B(x_i, r))) < t$, then

$$\sum_{i=1}^{p} d(\varphi(B(x_i, r))) \leqslant 2 \sum_{i=1}^{p} L_\varphi(x_i, r).$$

Let us denote the sum on the left-hand side by $\lambda^t(\varphi(F_k))$. From the Hölder inequality, we obtain

$$[\lambda^t(\varphi(F_k))]^n = 2^n p^{n-1} \sum_{i=1}^{p} L_\varphi^n(x_i, r) \leqslant 2^n H^n p^{n-1} \sum l_\varphi^n(x_i, r)$$
$$\leqslant \frac{2^n H^n (m(F) + \varepsilon)^{n-1}}{\Omega_n r^{n-1}} \sum m(\varphi(A_i)),$$

where Ω_n is the measure of the n-dimensional unit ball. Note that $\sum_{i=1}^{p} m(\varphi(A_i)) \leqslant 2m(\varphi(E_B)) \leqslant 2\eta(B)$. Since $F_k \subset F$,

$$[\lambda^t(\varphi(F_k))]^n \leqslant \frac{2^{n+1} H_n \Omega_{n-1} (m_1(F) + \varepsilon)^{n-1} \eta(B)}{\Omega_n m(B)}.$$

Let us carry out three limit transitions: for $\eta \to 0$, $\varepsilon \to 0$, and for $t \to 0$. As a result, for the 1-dimensional Hausdorff measure $\lambda_1(\varphi(B_k))$, we obtain the estimate

$$\lambda_1(\varphi(F_k)) \leqslant C\varphi'(y)[m_1(F)]^{n-1}.$$

Taking the limit for $k \to \infty$, we obtain

$$\lambda_1(\varphi(F)) \leqslant C\varphi^{-1}(y)[m_1(F)]^{n-1};$$

hence it follows that the mapping φ is absolutely continuous on E_y. Using the same argument for projections upon the rest of the coordinate axes, we obtain that $\varphi \in ACL(G)$. Consequently, the mapping φ has first generalized derivatives, therefore, it is differentiable almost everywhere and has K-differential almost everywhere. At the differentiability points of the mapping φ, it is obvious that the inequality

$$|\nabla\varphi|^n \leqslant K^{n-1}|J(x, \varphi)|$$

is valid. According to Corollary 4 of Theorem 3.5, due to the fact that the sign of the Jacobian is constant, the homeomorphism φ has the N-property, and we may apply the theorem for the change of variable to any compact domain $U \subset G$,

$$\int_U |\nabla\varphi(x)|^n dx \leqslant K^{n-1} \int_U |J(x, \varphi)| dx = K^{n-1}\mu(\varphi(U)) < \infty.$$

Thus, $\varphi \in L^1_{n,\mathrm{loc}}(G)$.

This completes the proof of the theorem.

Theorem 4.2. *Every quasiconformal homeomorphism $\varphi : G \to G'$ induces the* *bounded linear operator $\varphi^* : L_n^1(G) \to L_n^1(G')$ according to the rule $(\varphi^* u) = u \circ \varphi$, besides, $\|\varphi^*\| \leqslant K^{(n-1)/n}(\varphi)$.*

Proof. Let $u \in L_n^1(G')$ be a differentiable function, then the function $v = u \circ \varphi$ belongs to the class $ACL(G)$ because $\varphi \in ACL(G)$; consequently, it has generalized derivatives in the domain G. At differentiability points of the function $v(x)$ (i.e., almost at all points of the domain G), we have

$$|\nabla v|(x) \leqslant \overline{\lim_{y\to x}} \frac{|v(x)-v(y)|}{|x-y|}$$
$$\leqslant \overline{\lim_{y\to x}} \frac{|v(x)-v(y)|}{|\varphi(x)-\varphi(y)|} \overline{\lim_{y\to x}} \frac{|\varphi(x)-\varphi(y)|}{|x-y|},$$

i.e.,

$$|\nabla v|(x) \leqslant |\nabla u|(\varphi(x)) I_\varphi(x)$$

almost everywhere in G.

Since $I_\varphi(x) \leqslant K^{n-1}(J(x,\varphi))$ almost everywhere and φ has the N-property (Corollary of Theorem 3.5), then according to the theorem about change of variable,

$$\int_G |\nabla v(x)|^n \, dx \leqslant K^{n-1} \int_{G'} (|\nabla u|(\varphi(x)))^n |J(x,\varphi)| \, dx$$
$$\leqslant K^{n-1} \int_{G'} |\nabla u|^n \, dy.$$

Let us approximate an arbitrary function $u \in L_n^1(G)$ by smooth functions $u_m \in L_n^1(G)$. Since, according to the above, the operator φ^* is bounded on $L_n^1(G') \cap C^\infty(G)$, then the operator φ^* is uniquely extended on all $L_n^1(G')$ up to the bounded operator from $L_n^1(G')$ to $L_n^1(G)$. It is easy to see that due to uniqueness, the extension coincides with the operator defined according to the composition rule.

Corollary 1. *Since the mapping converse to the quasiconformal one is quasiconformal, the operator φ^* is the isomorphism of the Banach spaces $L_n^1(G')$ and $L_n^1(G)$.*

Corollary 2. *Let $\varphi : G \to G'$ be a quasiconformal homeomorphism. Then for any pair of closed sets $F_0, F_1 \subset G'$, the inequality*

$$C_n^1(F_0, F_1, G) \leqslant C_n^1(\varphi(F_0), \varphi(F_1), G') K^{n-1}$$

is valid.

Proof. The proof easily follows from the definition of the $[1,n]$-capacity and from Theorem 4.2, because the mapping φ transfers the admissible function to the admissible one due to the fact that the belonging of the function to the class L_n^1 is being preserved.

Theorem 4.3. *The homeomorphism $\varphi : G \to G'$ is quasiconformal iff for any pair of compact sets $F_0, F_1 \subset G$, the inequality*

$$Q^{-1}C_n^1(F_0, F_1, G) \leqslant C_n^1(\varphi(F_0), \varphi(F_1), G')$$

holds where, the constant Q does not depend on the choice of the compact sets F_0 and F_1.

Proof. The validity of the above inequality for quasiconformal mappings is proved in Corollary 2 of Theorem 4.2. Let the inequality formulated in the statement of the theorem be valid. Let us prove that in this case the mapping is quasiconformal. Let us fix the point $x_0 \in U$. Let us choose r_0 to be so small that $L_\varphi(x_0, r_0) < \rho(\varphi(x_0), \partial\varphi(g))$. Let $0 < r < r_0$. Let us consider a pair of closed sets $F_0 = B(\varphi(x_0, l_\varphi(x_0 r))$ and $F_1 = G \setminus B(\varphi(x_0), L_\varphi(x_0, r))$ and their pre-images $\mathcal{F}_0 = \varphi^{-1}(F_0), \mathcal{F}_1 = \varphi^{-1}(F_1)$. The point x_0 lies in $\mathcal{F}_0$, $\mathcal{F}_1$ contains the sphere $S(x_0, r)$. According to Proposition 6.3 of Chapter 3, $C_n^1(\mathcal{F}_0, \mathcal{F}_1, G) \geqslant \alpha^2(n) > 0$, and according to Proposition 6.2 of Chapter 3,

$$C_n^1(F_0, F_1, G) = \left(\log \frac{L_\varphi(x_0, r)}{l_\varphi(x_0, r)}\right)^{1-n} \Omega_{n-1}.$$

From the condition of the theorem, it follows that

$$\alpha^2(n) \leqslant C_n^1(\mathcal{F}_0, \mathcal{F}_1, G) \leqslant QC_n^1(F_0, F_1, G) \leqslant \Omega_{n-1}/(\log H_\varphi(x_0, r))^{n-1}.$$

Hence, we have

$$H_\varphi(x_0, r) \leqslant \exp\left[\left(\frac{\Omega_{n-1}}{\alpha^2(n)}\right)^{\frac{1}{n-1}}\right].$$

Thus, the mapping φ is quasiconformal.

The theorem is proved.

The smallest $K_n(\varphi)$ of the constants of Q, for which the inequalities from the statement of the theorem hold, is called the capacitance coefficient of quasiconformality.

A pair of closed sets F_0, F_1 with respect to G is called a hull in G if $F_1 \subset G$, $F_0 \supset \partial G$, and $\overline{\mathbf{R}^n \setminus F_0}$ is a compact set ($F_1 \cap F_0 = \emptyset$).

From the proof of the theorem, Theorem 4.3' obviously follows.

Theorem 4.3'. *A homeomorphism $\varphi : G \to G'$ is quasiconformal iff for any hull (F_0, F_1) in G, the inequality*

$$Q^{-1}C_n^1(F_0, F_1, G) \leqslant C_n^1(\varphi(F_0), \varphi(F_1), G') \leqslant QC_n^1(F_0, F_1, G)$$

is valid, where the constant Q does not depend on the choice of the hull.

The homeomorphism $\varphi : G \to G'$ (G, G' are domains in $\mathbf{R}^n$) is called quasiisometric if for any $x \in G$ and $y \in G'$ the inequalities

$$\varlimsup_{z \to x} \frac{|\varphi(x) - \varphi(z)|}{|x - z|} \leqslant M, \quad \varlimsup_{t \to y} \frac{|\varphi^{-1}(y) - \varphi^{-1}(t)|}{|y - t|} \leqslant M$$

are valid, where the constant M does not depend on the choice of the point $x \in G$ and $y \in G'$. The smallest $q_i(\varphi)$ of the constants M for which the *inequalities hold is called the metric coefficient of quasiisometricity. It is obvious that the quasiisometric homeomorphism is the quasiconformal one.*

Theorem 4.4. *A quasiisometric homeomorphism $\varphi : G \to G'$ belongs to the class $L_\infty^1(G)$.*

Proof. According to Theorem 4.1 $\varphi \in L^1_{n,\mathrm{loc}}(G)$. Consequently, φ is differentiable almost everywhere, and from the definition of the quasiisometric homeomorphism, it is clear that for every coordinate function φ_i, $i = 1, 2, \dots, n$, the inequality $|\nabla\varphi_i| \leqslant M$ holds. Consequently, $\varphi \in L_\infty^1(G)$.

The theorem is proved.

Theorem 4.4′. *Let $\varphi : G \to G'$ be a quasiisometric homeomorphism. Then the operator $\varphi_p^* : L_p^1(G') \to L_p^1(G)$ defined according to the rule $\varphi_p^* \circ u = u \circ \varphi$ is bounded. Besides, $\|\varphi_p^*\| \leqslant M^{p+n}$ for $p < \infty$ and $\|\varphi_\infty^*\| \leqslant M$.*

Proof. Since the mapping φ is the Lipschitz mapping, according to Corollary 4 of Theorem 3.5, it has the N-property.

The remainder of the proof repeats that of Theorem 4.2.

We only give the final calculations for $p < \infty$:

$$|\nabla v(x)| \leqslant |\nabla u|(\varphi(x))\, l_\varphi(x) \leqslant M\, |\nabla u|(\varphi(x)),$$

$$\int_G |\nabla v(x)|^p\, dx \leqslant M^p \int_G (|\nabla u|(\varphi(x)))^p \frac{|J(x,\varphi)|}{|J(x,\varphi)|}\, dx$$

$$\leqslant M^{p+n} \int_G (|\nabla u|(\varphi(x)))^p\, |J(x,\varphi)|\, dx \leqslant M^{p+n} \int_{G'} |\nabla u|^p\, dy.$$

Here $u \in L_p^1(G) \cap C^\infty(G)$, $v(x) = u(\varphi(x))$. We used the fact that the Jacobian $J(x,\varphi)$ exists almost everywhere and $M^{-n} \leqslant |J(x,\varphi)| \leqslant M^n$.

The remainder of the proof is the same as that of Theorem 4.2.

The case $p = \infty$. It has been proved that for any function $u \in L_\infty^1(G)$, $u \circ \varphi \in L_{p,loc}(G)$ for all $p < \infty$. It remains to prove that $|\nabla v| \leqslant |\nabla u| M$. This follows from Theorem 4.6 below for differentiation of composition.

Corollary 1. *The theorem is valid for the spaces $W_p^1(G)$ and $W_p^1(G')$.*

Proof. It suffices to verify that $\|\varphi_p^*(U)\|_{L_p(G)} \leqslant |||\varphi_p^*|||\, \|u\|_{L_p(G')}$ for any function $u \in L_p(G')$. We wrote $|||\varphi^*|||$ to distinguish the norm φ_p^* acting as the operator from $L_p(G') \to L_p(G)$ from the norm φ_p^* acting as the operator from $L_p^1(G') \to L_p^1(G)$.

From Lemma 3.2, it follows that for $p < \infty$,

$$\int_G |v(x)|^p\, dx \leqslant \int_G |(u \circ \varphi)(x)|^p\, dx$$

$$\leqslant M^n \int_G |(u \circ \varphi)(x)|^p\, |J(x,\varphi)|\, dx \leqslant M^n \int_G |u(y)|^p\, dy.$$

In the case $p = \infty$, it is obvious that

$$\operatorname*{ess\,sup}_{x\in G}|v(x)|\leqslant M\operatorname*{ess\,sup}_{y\in G}|u(y)|.$$

Corollary 2. *Since the homeomorphism inverse to the quasiisometric one is quasiisometric, the operator φ_p^* is the isomorphism of the Banach spaces $L_p^1(G')(W_p^1(G))$ and $L_p^1(G)(W_p^1(G))$.*

Corollary 3. *Let $\varphi : G \to G'$ be a quasiisometric homeomorphism. Then for any pair of closed sets $F_0, F_1 \in G'$, the inequality*

$$C_p^1(\varphi^{-1}(F_0), \varphi^{-1}(F_1); G)\leqslant Q_p C_n^1(F_0, F_1; G)$$

is valid. The constant Q_p does not depend on the choice of the pair F_0, F_1.

Proof. The proof is similar to that of Corollary 2 of Theorem 4.2.

Theorem 4.5. *Let $1 \leqslant p < \infty$, $p \neq n$, and let a homeomorphism $\varphi : G \to G'$ satisfy the condition: $Q_p^{-1}C_p^1(F_0, F_1, G) \leqslant C_p^1(\varphi(F_0), \varphi(F_1), \varphi(G)) \leqslant Q_p C_p^1(F_0, F_1 G)$ for any pair of closed sets $F_0, F_1 \subset G$, with the constant Q_p not depending on the choice of the pair F_0, F_1.*

Then the mapping φ is a quasiisometric homeomorphism.

Proof for the case $p > n$. Let us fix a point $a \in G$ and a ball $B(a,r)$ with the radius smaller than $\rho(a, \partial G)$. Let $\{b\}$ be any point on the sphere $S(a,r)$. According to Proposition 6.8 of Chapter 3, $C_p^1(\{a\},\{b\},B(a,r)) \geqslant \alpha^2(n)|a-b|^{n-p}$. Let us consider the image $\varphi(B(a,r))$ of the ball $B(a,r)$. In addition, let $L_\varphi(a,r) < \rho(\varphi(a), \partial G')$. Then, due to Proposition 6.8 of Chapter 3,

$$\begin{aligned}C_p^1(\{\varphi(a)\}, \{\varphi(b)\}; G')&\leqslant C_p^1(\{\varphi(a)\}, \{\varphi(b)\}, R^n\}\\ &\leqslant\beta^2(n)\,|\varphi(a)-\varphi(b)|^{n-p}.\end{aligned}$$

From the condition of the theorem, we directly obtain that

$$\begin{aligned}\alpha^2(n)\,|a-b|^{n-p}&\leqslant C_p^1(\{a\}, \{b\}, B(a, r))\\ &\leqslant C_p^1(\{a\}, \{b\}; G)\leqslant Q_p C_p^1(\{\varphi(a)\}, \{\varphi(b)\}; G')\\ &\leqslant Q_p C_p^1(\{\varphi(a)\}, \{\varphi(b)\}; R^n)\leqslant Q_p\beta^2(n)\,|\varphi(a)-\varphi(b)|^{n-p}.\end{aligned}$$

Since $p \neq n$, it follows that

$$|\varphi(a)-\varphi(b)|^{p-n}\leqslant\frac{\beta^2(n)\,Q_p}{\alpha^2(n)}\,|a-b|^{p-n},$$

that is,

$$\frac{|\varphi(a)-\varphi(b)|}{|a-b|}\leqslant\left(\frac{\beta^2(n)\;Q_p}{\alpha^2 n}\right)^{\frac{1}{p-n}}.$$

Taking the limit for $r \to 0$, we hence obtain

$$\overline{\lim_{b\to a}} \frac{|\varphi(b)-\varphi(a)|}{|b-a|} \leqslant M = \left(\frac{\beta^2(n)\,Q_p}{\alpha^2(n)}\right)^{\frac{1}{p-n}}.$$

Since the inverse mapping also satisfies the condition of the theorem,

$$\overline{\lim_{x\to y}} \frac{|\varphi^{-1}(x)-\varphi^{-1}(y)|}{|x-y|} \leqslant \left(\frac{\beta^2(n)\,Q_p}{\alpha^2(n)}\right)^{\frac{1}{p-n}},$$

and the quasiisometricity of the homeomorphism φ is proved.

Proof for $p < n$. The proof for $p < n$ is omitted, and the reader is referred to [92].

Let us denote by K_p the smallest one of the constants Q_p for which the inequalities from the condition of the theorem are valid.

As is seen from the proof of Theorem 4.4, the constants K_p do not coincide for different values of p. In the same way we can prove

Theorem 4.5′. *Let $1 \leqslant p < \infty$, $p \neq n$, and let a homeomorphism $\varphi : G \to G'$ satisfy the condition*

$$Q_p^{-1} C_p^1(F_0, F_1; G) \leqslant C_p^1(\varphi(F_0), \varphi(F_1); G') \leqslant Q_p C_p^1(F_0, F_1; G)$$

for any hull (F_0, F_1) in g. Here, the constant Q_p does not depend on the choice of the hull.

Then the mapping φ is a quasiisometric homeomorphism.

4.2. Differentiation of Composition

Theorem 4.6 (differentiation of composition). *Let a bounded operator $\varphi^* : L^1_p(G') \to L^1_p(G)$ be induced by a mapping $\varphi : G \to G'$ according to the rule $\varphi^* f = f \circ \varphi$, $f \in L^1_p(G')$; here the mapping φ has the N^{-1}-property. Then the formula about the differentiation of composition is valid: for every function $f \in L^1_p(G')$,*

$$\frac{\partial(f\circ\varphi)}{\partial x_i}(x) = \sum_{k=1}^{n} \frac{\partial f}{\partial y_k}(\varphi(x))\frac{\partial \varphi_k}{\partial x_i}(x).$$

Proof. If f is a smooth function, then the validity of the formula follows from Theorem 5.1 of Chapter 2. Let us take a sequence of smooth functions $f_m \in L^1_p(G')$ which converge to an arbitrary function f almost everywhere and in $L^1_p(G')$. Then

$$f_m \circ \varphi = \varphi^*(f_m) \to \varphi^* \circ \varphi = f \circ \varphi$$

in $L^1_p(G)$. One may assume that the partial derivatives

$$\frac{\partial(f_m\circ\varphi)}{\partial x_i}(x) \to \frac{\partial(f\circ\varphi)}{\partial x_i}(x) \quad \text{almost everywhere in } G,$$

and

$$\frac{\partial f_m}{\partial y_k}(y) \to \frac{\partial f}{\partial y_k}(y) \quad \text{almost everywhere in } G'.$$

Since the mapping φ has the N^{-1}-property, then, by the limit transition from the formula

$$\frac{\partial (f_m \circ \varphi)}{\partial x_i} = \sum_{i=1}^{n} \frac{\partial f_m}{\partial y_k}(\varphi(x)) \frac{\partial \varphi_k}{\partial x_i}(x),$$

the statement of the theorem follows.

Corollary. *For mappings with bounded distortion, the formula about the differentiation of composition is valid.*

In the space $L_p^1(G)$, we introduce the usual order. The function $u \in L_p^1(G)$ is greater than zero, $u > 0$, if $u(x) \geqslant 0$ almost everywhere, and there exists a set of positive measure on which $u(x) > 0$; $u > v$ if $u - v > 0$. The order in $W_p^1(G)$ is induced from $L_p^1(G)$. (The spaces $L_p^1(G)$ and $W_p^1(G)$ are K-lineals; see, for instance, [26].)

The sequence $\{u_m\} \in L_p^1(G)$ monotonically decreases if for $n \leqslant m$, $u_n < u_m$. If $\inf u_m = 0$ (i.e., if the inequality $w < u_m$, $w \in L_p^1(G)$ holds for all m, then it follows that $w \leqslant 0$), then we write $u_m \downarrow 0$. Let us give the definition of (o)-convergence in $L_p^1(G)$. The sequence $u_m \overset{(o)}{\to} u$, $u_m \in L_p^1(G)$ if there exists a monotone sequence $v_m \downarrow 0$, $v_m \in L_p^1(G)$ such that $|u_m - u| < v_m$ for all m.

Lemma 4.7. *Let $\sum_{\nu=1}^{\infty} u_\nu$ be an arbitrary absolutely converging series in the Banach space $W_p^1(G)$, and let u be the sum of this series. Then $\sum_{\nu=1}^{\infty} u_\nu \to u$, $m \to \infty$.*

Proof. It is obvious that

$$|u - \sum_{\nu=1}^{\infty} u_\nu| \leqslant \sum_{\nu=m+1}^{\infty} |u_\nu|.$$

Therefore, it remains to show that

$$g_m = \sum_{\nu=m+1}^{\infty} |u_\nu| \downarrow 0.$$

Firstly,

$$\|g_m\|_{W_p^1} \leqslant \sum_{\nu=m+1}^{\infty} \|u_\nu\|_{W_p^1}.$$

Thus, $\|g_m\|_{W_p^1} \to 0$ for $m \to \infty$.

Secondly, by applying Theorem 1.2 of Chapter 3 on the ball $B \subset G$, we see that g_m converges to zero on the ball B everywhere except for the set of zero

$(1,p)$-capacity. Covering the domain G by a countable number of balls, we obtain the convergence of the sequence $g_m \downarrow 0$ everywhere except for the set of zero $(1,p)$-capacity. Hence it follows that $g_m \downarrow 0$, since if $g_m \geqslant w > 0$ for all m, then for some $\varepsilon > 0$,

$$\mathrm{Cap}_{(1,p)}\{x \in G \mid w(x) > \varepsilon\} > 0.$$

Corollary. *From a sequence f_m converging to f in $W_p^1(G)$, one may extract a subsequence which (o)-converges to f.*

The operator $A : W_p^1(G') \to W_p^1(G)$ is said to be (o)-linear if from $u_m \overset{(o)}{\to} u$, it follows that $Au_m \overset{(o)}{\to} Au$.

Theorem 4.8. *If an operator $A : W_p^1(G') \to W_p^1(G)$ is (o)-linear, then it is bounded.*

Proof. Let us prove that the diagram of the operator A is closed, hence the continuity of the operator follows.

Let $u_m \to u$ in $W_p^1(G)$, and $w_m = Au_m \to w$ in $W_p^1(G)$. It is necessary to prove that $w = Au$. Following the corollary of the previous theorem, let us select the subsequence $w_{m_k} \overset{(o)}{\to} w$ and the subsequence $u_{m_{k_i}} \overset{(o)}{\to} u$ (the subsequence $\{u_{m_{k_i}}\}$ is singled out from the subsequence $\{u_{m_k}\}$). Since the operator A is (o)-linear, then $w_{m_{k_i}} = Au_{m_{k_i}} \overset{(o)}{\to} Au$. Due to the uniqueness of the (o)-limit, it follows that $Au = w$.

The theorem is proved.

4.3. Representation of Operators Preserving the Order

In this section, the domain G' is supposed to be bounded.

Theorem 4.9. *Let $\varphi : G \to G'$ be a mapping defined almost everywhere in G and let it induce an isomorphism $\varphi^* : L_p^1(G') \to L_p^1(G)$ according to the rule: for $f \in L_p^1(G')$, $\varphi^* f = f \circ \varphi$. Then the mapping φ is a quasiconformal homeomorphism for $p = n$, and φ is a quasiisometric homeomorphism for $p > n$; here the domains G' and $\varphi(G)$ are $(1,p)$-equivalent.*

The main difficulty in the proof of the theorem is to prove the continuity for $p = n$.

I. Due to the boundedness of the domain G', the coordinate functions $y_j(y) = y_j$ and $y = (y_1, \dots, y_n)$ belong to $L_p^1(G')$. Thus, φ is the mapping of the class L_p^1, because $\varphi_i(x) = y_i(\varphi(x)) = (\varphi^* y_i)(x)$ almost everywhere.

The continuity of the mapping φ will follow from Theorem 3.4 if we prove that the Jacobian of the mapping φ is distinct from zero almost everywhere in the domain G, and φ is a one-to-one mapping almost everywhere at least in some neighbourhood of the point $x \in G$.

Let us fix a ball $B(\bar{x},r) \subset G$ such that $B(\bar{x},2r) \subset G$. Let k be a smooth function which is equal to 1 on $B(\bar{x},r)$ and equal to 0 outside of $B(\bar{x},2r)$. The functions kx_i (x_i is the ith coordinate function) are bounded and belong to the class $L^1_n(G)$. Consequently, the function $\psi_i = \varphi^{*-1}_i(kx_i)$ are also bounded and belong to $L^1_n(G')$. Consider the mapping

$$\psi = (\psi_1, \psi_2, \dots, \psi_n).$$

Note that

$$(\psi \circ \varphi)(x) = x$$

for almost all $x \in B(\bar{x},r)$, since

$$(\psi_i \circ \varphi)(x) = \varphi^*(\varphi^{*-1}(kx_i)) = k(x)\, x_i,$$

and $k(x) = 1$ for $x \in B(\bar{x},r)$. Hence, almost everywhere it follows that φ is a one-to-one mapping on the ball $B(\bar{x},r)$.

Let us consider a sequence $\{\psi_m\}$ of smooth mappings of the class $W^1_n(G)$. This sequence converges in the metric of the space $W^1_n(G)$ to the mapping ψ (this implies that the ith coordinate function of the mapping ψ_m converges for $m \to \infty$ in the metric of the space $W^1_n(G)$ to the ith coordinate function of the mapping ψ ($i = 1, 2, \dots, n$). This function belongs to the space $W^1_n(G')$, since the domain G' is bounded). The function ψ_i is bounded and $\psi_i \in L^1_n(G')$. We state that the sequence of mappings $\varphi^*\psi_m = \psi_m \circ \varphi$ converges for $m \to \infty$ to the mapping $\psi \circ \varphi = I$ (I is the identical mapping).

First let us note that the isomorphism φ^* is (o)-linear. Indeed, since the isomorphism φ^* preserves the order, it suffices to establish that $\varphi^*(u_m) \downarrow 0$ if $u_m \downarrow 0$, $u_m \in L^1_n(G')$ $(m = 1, 2, \dots)$. If $w \leqslant \varphi^*(u_m)$ for all m, then $\varphi *^{-1}(w) \leqslant \varphi^{*-1}(\varphi^*(u_m)) = u_m$ for all m. (The converse isomorphism φ^{*-1} also preserves the order.) Hence it follows that $\varphi^{*-1}(w) \leqslant 0$, and consequently, $w \leqslant 0$. Since $0 \leqslant \varphi^*(u_m)$ for all m, then $\varphi^*(u_m) \downarrow 0$. Then, the restriction operator

$$\theta : L^1_n(G) \to W^1_n(B(\bar{x}, r))$$

is also (o)-linear. Thus, the operator

$$\theta \circ \varphi^* : W^1_n(G') \to W^1_n(B(\bar{x}, r))$$

is (o)-linear, and due to Theorem 4.8, it is bounded. Hence it follows that $\psi_m \circ \varphi \to I$ in $W^1_n(B(\bar{x},r))$.

According to the theorem for the differentiation of composition, by using the smoothness of ψ_m, we obtain

$$\frac{\partial(\psi_m \circ \varphi)}{\partial x_i}(x) = \sum_{k=1}^{n} \frac{\partial \psi_m}{\partial y_k}(\varphi(x)) \frac{\partial \varphi_k}{\partial x_i}(x)$$

almost everywhere in G. Taking the subsequence, we may assume that $\partial(\psi_m \circ \varphi)\backslash\partial x_i$ converges almost everywhere on $B(\bar{x},r)$ to $\partial I\backslash\partial x_i$. Thus,

$$\frac{\partial I}{\partial x_i} = \lim_{m\to\infty} \sum_{k=1}^{n} \frac{\partial \psi_m}{\partial y_k} (\varphi(x)) \frac{\partial \varphi_k}{\partial x_i}(x)$$

for almost all $x \in B(\bar{x}, r)$. Hence it follows that

$$\lim_{m\to\infty} J(\varphi(x), \psi_m)\cdot J(x, \varphi) = 1$$

for almost all $x \in B(\bar{x}, r)$; consequently, the Jacobian $J(x, \varphi)$ of the mapping φ is distinct from zero almost everywhere in $B(\bar{x}, r)$. It has been proved earlier that this is a one-to-one mapping almost everywhere.

According to Theorem 3.5, the mapping φ is monotone and, due to the corollary of this theorem, continuous.

II. In this section, we prove that in the domain G the mapping φ is a homeomorphism. Let us prove it by a chain of statements.

From Corollary 1 of Theorem 3.5, we obtain that for every compact domain U of all points $y \in \varphi(U)\setminus\partial(\varphi(U))$, the degree of the mapping φ is distinct from zero. According to Corollary 5 of the same theorem, φ has the N-property, and from Corollary 3 of Theorem 3.5, we obtain that φ has the N^{-1}-property.

Let us study the sets $\tilde{G} = \varphi^{-1}(\varphi(G) \cap G')$ and $\tilde{G}' = \varphi(\tilde{G})$. Let $x_0 \in \tilde{G}$, then $\varphi(x_0) \in \tilde{G}'$. Due to the continuity of φ there exists r_0 such that for all $r < r_0$ the ball $B(x_0, r)$ lies in G, and its image $\varphi(B(x_0, r))$ lies in G'. Since φ has the N^{-1}-property, then $m(\varphi(B(x_0, r))) > 0$. Let us recall that $\psi \circ \varphi = I$ almost everywhere. Thus, in any of the sets $\varphi(B(x_0, r))$, there exists a point y_r for which $\psi(y_r) \subset B(x_0, r)$. Due to the continuity of ψ, we obtain that $\psi(\varphi(x_0)) = x_0$. Since the point $x_0 \in G$ was chosen arbitrarily, it follows that $\psi \circ \varphi = I$ on $\tilde{G}$. Since $\tilde{G}$ is an open set, $\varphi|\tilde{G}$ is a homeomorphism, then $\tilde{G}'$ is an open set, and $\psi|\tilde{G}'$ is a homeomorphism. Also, $\tilde{G}' \subset G'$.

Proposition 4.10. *The set G' is connected.*

Proof. Let the set G' be nonconnected, and let A be its connected component. Consider the function $\tilde{v}(y) \equiv 1$ on A, $\tilde{v}(y) \equiv 0$ on $G'\setminus A$. Let us take a point $y_0 \in \partial A$. Since $\tilde{G}' \subset G'$, then $y_0 \in G'\setminus\tilde{G}'$. Let the ball $B(y_0, r) \subset G'$. We set $v(y) = \eta(y)\tilde{v}(y)$, where $\eta(y)$ is a smooth compactly supported function which is identically equal to 1 on $B(y_0, r/2)$, $\eta(y) \geqslant 0$, and $\eta(y) \equiv 0$ outside of $B(y_0, r)$.

It is obvious that v is a bounded smooth function in $\mathbf{R}^n\setminus(G'\setminus\tilde{G}')$, $v \in L_p^1(G')$. The function v is not extendable to the function of the class $L_p^1(G')$.[1] The composition $v \circ \varphi$ should belong to the class $L_p^1(G)$, because the mapping $\varphi \in Lp(G)$. It can easily be verified, because $v \circ \varphi \in ACL$ and $\nabla v \leqslant M < \infty$. Then $\varphi^{*-1}(v \circ \varphi) \in L_p^1(G')$ and it is the extension of the function v, which is impossible due to the construction of the function v.

The proposition is proved.

[1] Recall that $m(G'\Delta\tilde{G}') = 0$.

Proposition 4.11. *The operator $\varphi^* : W_p^1(G') \to L_p^1(G)$ is bounded, i.e., for any function $u \in W_p^1(G')$,*

$$\|\nabla(\varphi^* u)\|_{L_p(G)} \leqslant C\,[\|u\|_{L_p(G')} + \|\nabla u\|_{L_p(G')}],$$

where the constant C does not depend on the choice of the function $u \in W_p^1(G')$.

Proof. Let us recall that $\varphi \circ \varphi^{-1} = I$ almost everywhere. Consequently, $m(\varphi(\varphi^{-1}(G'))\backslash G') = 0$. Since $\varphi(G) \supset \varphi(\varphi^{-1}(G'))$, $m(\varphi(G)\backslash G') = 0$. According to Lemma 3.2, for every function $u \in W_p^1(G')$ and for any compact domain $U \subset G$, we have

$$\int_U |(u \circ \varphi)(x)|^p\,|J(x,\varphi)|\,dx = \int_{\varphi(U)} |u(y)|^p\,dy,$$

since, as was shown in the proof of Corollary 5 of Theorem 3.5,

$$N_\varphi(y,U) = \begin{cases} 1 & \text{for almost all } y \in \varphi(U), \\ 0 & \text{for } y \in \varphi(U). \end{cases}$$

From this formula by means of limit transition in compact domains $U_m \subset G$, $U_m \subset U_{m+1}$, $\cup U_m = G$, we obtain the equality

$$\int_G |(u \circ \varphi)(x)|^p\,|J(x,\varphi)|\,dx = \int_{\varphi(G)} |u(y)|^p\,dy = \int_{G'} |u(y)|^p\,dy$$

for any function $u \in W_p^1(G')$.

Denote by $W_{p,J}^1$ the completion by the norm

$$\|v\|_{\nabla^{-1}_{p,J}} = \left(\int_G |v(x)|^p\,|J(x,\varphi)|\,dx\right)^{1/p} + \|\nabla v\|_{L_p(G)}$$

of the space of functions from $L_p^1(G)$ for which this norm is bounded.

For the function $u(x) \in W_p^1(G')$, the function $u(\varphi(x)) = \varphi^* u(x) \in W_{p,J}^1$. Let us show that the diagram of the mapping $\varphi^* : W_p^1(G') \to W_{p,J}^1$ is closed. Let $u_m \to u$ in $W_p^1(G')$, and $v_m = \varphi^*(u_m) \to v$ in $W_{p,J}^1$. It is necessary to show that $v = \varphi^*(u)$. Taking the subsequence, one may assume $u_m \to u$ almost everywhere. Since according to Corollary 4 of Theorem 3.5, the mapping φ has the N^{-1}-property, $\varphi^* v_m = v_m \circ \varphi$ converges for $m \to \infty$ to $\varphi^* u$ almost everywhere. On the other hand, some subsequence of the sequence $\{|v_{m_k}(x)|^p |J(x,\varphi)|\}$ converges to $|v(x)|^p |J(x,\varphi)|$ almost everywhere. Since $|J(x,\varphi)| \neq 0$ almost everywhere, $v_{m_k}(x) = (\varphi^* u_{m_k})(x) \to v(x)$ actually almost everywhere, i.e., $v = \varphi^* u$.

According to the closed diagram theorem, the operator $\varphi^* : W_p^1(G') \to W_{p,J}^1$ is continuous.

The proposition is proved.

Let us recall that a pair of sets F_0, F_1 closed with respect to G' is called a hull in G' if $F_1 \subset G'$, $F_0 \supset \partial G'$ and if $\mathbf{R}^n \backslash F_0$ is a compact set. ($F_1 \cap F_0 = \emptyset$.)

To verify the quasiconformality (quasiisometricity) of the homeomorphism φ, it suffices to show that

$$C_p^1(\varphi^{-1}(F_0), \varphi^{-1}(F_1); \varphi^{-1}(G')) \leqslant KC_p^1(F_0, F_1, G')$$

with the constant K which does not depend on the choice of the hull (F_0, F_1) in the domain $\varphi(G)$. Note that any function admissible for the hull (F_0, F_1) is extended by zero on $\mathbf{R}^n$, and its belonging to the class L_p^1 is preserved.

Let us prove the inequality. Let u be an admissible function for the hull $(F_0, F_1) \subset \varphi(G)$, then $v = u \circ \varphi = \varphi^* u$ is the admissible function for the hull $(\varphi^{-1}(F_0), \varphi^{-1}(F_1)) \subset G$, and the estimate (Proposition 4.11)

$$\|v\|_{L_p^1(G)} \leqslant C\,[\|u\|_{L_p(G')} + \|u\|_{L_p^1(G')}] \leqslant C\,(C_1 d(G') + 1)\|u\|_{L_p^1(G')}$$

is valid.

Here C_1 is the constant from Theorem 4.2 of Chapter 2, $(\|u\|_{L_p(\mathbf{R}^n)} \leqslant C_1\|\nabla u\|_{L_p(\mathbf{R}^n)})$. The constants C and C_1 do not depend on the choice of the hull and on the admissible function. Therefore, the inequality for capacity is proved.

From Theorems 4.3 and 4.5 now follow the quasiconformality for $p = n$ and the quasiisometricity for $p > n$ of the homeomorphism $\varphi : G \to \mathbf{R}^n$. We used here the fact that the inequality

$$C_p^1(F_0, F_1) \leqslant QC_p^1(\varphi^{-1}(F_0), \varphi^{-1}(F_1))$$

for the hull F_0, F_1 is proved by applying the previous argument to the mapping $\varphi^{-1} = \psi$.

It remains to prove that the domains G' and $\varphi(G)$ are $(1, p)$-equivalent. Let us prove that the function g admits the unique extension to the domain G', with the class being preserved. First of all, let us recall that $m(\varphi(G \backslash G') = 0$. Due to the quasiconformality for $p = n$ and the quasiisometricity for $p > n$ from Theorems 4.3 and 4.5, it follows that $g \circ \varphi \in L_p^1(G)$. Therefore, $w = (\varphi^*)^{-1}(g \circ \varphi) \in L_p^1(G)$, and obviously, $\|w\|_{L_p^1(G')} = \|g\|_{L_p^1(\varphi(G))}$. The extension is constructed. It is unique, since $m(G' \backslash \varphi(G)) = 0$. Consequently, the restriction operator $\theta(\theta w = w|_{\varphi(G)}, w \in L_p^1(G))$ is an isometric isomorphism.

This completes the proof of Theorem 4.9.

Remark. Theorem 4.9 remains valid if we replace $L_p^1(G)$ and $L_p^1(G')$ by $W_p^1(G)$ and $W_p^1(G')$ (for $p > n$). The proof is literally the same.

Recall that in Theorem 3.1 of Chapter 4 it was proved that the closed set E, which is equal to the difference of $(1, p)$-equivalent domains G and $\tilde{G}$, is the NC_p-set.

In this subsection, we prove that NC_p-sets are removable for quasiconformal mappings. The theorems of the papers [4, 53, 113] about removability present the special case of this result. We also obtain the removability of the NC_p-set for quasiconformal mappings.

Theorem 4.12. *Let G be a domain in $\mathbf{R}^n$, and let E be an NC_p-set in the domain G. Then any quasiconformal homeomorphism φ of the domain $G \backslash E$*

onto a bounded domain $G' \subset \mathbf{R}^n$ is extended as far as the quasiconformal homeomorphism $\varphi : G \to \mathbf{R}^n$ without increase of the distortion coefficient.

Remark. The result is valid for $p = n$ and for an unbounded domain. For this it suffices to transfer the arguments below to the sphere.

Theorem 4.13. *Let G be a domain in $\mathbf{R}^n$, and let E be an NC_p-set in G, and $\varphi : G\backslash E \to \mathbf{R}^n$ is a mapping with bounded distortion. Suppose that for every point $x \in E$, there exists a ball $B(x,r)$ such that φ belongs to the class $L_n^1(B\backslash E)$. Then there exists the unique extension of the mapping φ up to the mapping with bounded distortion $\overline{\varphi} : G \to \mathbf{R}^n$ without the growth of distortion coefficient.*

Proof of Theorems 4.12 and 4.13. Theorem 4.12 follows from Theorem 4.13. Indeed, the condition with the ball holds for any quasiconformal mapping. According to Property 3.9 of Chapter 4, the set $G\backslash E$ is connected; consequently, the quasiconformal mapping φ is a mapping on $G\backslash E$ with bounded distortion. For every compact domain $V(\overline{V} \subset G)$, the mapping $\varphi \in L_n^1(V\backslash E)$. According to Property 3.6 and Theorem 3.1 of Chapter 4, φ is extended to the mapping $\varphi_V : V \to \mathbf{R}^n$ belonging to the class L_n^1. Thus, φ is extended to the mapping $\overline{\varphi} : G \to \mathbf{R}^n$ belonging to the class $L_{n,\mathrm{loc}}^1(G)$. Since E has the measure 0, φ is a mapping with bounded distortion [60]. If the mapping φ is a homeomorphism, then $\overline{\varphi}$ is also a homeomorphism. (This easily follows from the mapping φ being open.)

Remark. Theorem 4.13 contains the corresponding result of paper [4] as a special case, if E is a compact set.

Theorem 4.14. *Let G be a domain in $\mathbf{R}^n$, and let E be an NC_p-set ($p > 1$) in the domain G. Then any quasiisometric mapping $\varphi : G\backslash E \to \mathbf{R}^n$ is uniquely extended as far as the quasiisometric mapping $\overline{\varphi} : G \to \mathbf{R}^n$.*

Proof. By applying Property 3.3 of Chapter 4, we reduce the proof to the case $p > n$. The mapping $\theta_\varphi^{-1*} : L_p^1(G') \to L_p^1(G)$, where $G' = \varphi(G)$; $\varphi^* : L_p^1(G') \to L_p^1(G\backslash E)$, $\varphi^* f = f \circ \varphi$, $f \in L_p^1(G')$, $\theta : L_p^1(G) \to L_p^1(G\backslash E)$, $\theta(g) = g|_{G\backslash E}$, $g \in L_p^1(G)$, is a structural isomorphism of the spaces $L_p^1(G')$ and $L_p^1(G)$. According to Theorem 4.9, $\theta^{-1}\varphi^*$ is generated by the quasiisometric mapping $\overline{\varphi}$, which is obviously the extension by the continuity of the mapping φ.

CHAPTER 6

Extension of Differentiable Functions

The study of the extension problem for the Sobolev classes focuses mainly on the necessary extension conditions. Geometric interpretation of the necessary conditions of extension (the condition with the arc diameter and its refinements) concerns the use of nonlinear capacity methods. The extension conditions are first formulated and proved in the abstract variant, then they are interpreted for certain types of spaces. The proof of the estimates of capacity used in the arguments is based on the material of Chapter 3. In a plane, the arc diameter condition leads to the necessary and sufficient conditions of extension for some functional classes. For the Sobolev spaces with first generalized derivatives, the extension operator is constructed by means of quasiconformal mappings. For higher derivatives, the results for extension that are used were obtained by P. Jones and P. A. Schwartzman. In this case the proofs are not given.

The presentation is given for domains of Euclidean space.

§1 Arc Diameter Condition

1.1. Analysis of the Ahlfors Condition

Quasisphere. A surface in $\mathbf{R}^n$ is called a quasisphere if it is an image of a unit sphere under a quasiconformal homeomorphism of $\mathbf{R}^n$ onto itself.

The necessary and sufficient condition for a closed Jordan curve in a plane to be a quasicircle is the Ahlfors condition [2]:

The Ahlfors condition. Everywhere in the following, the neighbourhood $V(x_0)$ of a point x_0 belonging to a closed Jordan curve γ is supposed to be chosen so small that the connected component γ_V is a Jordan arc noncoinciding with the whole of the curve γ. The component γ_V belongs to the intersection of the curve γ and the circle $V(x_0)$ and contains the point x_0.

The closed Jordan curve $\gamma \subset \mathbf{R}^2$ satisfies the Ahlfors condition if for any point $x_0 \in \gamma$ there exists a neighbourhood $V(x_0)$ in which the following condition is satisfied: any triple of points $\xi_1, \xi_2, \xi_3 \subset \gamma_V$ satisfies the inequality $|\xi_1 - \xi_3| \leqslant C|\xi_1 - \xi_2|$ if the point ξ_3 lies between the points ξ_1 and ξ_2 on the arc γ_V. The constant C does not depend on the choice of the point $x_0 \in \gamma$ and on the choice of the triple of points ξ_1, ξ_2, ξ_3.

A simply-connected domain $G \subset \mathbf{R}^2$ satisfies the Ahlfors condition if its boundary is a Jordan curve satisfying the Ahlfors condition.

Let us give the condition which is equivalent to the Ahlfors condition for bounded simply-connected domains in $\mathbf{R}^2$. It should be noted at once that the constant in the new condition depends on the constant in the Ahlfors condition and on the shape of the domain as well.

Let G be a simply-connected plane domain whose boundary is a closed Jordan curve. Let us choose two points, x and y, belonging to ∂G. Let us consider the arc $\gamma_{x,y} \subset \partial G$ which connects the points x and y and has the least diameter. If there exists a constant C (independent of the choice of the pair of points x, y) for which the inequality

$$\operatorname{diam} \gamma_{x,y} \leqslant C|x - y|$$

is valid, then we say that for the domain G, the preliminary arc diameter condition is satisfied.

It is obvious that for bounded simply-connected plane domains the preliminary arc diameter condition is equivalent to the Ahlfors condition.

1.2. The Arc Diameter Condition

The domain $G \subset \mathbf{R}^n$ satisfies the condition with the arc diameter if for every pair of points $x, y \in G$ the inequality

$$\operatorname{diam}(x, y, G) \leqslant C|x - y|$$

is valid.

Here $\operatorname{diam}(x, y, G) \leqslant C|x-y|$ is the minimum of diameters of smooth curves connecting the points x and y in the domain G. The open set $G \subset \mathbf{R}^n$ satisfies the arc diameter condition if each of its connected components satisfies this condition.

The domain G satisfies the bilateral condition with the arc diameter if it and the domain $G^* = \operatorname{Int}(\mathbf{R}^n \backslash G)$ satisfy the arc diameter condition.

Let us give typical examples of domains satisfying the arc diameter condition.

Example 1. (Domains with nonzero angles). Let us consider in a plane the domains

$$G = \{(x, y) : 0 < x < 1, |y| < x^2\}, \qquad G^* = \operatorname{Int}(\mathbf{R}^2 \backslash G).$$

The domain G satisfies the arc diameter condition. The domain G^* does not satisfy the condition with the arc diameter, i.e., the domain G does not satisfy the bilateral arc diameter condition.

Example 2. (Domains with peaks). Let us give the description of domains with peaks, exactly following Gehring and Väisälä. The example is given for dimension 3 only for the sake of clarity and can easily be modified for any large dimension.

A set in $\mathbf{R}^3$ is said to be a peak if by the isometry φ of the space $\mathbf{R}^3$ it can be mapped onto $S = \{x = (r, \theta, x_3) : r = g(x_3), 0 < x_3 < a\}$, where (r, θ, x_3) are cylindric coordinates, $a < \infty$ and $g(x)$ is a smooth function satisfying the following conditions: a) $g(a) = 0$; b) $g'(u)$ is continuous and decreases at $0 < u < a; c) \lim_{u \to a} g'(u) = 0$.

The image of the point $Q = (0, 0, a)$ for the isometry φ^{-1} is called the vertex of the peak, $e_3 = (0, 0, 1)$ is the direction of the peak, the image of the disc $B = \{x : 0 \leqslant r \leqslant g(0), \hat{x}_3 = 0\}$ is the foundation of the peak.

The domain $G \subset \mathbf{R}^3$ is called a domain with peak if on its boundary there exist a point x and a neighbourhood $U(x)$ such that $\partial G \cap U$ is the peak with the vertex x. Let n be the direction of the peak. If the points $x + u\overline{n}$ $(u > 0)$ do not belong to G, then the domain is said to have the peak directed outward. If the points $x + u\overline{n}$ belong to G, then G is said to have the peak directed inward. The verification of whether $x + u\overline{n}$ belongs to G is realized at sufficiently small $u > 0$.

It is immediately verified that, unlike in the planar case (i.e., unlike domains with zero angles), both the domains with peaks directed inward and those with peaks directed outward may satisfy the arc diameter condition, both unilateral and bilateral ones, if the boundary of a domain is a smooth surface outside of the peak vertex.

1.3. Properties of Domains Satisfying the Arc Diameter Condition

Property 1. If a domain G satisfies the arc diameter condition, then it is connected at every boundary point.

Let us preliminarily recall the following:

Definition. A domain G is connected at a point $x \in \partial G$ if for every ball $B(x, R)$ with the centre at the point x there exists a ball $B(x, r)$, $r < R$ such that any two points of the intersection $B(x, r) \cap G$ may be connected by a curve in $B(x, R) \cap G$.

Proof. Suppose that a domain in G is not connected at a point $x \in \partial G$. In this case there exists a ball $B(x, r)$ such that in every ball $B(x, r)$ of a smaller radius there is at least a pair of points $(x_r, y_r) \subset G$ which cannot be connected by a curve in $B(x, R) \cap G$, i.e., $|x_r - y_r| < r$, and

$$\operatorname{diam}(x_r, y_r, G) > R - r.$$

Since $r < R$ is chosen arbitrarily, the domain G does not satisfy the arc diameter condition.

Corollary. *If a domain $G \subset \mathbf{R}^2$ satisfies the arc diameter condition, then the closures of any two connected components of G_1^* and G_2^* of the open set $G^* = \mathrm{Int}(\mathbf{R}^2 \backslash G)$ do not intersect.*

Property 2. If a domain $G \subset \mathbf{R}^2$ satisfies the arc diameter condition, then $\partial G \backslash \partial G^*$ has topological dimension not exceeding $n-2$.

Proof. Let us recall that the set $A \subset \mathbf{R}^n$ has a topological dimension not exceeding $n-2$ if for every ball $B(x,r)$, the set $B(x,r)\backslash A$ is connected.

Suppose that $\partial G \backslash \partial G^*$ has a topological dimension greater than $n-2$. Then there exists a point $x \in \partial G \backslash \partial G^*$ and a ball $B(x,R)$ which satisfy the conditions: $B(x,R) \cap G^* \neq \emptyset$; in any ball $B(x,r)$ for $r < R$, there exist two points, x_r and y_r, which cannot be connected by a curve in the set $A = B(x,R)\backslash \partial G$. Consequently, any curve connecting the points x_r and y_r in the domain G has a diameter greater than $R-r$. The radius r may be chosen arbitrarily small. Hence it follows that

$$\mathrm{diam}(x_r, y_r, G) > R,$$

while $|x_r - y_r| < 2r$.

We obtained the contradiction to the condition with the arc diameter for the domain G.

This completes the proof.

The example of the ball $B(0,1) \subset \mathbf{R}^3$ without a segment $[0,1]$ on the x-axis shows that the topological dimension $n-2$ for $\partial G \Delta \partial G^*$ is realized in the domain satisfying even the bilateral arc diameter condition.

Moreover, if one removes a relatively closed set of topological dimension $n-2$ from the ball, then the obtained domain satisfies the arc diameter condition.

Corollary 1. *If a plane domain G satisfies the arc diameter condition, then each of its connected components W of its complement, for which* $\mathrm{Int}\, W \neq \emptyset$, *is the closure of the domain. The remaining part of the complement is connected nowhere.*

This easily follows from Properties 1 and 2.

Corollary 2. *If a plane finitely-connected domain G satisfies the bilateral arc diameter condition, then each component W of its complement is either an isolated point or a closure of the domain satisfying the condition with the arc diameter.*

The proof easily follows from the definition of the bilateral arc diameter condition and from the previous corollary.

Property 3. If a domain G satisfies the arc diameter condition, then for any two points x, y of its boundary and for any $\varepsilon > 0$, there exists a curve

$\gamma_{x,y} : [0,1] \to G$ such that $\gamma_{x,y}(0) = x$, $\gamma_{x,y}(1) = y$, $\gamma_{x,y}(t) \subset G$ for all $t \in (0,1)$, and the diameter of the curve γ satisfies the inequality

$$\operatorname{diam} \gamma_{x,y} \leqslant (1+2\varepsilon)C|x-y|.$$

Here the constant C is the same as in the arc diameter condition for the domain G.

For brevity's sake, we say that the curve $\gamma_{x,y}$ connects two points of the boundary of ∂G in the domain G.

Proof. Let us fix the points $x, y \in \partial G$. Choose two sequences $\{x_m \in G\}$ and $\{y_m \in G\}$, $m = 1, 2, \ldots$, such that $x_m \to x$, $y_m \to y$ for $m \to \infty$,

$$|x_m - x_{m+1}| < (\varepsilon/2^{m+1})|x-y|,$$
$$|x_1 - y_1| < (1+\varepsilon)|x-y|,$$
$$|y_m - y_{m+1}| < (\varepsilon/2^{m+1})|x-y|$$

for all m. According to Property 1, the domain G is connected at each of its boundary points x and y. Therefore, one may construct two sequences of balls $\{B(x,r_m)\}$ and $\{B(y,r_m)\}$ which satisfy the following conditions: $B(x,r_0) \cap B(y,r_0) = \emptyset$, $B(x,r_m) \supset B(x,r_{m+1})$, and $B(y,r_m) \supset B(y,r_{m+1})$ for all m, $r_m \to 0$ for $m \to \infty$; any two points from the set $B(x,r_{m+1}) \cap G$ may be connected by the curve lying in$B(x,r_m) \cap G$; any two points from $B(y,r_{m+1}) \cap G$ may be connected by the curve lying in $B(y,r_m) \cap G$. Without loss of generality, one may put $x_m \in B(x,r_m)$, $y_m \in B(y,r_m)$ for all m. Due to the fact that the domain G satisfies the condition with the arc diameter, there exists a curve $\gamma_1 : [1/4, 3/4] \to G$ connecting the points x_1 and y_1, and $\operatorname{diam} \gamma_1 \leqslant C|x_1 - y_1|$. For any pair x_{m-1}, x_m $(m \geqslant 2)$, due to the arc diameter condition and due to the choice of the sequence of balls $B(x,r_m)$, there exists a curve

$$\gamma_m : [\tfrac{1}{2}^{m+1}, \tfrac{1}{2}^{m}] \to B(x, r_{m-1}) \cap G$$

connecting the points x_m, x_{m-1} for which

$$\operatorname{diam} \gamma_m \leqslant C|x_m - x_{m-1}| \leqslant \frac{\varepsilon^C}{2^m}|x-y|.$$

Similarly, for any pair y_{m-1}, y_m, there exists a curve

$$\overline{\gamma}_m : [1 - \tfrac{1}{2}^{m}, 1 - \tfrac{1}{2}^{m+1}] \to \overline{B}(y, r_{m-1}) \cap G$$

connecting the points y_{m-1} and y_m, for which

$$\operatorname{diam} \overline{\gamma} \leqslant C|y_m - y_{m-1}| < \frac{\varepsilon C}{2^m}|x-y|.$$

Let us consider the curve $\gamma : (0,1) \to G$ coinciding on the segment $[1/4, 3/4]$ with γ_1, coinciding on any of the segments $[1/2^{m+1}, 1/2^m]$ with γ_m, and on any of the segments $[1-1/2^m, 1-1/2^{m+1}]$ with the curve $\overline{\gamma}_m$. By construction, $\lim_{t\to 0}\gamma(t) = x$, and $\lim_{t\to 1}\gamma(t) = y$. Consequently, the curve γ connects the points $x, y \in \partial G$ in the domain G. From the construction of the curve $\gamma_m, \overline{\gamma}_m$, it follows that

$$\begin{aligned}\operatorname{diam}\gamma &\leqslant \operatorname{diam}\gamma_1 + \sum_{m=2}^{\infty}\operatorname{diam}\gamma_m + \sum_{m=2}^{\infty}\overline{\gamma}_m \\ &\leqslant (1+\varepsilon)C|x-y| + \tfrac{\varepsilon}{2}C|x-y| + \tfrac{\varepsilon}{2}C|x-y| \leqslant (1+2\varepsilon)C|x-y|.\end{aligned}$$

This completes the proof.

Property 4. If the domain $G \subset \mathbf{R}^n$ satisfies the arc diameter condition, then for any two points $x, y \in \overline{G}$, the inequality

$$\operatorname{diam}(x, y, G) \leqslant C|x-y|$$

is valid. Here C is the constant form of the arc diameter condition, and $\operatorname{diam}(x, y, G)$ is the minimum of the diameters of all curves connecting the points x, y in the domain G.

The proof obviously follows from Property 3.

Property 5. If a plane bounded domain G satisfies the bilateral arc diameter condition, then the boundary of each connected component of the set $G^* = \operatorname{Int}(\mathbf{R}^2\backslash G)$ is a curve. This curve satisfies the Ahlfors condition and has a constant which only depends on constants in the bilateral arc diameter condition. The remaining part of the boundary is completely nonconnected.

Proof. Every connected component W_i of the complement $\operatorname{Int}(\mathbf{R}^2\backslash G)$ is a simply-connected domain which is locally connected at each boundary point (Property 1). Consequently, the conformal mapping of the circle B onto W_i is extended by continuity up to the topological mapping of the closed circle $\overline{B}$ onto $\overline{W}_i \subset \overline{\mathbf{R}}^2$, i.e., the boundary $\partial W_i = \Gamma_i$ is a closed Jordan curve.

Now let us prove that Γ_i satisfies the Ahlfors condition. Let us choose on Γ_i any two points x and y. According to Property 4, there exist arcs $\gamma_{x,y}$ and $\gamma^*_{x,y}$, connecting the points x and y in G and W_i, respectively, such that

$$\max(\operatorname{diam}(x, y, G), \operatorname{diam}(x, u, W_i)) \leqslant C|x-y|,$$

where C is the constant from the bilateral arc diameter condition. This inequality, obviously, is equivalent to the Ahlfors condition.

If the connected component of the boundary ∂G of the domain G is not a boundary of some of the domains W_i, then, due to Corollary 2 of Property 2, it belongs to the completely nonconnected set $\partial G\backslash\partial G^*$.

This completes the proof.

The Ahlfors condition for unbounded Jordan curves. *An unbounded Jordan curve γ satisfies the Ahlfors condition if for any triple of points $\xi_1, \xi_2, \xi_3 \subset \gamma$, the inequality*

$$|\xi_1 - \xi_3| < C|\xi_1 - \xi_2|$$

holds if the point ξ_3 lies between the points ξ_1 and ξ_2 on the arc γ. The constant C does not depend on the choice of the triple of the points.

Property 6. If a plane domain G satisfies the bilateral arc diameter condition, then any unbounded component of its boundary ∂G is a Jordan curve satisfying the Ahlfors condition.

The proof is similar to that of Property 5.

§2 Necessary Extension Conditions for Seminormed Spaces

2.1. The Extension Operator. Capacitary Extension Condition

An operator θ acting from a seminormed space of functions $\mathcal{F}(G)$ defined in a domain $G \subset \mathbf{R}^n$ into a seminormed space of functions $\mathcal{F}_1(\mathbf{R}^n)$ is called an extension operator if it is bounded and if $(\theta u)/G = u$ for any function $u \in \mathcal{F}(G)$.

Theorem 2.1. *Let G be a domain in $\mathbf{R}^n$, and let $\mathcal{F}(G)$, $\mathcal{F}_1(\mathbf{R}^n)$ be seminormed spaces of functions. If there exists an extension operator $\theta : \mathcal{F}(G) \to \mathcal{F}_1(\mathbf{R}^n)$, then for every pair of closed sets $(F_0, F_1) \subset G$, the inequality*

$$\operatorname{Cap}_{\mathcal{F}_1}(F_0, F_1, \mathbf{R}^n) \leqslant \|\theta\| \operatorname{Cap}_{\mathcal{F}}(F_0, F_1, G)$$

is valid.

Proof. Let us choose an arbitrary function $u \in \mathcal{F}(G)$ admissible for variational $\mathcal{F}$-capacity of a pair of closed sets $F_0, F_1 \subset G$. According to the definition of an admissible function, there exists a neighbourhood $U(F_0) \subset G$ of a set F_0, in which the function u vanishes, and there exists a neighbourhood $V(F_1)$ of a set F_1 in which the function u turns into 1. The extension operator θ does not change the function on the domain G. Consequently, the function θu is admissible for the variational $\mathcal{F}_1$-capacity of the pair (F_0, F_1) in $\mathbf{R}^n$. From the boundedness of the extension operator θ and from the definition of capacity, it follows that

$$\operatorname{Cap}_{\mathcal{F}_1}(F_0, F_1, \mathbf{R}^n) \leqslant \|\theta u\|_{\mathcal{F}_1(\mathbf{R}^n)} \leqslant \|\theta\| \, \|u\|_{\mathcal{F}(G)}.$$

Due to arbitrariness in the choice of $\mathcal{F}$-admissible function for the pair (F_0, F_1), we immediately obtain

$$\operatorname{Cap}_{\mathcal{F}_1}(F_0, F_1, \mathbf{R}^n) \leqslant \|\theta\| \operatorname{Cap}_{\mathcal{F}}(F_0, F_1, G).$$

This completes the proof.

2.2. Additional Properties of Capacity

The traditionally investigated classes of functions L_p, W_p^l, L_p^l, $C^k, \ldots$ are defined in the same way in any domain G of the Euclidean space. One may consider the trace classes $L_p(\mathbf{R}^n)|_G$, $W_p^l(\mathbf{R}^n)|_G$, $L_p^l(\mathbf{R}^n)|_G$ together with the classes $L_p(G)$, $W_p^l(G)$, $L_p^l(G)$. Trace classes may or may not coincide with classes in the domain, depending on the class under consideration and on the structure of a domain.

A homeomorphism $\varphi : G \to G'$ is called the change of variables for the pair of seminormed spaces of functions $\mathcal{F}_0(G)$ and $\mathcal{F}_1(G')$ if for every function $u \in \mathcal{F}_1(G')$ the function $u \circ \varphi^{-1} \in \mathcal{F}_0(G)$, and for every function $v \in \mathcal{F}_0(G)$ the function $v \circ \varphi^{-1} \in \mathcal{F}_1(G')$. In this case, the operators $\varphi^* : \mathcal{F}_1(G') \to \mathcal{F}_0(G)(\varphi^* u = u \circ \varphi)$ and $(\varphi^*)^{-1} : \mathcal{F}_0(G) \to \mathcal{F}_1(G')((\varphi^*)^{-1} v = v \circ \varphi^{-1})$ should be bounded operators.

For brevity's sake, we may say the following: the homeomorphism $\varphi : G \to G'$ is called the change of variables for the pair of seminormed spaces of functions $\mathcal{F}_0(G), \mathcal{F}_1(G')$ if the operator $\varphi^* : \mathcal{F}_1(G') \to \mathcal{F}_0(G)$ induced by this homeomorphism is the isomorphism of the spaces $\mathcal{F}_1(G')$ and $\mathcal{F}_0(G)$.

A seminormed space of functions $\mathcal{F}(\mathbf{R}^n)$ is called invariant with respect to isometries if the operator τ^* induced by the transition $x \mapsto x + \tau$ is the isometry of the space $\mathcal{F}(\mathbf{R}^n)$ onto itself.

A seminormed space of functions $\mathcal{F}(\mathbf{R}^n)$ is called invariant with respect to isometries if the operator φ induced by the isometry of $\mathbf{R}^n$ is the isometry of the space $\mathcal{F}(\mathbf{R}^n)$ onto itself.

Lemma 2.2. *If a seminormed space of functions $\mathcal{F}(\mathbf{R}^n)$ is invariant with respect to isometries, then*

$$\mathrm{Cap}_{\mathcal{F}}(\{x\}, \{y\}, \mathbf{R}^n) \qquad \textit{and} \qquad \mathrm{Cap}_{\mathcal{F}}(\{x\}, S(x, r), \mathbf{R}^n)$$

are the functions of the distance between the points $|x - y|$ or of the radius r of the sphere $S(x, r)$ respectively.

The proof is obvious since, by isometry, any pair of points x, y may be transformed to the pair $(0, \ldots, 0), (0, \ldots, 0, |x - y|)$.

Let us denote a function $\mathrm{Cap}_{\mathcal{F}}(\{x\}, \{y\}, \mathbf{R}^n)$ by $\gamma_{\mathcal{F}}(|x-y|)$ and $\mathrm{Cap}_{\mathcal{F}}(\{x\}, \mathbf{R}^n \backslash B(x, r), \mathbf{R}^n)$ by $\delta_{\mathcal{F}}(r)$. It is natural that the main interest is with the case when $\gamma_{\mathcal{F}}$ and $\delta_{\mathcal{F}}$ are distinct from zero. Let us recall that the inequality $\gamma_{\mathcal{F}}(|x-y|) > a^2 > 0$ for $|x-y| < 1$ is the necessary condition for the existence of the imbedding operator $I : \mathcal{F}(\mathbf{R}^n)|_{B(0,1)} \to C(B(0,1))$, provided $\mathcal{F}(\mathbf{R}^n)$ is normed. The norm in $\mathcal{F}(\mathbf{R}^n)|_{B(0,1)}$ may be defined as the norm in the factor space $\mathcal{F}(\mathbf{R}^n) \backslash \mathcal{F}_0$, where $\mathcal{F}_0$ is a subspace of all functions from $\mathcal{F}(\mathbf{R}^n)$ vanishing inside the unit circle.

2.3. The Invisibility Condition

The union of all points from G which may be connected with some point from

A by a segment (lying entirely in G) is a conic hull $C_G(A)$ of a set $A \subset G$ in the domain G.

Let us note that the conic hull of an open set is open.

Let us consider a seminormed space of functions $\mathcal{F}_0(G)$ in the domain $G \subset \mathbf{R}^n$. The space $\mathcal{F}_0(G)$ satisfies the invisibility condition if for any open set $V \subset G$ and for any function $u \in \mathcal{F}_0(G)$ equal to zero in some neighbourhood of the set $G \cap (C_G(V) \backslash V)$, the function

$$\tilde{u}(x) = \begin{cases} u(x), x \in C_G(V) \\ 0, x \notin C_G(V) \end{cases}$$

belongs to the class $\mathcal{F}_0(G)$. In addition, the inequality

$$\|\tilde{u}\|_{\mathcal{F}_0(G)} \leqslant K \|u\|_{\mathcal{F}_0(G)}$$

is valid. Here the constant K does not depend on the choice of the set V and on the function u.

For the spaces $B^l_{p,\theta}(G)$, $L^l_p(G)$, $W^l_p(G)$, and $C^K(G)$, the invisibility condition holds with the constant $K = 1$.

Lemma 2.3. *Let V be an open subset of a domain G, and let $F_1 \subset V$ be a closed subset of G. If the space $\mathcal{F}_0(G)$ satisfies the invisibility condition, then*

$$K \operatorname{Cap}_{\mathcal{F}_0(G)} \overline{(G \cap (C_G(V) \backslash V)}, F_1, G) \geqslant \operatorname{Cap}_{\mathcal{F}_0(G)}(G \backslash V, F_1, G),$$

where K is the constant from the invisibility condition.

Proof. Any function $u \in \mathcal{F}_0(G)$, admissible for the pair $\{G \cap (C_G(V) \backslash V), F_1\}$, vanishes in some neighbourhood of the first of sets of this pair. In this case, the function $u(x)$ (see the definition of the invisibility condition) is admissible for the pair $G \backslash V, \mathcal{F}_1$ in the domain G due to the definition of capacity and of the invisibility condition. Due to the invisibility condition,

$$\|\tilde{u}\|_{F_0(G)} \leqslant K \|u\|_{F_0(G)}.$$

The arbitrariness in the choice of the admissible function allows one to obtain the statement of the lemma from this inequality.

In a number of cases, it will be more convenient to verify the fulfillment of a stronger condition than the invisibility condition, because the former is simpler from the technical point of view.

Strong invisibility condition. *Let us consider a seminormed space of functions $\mathcal{F}_0(G)$ in a domain $G \subset \mathbf{R}^n$. Let open sets V_1, V_2, V_3 be such that $V_1 \cup V_2 \cup V_3 = G$ and $\partial V_1 \cap \partial G$ do not intersect with $\partial V_3 \cap \partial G$. Let us consider a function $u \in \mathcal{F}_0(G)$ which equals zero in V_2.*

The space $\mathcal{F}_0(G)$ satisfies the strong invisibility condition if for every such function u, the function

$$\tilde{u}(x) = \begin{cases} u(x), x \in V_1 \\ 0, x \notin V_1 \end{cases}$$

belongs to the class $\mathcal{F}_0(G)$, and the inequality

$$\|\tilde{u}\|_{\mathcal{F}_0(G)} \leqslant K\|u\|_{\mathcal{F}_0(G)}$$

is valid, where the constant $K \geqslant 1$ does not depend on the choice of the function and on the triple of the sets V_1, V_2, V_3.

For the spaces L_p, W_p^l, L_p^l, C^K, the strong invisibility condition is obviously satisfied by the constant $K = 1$.

Lemma 2.4. *Let S be a closed subset of a domain G. If the space $\mathcal{F}_0(G)$ satisfies the strong invisibility condition, then for every compact set $F_1 \subset G$, $F_1 \cap S = \emptyset$,*

$$K \operatorname{Cap}_{\mathcal{F}_0(G)}(\partial S, F_1, G) \geqslant \operatorname{Cap}_{\mathcal{F}_0(G)}(S, F_1, G),$$

where K is the constant from the invisibility condition.

Proof. If $S = \emptyset$, then the inequality is obvious. Let us consider the case $S \neq \emptyset$. Choose an arbitrary function u which is admissible for the pair $\partial S, F_1$. It vanishes in some neighbourhood $U(\partial S)$ of the set S. From the invisibility condition, it follows that

$$\|\tilde{u}\|_{\mathcal{F}_0(G)} \leqslant K\|u\|_{\mathcal{F}_0(G)}.$$

By construction, the function $\tilde{u}$ is admissible for the pair (S, F_1) in the domain G. Hence, due to arbitrariness in the choice of the function u, the statement of the lemma follows.

2.4. The Extension Theorem

Theorem 2.5. *Let G be a domain in $\mathbf{R}^n$, $\mathcal{F}_0(G)$ be a seminormed space of functions which satisfies the invisibility condition, and $\mathcal{F}_1(G)$, $\mathcal{F}_2(G)$ be seminormed spaces of functions invariant with respect to isometries.*

Suppose that

1) *$\mathcal{F}_2(\mathbf{R}^n)|_G \subset \mathcal{F}_0(G)$ and $\|u|_G\|_{\mathcal{F}_0(G)} \leqslant K\|u\|_{\mathcal{F}_2(\mathbf{R}^n)}$ for any function $u \in \mathcal{F}_2(\mathbf{R}^n)$. The constant K is independent of the choice of the function u, and*
2) *$\gamma_{\mathcal{F}_1}(t) \geqslant \alpha(C)\delta_{\mathcal{F}_2}(Ct) > 0$, where $\lim_{C\to\infty} \alpha(C) = \infty$.*

If there exists an extension operator $\theta : \mathcal{F}_0(G) \to \mathcal{F}_1(\mathbf{R}^n)$, then the domain G satisfies the arc diameter condition.

Proof. Suppose that the domain G does not satisfy the arc diameter condition. Thus, for any $C > 0$, there exists a pair of points $x_C, y_C \in G$ for which

$$\operatorname{diam} \gamma_C \geqslant C|x_C - y_C|$$

for any curve γ_C which connects the points x_C and y_C in the domain G, i.e.,

$$\operatorname{diam}(x_C, y_C, G) \geqslant C|x_C - y_C|. \tag{2.1}$$

This inequality implies that the connected component A_C of the intersection of the open ball $B_C = B(x_C, C|x_C - y_C|)$ with the domain G contains x_C and does not contain the point y_C. Due to convexity of the ball B_C, the intersection of the conic hull $C_G(A_C)$ with the connected component A_C^1 of the ball B_C is empty (the connected component contains the point y_C). Indeed, if $C_G(A_C) \cap A_C^1$ contains at least one point z_C, then by the definition of $C_G(A_C)$, the point z_C may be connected with a point from A_C. To do this, we use the segment entirely lying in the domain G and in the ball B_C. We obtain $A_C \cap A_C^1 \neq \emptyset$, which contradicts the inequality

$$\operatorname{diam}(x_C, y_C, G) \geqslant C|x_C - y_C|.$$

From the monotonicity property for variational capacity, we obtain

$$\operatorname{Cap}_{\mathcal{F}_0(G)}(\{y_C\}, \{x_C\}, G) \leqslant \operatorname{Cap}_{\mathcal{F}_0(G)}(G \backslash A_C, \{x_C\}, G).$$

From Lemma 2.3, due to the invisibility condition's being fulfilled for $\mathcal{F}_0(G)$, we obtain

$$\operatorname{Cap}_{\mathcal{F}_0(G)}(G \backslash A_C, \{x_C\}, G) \leqslant K_1 \operatorname{Cap}_{\mathcal{F}_0(G)} \overline{(G \cap (C_G(A_C) \backslash A_C)}, \{x_C\}, G),$$

where K_1 is the constant from the invisibility condition.

Due to monotonicity of variational capacity,

$$\operatorname{Cap}_{\mathcal{F}_0(G)} \overline{(G \cap (C_G(A_C) \backslash A_C)}, \{x_C\}, G) \leqslant \operatorname{Cap}_{\mathcal{F}_0(G)}(G \backslash B_C, \{x_C\}, G).$$

Finally, we obtain

$$\operatorname{Cap}_{\mathcal{F}_0(G)}(\{y_C\}, \{x_C\}, G) \leqslant K_1 \operatorname{Cap}_{\mathcal{F}_0(G)}(G \backslash B_C, \{x_C\}, G).$$

Applying Condition 1 of the theorem to the right-hand side of the latter inequality, we obtain

$$\operatorname{Cap}_{\mathcal{F}_0}(G \backslash B_C, \{x_C\}, G) \leqslant K_1 K \operatorname{Cap}_{\mathcal{F}_2}(G \backslash B_C, \{x_C\}, \mathbf{R}^n).$$

Due to monotonicity of capacity,

$$\operatorname{Cap}_{\mathcal{F}_2}(G \backslash B_C, \{x_C\}, \mathbf{R}^n) \leqslant \operatorname{Cap}_{\mathcal{F}_2}(\mathbf{R}^n \backslash B_C, \{x_C\}, \mathbf{R}^n),$$

that is,

$$\operatorname{Cap}_{\mathcal{F}_0(G)}(\{y_C\}, \{x_C\}, G) \leqslant K_1 K \delta_{\mathcal{F}_2}(C|x_C - y_C|).$$

From the existence of the extension operator $\theta : \mathcal{F}_0(G) \to \mathcal{F}_1(\mathbf{R}^n)$ and from Theorem 2.1, it follows that

$$\mathrm{Cap}_{\mathcal{F}_1}(\{y_C\}, \{x_C\}, \mathbf{R}^n) \leqslant \|\theta\| \, \mathrm{Cap}_{\mathcal{F}_0(G)}(\{y_C\}, \{x_C\}, G).$$

Recalling the definition of the function $\gamma_{\mathcal{F}_1}$ for the space $\mathcal{F}_1$, which is invariant relative to isometries, and comparing the last two inequalities, we have

$$\alpha(C) = \frac{\gamma_{\mathcal{F}_1}(|x_C - y_C|)}{\delta_{\mathcal{F}_2}(C|x_c - y_c|)} \leqslant \|\theta\| K_1 K \tag{2.2}$$

for all $C > 0$.

This contradicts the second condition of the theorem. Thus, our domain satisfies the arc diameter condition.

This completes the proof.

Remark 1. As it follows from the necessary conditions of imbedding [34], for the normed space $\mathcal{F}_2$ the condition $\delta_{\mathcal{F}_2}(t) > 0$ may be substituted by a stronger requirement of the existence of bounded imbedding operators $I_r : \mathcal{F}_2(B(0,r)) \to C(B(0,r))$ on every ball $B(0,r)$.

The theorem acquires a simpler form if $\mathcal{F}_1(\mathbf{R}^n) = \mathcal{F}_2(\mathbf{R}^n)$. In this case, Condition 1 follows from the existence of the extension operator.

Remark 2. Under the conditions of Theorem 2.5, the domain G satisfies the arc diameter condition

$$\mathrm{diam}(x, y, G) \leqslant C|x - y|$$

for any two points $x, y, \in G$. Here the constant C does not exceed the lowest upper bound of C^*, for which

$$\alpha(C^*) \leqslant K_1 K \|\theta\|.$$

Let us recall that K_1 is the constant from the invisibility condition and that K is the constant from Condition 1 of Theorem 2.5.

The proof obviously follows from inequalities (2.1) and (2.2).

Remark 3. Under the conditions of Theorem 2.5, the domain G is locally connected at every boundary point of the set $\partial G \backslash \partial G^*$, and it has the topological dimension not exceeding $n - 2$; i.e., it divides no ball into connected components. If $n = 2$, then no two connected components of the open set $G^* = \mathrm{Int}(\mathbf{R}^n \backslash G)$ have common boundary points.

The proof follows from the properties of domains which satisfy the arc diameter condition.

Remark 4. Let us weaken Condition 2 of Theorem 2.5. Suppose that for the function $\alpha(C,t) = \gamma_{\mathcal{F}_1}(t)\backslash\delta_{\mathcal{F}_2}(Ct)$, the relation

$$\lim_{C\to\infty} \inf_{t<t_0} \alpha(C,t) = \infty$$

is valid for all $t_0 > 0$.

Then the statement of the theorem is valid as follows: for any two points $x, y \in G$ for $|x-y| < t_0$, the inequality

$$\operatorname{diam}(x,y,G) \leqslant C|x-y|$$

holds, where the constant C does not exceed the lowest upper bound of $C^*_{t_0}$, for which

$$\inf_{t<t_0} \alpha(C^*,t) \leqslant K_1, K\|\theta\|.$$

The proof follows from inequalities (2.1) and (2.2), from the proof of Theorem 2.5, and from Remark 2.

In Theorem 2.5 it was necessary that $\gamma_{\mathcal{F}_1}(t) \geqslant \alpha(C)\delta_{\mathcal{F}_2}(Ct)$ for all t, and the function $\alpha(C) \to \infty$ for $C \to \infty$. If the inequality only holds for t close to zero, not for all t, then Theorem 2.5 remains valid for bounded domains. Let us give this variant of the main theorem.

Theorem 2.6. *Let G be a bounded domain in $\mathbf{R}^n$, $\mathcal{F}_0(G)$ be a seminormed space of functions which satisfy the invisibility condition, and $\mathcal{F}_1(\mathbf{R}^n), \mathcal{F}_2(\mathbf{R}^n)$ be seminormed spaces of functions, invariant with respect to isometries.*

Suppose that:

1) $\mathcal{F}_2(\mathbf{R}^n)|_G \subset \mathcal{F}_0(G)$, and $\|u|_G\|_{\mathcal{F}_0(G)} \leqslant K\|u\|_{\mathcal{F}_2(\mathbf{R}^n)}$;

2) there exists $t_0 > 0$ such that

$$\gamma_{\mathcal{F}_1}(t) \geqslant \alpha(C)\delta_{\mathcal{F}_2}(Ct) > 0$$

for all $t < t_0$. Here $\alpha(C) \to \infty$ for $C \to \infty$.

If there exists an extension operator $\theta : \mathcal{F}_0(G) \to \mathcal{F}_1(G)$, then the domain G satisfies the condition with the arc diameter.

Proof. Suppose that the domain G does not satisfy the arc diameter condition. Thus, for any $C > 0$, there exists a pair of points $x_C, y_C \in G$ for which $\operatorname{diam}\gamma_C$ of any curve γ_C connecting x_C and y_C exceeds $C|x_C - y_C|$. Due to boundedness of the domain G, one may consider the sequences $\{x_C\}$, $\{y_C\}$ to converge in $\mathbf{R}^n$. Let us prove that $\lim_{C\to\infty} x_C = \lim_{C\to\infty} y_C$. If the limits are not equal, then $x_0 = \lim_{C\to\infty} x_C$ does not coincide with the point $y_0 = \lim y_C$. Then we have a chain of obvious inequalities

$$\operatorname{diam}(x_0,y_0,G) \leqslant \operatorname{diam} G \leqslant \frac{\operatorname{diam} G}{|x_0-y_0|}|x_0-y_0|;$$

i.e., for the points x_0, y_0 in the domain G, the arc diameter condition holds for the constant $C = \alpha(C)/|x_0 - y_0|$. Then for the points x_C and y_C, for sufficiently large C (due to convergence of $x_C \to x_0$ and $y_C \to y_0$) the inequality

$$\operatorname{diam}(x_C, y_C, G) \leqslant \frac{2\operatorname{diam}(G)}{|x_0 - y_0|}|x_C - y_C|$$

is valid. This contradicts the construction of sequences $\{x_C\}$, $\{y_C\}$. We have proved that

$$\lim_{C\to\infty} |x_C - y_C| = 0.$$

Let us choose C_0 to be so large that $|x_C - y_C| < t_0$ for all $C > C_0$. From here the proof of Theorem 2.6 exactly coincides with that of Theorem 2.5.

2.5. Verification of the Conditions of the Theorem for the Spaces $L_p^l(G)$, $W_p^l(G)$.

We consider the case $\mathcal{F}_1(\mathbf{R}^n) = L_p^l(\mathbf{R}^n)$, $\mathcal{F}_2(\mathbf{R}^n) = L_p^l(\mathbf{R}^n)$, $\mathcal{F}_0(G) = L_p^l(G)$. All the conditions of Theorem 2.5, except for Condition 2, are obviously fulfilled. Moreover, for the space $L_p^l(G)$, the strong invisibility condition holds.

For $lp > n$, Condition 2 of Theorem 2.5 is satisfied as follows.

Proposition 2.7. *For the spaces $L_p^l(\mathbf{R}^n)$ for $lp > n$, the function $\delta_{L_p^l}(t)$ is positive for all $t > 0$, and the relation*

$$\gamma_{L_p^l}(t) = \frac{\gamma(1)}{\delta(1)}\delta_{L_p^l}(Ct)C^{l-\frac{n}{p}}$$

is valid, where C is the positive constant which only depends on n, l, and p.

Proof. Let us consider the $[l,p]$-capacities

$$\gamma(a) = C_p^l(\{0\}, \{x_a\}, \mathbf{R}^n)(x_a = (a, 0, \dots, 0))$$

and

$$\delta(a) = C_p^l(\{0\}, S_a, \mathbf{R}^n)(S_a = S(0,a)).$$

Let us prove that the relation $\gamma(a)/\delta(a)$ does not depend on the choice of the number a. First recall that, as was proved in Proposition 6.1 of Chapter 3, $\gamma(a) > 0$ and $\delta(a) > 0$ for all a. In the transformation of the similarity $\tau_a : \mathbf{R}^n \to \mathbf{R}^n$, $\tau_a(x) = ax$, the seminorm of any function $u \in L_p^l(\mathbf{R}^n)$ is transformed according to the law

$$a^{\frac{n}{p}-l}\|u \circ \tau_a\|_{L_p^l(\mathbf{R}^n)} = \|u\|_{L_p^l(\mathbf{R}^n)}.$$

Consequently, for $[l,p]$-capacity the relations

$$\gamma(a) = a^{\frac{n}{p}-l}\gamma(1), \qquad \delta(a) = a^{\frac{n}{p}-l}\delta(1)$$

are valid, i.e., $\gamma(a)/\delta(a) = \gamma(1)/\delta(1) = \lambda_0$, where the number λ_0 only depends on n, l, p.

It remains to perform simple calculations:

$$\frac{\gamma(t)}{\delta(Ct)} = \frac{\gamma(t)}{\lambda_0\gamma(Ct)} = \frac{t^{\frac{n}{p}-l}}{\lambda_0(Ct)^{\frac{n}{p}-l}} = \frac{1}{\lambda_0}C^{l-\frac{n}{p}},$$

from which follows the inequality formulated in the proposition. (The positiveness of the function $\delta_{L^l_p}(t)$ for $lp > n$ may also be obtained from the theorem of imbedding of $W^l_p(\mathbf{R}^n)$ into $C(\mathbf{R}^n)$ in the variant studied by Maz'ya [48].)

From Proposition 2.7 and Theorem 2.5, there follows

Theorem 2.8. *Let G be a domain in $\mathbf{R}^n$. If there exists an extension operator*

$$\theta : L^l_p(G) \to L^l_p(\mathbf{R}^n)$$

for $lp > n$, then the domain G satisfies the arc diameter condition.

Remark. Under the conditions of Theorem 2.8, the domain G satisfies the arc diameter condition

$$\operatorname{diam}(x, y, G) \leqslant C|x - y|$$

for any two points $x, y \in G$. In this case, the constant C satisfies the inequality

$$C \leqslant \left(\|\theta\| \frac{\gamma(1)}{\delta(1)}\right)^{\frac{l}{lp-n}}$$

where $\delta(1) = C^l_p(\{0\}, S(0,1), \mathbf{R}^n)$, $\gamma(1) = C^l_p(\{0\}, \{1\}, \mathbf{R}^n)$.

The proof follows from Proposition 2.7 and from Remark 2 of Theorem 2.5. Let us recall that for L^l_p, the constant in the invisibility condition was equal to 1; this explains its absence in the estimate.

Recall that the norm in the space $W^l_p(G)$ is a sum of norms:

$$\|f\|_{W^l_p(G)} = \|f\|_{L_p(G)} + \|f\|_{L^l_p(G)}.$$

Therefore, for every pair of the sets$F_0, F_1 \subset G$ closed with respect to G, the inequality

$$C^l_p(F_0, F_1, G) \leqslant \operatorname{Cap}_{W^l_p(G)}(F_0, F_1, G) \tag{2.3}$$

is valid.

Proposition 2.9. *The inequality*

$$\mathrm{Cap}_{W_p^l(\mathbf{R}^n)}\left(S(0,R),\overline{B(0,r)},\mathbf{R}^n\right) \leqslant [1+PR^l]_{C_p^l}\left(S(0,R),\overline{B(0,r)},\mathbf{R}^n\right)$$

is valid, where the constant p does not depend on r.

Proof. Since for the spaces $W_p^l(\mathbf{R}^n)$ and $L_p^l(\mathbf{R}^n)$, the strong invisibility condition with the constant $K = 1$ holds, then, due to Lemma 2.4, each of the capacities equals the same capacity for the pair $\mathbf{R}^n \setminus B(0,R), \overline{B(0,r)}$. Therefore, any of the admissible functions for $[l,p]$-capacity may be considered to be equal to zero outside of the circle $B(0,R)$. The Sobolev inequality (Chapter 2, Theorem 4.3), applied to an arbitrary admissible function f yields the estimate

$$\|f\|_{L_p(\mathbf{R}^n)} \leqslant PR^l \|f\|_{L_p^l(\mathbf{R}^n)}.$$

Therefore,

$$\begin{aligned}\mathrm{Cap}_{W_p^l(\mathbf{R}^n)}(S(0,R),\overline{B(0,r)},\mathbf{R}^n) &\leqslant \|f\|_{W_p^l(\mathbf{R}^n)} \\ &= \|f\|_{L_p(\mathbf{R}^n)} + \|f\|_{L_p^l(\mathbf{R}^n)} \leqslant [1+PR^l]\,\|f\|_{L_p^l(\mathbf{R}^n)}.\end{aligned}$$

We make use of an obvious fact by stating that every $[l,p]$- admissible function for the pair of sets under consideration, is W_p^l-admissible. Due to the fact that the admissible function was chosen arbitrarily, the proposition is proved.

Proposition 2.10. *For the spaces $W_p^l(\mathbf{R}^n)$ for $lp > n$ the function $\delta_{W_p^l}(t)$ is positive for all $t > 0$, and for any $R_0 > 0$, the relation*

$$\gamma_{W_p^l}(t) \geqslant A_1 \delta_{L_p^l}(Ct)C^{\frac{n}{p}-l}$$

holds for all $t < R_0$. Here $A_1 = (\delta_{L_p^l}(1))/(\gamma_{L_p^l}(1)(1+PR_0^l))$, where P is the constant from the Sobolev inequality that only depends on n, l, p.

Proof. Due to inequality (2.3),

$$\gamma_{W_p^l}(t) \geqslant \gamma_{L_p^l}(t).$$

From Propositions 2.7 and 2.9, we obtain for

$$\gamma_{L_p^l}(t) \leqslant A\delta_{L_p^l}(Ct)C^{\frac{n}{p}-l} \geqslant \left(\frac{A}{1+PR_0^l}\right)\delta_{W_p^l}(Ct)C^{\frac{n}{p}-l},$$

where $A = \delta_{L_p^l}(1)/\gamma_{L_p^l}(1)$

This completes the proof.

From Proposition 2.10 and Theorem 2.6, there follows

Theorem 2.11. *Let G be a bounded domain in $\mathbf{R}^n$. If there exists an extension operator $\theta : W_p^l(G) \to W_p^l(\mathbf{R}^n)$ for $lp > n$, then the domain G satisfies the arc diameter condition.*

From Proposition 2.10 and Remark 4 of Theorem 2.5 for unbounded domains, we obtain

Theorem 2.12. *Let G be a domain in $\mathbf{R}^n$ for which there exists a bounded extension operator*

$$\theta : W_p^l(G) \to W_p^l(\mathbf{R}^n)$$

for $lp > n$.

Then for any two points $x, y \in G$, $|x - y| < R_0$, the inequality

$$\operatorname{diam}(x, y, G) \leqslant A(R_0)|x - y|$$

is valid, where the constant $A(R_0)$ only depends on n, l, p, R_0, and is the monotone increasing function of R_0.

Proof. Except for Condition 2, all the conditions of Theorem 2.5 are satisfied for the domain G. Let us consider the function

$$\alpha(C, t) = \gamma_{W_p^l}(t)/\delta_{W_p^l}(Ct).$$

From Proposition 2.10, it follows that

$$\alpha(C, t) = \frac{\delta_{L_p^l}(1)}{\gamma_{L_p^l}(1)}(1 + Pt^l)^{-1}C^{l - \frac{n}{p}},$$

where the constant P only depends on n, l, p. The function $\alpha(C, t)$ monotonically decreases by t. Therefore,

$$\inf_{t < R_0} \alpha(C, t) = \alpha(C, R_0).$$

From Remark 4 of Theorem 2.5, it follows that for all $x, y \in G$, $|x - y| < R_0$

$$\operatorname{diam}(x, y, G) \leqslant A(R_0)|x - y|,$$

where A may be taken to be equal to

$$A(R_0) = \left(\frac{\gamma_{L_p^l}(1)}{\delta_{L_p^l}(1)} \|\theta\| (1 + Pr_0^l) \right)^{\frac{p}{lp - n}}$$

For large R_0 it is useful to recall that

$$A(R_0) \leqslant A_1 R_0^{\frac{lp}{lp - n}}.$$

Corollary. *Under the conditions of Theorem 2.12, the domain G is locally connected at every boundary point of the set $\partial G \backslash \delta G^*$, and has topological dimension not exceeding $n-2$. For $n = 2$, the connected components of the set $G^* = \mathrm{Int}(\mathbf{R}^n \backslash G)$ have no common boundary points.*

It follows from the fact that the domain G locally satisfies the arc diameter condition (for the properties of this class of domains, see §1).

Remark. The results similar to Theorems 2.8 and 2.11 are also valid for the spaces $B^l_{p,\theta}$, but we will touch upon them later since the necessary estimates for capacity have not been obtained yet.

Let us formulate several obvious corollaries of Theorems 2.7 and 2.11.

Theorem 2.13. *Let a domain $G \subset \mathbf{R}^n$ be such that the set $G^* = \mathrm{Int}(\mathbf{R}^n \backslash G)$ is also a domain.*

Suppose that there exist two extension operators

$$\theta_1 : L^{l_1}_{p_1}(G) \to L^{l_1}_{p_1}(\mathbf{R}^n), \qquad n < l_1 p_1,$$
$$\theta_2 : L^{l_2}_{p_2}(G^*) \to L^{l_2}_{p_2}(\mathbf{R}^n), \qquad n < l_2 p_2.$$

Then the domain G satisfies the bilateral arc diameter condition.

Theorem 2.14. *Let a bounded domain $G \subset \mathbf{R}^n$ be such that the set $G^* =$ $\mathrm{Int}(\mathbf{R}^n \backslash G)$ is also a domain. Suppose that there exist two extension operators*

$$\theta_1 : W^{l_1}_{p_1}(G) \to W^{l_1}_{p_1}(\mathbf{R}^n), \qquad n < l_1 p_1,$$
$$\theta_2 : W^{l_2}_{p_2}(G^*) \to W^{l_2}_{p_2}(\mathbf{R}^n), \qquad n < l_2 p_2.$$

Then the domain G satisfies the bilateral arc diameter condition.

Proof. Let us consider the ball $B(0, 2\,\mathrm{diam}\,G)$. From the existence of the extension operator $\theta_2 : W^{l_2}_{p_2}(G^*) \to W^{l_2}_{p_2}(\mathbf{R}^n)$ there easily follows the existence of the extension operator

$$\theta_3 : W^{l_2}_{p_2}(G^* \cap B(0, 2\,\mathrm{diam}\,G)) \to W^{l_2}_{p_2}(\mathbf{R}^n).$$

Now it only remains to apply Theorem 2.5.

Let us remind that a simply-connected domain G satisfying the bilateral arc diameter condition satisfies the Ahlfors condition.

Taking this fact into account, let us formulate the two-dimensional variant of Theorems 2.11 and 2.12.

Theorem 2.11′. *Let a simply-connected domain G be such that G^* is a domain.*

Suppose that there exist two extension operators

$$\theta_1 : L^{l_1}_{p_1}(G) \to L^{l_1}_{p_1}(\mathbf{R}^n), \qquad 2 < l_1 p_1,$$
$$\theta_2 L^{l_2}_{p_2}(G^*) \to L^{l_2}_{p_2}(\mathbf{R}^n), \qquad 2 < l_2 p_2.$$

Then the boundary ∂G of the domain G is a Jordan curve and satisfies the Ahlfors condition.

Theorem 2.12′. *Theorem 2.12′ is formulated in a similar way under the additional requirement of boundedness of the domain G and with the replacement of L_p^l for W_p^l in the statement of the theorem.*

Remark 1. As it has been proved by Ahlfors, every Jordan curve satisfying the Ahlfors condition is an image of circle at quasiconformal mapping of a plane onto itself.

Remark 2. The conditions formulated in Theorems 2.11 and 2.12 are necessary and sufficient for the extension operators θ_1, θ_2 to exist in the case of the spaces $L_p^l (l = 1)$ and W_p^l. For details, see §5 of this chapter.

§3 Necessary Extension Conditions for Sobolev Spaces

Necessary conditions of extension for the spaces $L_p^l(G)$ $(lp > n)$ in arbitrary domains and for the spaces $W_p^l(G)$ $(lp > n)$ in bounded domains were obtained in the previous subsection as corollaries of the theorem about extension for semi-normed spaces. In this subsection we investigate the case $lp = n$ for domains from $\mathbf{R}^n$ and $1 \leqslant lp \leqslant 2$ for plane domains. Additionally, some refinements are possible in the plane case for the spaces $W_p^l(G)$.

3.1. Necessary Extension Conditions for L_p^l, W_p^l at $lp = n$

Theorem 3.1. *Let G be a domain in $\mathbf{R}^n$. If there exists an extension operator*

$$\theta : L_p^l(G) \to L_p^l(\mathbf{R}^n, \qquad lp = n,$$

then the domain G satisfies the condition with the arc diameter.

Let us prove Lemma first.

Lemma 3.2. *If for the domain G there exists an extension operator*

$$\theta : L_p^l(G) \to L_p^l(\mathbf{R}^n),$$

then there exists an extension operator

$$\theta_k : L_p^l(\varphi(G)) \to L_p^l(\mathbf{R}^n)$$

for the domain $\varphi(G)$ obtained from G by transforming the similarity $\varphi(x) = kx$. Here $\|\theta_k\| = \|\theta\|$.

Proof. The construction of the operator θ_k is clear from the diagram

$$\begin{array}{ccc} L_p^l(G) & \xrightarrow{\theta} & L_p^l(\mathbf{R}^n) \\ \uparrow \varphi^* & & \downarrow (\varphi^{-1})^* = (\varphi^*)^{-1} \\ L_p^l(\varphi(G)) & \xrightarrow{\theta_k} & L_p^l(\mathbf{R}^n) \end{array}$$

Recalling the transformation of the semi-norm $\|\cdot\|_{L^l_p}$ for similarities, we obtain that $\|\varphi^*\| = K^{l-n/p}$, $\|(\varphi^{-1})^*\| = K^{n/p-l}$. Consequently, $\|\theta_k\| \leqslant \|\varphi^*\|\,\|\theta\|\,\|(\varphi^{-1})^*\| \leqslant \|\theta\|$. Since $\theta = \varphi^*\theta_k(\varphi^*)^{-1}$, $\|\theta\| \leqslant \|\theta_k\|$. The lemma is proved.

Proof of the theorem. Suppose that the domain G does not satisfy the arc diameter condition. Thus, for every $C > 0$, there exists a pair of points $x_C, y_C \in G$ for which $\operatorname{diam}(\gamma_C)$ of any curve γ_C connecting the points x_C and y_C is larger than $C|x_C - y_C|$. That is, the connected component A_C of the intersection of the ball $B\left(\frac{x_C+y_C}{2}, C|x_C - y_C|\right)$ with G contains the point x_C and does not contain the point y_C. Let us perform the transformation of similarity φ_C of the domain G. The transformation φ_C has the similarity coefficient $1/C|x_C - y_C|$. Recall that according to Proposition 6.2. of Chapter 3, the $[l,p]$-capacity is invariant for similarities for $lp = n$.

Let us choose in the set A^0_C a curve $\gamma_{0,C}$ connecting the point x_C with the sphere $S_C = S\left(\frac{x_C+y_C}{2}, C|x_C - y_C|/2\right)$. Similarly, in the connected component A^1_C of the set $B\left(\frac{x_C+y_C}{2}, \frac{C|x_C-y_C|}{2}\right)$ containing the point y_C, let us choose the curve $\gamma_{1,C}$ which connects y_C with the sphere S_C.

From the existence of the extension operator, due to Theorem 2.1, we obtain

$$C^l_p(\gamma_{0,C}, \gamma_{1,C}, \mathbf{R}^n) \leqslant \|\theta\| C^l_p(\gamma_{0,C}, \gamma_{1,C}, G). \tag{3.1}$$

According to the lemma on similarity (Lemma 3.2),

$$C^l_p(\varphi_C(\gamma_{0,C}), \varphi_C(\gamma_{1,C}), \mathbf{R}^n) \leqslant \|\theta\| C^l_p(\varphi_C(\gamma_{0,C}), \varphi_C(\gamma_{1,C}), \varphi(G)). \tag{3.2}$$

The curves $\varphi_C(\gamma_{0,C})$, $\varphi_C(\gamma_{1,C})$ by construction connect the spheres

$$S\left(\varphi\left(\frac{x_c+y_c}{2}\right), 1/C\right) \quad \text{and} \quad S\left(\varphi\left(\frac{x_C+y_C}{2}\right), 1/2\right).$$

Consequently, by Proposition 6.8 of Chapter 3,

$$\lim_{C\to\infty} C^l_p(\varphi_C(\gamma_{0,C}), \varphi_C(\gamma_{1,C}), \mathbf{R}^n) = \infty; \tag{3.3}$$

i.e., in inequality (3.2) the left-hand side tends to infinity if $C \to \infty$.

On the other hand, since the sets A^0_C and A^1_C do not intersect, due to monotonicity of capacity,

$$\begin{aligned} &C^l_p(\varphi(\gamma_{0,C}), \varphi(\gamma_{1,C}), \varphi(G)) \\ &\quad\leqslant C^l_p\left(\varphi(\gamma_{0,C}), \varphi(\overline{A^1_C}), \varphi(G)\right) \\ &\quad\leqslant C^l_p\left(\varphi(\overline{A^0_C}), \varphi(\overline{A^1_C}), \varphi(G)\right). \end{aligned} \tag{3.4}$$

Recall that the connected component B_C^1 of the ball $B((x_C+y_C)/2, C|x_C-y_C|)$, containing the point y_C does not contain the point x_C. Consequently, $A_C^0 \cap B_C^1 = \emptyset$. Due to monotonicity of capacity,

$$C_p^l(\varphi(\overline{A_C^0}), \varphi(\overline{A_C^1}), \varphi(G)) \leqslant C_p^l(\varphi(\overline{A_C^0}), \varphi(\overline{B_C^1}), \varphi(G)). \tag{3.5}$$

From the fact that $[l,p]$-capacity is symmetric with respect to order in the pair (F_0, F_1), we obtain

$$C_p^l(\varphi(\overline{A_C^0}), \varphi(\overline{A_C^1}), \varphi(G)) = C_p^l(\varphi(\overline{A_C^1}), \varphi(\overline{A_C^0}), \varphi(G)),$$
$$C_p^l(\varphi(\overline{A_C^0}), \varphi(\overline{A_C^1}), \varphi(G)) = C_p^l(\varphi(\overline{B_C^1}), \varphi(\overline{A_C^0}), \varphi(G)),$$

i.e.,

$$C_p^l(\varphi(\overline{A_C^1}), \varphi(\overline{A_C^0}), \varphi(G)) \leqslant C_p^l(\varphi(\overline{B_C^1}), \varphi(\overline{A_C^0}), \varphi(G)). \tag{3.6}$$

From the strong invisibility condition, it follows that

$$C_p^l(\varphi(\overline{B_C^1}), \varphi(\overline{A_C^0}), \varphi(G)) \leqslant C_p^l(\partial\varphi(\overline{B_C^1}), \varphi(\overline{A_C^0}), \varphi(G)).$$

Finally, from (3.4)–(3.6) and from the latter inequality, we obtain

$$C_p^l(\varphi(\gamma_{0,C}), \varphi(\gamma_{1,C}), \varphi(G)) \leqslant C_p^l(\partial\varphi(\overline{B_C^1}), \varphi(\overline{A_C^0}), \varphi(G)). \tag{3.7}$$

Recall that $\partial\varphi(\overline{B_C^1}) \subset S((x_C+y_C)/2, 1)$, and

$$\varphi(\overline{A_C^0}) \subset B\left(\frac{x_C+y_C}{2}, \frac{1}{2}\right).$$

Therefore, from the monotonicity of capacity with respect to the definition domain, we obtain

$$\begin{aligned} C_p^l(\partial\varphi(\overline{B_C^1}), \varphi(\overline{A_C^0}), \varphi(G)) &\leqslant C_p^l(\partial\varphi(\overline{B_C^1}), \varphi(\overline{A_C^0}), \mathbf{R}^n) \\ &\leqslant C_p^l(S(0,1), B(0,1/2), \mathbf{R}^n) \\ &= C_p^l(S(0,1/2), S(0,1), \mathbf{R}^n). \end{aligned} \tag{3.8}$$

According to Proposition 6.2 of Chapter 3,

$$P = C_p^l(S(0,1/2), S(0,1), \mathbf{R}^n) < \infty.$$

From (3.6), (3.7) and from the latter inequality, we obtain

$$C_p^l(\varphi(\gamma_{0,C}), \varphi(\gamma_{1,C}), \varphi(G)) \leqslant P,$$

where the constant P is independent of C. Consequently, the right-hand side of inequality (3.2) is bounded while the left-hand side tends to infinity. The obtained contradiction proves the theorem.

Theorem 3.3. *Let G be a bounded domain in $\mathbf{R}^n$. If there exists an extension operator*

$$\theta : W_p^l(G) \to W_p^l(\mathbf{R}^n), \qquad lp = n,$$

then the domain G satisfies the condition with the arc diameter.

Proof. The beginning of the proof is the same as in the previous theorem. The same arguments as in the proof of Theorem 2.6 allow one to set $|x_C - y_C| \to 0$.

From the inequality

$$C_p^l(\gamma_{0,C}, \gamma_{1,C}, \mathbf{R}^n) \leqslant \mathrm{Cap}_{W_p^l}(\gamma_{0,C}, \gamma_{1,C}, \mathbf{R}^n)$$

and from equality (3.3), it follows that

$$\lim_{C \to \infty} \mathrm{Cap}_{W_p^l}(\gamma_{0,C}, \gamma_{1,C}\mathbf{R}^n) = \infty.$$

Following the proof of the previous theorem, let us consider the sets A_C^0, A_C^1, B_C^1. As in the proof of Theorem 3.1, from the monotonicity of W_p^l-capacity and from the invisibility condition, we obtain

$$\begin{aligned}
\mathrm{Cap}_{W_p^l}(\gamma_{0,C}, \gamma_{1,C}, G) &\leqslant \mathrm{Cap}_{W_p^l}(\overline{A_C^1}, \overline{A_C^0}, G) \\
&\leqslant \mathrm{Cap}_{W_p^l(G)}(\overline{B_C^1}, \overline{A_C^0}, G) \\
&\leqslant \mathrm{Cap}_{W_p^l}(\partial \overline{B_C^1}, \overline{A_C^0}, G) \\
&\leqslant \mathrm{Cap}_{W_p^l}(S(0, C|x_C - y_C|), S(0, \tfrac{C}{2}|x_C - y_C|), \mathbf{R}^n).
\end{aligned}$$

Since every function admissible for this pair of spheres may be considered to vanish outside the ball $S(0, C|x_C - y_C|)$, and due to the obvious inequality $C|x_C - y_C| < \operatorname{diam} G$, from the Sobolev inequality we have

$$\begin{aligned}
&\mathrm{Cap}_{W_p^l}(S(0, C|x_C - y_C|), S(0, \tfrac{C}{2}|x_C - y_C|), \mathbf{R}^n) \\
&\qquad \leqslant (1 + A(\operatorname{diam} G)^l) C_p^l(S(0, C|x_C - y_C|), S(0, \tfrac{C}{2}|x_C - y_C|), \mathbf{R}^n),
\end{aligned}$$

where the constant A only depends on n, l, p. Due to invariance of $[l,p]$-capacity for similarities ($lp = n$), we obtain from the two previous inequalities,

$$\mathrm{Cap}_{W_p^l}(\gamma_{0,C}, \gamma_{1,C}, G) \leqslant A_1 C_p^l(S(0, \tfrac{1}{2}), S(0,1), \mathbf{R}^n).$$

The end of the proof is the same as that in the previous theorem after inequality (3.8).

3.2. Necessary Conditions for L^l_p, W^l_p at $1 \leqslant lp \leqslant 2$ in Plane Domains

Theorem 3.4. *Let G be a domain in $\mathbf{R}^n$. Suppose that there exists an extension operator*

$$\theta : L^l_p(G) \to L^l_p(\mathbf{R}^n), \qquad 1 \leqslant lp \leqslant 2.$$

Then every connected component of the set $G^* = \operatorname{Int}(\mathbf{R}^2 \backslash G)$ satisfies the arc diameter condition.

Proof. Suppose that some connected component W of the open set G^* does not satisfy the condition with the arc diameter. Consequently, for every C there exists a pair of points $x_C, y_C \in W$ such that every curve γ_C connecting x_C and y_C has $\operatorname{diam}\gamma_C \geqslant C|x_C - y_C|$. Let us connect the points x_C and y_C by a segment. Denote the middle of this segment by z_C. Let us consider three circles:

$$\begin{aligned} B^0_C &= B(z_C, |x_C - y_C|), \\ B^1_C &= B(z_C, \tfrac{C}{2}|x_C - y_C|), \\ B^2_C &= B(z_C, C|x_C - y_C|). \end{aligned}$$

By construction, the open set $G \backslash B^0_C$ is not connected. Among its connection components there exist at least two components V^0_C, V^1_C intersecting each of the circles $S(z_C, t)$ for $|x_C - y_C| \leqslant t \leqslant C|x_C - y_C|$. Let us choose an arbitrary curve $\gamma_{0,C}$ connecting the circles $S^1_C = \partial B^1_C$, $S^2_C = \partial B^2_C$ in V^0_C, and the curve $\gamma_{1,C}$ connecting the circles S^1_C and S^2_C in V^1_C.

Due to the existence of the extension operator θ, it follows from Theorem 2.1 that

$$C^l_p(\gamma_{0,C}, \gamma_{1,C}, \mathbf{R}^2) \leqslant \|\theta\| C^l_p(\gamma_{0,C}, \gamma_{1,C}, G).$$

From Proposition 6.6 of Chapter 3, it follows (for $lp < 2$) that

$$\begin{aligned} C^l_p(\gamma_{0,C}, \gamma_{1,C}, \mathbf{R}^2) &\geqslant A(C^{2-lp}|x_C - y_C|^{2-lp} - \left(\tfrac{C}{2}\right)|x_C - y_C|^{2-lp})^{1/p} \\ &= A_1 C^{\frac{2-lp}{p}} |x_C - y_C|^{\frac{2-lp}{p}}, \end{aligned} \tag{3.9}$$

where the constants $A > 0$ and $A_1 > 0$ only depend on l and p.

For $lp = 2$, it follows from the same proposition that

$$C^l_p(\gamma_{0,C}, \gamma_{1,C}, \mathbf{R}^2) \geqslant A_2 > 0, \tag{3.10}$$

where the constant A_2 only depends on l and p.

Let us estimate the capacity of the pair $\gamma_{0,C}, \gamma_{1,C}$ in the domain G. Due to monotonicity of $[l,p]$-capacity,

$$C^l_p(\gamma_{0,C}, \gamma_{1,C}, G) \leqslant C^l_p(\overline{\gamma^1_C \cup B^0_C}) \cap G, G).$$

Due to the fact that $\partial(\overline{(V_C^1 \cup B_C^0)} \cap G) \subset S_C^0 = \partial B_C^0$, due to the strong invisibility condition and monotonicity of capacity,

$$\begin{aligned} C_p^l(\gamma_{0,C}\overline{(V_C^1 \cup B_C^0)} \cap G, G) &\leqslant C_p^l(\gamma_{0,C}, S_C^0 \cap G, G) \\ &\leqslant C_p^l(S_C^1 \cap G, S_C^0 \cap G, G). \end{aligned} \tag{3.11}$$

By using the monotonicity of capacity with respect to the domain, we obtain

$$C_p^l(S_C^1 \cap G, S_C^0 \cap G, G) \leqslant C_p^l(S_C^1 \cap G, S_C^0 \cap G, \mathbf{R}^2). \tag{3.12}$$

Due to monotonicity of capacity,

$$C_p^l(S_C^1 \cap G, S_C^0 \cap G, \mathbf{R}^2) \leqslant C_p^l(S(0, C|x_C - y_C|/2), S(0, |x_C - y_C|), \mathbf{R}^2). \tag{3.13}$$

From (3.10)–(3.13), we finally obtain

$$C_p^l(\gamma_{0,C}, \gamma_{1,C}, G) \leqslant C_p^l(S(0, C|x_C - y_C|/2, S(0, |x_C - y_C|), \mathbf{R}^2).$$

According to Proposition 6.3 of Chapter 3, for $lp < 2$,

$$C_p^l(S(0, C|x_C - y_C|/2), S(0, |x_C - y_C|/2), \mathbf{R}^2) \sim |x_C - y_C|^{(2-lp)/p}.$$

Finally, we have

$$C_p^l(\gamma_{0,C}, \gamma_{1,C}, G) \leqslant A_3 |x_C - y_C|^{(2-lp)/p},$$

where the constant A_3 only depends on l and p. Comparing it with (3.9), we have

$$A_1 C^{(2-lp)/p} |x_C - y_C|^{(2-lp)/p} \leqslant A_3 |x_C - y_C|^{(2-lp)/p},$$

which is impossible for sufficiently large C.

The obtained contradiction proves the theorem for $lp < 2$.

According to Proposition 6.2 of Chapter 3, for $lp = 2$,

$$\begin{aligned} &C_p^l(S(0, C|x_C - y_C|/2), S(0, |x_C - y_C|/2), \mathbf{R}^2) \\ &\quad = C_p^l(S(0, C/2), S(0,1), \mathbf{R}^2) \leqslant A_4 (\ln \tfrac{C}{2})^{-(1/p)}, \end{aligned}$$

where the constant A_4 only depends on l and p.

Comparing this with (3.10), we obtain

$$A_2 \leqslant A_4 (\ln \tfrac{C}{2})^{-1/p},$$

which is impossible for large C.

The obtained contradiction proves the theorem for $lp = 2$.

Theorem 3.5. *Let G be a bounded domain in $\mathbf{R}^2$. Suppose that there exists an extension operator*

$$\theta : W_p^l(G) \to W_p^l(\mathbf{R}^2), 1 \leqslant lp \leqslant 2.$$

Then every connected component $G^* = \operatorname{Int}(\mathbf{R}^2 \backslash G)$ satisfies the arc diameter condition.

The proof is obtained by modifying the proof of Theorem 3.4, just as Theorem 3.3 is obtained from Theorem 3.1.

3.3. Necessary Conditions Different from the Arc Diameter Condition

For bounded plane domains, the necessary conditions of extension for L_p^l at $1 \leqslant lp \leqslant 2$ may be refined. From this refinement, in particular, there follows the connection of the domain G satisfying the extension condition at the point ∞.

Note that the band $-1 < x < 1$ has a complement whose every connected component satisfies the arc diameter condition although the band is not connected at ∞; i.e., from Theorem 3.4, the connectedness of the domain G at ∞ does not follow.

Theorem 3.6. *Let $1 \leqslant lp \leqslant 2$, and for a domain G, let there exist an extension operator $\theta : L_p^l(G) \to L_p^l(\mathbf{R}^2)$.*

Let the circle $B(a,r)$ be situated so that $G \backslash B(a,R)$ should contain at least two nonempty connected components V_0 and V_1, and the circle $S(a,R)$ should have nonempty intersections with V_0 and $V_1 (R > r)$.

Then there exists a constant C only depending on the norm of the operator θ and on the numbers p and l, such that $R \leqslant Cr$.

Proof. One may assume that $2r < R$, otherwise the theorem is proved with the constant $C = 2$. In this case, the intersections of the circle $S(a,2r)$ with the sets V_0 and V_1 are nonempty. Let us consider the continuums $F_0 \subset \bar{V}_0 \cap \bar{D}_{R,2r}(a)$ and $S(a,R)(D_{R,2r}(a)$ is the ring $B(a,R) \backslash B(a,2r))$. According to Theorem 2.1, for the pair of continuums (F_0, F_1), the estimate

$$C_p^l(F_0, F_1, \mathbf{R}^2) \leqslant \|\theta\| C_p^l(F_0, F_1, G)$$

is valid. From the properties of capacity and from Proposition 6.6 of Chapter 3, the inequality

$$C(\ln \tfrac{2R}{r})^{1/p} \leqslant C_p^1(F_0, F_1, D_{r,2r}(a)) \leqslant C_p^1(F_0, F_1, \mathbf{R}^2)$$

follows for $lp = 2$, and

$$\begin{aligned} &C(2-lp)^{-1/p}\left[R^{(2-lp)/p} - \left(\tfrac{r}{2}\right)^{(2-lp)/p}\right] \\ &\qquad \leqslant C_p^l(F_0, F_1, D_{R,2r}(a)) \leqslant C_p^l(F_0, F_1, \mathbf{R}^2) \end{aligned}$$

follows for $1 \leqslant lp < 2$.

On the other hand, using the properties of capacity and Propositions 6.1 and 6.2 of Chapter 1, we obtain the estimate from above for $C_p^l(F_0, F_1, G)$

$$\begin{aligned} C_p^l(F_0, F_1, G) &\leqslant C_p^l(\overline{V_0 \cap B(a,r)}, F_1, G) \leqslant C_p^l(\overline{V_0} \cap \overline{B(a,r)}, V_1 \backslash B(a,r), G) \\ &\leqslant C_p^l(\bar{G} \cap \bar{B}(a,r), \bar{G} \backslash B(a,2r), G) \\ &\leqslant \begin{cases} Cr^{(2-lp)/p} & \text{for } 1 \leqslant lp \leqslant 2, \\ C & \text{for } lp = 2. \end{cases} \end{aligned}$$

Comparing the two inequalties above, we obtain the relations

$$\ln \frac{2R}{r} \leqslant \|\theta\| C$$

for $lp = 2$,

$$\frac{R^{2-lp} - (\frac{r}{2})^{2-lp})^{1/p}}{(2-lp)^{1/p}} \leqslant \|\theta\| C r^{(2-lp)/p}$$

for $1 \leqslant lp \leqslant 2$.

Proposition 3.7. *Let $1 \leqslant lp \leqslant 2$, and let there exist a bounded extension operator $\theta : L_p^l(G) \to L_p^l(\mathbf{R}^2)$.*

Then for any circle $B(a, R)$ and for any points $x, y \in G \backslash \bar{B}(a, R)$ there exists a curve $\gamma \subset G \backslash B(a, R/\lambda C)$ connecting the points x and y. Here C is the constant from Theorem 3.6, and l is an arbitrary number exceeding 1.

Proof. Suppose that $G \backslash \bar{B}(a, R)$ is nonempty and there exist points $x, y \in G \backslash \bar{B}(a, R)$ that cannot be connected by a curve in $G \backslash B(a, R/\lambda C)$. Then these points lie in different connection components of the set $G \backslash B(a, R/(\lambda C))$. The circle $S(a, R)$ has the nonempty intersection with each of these connection components. For the disc $B(a, R/\lambda C)$ and for the circle $S(a, R)$, all conditions of Theorem 3.6 are satisfied; therefore, $R \leqslant CR/\lambda C$, i.e., $\lambda \leqslant 1$, which contradicts the choice of λ.

The domain $G \subset \mathbf{R}^n$ is said to be locally connected in the neighbourhood of the point $\infty \in \partial G$ if for any neighbourhood $W \ni \infty$ there exists a neighbourhood $V \subset W$ of an infinitely distant point, such that any two points $x, y \in V \cap G$ may be connected by the curve $\gamma \in W \cap G$.

Corollary. *Let a domain $G \subset \mathbf{R}^2$ be unbounded, and let there exist a bounded extension operator $\theta : L_p^l(G) \to L_p^l(\mathbf{R}^2)$, $1 \leqslant lp \leqslant 2$. Then the domain G is locally connected in the neighbourhood of the point $\infty \in \partial G$, and in addition, $\mathbf{R}^2 \backslash G$ may only contain one infinitely connected component which is also locally connected at the point ∞.*

Proof. Local connectedness of the domain G in the neighbourhood of an infinitely distant point directly follows from Proposition 3.7. To prove the second statement, let us suppose that $\mathbf{R}^2 \backslash \bar{G}$ contains two infinite connected components W_0 and W_1. Let the circle $B(a, r)$ intersect each of these components. Then $G \backslash \bar{B}(a, r)$ also contains two infinite nonempty connected components V_0 and V_1, which intersect any of the circles $S(a, R), R > r$. This contradicts the first statement of the corollary.

This completes the proof.

If Γ is a closed Jordan curve in $\mathbf{R}^2$ and $x, y \neq \infty$ are two different points on Γ, then the complement to $x \cup y$ in Γ consists of two nonintersecting arcs. The arc with the smaller diameter is called the smaller arc. If there exists a constant $M < \infty$, the curve Γ is said to be the Ahlfors curve or the curve satisfying the Ahlfors condition. The constant $M < \infty$ does not depend on the choice of $x, y \in \Gamma$ and is such that $|x - z| \leqslant M|x - y|$ for any point z on the smaller arc connecting the points x, y in Γ.

The set $A \subset G$ closed with respect to the domain $G \subset \mathbf{R}^n$ is said to be connected nowhere in the domain G if A does not divide any ball $B \subset G$ into nonempty connected components; i.e., for any ball $B \subset G$ the set $B \backslash A$ is connected.

Theorem 3.8. *For a domain $G \subset \mathbf{R}^2$, let there exist two bounded extension operators*

$$\theta_1 : L^{l_1}_{p_1}(G) \to L^{l_1}_{p_1}(\mathbf{R}^2), \quad 1 \leqslant l_1 p_1 \leqslant 2,$$
$$\theta_2 : L^{l_2}_{p_2}(G) \to l^{l_2}_{p_2}(\mathbf{R}^2), \quad 2 \leqslant l_2 p_2 \leqslant \infty.$$

Then:

1) the boundary $\Gamma_i = \partial W_i$ of any connected component of the set $G^* = \mathbf{R}^2 \backslash G$ is a Jordan curve satisfying the Ahlfors condition with the constant M, which depends only on norms of the operators θ_1 and θ_2 and on differences $2 - l_1 p_1$ and $l_2 p_2 - 2$;

2) here the curves Γ_i do not intersect;

3) in the domain $\tilde{G} = \mathbf{R}^2 \backslash G^*$, the set $\partial G \backslash \partial G^*$ is connected nowhere;

4) the domain G is locally connected at ∞.

Proof. Every connected component W_i of the complement $\mathbf{R}^2 \backslash \bar{G}$ is a simply-connected domain which is locally connected at every boundary point $a \in \partial W_i$ (Corollary 1 of Proposition 3.7). Consequently, the conformal mapping of the circle B onto W_i is extended by infinity up to the topological mapping of a closed disc $\bar{B}$ on $\bar{W}_i \subset \bar{\mathbf{R}}^2$; i.e., the boundary $\partial W_i = \Gamma$ is a Jordan curve.

If a is a common point of Γ_i and $\Gamma_j (i \neq j)$, then obviously at this point, the domain G is not locally connected. But due to Theorem 2.8 for $2 < lp \leqslant \infty$ or due to Theorem 3.1 for $lp = 2$, the domain G satisfies the arc diameter condition. Thus (Property 1 from Section 1), the domain G must be locally connected at every boundary point; i.e., $\Gamma_i \cap \Gamma_j = \varnothing$. According to the corollary of Proposition 3.7, only one curve of Γ_i may be unbounded; therefore, the curves Γ_i do not intersect in $\overline{\mathbf{R}}^2$.

From Theorems 2.8 and 3.3, it follows that the domain G satisfies the bilateral arc diameter condition. Therefore (Property 5 of Section 1), each of the curves satisfies the Ahlfors condition. Due to Property 3 of Section 1, the set $\partial G \backslash \partial G^*$ has topological dimension not exceeding zero, i.e., is connected nowhere.

3.4. Refinement for the Space W_p^l

Theorem 3.9. *Let $l \leqslant lp \leqslant 2$, and let there exist a bounded extension operator*

$$\theta : W_p^l(G) \to W_p^l(\mathbf{R}^2).$$

Let a disc $B(a,r)$ be situated in such a way that $G\backslash\bar{B}(a,r)$ contains at least two nonempty connected components V_0 and V_1, and the intersections of the circle $S(a,R), r < R < R_0$ with the sets $\bar{V}_0$ and $\bar{V}_1$ are nonempty.

Then there exists a constant C which depends only on the norm of the operator θ and on the numbers $p, l,$, and R_0, such that $R < Cr$.

The proof exactly repeats that of Theorem 3.6, except that Proposition 2.9 is used instead of the corresponding results for L_p^l.

Proposition 3.10. *Let $1 \leqslant lp \leqslant 2$, and let there exist an extension operator*

$$\theta : W_p^l(G) \to W_p^l(\mathbf{R}^2).$$

Then for any circle $B(a,R)$, $R < R_0$, any two points $x, y \in G\backslash\overline{B}(a,R)$ may be connected by a curve $\gamma \subset G\backslash B(a, R/\lambda C)$. Here R_0, C are constants from Theorem 3.9, and λ is an arbitrary number exceeding 1.

Corollary 1. *Under the conditions of Proposition 3.10, the domain G is locally connected at every boundary point.*

Corollary 2. *Under the conditions of Proposition 3.10, any connected component W_i of the set $G^* = \mathbf{R}^2\backslash\overline{G} = \cup W_i$ is locally connected at every boundary point.*

The domain G need not be connected at ∞. From the sufficient extension conditions given in the next subsection, it is clear that the space W_p^l may be extended from the band, i.e., from the domain $G = \{(x,y) \in \mathbf{R}^2 : a < y < b\}$. It is obvious that $\mathbf{R}^2\backslash G$ consists of two unbounded connected components.

Corollary 3. *Let the conditions of Proposition 3.10 be satisfied, and let Γ_i, Γ_j be boundaries of different connected unbounded components W_i, W_j of the set $G^* = \mathbf{R}^2\backslash G$.*

Then $\rho(\Gamma_i, \Gamma_j) \geqslant 2R_0/C$, where C is the constant from Theorem 3.6.

Proof. Let $x \in \Gamma_i$, $y \in \Gamma_j$ and $|x - y| < 2R_0/M$. Let us fix the disc $B(a,R)$ with the centre at the point $a = (x+y)/2$ and with the radius $m|x-y| < R \leqslant R_0$. Then $G\backslash\bar{B}(a,r)$, $r = |x-y|/2$ contains at least two unbounded connected components which intersect the sphere $S(a, R_0)$. According to Theorem 3.6,

$$R_0 \leqslant \tfrac{C}{2}|x-y| \quad \text{or} \quad |x-y| \geqslant \tfrac{2R_0}{C}.$$

Q. E. D.

Remark 1. Since C depends on R_0, the relation might be chosen to be maximal, then $\rho(\Gamma_i, \Gamma_j) \geqslant M$ where M only depends on the norm of the extension operator.

We say that a Jordan curve Γ locally satisfies the Ahlfors conditions if there exist numbers $R_0 > 0$ and $M > 0$ such that for the points $x, y \in \Gamma$ for $|x - y| < R_0/M_0$, the inequality

$$\left|z - \frac{x+y}{2}\right| \leqslant M_0|x - y|$$

is valid for any point z lying on the smaller arc which connects the points x and y.

Theorem 3.11. *For a domain $G \subset \mathbf{R}^2$, let there exist two extension operators*

$$\theta : W_{p_1}^{l_1}(G) \to W_{p_1}^{l_1}(\mathbf{R}^2), \qquad 1 \leqslant l_1 p_1 \leqslant 2,$$

and

$$\theta : W_{p_2}^{l_2}(G) \to W_{p_2}^{l_2}(\mathbf{R}^2), \qquad 2 \leqslant l_2 p_2 \leqslant \infty.$$

Then:

1) the boundaries $\Gamma_i = \partial W_i$ of the connected components W_i of the set $G^* = \mathbf{R}^2 \backslash G = \cup W_i$ are pairwise nonintersected, and each of them is a Jordan curve locally satisfying the Ahlfors condition;

2) for the boundaries Γ_i and Γ_j of different unbounded connected components W_i and W_j, the inequality

$$\rho(\Gamma_i, \Gamma_j) \geqslant \tfrac{R_0}{M}$$

is valid. Here M is the constant from the Ahlfors local condition;

3) in the domain $\tilde{G} = \mathbf{R}^2 \backslash G^*$, the set $\partial G \backslash \partial G^*$ is connected nowhere.

This theorem is proved just as Theorem 3.8 was, except that Proposition 3.10 replaces Proposition 3.7.

Remark. The Ahlfors condition, in its local variant, may be written in an asymmetric form; for the points x, y, and z, chosen according to the definition of the Ahlfors local condition, the inequality

$$|x - z| \leqslant \tfrac{1}{2}(C + 1)|x - y|$$

is valid where C is the constant of Theorems 3.6 or 2.12.

§4 Necessary Extension Conditions for Besov and Nickolsky Spaces

4.1. Extension Theorem for $lp > n$

For a space $b^l_{p,\theta}(G)$, due to the definition of the seminorm $\|\cdot\|^l_{p,\theta,G,h}$, the invisibility condition is satisfied with the constant $K = 1$. This obviously follows from the definition of the space $b^l_{p,\theta,}(G)$. By setting $\mathcal{F}_0(G) = b^l_{p,\theta}(G)$, $\mathcal{F}_1(\mathbf{R}^n) = b^l_{p,\theta}(\mathbf{R}^n) = \mathcal{F}_2(\mathbf{R}^n)$, we obtain the extension theorem for $b^l_{p,\theta}$ from Theorem 2.5 and Lemma 7.1 in Chapter 3.

Theorem 4.1. *G is a domain in* $\mathbf{R}^n$. *If there exists an extension operator*

$$T : b^1_{p,\theta}(G, h) \to b^l_{p,\theta}(\mathbf{R}^n, \infty)$$

for $lp > n$, *then the domain G satisfies the arc diameter condition.*

For the spaces $B^l_{p,\theta}(G)$, the same arguments made in the proof of Theorem 2.11 lead to Theorem 4.2.

Theorem 4.2. *Let G be a bounded domain in* $\mathbf{R}^n$ *for which there exists a bounded extension operator*

$$\theta : B^l_{p,\theta}(G, h) \to B^l_{p,\theta}(\mathbf{R}^n)$$

for $lp > n$. *Then the domain G satisfies the arc diameter condition.*

Remark. Instead of the Sobolev inequality used in the proof of Theorem 2.11, in the proof of Theorem 4.2 we use the inequality [10]

$$\|f\|_{L_p(B(0,R_0))} \leqslant C\|f\|_{b^l_{p,\theta}(B(0,R_0))}$$

for the functions $f \in b^l_{p,\theta}(\mathbf{R}^n)$ which vanish outside of the ball $B(0, R_0)$. The constant C only depends on l, n, p, R_0.

Theorem 4.3. *Let a bounded domain* $G \in \mathbf{R}^n$ *be such that the set* $G^* = \operatorname{Int}(\mathbf{R}^n \backslash G)$ *is also a domain. Suppose that there exist two extension operators*

$$\theta_1 : B^{l_1}_{p_1,\theta_1}(G, h) \to B^{l_1}_{p_1,\theta_1}(\mathbf{R}^n), \qquad n < l_1 p_1,$$
$$\theta_2 : B^{l_2}_{p_2,\theta_2}(G^*, h) \to B^{l_2}_{p_2,\theta_2}(\mathbf{R}^n), \qquad n < l_2 p_2.$$

Then the domain G satisfies the bilateral arc diameter condition.

The proof repeats that of Theorem 2.12 exactly.

For dimension 2 from Property 4 of domains satisfying the arc diameter condition, it follows that under the conditions of Theorem 4.3, the boundary of the domain is a quasicircle.

For the spaces $b^l_{p,\theta}$, Theorem 4.3 is valid for unbounded domains as well.

4.2. Extension Conditions for $lp = n$

Theorem 4.4. *Let G be a domain in $\mathbf{R}^n$. If there exists an extension operator*

$$\theta_h : b^l_{p,\theta}(G,h) \to b^l_{p,\theta}(\mathbf{R}^n,\infty) \qquad \begin{array}{l} lp = n, \\ 1 \leqslant p < \infty, \\ 1 \leqslant \theta \leqslant \infty, \end{array}$$

then the domain G satisfies the condition with the arc diameter.

Proof. Note that $b^l_{p,\theta}(G,\infty) \subset b^l_{p,\theta}(G,h)$ and $\|u\|^{(l)}_{p,\theta,G,\infty} \geqslant \|u\|^{(l)}_{p,\theta,G,h}$ for all $u \in b^l_{p,\theta,G,\infty}$. Due to the existence of the extension operator θ_h, we obtain

$$\|u\|^{(l)}_{p,\theta,G,h} \geqslant \frac{1}{\|\theta_h\|}\|\theta_h u\|^{(l)}_{p,\theta,\mathbf{R}^n,\infty}.$$

Therefore, for any function $u \in b^l_{o,\theta}(G,\infty)$, we have

$$\|\theta_h\|\,\|u\|^{(l)}_{p,\theta,G,\infty} \geqslant \|\theta_h u\|^{(l)}_{p,\theta,\mathbf{R}^n,\infty}; \tag{4.1}$$

i.e., the operator θ_k is also the bounded extension operator for $b^l_{p,\theta,G,\infty}$. The seminorm $\|\cdot\|^{(l)}_{p,\theta,G,\infty}$ is invariant for similarities since $lp = n$.

Now the proof repeats that of Theorem 3.3 with minor variations. The difference is that, instead of the strong invisibility condition, one should use the invisibility condition, just as in Section 2 when proving extension theorems. Note that the vanishing of the capacity of a point with respect to the complement to a sphere in the proof of Theorem 3.3 is a redundant requirement. It suffices to know that it is smaller than ∞, since the contradiction was obtained when comparing this capacity to that of the pair of approaching continuums which tends to ∞.

Theorem 4.5. *Let G be a bounded domain in $\mathbf{R}^n$. If there exists an extension operator*

$$\theta : B^l_{p,\theta}(G,h) \to B^l_{p,\theta}(\mathbf{R}^n) \qquad \begin{array}{l} lp = n, \\ 1 \leqslant p < \infty, \\ 1 \leqslant \theta \leqslant \infty, \end{array}$$

then the domain G satisfies the arc diameter condition.

It is obtained from Theorem 4.4 by means of the same arguments as those used for Theorem 4.2 and Theorem 4.1.

Theorem 4.6. *Let G be a bounded domain in $\mathbf{R}^2$. Suppose that there exists an extension operator*

$$\theta : b^l_{p,\theta}(G,h) \to b^l_{p,\theta}(\mathbf{R}^2,\infty) \qquad \begin{array}{l} 1 \leqslant lp \leqslant 2, \\ 2 \leqslant p \leqslant \infty, \\ p \leqslant \theta. \end{array}$$

Then every connected component of the set $G^* = \operatorname{Int}(\mathbf{R}^2 \setminus G)$ *satisfies the arc diameter condition.*

Proof. For the same reasons as those in the proof of Theorem 4.4, one may assume h to be equal to ∞.

Suppose that some connected component of the set G^* does not satisfy the arc diameter condition. Consequently, for any $C > 0$, there exists a pair of points $X_c, Y_c \in W$ such that every curve γ_C connecting x_C and y_C has $\operatorname{diam} \gamma_C \geqslant C|x_C - y_C|$. Let us connect the points x_C and y_C by a segment and denote the middle of this segment by z_C. The segment divides the domain G into several connected components. Among them there are at least two components, G_1 and G_2, whose diameters exceed $C|x_C - y_C|$. Let G_2 be the component of the smaller diameter.

Let us denote $\sup_{x \in G} \frac{|z_C - x|}{|x_C - y_C|}$ by C_1. Let us consider three circles:[1]

$$\begin{aligned} B^0 &= B^0_{C_1} = B(z_C, C_1^{(1-\varepsilon)/lp}|x_C - y_C|) \\ B^1 &= B^1_{C_1} = B(z_C, \tfrac{C_1}{2}|x_C - y_C|), \\ B^2 &= B^2_{C_2} = B(z_C, C_1|x_C - y_C|). \end{aligned}$$

By construction, the open set $G \setminus B^0$ is not connected. Among its connection components there exist at least two, V_0 and V_1, that intersect the circles $S(z_C, t)$ for all

$$t \in (C_1^{(1-\varepsilon)/lp}|x_C - y_C|, C_1|x_C - y_C|).$$

Let us choose an arbitrary curve γ_0 connecting the circles $S^1 = \partial B^1$ and $S^2 = \partial B^2$ in V_0, and a curve γ_1 connecting the circles S^1_C and S^2_C in V_1.

Due to the existence of the extension operator, it follows from Theorem 2.1 that

$$C^l_{p,\theta}(\gamma_0, \gamma_1, \mathbf{R}^2, \infty) \leqslant \|\theta\| C^l_{p,\theta}(\gamma_0, \gamma_1, G, \infty).$$

From Chapter 3's Proposition 7.6 and Lemma 7.1 for $lp < 2$, it follows that

$$C^l_{p,\theta}(\gamma_0, \gamma_1, \mathbf{R}^2, \infty) \geqslant A C_1^{(2-lp)/p}|x_C - y_C|^{(2-lp)/p}, \tag{4.2}$$

where the constant A depends on l, p, θ.

For $lp = 2$,

$$C^l_{p,\theta}(\gamma_0, \gamma_1, \mathbf{R}^2, \infty) \geqslant A_2 > 0$$

where the constant A_2 only depends on l and p.

Let us estimate the capacity of the pair γ_0, γ_1 in the domain G. Due to monotonicity of capacity,

$$C^l_{p,\theta}(\gamma_0, \gamma_1, G, \infty) \leqslant C^l_{p,\theta}(\gamma_0, (\overline{V_1, \cup B_0}) \cap G, G, \infty). \tag{4.3}$$

[1] ε is sufficiently small. It will be chosen later.

Suppose the estimate

$$C^l_{p,\theta}(\gamma_0, (\overline{V_1 \cup B_0}) \cap G, G, \infty) \leqslant o(C_1^{(2-lp)/p}|x_C - y_C|^{(2-lp)/p}) \tag{4.4}$$

to be valid. Then we may complete the proof of this theorem as we did the proof of Theorem 3.4. From the previous inequalities, it follows that

$$\begin{aligned}\|\theta\| o(C_1^{(2-lp)/p}|x_C - y_C|^{(2-lp)/p}) &\geqslant \|\theta\| C^l_{p,\theta}(\gamma_0, \gamma_1, G, \infty)\\ &\geqslant C^l_{p,\theta}(\gamma_0, \gamma_1, \mathbf{R}^2, \infty)\\ &\geqslant AC_1^{(2-lp)/p}|x_C - y_C|^{(2-lp)/p}.\end{aligned}$$

The contradiction obtained in the inequality proves the theorem. It only remains to prove estimate (4.4). Due to imbedding theorems for $b^l_{p,\theta}$,

$$C^l_{p,\theta}(\gamma_0, (\overline{V_1 \cup B_0}) \cap G, G, \infty) \leqslant C^l_{p,p}(\gamma_0, (\overline{V_1 \cup B_0}) \cap G, G, \infty),$$

since $b^l_{p,p}(G)$ is imbedded into $b^l_{p,\theta}(G)$.

Therefore, it suffices to obtain estimate (4.4) for $\theta = p$ only.

Let us take an arbitrary function f, admissible for the $b^l_{p,p}$-capacity of the pair $\overline{B}_0 \cap G, \gamma_0$. (We use the symmetry of the $C^l_{p,p}$-capacity relative to the change of order in the pair of sets in question.) Let us consider a function $\tilde{f}(x) = f(x)$ outside of G_2 and equal to zero in G_2.

For $x \in G_1$, the difference of $\Delta^m(t, G)f(x)$ differs from $\Delta^m(t, G)\tilde{f}(x)$ for $|t| > C_1^{(1-\varepsilon)/lp}$. The angle between the two vectors t_1 and t_2, for which $m-l$-differences of the functions f and $\tilde{f}$ are distinct at the point x, does not exceed $1/C_1^{(1-\varepsilon)/lp}$. That is, to each point we may associate two angles φ^1_x and φ^2_x such that for $\varphi^1_x < \varphi < \varphi^2_x$ for the vector $(r, \varphi) = t$ $(r = |t|)$, the differences $\Delta^m(t, G)f(x)$ and $\Delta^m(t, G)\tilde{f}(x)$ are distinct; for all the remaining φ they coincide.

First let us estimate L_p-norms. Let us fix arbitrarily $t > 0$. Let Γ_C be the intersection of the domain G with the band lying between the straight lines which are parallel to the vector t and pass through the points x_C and y_C. Let us consider a sequence of circular layers

$$\begin{aligned}D_1 &= B^1 \backslash B^0,\\ D_2 &= B(z_C, 2C_1|x_C - y_C|) \backslash B^1,\\ D_3 &= B(z_C, 3C_1|x_C - y_C|) \backslash B(z_C, 2C_1|x_C - y_C|),\\ D_S &= B(z_C, SC_1|x_C - y_C|) \backslash B(z_C, (S-1)C_1|x_C - y_C|).\end{aligned}$$

The area of the intersection $\Gamma_S = D_S \cap \Gamma_C$ does not exceed $C_1|x_C - y_C|$. Note that outside of the set Γ_C, the differences $\Delta^m(t, G)f(x)$ and $\Delta^m(t, G)\tilde{f}(x)$ coincide.

Thus,

$$\int_{G_1} |\Delta^m(t,G)(f-\tilde{f})(x)|^p\,dx = \sum_{S=1} \int_{\Gamma_S} |\Delta^m(t,G)(f-\tilde{f})(x)|^p\,dx. \qquad (4.5)$$

Since $0 \leqslant f(x) \leqslant 1$ and $\tilde{f}(x) \leqslant 1$ for all $x \in G$ (see the definition of function admissible for capacity), then

$$\int_{\Gamma_S} |\Delta^m(t,G)(f-\tilde{f})|^p \leqslant m(\Gamma_S)2^m \leqslant 2^m C_1 |x_C - y_C|^2.$$

Now let us estimate the difference $f - \tilde{f}$ in the norm of the space $b^l_{p,\theta}$:

$$\begin{aligned}
&\int_{|t|<h} \frac{\|\Delta^m(t,G)(f-\tilde{f})(x)\|^p_{Lp(G)}}{|t|^{lp+2}}\,dt \\
&= \iint\limits_{|t|<h} \frac{\|\Delta^m(t,G)(f-\tilde{f})(x)\|^p_{Lp(G)}}{r^{lp+1}}\,dr\,d\varphi \\
&= \sum_r \iint\limits_{|t|<h} \frac{\|\Delta^m(t,G)(f-\tilde{f})(x)\|^p_{Lp(\Gamma_s)}}{r^{lp+1}}\,dr\,d\varphi.
\end{aligned}$$

Let us recall that $\Delta^m(t,G)(f-\tilde{f})(x) \neq 0$ only for $r > C_1^{(1-\varepsilon)/lp}|x_C - y_C|$, $\varphi_x^1 < \varphi < \varphi_x^2$; and by using (4.5), let us estimate each of the integrals in this sum:

$$\begin{aligned}
&\iint\limits_{|t|<h} \frac{\|\Delta^m(t,G)(f\tilde{f})(x)\|^p_{Lp(\Gamma_1)}}{r^{lp+1}}\,dr\,d\varphi \\
&\leqslant \int_0^{1/C_1^{(1-\varepsilon)/lp}} \left(\int_{C_1^{(1-\varepsilon)/lp}}^{\infty} \frac{2^m C_1 |x_C - y_C|^2}{r^{lp+1}}\,dr \right) d\varphi \\
&\leqslant P C_1^{\varepsilon(1+(1/lp))-(1/lp}|x_C - y_C|^{2-lp}.
\end{aligned}$$

It is clear that for sufficiently small ε, the right-hand side tends to zero for $C_1 \to \infty$.

Let us estimate the domains Γ_S. Note that in Γ_S, the difference $\Delta^m(t,G)$ $(f-\tilde{f})(x)$ is distinct from zero only for $|t| > SC_1|x_C - y_C|$ $(S > 1)$.

By repeating the estimate, we obtain

$$\iint\limits_{|t|<h} \frac{\|\Delta^m(t,G)(f-\tilde{f})(x)\|^p_{Lp(\Gamma_s)}}{|t|^{lp+2}}\,dt \leqslant \frac{1}{\varphi^{lp}} C_1^{(-1)/lp}|x_C - y_C|^{2-lp}.$$

By summing up all the estimates for Γ_s, we obtain

$$J_{G_1}(f-\tilde{f}) = \iint\limits_{|t|<h} \frac{\|\Delta^m(t,G)(f-\tilde{f})(x)\|^p_{Lp(G_1)}}{|t|^{lp+2}}\, dt \leqslant C_1^{\varepsilon(1+\frac{1}{lp})-\frac{1}{lp}} \sum \frac{1}{m^{lp}} |x_C - y_C|^{2-lp}. \tag{4.6}$$

Hence it follows that

$$J_{G_2}(F) \leqslant J_{G_1}(f) + K_1 C_1^{\varepsilon(1+\frac{1}{lp})-\frac{1}{lp}} |x_C - y_C|^{2-lp}$$

Here J_{G_2} is the integral obtained from J_{G_1} as a result of substitution of the domain G_1 for G_2. For the ball B^0, the estimate

$$J_{B^0\cap G}(\tilde{f}) \leqslant J_{B^0\cap G}(f)$$

is obvious, since $|\Delta^m(t,G)\tilde{f}(x)| \leqslant |\Delta^m(t,G)f(x)|$ for $x \in B^0 \cap G$.

Recalling that $(\|\tilde{f}\|^l_{p,p,G,\infty})^p = J_{G_1}(\tilde{f}) + J_{G_2}(\tilde{f}) + J_{B^0\cap G}(\tilde{f})$ and that the same representation is valid for f, we obtain

$$\|\tilde{f}\|^{(l)}_{p,p,G,\infty} \leqslant \|f\|^{(l)}_{p,p,G,\infty} + 0\left(\tfrac{1}{C_1}\right).$$

In terms of capacity, this inequality means that

$$C^l_{p,\theta}(\gamma_0, (\overline{V_1 \cup B^0}) \cap G, G, \infty) \leqslant C^l_{p,\theta}(\gamma_0, \overline{B}^0 \cap G, G, \infty) + 0\left(\tfrac{1}{C}\right).$$

Due to monotonicity of capacity, it follows that

$$C^l_{p,\theta}(\gamma_0, (\overline{V_1 \cup B^0}) \cap G, G, \infty) \leqslant C^l_{p,\theta}(\mathbf{R}^2 \backslash B^2, B^0, \mathbf{R}^2, \infty) + 0\left(\tfrac{1}{C_1}\right). \tag{4.7}$$

From Lemma 4.3, it follows that

$$C^l_{p,\theta}(\mathbf{R}^2 \backslash B^2, B^0, \mathbf{R}^2, \infty) \leqslant (C_1 |x_C - y_C|)^(2 - lp)/p\alpha(C_1),$$

where

$$\alpha(C_1) = C^l_{p,\theta}\left(B(0,1), B\left(0, C_1^{1/lp-\varepsilon/lp-1}\right), \mathbf{R}^2, \infty\right).$$

According to Proposition 4.8, $\lim \alpha(t) = 0$ for $t \to \infty$.

Finally, we obtain estimate (4.4) from (4.7). This completes the proof.

Remark. The theorem remains valid for the spaces $B^l_{p,\theta}$ as well.

Let us formulate the results similar to Theorems 3.8 and 3.12 of Chapter 2.

Theorem 4.7. *For a bounded domain $G < \mathbf{R}^2$, let there exist two bounded extension operators*

$$\theta_1 : B^{l_1}_{p_1,\theta_1}(G) \to B^{l_1}_{p_1,\theta_1}(\mathbf{R}^2) \qquad \begin{array}{l} 1 \leqslant l_1p_1 \leqslant 2, \\ 2 \leqslant p \leqslant \theta, \end{array}$$

$$\theta_2 : B^{l_2}_{p_2,\theta_2}(G) \to B^{l_2}_{p_2,\theta_2}(\mathbf{R}^2) \qquad 2 \leqslant l_2p_2 \leqslant \infty.$$

Then:

1) the boundary $\Gamma_i = \partial W_i$ of any connected component W_i of the set $G^* = \mathbf{R}^2 \backslash \bar{G} = \cup W_i$ is a quasicircle;

2) here the curves Γ_i do not intersect;

3) in the domain $\tilde{G} = \mathbf{R}^2 \backslash G^*$, the set $\partial \backslash \partial G^*$ is connected nowhere.

Remark. In the case $lp = 2$, the theorem is valid if only one extension operator is supposed to exist.

§5 Sufficient Extension Conditions

5.1. Quasiconformal Extension

We follow the book by Ahlfors [1] in our presentation.

Let us consider a quasiconformal mapping φ of the entire plane onto itself having a distortion coefficient K. In this case, the real axis L is mapped on an unbounded curve $\varphi(L)$ satisfying the Ahlfors condition. The curve $\varphi(L)$ divides the plane into two parts, Ω and Ω', which correspond to the upper and lower half-planes H and $H^*(\varphi(H) = \Omega, \varphi(H^*) = \Omega^*)$. Let j denote a mapping interchanging H and H^*. Then $\varphi \circ j \circ \varphi^{-1}$ is a quasiconformal mapping interchanging Ω and Ω^* and leaving the points $\varphi(L)$ immovable. $\varphi(L)$ is said to admit K^2-quasiconformal mapping.

A boundary γ of an unbounded domain G is called a quasistraight line if γ is an image of the real axis for some quasiconformal homeomorphism φ of the plane onto itself.

Recall that an unbounded Jordan curve is a quasistraight line iff it satisfies the Ahlfors condition.

The constant in the Ahlfors condition only depends on the distortion coefficient of a quasiconformal homeomorphism φ which assigns the quasistraight line.

Theorem 5.1. *Let a boundary γ of an unbounded simply-connected domain $G \subset \mathbf{R}^2$ satisfy the Ahlfors condition. Then there exists a homeomorphism $\varphi : \bar{G} \to \mathbf{R}^2 \backslash G$ which is quasiconformal, differentiable in the domain G and identical on the boundary. Besides, the homeomorphism satisfies the condition*

$$P^{-1}(C) < \frac{|\varphi(x) - \varphi(y)|}{|x-y|} < P(C)$$

for any points $x, y, \in \gamma$. The constant $P(C)$ only depends on the constant C in the Ahlfors condition.

Similar to the notion of a quasiconformal mapping for a quasistraight line, one can introduce the notion of a quasiconformal mapping for a quasicircle.

Let us consider on a plane disc $B(0,1)$ of radius 1 with the centre at the point 0. Let j be inversion with respect to the unit circle $S(0,1)$. If the boundary γ of a domain G is a quasicircle, then there exists a quasiconformal homeomorphism $\varphi : \mathbf{R}^2 \to \mathbf{R}^2$ mapping $B(0,1)$ on the domain G. Then $\varphi \circ j \circ \varphi^{-1}$ is a quasiconformal homeomorphism which transfers the domain G into the domain $\mathbf{R}^2 \backslash G$ and leaves the points of the curve γ immovable. We say that γ admits quasiconformal mapping. The converse is also valid: if γ admits quasiconformal mapping, then γ is a quasicircle [1, 100].

Theorem 5.2. *Let a boundary of a bounded domain $G \subset \mathbf{R}^2$ be a Jordan curve γ locally satisfying the Ahlfors condition. Then there exists a differentiable homeomorphism $\varphi : \bar{G} \to \mathbf{R}^2 \backslash G$, quasiconformal in G, which leaves the points of the boundary immovable. The boundary γ has a neighbourhood $V(\gamma)$ such that for any pair of points $x, y \in V(\gamma) \backslash G$, the inequality*

$$P^{-1}C < \frac{|\varphi(x) - \varphi(y)|}{|x - y|} < P(C)$$

is valid. The constant $P(C)$ only depends on the constant C in the Ahlfors condition.

This theorem was actually proved in [1,100].

In Section 1, it was proved that every simply-connected domain whose boundary satisfies the Ahlfors condition also satisfies the bilateral arc diameter condition. The converse is also valid: if a simply-connected plane domain satisfies the bilateral condition with the arc diameter, then its boundary satisfies the Ahlfors condition (for bounded domains it locally satisfies the Ahlfors condition). It follows from Property 5 of Section 1.

Therefore, in Theorem 5.1, the first sentence of the statement may be written as follows: "Let an unbounded simply-connected domain satisfy the bilateral condition with the arc diameter," and in Theorem 5.2: "Let a bounded simply-connected domain satisfy the bilateral condition with the arc diameter."

5.2. Extension Conditions for Sobolev Classes

Theorem 5.3. *The bounded extension operators*

$$\theta_1 : L^{l_1}_{p_1}(G) \to L^{l_1}_{p_1}(\mathbf{R}^2), \qquad 1 \leqslant l_1 p_1 \leqslant 2,$$
$$\theta_2 : L^{l_2}_{p_2}(G) \to L^{l_2}_{p_2}(\mathbf{R}^2), \qquad 2 \leqslant l_2 p_2 \leqslant \infty$$

from a bounded finitely connected domain $G \subset \mathbf{R}^2$ exist iff the domain G satisfies the bilateral condition with the arc diameter.

Theorem 5.4. *Bounded extension operators*

$$\theta_1 : W_{p_1}^{l_1}(G) \to W_{p_1}^{l_1}(\mathbf{R}^2), \qquad 1 \leqslant l_1 p_1 \leqslant 2,$$
$$\theta_2 : W_{p_2}^{l_2}(G) \to W_{p_2}^{l_2}(\mathbf{R}^2), \qquad 2 \leqslant l_2 p_2 \leqslant \infty$$

from a bounded finitely connected domain G exist iff the domain G satisfies the bilateral condition with the arc diameter.

Proof of the theorems. Necessity of the conditions was obtained in Theorems 3.8 and 3.11. Sufficiency for $l = 1$ is proved as in the theorem below about necessary and sufficient extension conditions for $L_p^1(G)$ classes from unbounded finitely connected domains.

Sufficient conditions for $l_1 = l_2 = 1$ were proven in one of the author's papers [25]. Sufficiency of conditions of Theorem 5.4 for arbitrary l_1, l_2 can be drawn from a later paper by Jones [99]. From the work of Schwartzman [121], the sufficiency of conditions of Theorem 5.3 follows for arbitrary l_1 and l_2. In fact, the sufficiency of conditions of Theorem 5.3 and 5.4 for arbitrary l_1 and l_2 follows from an earlier paper of Schwartzman [120]. The result about extension for the Sobolev spaces was not, however, formulated in this paper.

Remark 1. The extension algorithm proposed in papers [120] and [99] gives sufficient conditions for a wide class of domains in space. These conditions are stronger than the condition with the arc diameter.

Remark 2. In the case $l_1 p_1 = l_2 p_2 = 2$, Theorems 5.3 and 5.4 remain valid if only one extension operator exists.

Let us give the theorem about extension from [121]. Note that the extension conditions formulated in paper [121] (the Lichtenstein condition) contain additional requirement as compared to (ε, δ)-condition given in Jones's paper [99]. This is due to a minor inaccuracy of paper [99]. In fact, this additional requirement was used in the proof of extension theorems in [99]. The equivalence, in plane-bounded finitely-connected domains, of the Lichtenstein and Ahlfors conditions upon the curves which are boundaries of complement components nondegenerating to a point was proved in [99]. That is, in plane-bounded finitely-connected domains, the bilateral arc diameter condition and the Lichtenstein condition are equivalent.

The Lichtenstein condition (see [121]). *An open set G satisfies the Lichtenstein condition if there exist constants $\varepsilon = \varepsilon(G)$ and $\delta = \delta(G)$ such that:*

1) for every cube Q with the centre in G and with $\operatorname{diam} Q < \delta$, *the inequality*

$$\operatorname{diam}(Q \cap G) > \varepsilon \operatorname{diam} Q$$

holds.

2) for any $x,y \in G$ with $|x-y| < \delta$ in G there exists a continuous curve Γ connecting x and y and such that for any point $z \in \Gamma$ the inequalities

$$|x-z| + |z-y| < \tfrac{1}{\varepsilon}|x-y|, \qquad \operatorname{dist}(z, \mathbf{R}^n \backslash G) \geqslant \tfrac{1}{\varepsilon}\min(|x-z|.|z-y|)$$

are valid. Let us denote this class of open sets by Arc.

Theorem 5.5 [121]. *If $G \in$ Arc is a bounded domain in $\mathbf{R}^n$, then for $1 \leqslant p \leqslant \infty$, there exists a linear extension operator $\theta : L_p^l(G) \to L_p^l(\mathbf{R}^n)$.*

The class L_p^l may be replaced by W_p^l or by $B_{p,\theta}^l$ ($p > 1$). For the class W_p^l, the domain G may be unbounded [121]. For instance, the band $|x| < 1$ in $\mathbf{R}^2$.

Remark 1. In unbounded domains, analogous results for the spaces L_p^l, $B_{p,\theta}^l$ are not known.

Remark 2. Condition 1 from the Lichtenstein condition is automatically satisfied.

Theorem 5.6. *Bounded extension operators*

$$\theta_1 : L_{p_1}^1(G) \to L_{p_1}^1(\mathbf{R}^2), \qquad 1 \leqslant p_1 \leqslant 2,$$
$$\theta_2 : L_{p_2}^1(G) \to L_{p_2}^1(\mathbf{R}^2), \qquad 2 \leqslant p_2 \leqslant \infty$$

from an unbounded finitely-connected domain $G \subset \mathbf{R}^2$ exist iff the domain G satisfies the bilateral condition with the arc diameter and is locally connected at ∞.

Proof. Necessity was proved in Theorem 3.8.

Sufficiency. Since the domain G satisfies the bilateral condition with the arc diameter, then, due to Property 5 from Section 1, the boundary of each component of the set $G^* = \operatorname{Int}(\mathbf{R}^2 \backslash G)$ is a curve satisfying the Ahlfors condition. Due to connection of the domain G at infinity, there exists only one unbounded component of all such curves. Let us denote the unbounded curve by Γ_∞, the bounded ones by $\Gamma_1, \Gamma_2, \ldots, \Gamma_k$.

The remaining part of the boundary of the domain G, due to Property 5 from Section 1, and due to the fact that G is finitely connected, consists of a finite number of isolated points. Since an isolated point is a removable singularity for any of the $L_p^1(G)$ spaces, then we extend the $L_p^1(G)$ class to these points.

Extension through the Unbounded Component of the Boundary. Recall that according to Corollary 1 of Property 5 (Section 1), the curves $\Gamma_1 m, \ldots, \Gamma_K, \Gamma_\infty$ do not intersect pairwise. The curve Γ_∞ is a quasistraight line. It divides the plane $\mathbf{R}^2$ into two domains: Ω and Ω^*. Let Ω be the domain containing the domain G. Let us consider a conformal mapping $\varphi : \mathbf{R}_+^2 \to \Omega$. According to the theorem about extension across a quasistraight line [1], it may also be extended up to a quasiconformal homeomorphism $\overline{\varphi} : \mathbf{R}^2 \to \mathbf{R}^2$ with distortion coefficient $q(\overline{\varphi})$, which depends only on the constant in the Ahlfors condition for the curve Γ_∞. That is, it depends only on the constant in the bilateral arc diameter condition for the domain G.

The images γ_m of the straight lines $y = 1/m$ for the mapping φ are smooth curves satisfying the Ahlfors condition with constants C_m, which are bounded from above by C_0. The constant C_0 depends on the distortion coefficient $q(\overline{\varphi})$ only; i.e., it only depends on the constant in the bilateral arc diameter condition [1]. At sufficiently large m_0, the curves γ_m $(m > m_0)$ and $\Gamma_1 . \Gamma_2, \dots, \Gamma_k$ do not intersect. Further, we set $m > m_0$.

Each of the curves γ_m divides the domain G into two connected components. Let us denote by G_m the component whose boundary contains γ_m and $\Gamma_1, \Gamma_2, \dots, \Gamma_k$. Theorem 5.1 allows us to construct a sequence of quasiisometries

$$\varphi_m : G_m \cup \gamma_m \to \mathbf{R}^2 \backslash G_m,$$

which are immovable on the curves γ_m. According to the same theorem, for quasiisometricity coefficients $qi_m = qi(\varphi_m)$ of these mappings the estimate $qi_m < P(C)$ is valid where the constant C only depends on C_0, i.e., only depends on the constant in the bilateral condition with the arc diameter.

The sequence of quasiisometries $\{\varphi_m\}$ uniformly converges locally to the quasiisometry φ_0 constructed by Theorem 5.1 for the curve Γ_∞ (this can easily be seen from the algorithm of construction of quasiisometries in [1]). Besides, $G_m \subset G_{m+1}$ for all m.

Let us take a continuous function $u \in L_p^1(G)$. The function u_m^* equals u on G_m and equals $u \circ \varphi^{-1} m$ on $\mathbf{R}^2 \backslash G_m$ and belongs to the class $L_p^1(G_m \cup (\mathbf{R}^2 \backslash \varphi(G_m)))$. By straightening the locally smooth curve γ_m, one can easily show that the function u_m^* has generalized derivatives, i.e., $u_m^* \in L_p^1(\bar{G}_m \cup (\mathbf{R}^2 \backslash \varphi(G_m)))$. Let us denote the set $\bar{G}_m \cup (\mathbf{R}^2 \backslash \varphi(G_m))$ by G_m^*, and the set $\bar{G} \backslash (\mathbf{R}^2 \backslash \varphi(G_m))$ by G^*. Due to quasiisometricity of mapping φ_m and to the uniform estimate $qi_m \leqslant P(C)$,

$$\|u_m^*\|_{L_p^1(G_m^*)} \leqslant M \|u\|_{L_p^1(G)}, \tag{5.1}$$

where the constant M depends on P and on the constant in the bilateral condition with the arc diameter.

Let $1 < p < \infty$. Let us fix $m_1 > m_0$. Due to locally uniform convergence of φ_m to φ_0, the functions u_m^* uniformly converge locally to the function u^*. Due to inequality (5.1), the sequence u_{m_1+k} is uniformly bounded in the space $L_p^1(G_{m_1})$. Therefore, due to reflexivity of the space $L_p^1(G_{m_1}^*)$ for $p > 1$, one can extract a weakly converging subsequence from it. To avoid cumbersome denotations, we assume that the sequence u_{m+k}^* itself weakly converges in $L_p^1(G_{m_1}^*)$. It is obvious that the function u^* is the weak limit of this sequence. Due to inequality (5.1) and weak convergence of u_{m+k}^* to u^* in $L_p^1(G_{m_1}^*)$, we have

$$\|u^*\|_{L_p^1(G_{m_1}^*)} \leqslant \lim_{k \to \infty} \|u_{m+k}^*\|_{L_p^1(G_{m_1}^*)} \leqslant M \|u\|_{L_p^1(G)}.$$

By taking the limit by m, we obtain

$$\|u^*\|_{L_p^1(G^*)} \leqslant M\|u\|_{L_p^1(G)}. \tag{5.2}$$

In the case $p=\infty$, it follows directly from Theorem 5.1 that the function u^* is the Lipschitz one in G^* with the Lipschitz constant which increased at most by $P(C)$ times. The case $p=1$ is obtained from inequality (5.2) by limit transition over p.

We extended the function u to the domain G^* whose boundary consists of a finite number of quasicircles. Applying the same method as in the case Γ_∞ to each of these quasicircles, and using Theorem 5.2 instead of Theorem 5.1, one can construct an extension operator to some domain in $G_1^* \supset G^*$. Multiplication of each of the extended functions by the function of the class $C_0^\infty(G_1^*)$ equal to 1 in $\bar{G}^*$ completes the construction of the bounded extension operator.

This completes the proof.

Remark 1. It is easy to see that the constructed extension operator is linear.

Remark 2. In the case $p_1=p_2=2$, the theorem remains valid if only one extension operator exists.

5.3. Example of Estimating the Norm of an Extension Operator

Let a boundary of a simply-connected domain in G be a quasistraight line. Let us consider in this domain the spaces $L_2^1(G)$. The extension operator

$$\theta : L_2^1(G) \to L_2^1(\mathbf{R}^2)$$

may be constructed according to the diagram

$$\begin{array}{ccc} L_2^1(\mathbf{R}_+^2) & \xrightarrow{\ \overline{\theta}\ } & L_2^1(\mathbf{R}^2) \\ \uparrow \varphi^* & & \downarrow (\varphi^{-1})^* \\ L_2^1(G) & \overset{\theta}{\dashrightarrow} & L_2^1(\mathbf{R}^2) \end{array}$$

Here $\overline{\theta}$ is any extension operator, for instance, that by symmetry φ^* is a structural isomorphism induced by the quasiconformal homeomorphism $\varphi : \mathbf{R}^2 \to \mathbf{R}^2$ which maps the real axis onto the curve ∂G, and it maps the upper half-plane $\mathbf{R}_+^2$ onto G.

One can easily calculate that $\|\varphi^*\| \leqslant K$, $\|(\varphi^{-1})^*\| \leqslant K$ where K is the distortion coefficient of the mapping φ. Then $\|\theta\| \leqslant K^2\|\overline{\theta}\|$.

Remark 1. The distortion coefficient φ may be calculated if we know the constant in the Ahlfors condition for ∂G [1].

Remark 2. Since the quasiisometricity coefficient for quasiisometries, which was used in the proof of Theorem 5.6, may be calculated in terms of the distortion coefficient of the mapping φ, then the norm of the extension operator may be calculated in this theorem as well.

Remark 3. The example may easily be transferred to the n-dimensional case for domains whose boundaries are quasiplanes, if the function u belongs to the class L_n^1.

Remark 4. Instead of the quasistraight line, one may consider a quasicircle or (in the n-dimensional case) a quasisphere.

5.4. The Extension Condition for Nickolsky–Besov Spaces

Theorem 5.7. *Bounded extension operators*

$$\Lambda_1 : B^l_{p,\theta}(G) \to B^l_{p,\theta}(\mathbf{R}^2), \qquad 1 \leqslant lp \leqslant 2,\ 2 \leqslant p,\ p \leqslant \theta$$

$$\Lambda_2 : B^{l_1}_{p_1,\theta_1}(G) \to B^{l_1}_{p_1,\theta_1}(\mathbf{R}^2) \qquad 2 \leqslant l_1 p_1$$

from an unbounded finitely-connected domain $G \subset \mathbf{R}^2$ exist iff the domain G satisfies the bilateral condition with the arc diameter.

Necessity is proved in Theorems 4.6, 4.14, and 4.17. Sufficiency follows from Theorem 5.5 in the variant for $B^l_{p,\theta}$ [120, 121]. In the case $l < 1$, $\theta = p$, the proof is possible using the method of Theorem 5.6.

Let us give one more variant of necessary and sufficient extension conditions following from Theorems 4.6, 4.14, and 5.5.

Theorem 5.8. *Bounded extension operators*

$$\Lambda_1 : B^l_{p,\theta}(G) \to B^l_{p,\theta}(\mathbf{R}^2), \qquad 2 \leqslant lp \leqslant \infty,$$

$$\Lambda_2 : B^{l_1}_{p_1,\theta_1}(G^*) \to B^{l_1}_{p_1,\theta_1}(\mathbf{R}^2), \qquad 2 \leqslant l_1 p_1,$$

from a plane simply-connected domain $G \subset \mathbf{R}^2$ exist iff the domain G satisfies the bilateral condition with the arc diameter (i.e., if the boundary of the domain is a quasircle).

For $lp = 2$, we obtain

Theorem 5.9. *A bounded extension operator*

$$\Lambda : B^l_{p,\theta}(G) \to B^l_{p,\theta}(\mathbf{R}^2), \qquad lp = 2,$$

from an unbounded finitely-connected domain $G \subset \mathbf{R}^2$ exists iff the domain G satisfies the bilateral condition with the arc diameter.

Remark. Theorems 5.7 and 5.9 remain valid for the classes $B^l_{p,\theta}(G)$ as well.

Comments

Chapter 2

The classes W_p^l of functions were introduced by S. L. Sobolev [72, 74, 75], who established the basic properties of these classes. The number of papers studying the W_p^l classes, their various generalizations and analogies, is rather great at present, and we do not even mention the most important ones here. However, we refer the reader to the monographs [57, 8, 71], which contain extensive references to the literature on the W_p^l classes.

The main tool for the investigation of the functions of the class W_p^l in the papers of Sobolev (just as in this book) are integral representations of a function in terms of its derivatives of the order l. The methods of constructing such representations were further developed in the papers of Il'yin [32, 33], Nickolsky [55, 56], and others.

Integral representations of a vector-function f in terms of the magnitudes $L_j f$, $j = 1, 2, \ldots, n$, where $L_1, L_2, \ldots, L_n$ is a system of differential operators with the complete integrability condition, were obtained in the paper of Reshetnyak [64]. This paper also shows how integral formulae of Sobolev (for the domains that are starlike with respect to a ball) and some others can be deduced by means of the above representations. In particular, in this way one can obtain integral representations established earlier by Il'yin [34] and the representations of a vector-function in terms of the values of the differential operators

$$\varepsilon_{i,j}(f) = \frac{1}{2}\left(\frac{\partial f_i}{\partial x_j} + \frac{\partial f_j}{\partial x_i}\right)$$

$$w_{i,j}(f) = \varepsilon_{i,j}(f) - \frac{1}{n}(\operatorname{div} f)\delta_{i,j}, \qquad i, j = 1, 2, \ldots, n,$$

obtained in Reshetnyak's paper [62]. The differential operator ε_{ij} and some estimates of the function by means of this operator obtained in [62] are known as the Corn inequality and have applications in the elasticity theory. The differential operator w arises when studying spatial quasiconformal mappings [59, 63, 66].

The method of constructing integral representations of functions in terms of the magnitudes $L_j f, j = 1, 2, \ldots, n$, where $L_1, L_2, \ldots, L_n$ is a system of operators with the complete integrability condition, based on the use of the relations

$$L_j = Y \circ \frac{\partial}{\partial x_j} \circ Y^{-1}, \qquad j = 1, 2, \ldots, n,$$

was proposed by Reshetnyak. By applying this relation, as shown in Sections 2 and 4, we may reduce the general case to the problem of the integral representation of a function in terms of its first-order derivatives.

The relations between the classes $W_p^l(U)$, $L_q(U)$ and $C(U)$ established in Theorems 4.2, 4.3, 4.8, and 4.9 were only known before for the case where a domain U belongs to the class S, i.e., where U is either star-like with respect to a ball, or is a union of a finite number of domains that are star-like with respect to a ball. The class J is wider than the class S and contains domains that have a rather pathological arrangement. One can prove that the boundary of a domain of the class S is a set of finite area (taken as an $(n-1)$-dimensional Hausdorff measure). One can easily construct examples of domains of the class J whose boundary area is infinite. The possibility of constructing an integral representation of the form (4.3) in the domains of the class J for the case of smooth functions was established by Reshetnyak [67]. The deduction of these representations given in Sections 2 and 4 is a modification of the arguments given in [67].

The classes of functions $\bar{W}_1^1(U)$ seem to be considered for the first time in papers [90, 100, 107]. Note that a continuous function f belongs to the class $\bar{W}_1^1(U)$ if its diagram is an $(n-1)$dimensional surface of finite area in the sense of Lebesque, i.e., if there exists a sequence of polygonal functions $\{f_m\}$, $m = 1, 2, \ldots, n$, which converges to f uniformly on every compact subset U and such that the diagram areas of the functions of $\{f_m\}$ are uniformly bounded.

The ways to introduce norms in the spaces $W_p^l(U)$ considered in Subsection 4.4 for the domains of the type S were first given by Sobolev [74].

The approximation theorem for functions from $W_p^l(U)$ by smooth functions was obtained in paper [89] (see also paper [15] in which the analogous result was established for a wide class of spaces). The change of variable theorem (Theorem 4.7) for functions with generalized derivatives was proved by Nickolsky [54].

With respect to an essentially more general class of mappings than that considered in Lemma 4.4, the domains of the class J are invariant to spatial quasiconformal mappings. This was established in [102].

Compactness in the imbedding theorems for the space W_p^l was obtained by Kondrashov [37]. The estimates of the form considered in Subsection 4.8 may be applied when studying elliptic equations. Theorem 4.10 was proved by Il'yin [31] for the case where U is a domain of the class S.

Theorem 4.12 belongs to Sobolev. Theorem 4.11 cannot be called original, although the authors cannot give the primary source. The fact that every convex function belongs to the class $\bar{W}_{1,\mathrm{loc}}^2$ was established by Bakel'man in his doctoral thesis. The proof of this fact and the convexity criterion given in Theorem 4.13 were first published in [61]. The proof given in that paper, which coincides with the one given in this book, was first proposed by Borovsky instead of the original, more complicated proof of Reshetnyak.

A great number of investigations deal with the problem of differentiability almost everywhere of functions of many variables under certain conditions. Since we are unable to give a full survey of results concerning this problem, let us refer the reader to the papers by Stepanov [69], Calderon [84, 86], and Aleksandrov [1]. Theorem 5.1 is published in the papers of Reshetnyak [59, 61]. Note that by letting $l = 1$, $p = n$ in Theorem 5.2, we obtain the result which is a special case of a theorem by Calderon [86]. Theorem 5.3 was established by Aleksandrov [1]. Theorem 5.4 belongs to Väisälä [112].

The theorem for the behaviour of functions of the classes W_p^l on almost all k-dimensional planes (Theorem 5.5) and the characteristic of a function of the class W_p^1 in terms of its behaviour on almost all straight lines parallel to the coordinate axes (Theorem 5.6) belong to Nickolsky [55].

Chapter 3

The notion of capacity appeared in potential theory, based on the study of properties of harmonic functions and analytical functions of one complex variable. For the history of the problem see, for instance, the book [43]. Presentation of the basic facts of classical potential theory is contained in monographs [43, 12]. Within the concept presented here, the classical case corresponds to the value $p = 2$.

The main contents of Chapter 3 deals with the revision of article [65]. This article contains arguments for special integral operators, namely, Bessell potentials. The notion of capacity with respect to an arbitrary nonnegative operator was introduced by Gol'dshtein. This notion proves to be useful in the investigations of the most general properties of capacity.

In connection with the notion of nonlinear capacity, let us mention the papers of Fuglede [91], Aronszajn and Smith [82], and Maz'ya [45]. In [91] the notion of a p-module of a family of measures is introduced. Let (X, S) be a measurable space where S is a σ-ring, and $m \geqslant 0$ is some fixed measure on S. Let a family E of nonnegative measures be defined onto the set S. A function $f \geqslant 0$ defined in X is said to be admissible with respect to the family of measures E if $\int f(x)\mu(dx) \geqslant 1$ for all $\mu \in E$. The greatest lower bound $M_p(E)$ of the magnitude $\int [f(x)]^p m(dx)$ on the set of all functions f admissible with respect to the family of measures E is called the p-module of the family of measures E. Let T be a nonnegative operator of the form

$$TU(x) = \int K(x,y)U(y)m(dy)$$

and let $E \subset X$. Setting for $x \in E$, $A \in S$,

$$\mu_x(A) = \int_A K(x,y)m(dy),$$

we obtain a family of measures. It is easy to show that the p-module of this family coincides with $\mathrm{Cap}_{T,p}(E)$. In the general case, the notion of (T,p)-capacity is not reduced to the notion of p-module of a system of measures.

The systematic investigation of functions representable by Bessel potentials is contained in the papers of Aronszajn and Smith [82, 83]. These papers study in detail the sets of zero (l,p)-capacity (in the terminology of the present book) and give their applications to the investigation of properties of Bessel potentials.

The papers of Maz'ya [45, 48] study various characteristics of sets and domains in the Euclidean space which were constructed on the basis of the capacity concept. Various applications of methods concerning nonlinear capacity in the theory of functions and in differential equations are contained in Maz'ya's monographs [50, 51]. These books, in particular, investigate in detail the relation between capacity and the existence of imbedding operators into the spaces C and L_p, and give situations in which one can obtain necessary and sufficient conditions of imbedding in terms of capacity.

It should also be noted that soon after the paper of Reshetnyak [65] was published, the paper of Maz'ya and Khavin [47] appeared. The latter presented the theory of nonlinear capacity on the basis of the notion of the Newton potential. In addition, the papers by Meyers [103, 104] were published, which are closer to the paper of Reshetnyak. In Meyers' papers, the theory of capacity is also constructed by means of Bessel potentials.

Chapter 4

The theorem about monotonicity of extremal functions for $(1,p)$-capacity in the case $p=n$ is well known (see, for instance, the paper by Mostow [54]). Continuity of the extremal function is the corollary of the well-known facts from the theory of elliptic equations (see the papers by Ladyzhenskaya, Ural'tseva [41, 42]).

In the presentation of the approximation theorem and corollaries for the function of the class W_p^1 by extremal functions, we follow the paper of Vodop'janov and Gol'dshtein [21] (see also the paper of Vodp'janov, Gol'dshtein and Reshetnyak [23]). Property 3.9 in the case $p=n$ was obtained in the paper of Väisälä [113].

Chapter 5

The material of Section 1 on the multiplicity function of an imbedding, on the change of variable theorem for an imbedding, and on the change of variable theorem for an integral is classical. In the presentation, we mainly use the monograph of Rado and Reichelderfer [108]. The given properties for the degree of mapping may be found in the monograph of Nirenberg [106]. For the necessary information concerning the linking index, see the article by Boltyansky [11]. The change of variable theorem for the integral for mappings of the class L_n^1 (not necessarily continuous ones) almost without changes follows the paper of Vodop'janov and Gol'dshtein [22]. The material of the third section, except for the theorem on the continuity of a monotone function of the class L_n^1 (which belongs to so-called folklore information), is contained in the paper of Vodop'janov and Gol'dshtein [22]. The theorems for the change of independent variables for the classes L_p^1 are given in the papers of Vodop'janov and Gol'dshtein [19, 20, 22]. It should be noted that related problems using the methods of Banach algebras were studied for Royden algebras in the papers by Nakai [105] and Lewis [101].

The Royden algebra $M_p(G)$ is the totality of bounded continuous functions $u \in L_p^1(G)$. Algebraic operations in $M_P(G)$ are the ordinary addition and multiplication of functions. The algebra $M_p(G)$ is considered with the norm

$$\|u\|_{M_p(G)} = \sup |u(x)| + \|u\|_{L_p^1(G)}.$$

The fact that the algebras $M_p(G')$ and $M_p(G)$ are algebraically isomorphic (G, G' are domains of the Euclidean space) results in the existence of a quasiformal (for $p = n$) and quasiisometric (for $1 < p < n$) homeomorphism $\varphi : G \to G'$. The result for $p = 2$, $n = 2$, belongs to Nakai [105]; the case $1 < p \leqslant n$, $n > 2$, was generalized by Lewis. The case $p > n$ cannot be investigated in papers [105, 101], since the algebra $M_p(G)$ for $p > n$ makes it impossible to distinguish inner points of the domain G from boundary ones (in compactification). The simplest example: a ball and a ball with punctured centre on which the algebras $M_p(G)$ coincide. For the case $p > n$, the result for the algebras $M_p(G)$ remains valid after the following refinement: the domains G' and $\varphi(G)$ need not coincide but should be $(1,p)$-equivalent. Due to the imbedding theorems, this follows from Theorem 4.9. Another variant of the proof of this fact may be found in the paper by Vodop'janov and Gol'dshtein [20].

Let us comment on the results of Subsection 4.4 concerning removable singularities for quasiconformal and quasiisometric mappings. The problem of describing sets of removable singularities for quasiconformal mappings originates from the paper of Ahlfors and Beurling [81]. In this paper, the description of sets of removable singularities is given in terms of moduli of families of curves. By taking into account the result of the paper by Hasse [95] on the coincidence of p-moduli and p-capacities and by using our terminology, the result of the paper by Ahlfors and Beurling states the removability of NC_2-sets (in the terminology of Ahlfors and Beurling—NED-sets) for plane quasiconformal homeomorphisms. The same class of sets is removable for analytical functions of the class AD [81], for quasiisometric mappings (Theorem 4.14), and for the space L_2^1 (Theorem 3.13 of Chapter 3). In the case of dimension $n > 2$ for a class analogous to AD, there are fewer removable sets than for quasiconformal homeomorphisms and for the spaces L_n^1 (see the paper by Hedberg [96]). The well-known theorems for the description of sets of removable singularities for quasiconformal mappings are special cases of Theorem 4.14 (see the papers of Väisälä [113], Miklyukov [53], Aseev and Sychev [4]). The theorems for removable singularities given in Chapter 5 were published in the paper by Vodop'janov and Gol'dshtein [21].

Chapter 6

The general presentation in this chapter follows the paper by Vodop'janov and Gol'dshtein [25] and the papers of Gol'dshtein [28, 29, 116–118]. The invisibility principle was used in the paper of Gol'dshtein [118] in the form close to that given in Chapter 6. For anisotropic Sobolev classes and anisotropic Besov–Nickolsky classes, the extension theorems were obtained by Vodop'janov [114, 115]. He uses the refined variant of the invisibility principle.

Let us briefly give the basic stages of studying the problem concerning the extension of functions in the Sobolev classes and the Nickolsky–Besov classes across the boundary of the definition domain.

For domains with sufficiently smooth boundary, the extension theorem was proved by Babich [5] and Nickolsky [55], [56]; for domains satisfying the strong cone condition, it was proved in the papers by Calderon [87], Besov [7,8], Besov and Il'yin [9]; for domains of the class $\widetilde{\mathrm{Lip}1}$—in the papers by Burenkov [16, 17]. In the paper by Burago and Maz'ya [13], where necessary and sufficient extension conditions for functions of the class BV were obtained, these conditions are formulated in terms of isoperimetric inequality. The extension theorem for domains satisfying the Lichtenstein condition (the ε,δ-condition in the terminology of P. Jones) was obtained in the papers by Schwartzman [120, 121] and Jones [99].

The original idea of extension using the structural isomorphism theorem is very simple: let $\varphi : R^n \to \mathbf{R}^n$ be a quasiconformal homeomorphism and let G be a standard domain in $\mathbf{R}^n$, for instance, a ball. Let us construct an extension operator θ according to the diagram

$$\begin{array}{ccc} L^1_n(G) & \xrightarrow{\bar\theta} & L^1_n(\mathbf{R}^n) \\ \uparrow{\scriptstyle \varphi^*} & & \downarrow{\scriptstyle (\varphi^{-1*})} \\ L^1_n(\varphi(G)) & \overset{\theta}{\dashrightarrow} & L^1_n(\mathbf{R}^n) \end{array}$$

Here $\varphi^*, (\varphi^{-1})^*$ are structural isomorphisms of spaces of functions, induced by the quasiconformal mappings φ and φ^{-1}; θ is any of the known extension operators. For the spaces $L^1_p(W^1_p)$, one can use quasiisometries. There arises a wider class of domains than $\widetilde{\mathrm{Lip}1}$. These are domains whose boundaries are, respectively, quasiconformal and Lipschitz manifolds. However, even in a plane $p \neq 2$, we do not obtain necessary and sufficient extension conditions.

The examples of quasiconformal curves that are nowhere locally rectifiable [9] show that necessary and sufficient extension conditions should be of geometric or, maybe, capacitance nature.

For $l = 1$, the extension method used in the chapter has the advantage: by means of this method, one can easily obtain relatively exact estimates of the norm of an extension operator. It should also be noted that sufficient conditions obtained in the paper of Schwartzman [120, 121] do not coincide in a space with the arc diameter condition. For $p \neq 2$, Maz'ya constructed an example [52] showing that the bilateral arc diameter condition is not necessary if either θ_1 or θ_2 does not exist.

The restriction $p \geqslant 2$, $\theta \geqslant p$ in Theorem 4.6 is superfluous as is shown in the paper of Vodop'janov [115].

REFERENCES

[1] Aleksandrov, A. D. "The existence almost everywhere of the second differential and the related convex surfaces" (In Russian), *Uchyonye zapiski LGU, matematika* **6** (1939), 3–35.

[2] Ahlfors, L. V. *Lectures on Quasiconformal mappings.* Van Nostrand, Princeton, N. J., 1966.

[3] Aseev, V. V. "The example of an NC_p-set in an n-dimensional space, having positive $(n-1)$-dimensional Hausdorff's measure" (In Russian), *Doklady Akademii Nauk SSSR* **216** (1974), 717–719.

[4] Aseev, V. V., and Sychev, A. V. "On removable sets for spatial quasiconformal mappings" (In Russian), *Siberian Math. Journal* **15** (1974), 1213–1227.

[5] Babich, V. M. "On the distribution of functions" (In Russian), *Uspekhi matem. nauk* **8** (1953), 111–113.

[6] Bakel'man, I. Ya. *Geometric Methods of Solving Elliptic Equations* (In Russian). Nauka, Moscow, 1955.

[7] Besov, O. V. "Extension of functions from the spaces L_p^1 and W_p^1" (In Russian), *Trudy MIAN SSSR* **89** (1967), 5–17.

[8] Besov, O. V. "On coercitivity in non-isometric space of S. L. Sobolev" (In Russian), *Matematicheskii sbornik* **3** (1967), 585–599.

[9] Besov, O. V., and Il'yin, V. P. "Extension of a class of domains in the imbedding theorems" (In Russian), *Matematicheskii sbornik* **75** (1968), 483–595.

[10] Besov, O. V., Il'yin, V. P., and Nickolsky, S. M. *Integral representations of functions and imbedding theorems* (In Russian). Nauka, Moscow, 1975.

[11] Boltyanskii, V. G. "Homotopic theory of continuous mappings and vector fields" (In Russian), *Trudy MIAN SSSR* **47** (1955), 89–120.

[12] Brelot, M. *Élements de théorie classique du potential.* C.D.U., Paris, 1969.

[13] Burago, Yu. D., and Maz'ya, V. G. "Some questions of the potential theory and theory of functions for domains with irregular boundaries" (In Russian), *Zapiski nauchnyh seminarov LOMI* **3** (1967), 1–67.

[14] Bourbaki, N. *Elements de Mathématique.* Livre VI: *Intégration.* Hermann, Paris, 1964.

[15] Burenkov, V. I. "On density of infinitely differentiable functions in the space of functions assigned on an arbitrary open set" (In Russian), in:

Teoriya kubaturnyh formul i prilozheniya funktsional'nogo analiza k nekotorym zadacham matematicheskoj fiziki. Institute of Mathematics, Novosibirsk, 1975, pp. 63–71.

[16] Burenkov, V. I. "On a method of extending differentiable functions" (In Russian), *Trudy MIAN SSSR* **140** (1976), 27–67.

[17] Burenkov, V. I. "On the extension of functions with seminorm being preserved" (In Russian), *Doklady Akademii nauk SSSR* **228**, 779–782.

[18] Belinskii, P. P. *General Properties of Quasiconformal Mappings* (In Russian). Nauka (Siberian Division), Novosibirsk, 1974.

[19] Vodop'janov, S. K., and Gol'dshtein, V. M. "Structural isomorphisms of the spaces W_n^1 and quasiconformal mappings" (In Russian), *Siberian Matemat. Journal* **16** (1975), 224–246.

[20] Vodop'janov, S. K., and Gol'dshtein, V. M. "Functional characteristics of quasiisometric mappings" (In Russian), *Siberian Math. J.* **17** (1976), 768–773.

[21] Vodop'janov, S. K., and Gol'dshtein, V. M. "The removability criterion of sets for the spaces W_p^1, for quasiconformal and quasiisometric mappings" (In Russian), *Siberian Math. J.* **18** (1977), 48–68.

[22] Vodop'janov, S. K., and Gol'dshtein, V. M. "Quasiconformal mappings and spaces of functions with first generalize derivatives" (In Russian), *SIberian Math. J.* **17** (1976), 515–531.

[23] Vodop'janov, S. K., Gol'dshtein, V. M., and Reshetnyak, Yu. G. "On geometric properties of functions with first generalized derivatives" (In Russian), *Uspekhi matematicheshih nauk* **34** (1979), 17–62.

[24] Vodop'janov, S. K., Gol'dshtein, V. M., and Latfullin, T. G. "The extension criterion of functions of the class L_2^1 from plane unbounded domains" (In Russian), *Siberian Math. J.* **20** (1979), 416–420.

[25] Vodop'janov, S. K., and Gol'dshtein, V. M. *The Sobolev spaces and special classes of mappings* (In Russian). Novosibirsk State University, Novosibirsk, 1981.

[26] Vulikh, B. Z. *Foundations of the Theory of Semiordered Spaces* (In Russian). Fizmatgiz, Moscow, 1961.

[27] Gel'fand, I. M., and Shilov, G.E. *Generalized Functions and Operations with them* (In Russian). Fizmatgiz, Moscow, 1958.

[28] Gol'dshtein, V. M. "Extension of differentiable functions preserving the class from the plane domains" (In Russian), *Doklady Akademii nauk SSSR* **257** (1981), 451–454.

[29] Gol'dshtein, V. M. "Extension of functions of the classes $B_{p,q}^l$ across quasiconformal boundaries" (In Russian), in: *Teoriya kubaturnyh formul i prilozheniya funktsional'nogo analiza k zadacham matematicheskoj fiziki.* Novosibirsk, 1979, **1**, pp. 12–32.

[30] Il'yin, V. P. "Some inequalities in functional spaces and their application in the investigations of convergence of variational processes" (In Russian), *Trudy MIAN SSSR* **53** (1953), 64–127.

[31] Il'yin, V. P., and Solonnikov, V. A. "On some properties of differentiable functions of many variables" (In Russian), *Trudy MIAN SSSR* **66** (1962), 205–226.

[32] Il'yin, V. P. "Properties of some classes of differentiable functions of many variables assigned in an n-dimensional domain" (In Russian), *Trudy MIAN SSSR* **66** (1962), 227–363.

[33] Il'yin, V. P. "On some properties of classes of differentiable functions assigned in a domain" (In Russian), *Trudy MIAN SSSR* **84** (1965), 93–143.

[34] Il'yin, V. P. "Integral representations of differentiable functions and their application to the problems of extension of functions from $W_p^{(l)}$" (In Russian), *Siberian Math. J.* **11** (1970), 573–586.

[35] Kantorovich, L. V., and Akilov, G. P. *Functional Analysis* (In Russian). Nauka, Moscow, 1977.

[36] Cartan, H. *Calcul Differentiel. Formes Differentielles.* Hermann, Paris, 1967.

[37] Kolmogorov, A. N., and Fomin, S. V. *Elements of the Theory of Functions and Functional Analysis* (In Russian). Nauka, Moscow, 1967.

[38] Kondrashov, V. I. "On some properties of functions from the space L_p" (In Russian), *Doklady AN SSSR* **48** (1945), 563–566.

[39] Kondrat'jev, V. A. "On the solvability of the first boundary-value problem for strongly elliptic equations" (In Russian), *Trudy Moskovskogo matematicheskogo obshchestva* **16** (1967), 293–318.

[40] Kuratowski, K. *Topology.* Vol. 1. Academic Press, New York and London, 1966.

[41] Ladyshenskaya, O. A., and Ural'tseva, N. N. "Quasilinear elliptic equations and variational problems with many variables" (In Russian), *Uspekhi matem. nauk* **16** (1961), 19–20.

[42] Ladyshenskaya, O. A., and Ural'tseva, N. N. *Linear and Quasilinear Elliptic Equations* (In Russian). Nauka, Moscow, 1964.

[43] Landkoff, N. S. *Foundations of the Modern Potential Theory* (In Russian). Nauka, Moscow, 1966.

[44] Lizorkin, P. I. "Characteristics of boundary values of a function from $L_p^{(l)}(E^n)$ on hyperplanes" (In Russian), *Doklady AN SSSR* **150** (1963), 984–986.

[45] Maz'ya, V. G. "Polyharmonic capacity in the theory of the first boundary-value problem" (In Russian), *Siberian Math. J.* **6** (1965), 127–148.

[46] Maz'ya, V. G. "On the continuity at the boundary point of solutions of quasilinear elliptic equations" (In Russian), *Vestnik Leningradskogo gosudarstvennogo universiteta. Matematika, mekhanika i astronomiya* **13** (1970), 42–55.

[47] Maz'ya, V. G., and Khavin, V. P. "Nonlinear analogue of Newton potential and metric properties of (p, l)-capacity" (In Russian), *Doklady AN SSSR* **194** (1970), 770–773.

[48] Maz'ya, V. G. "Classes of sets and measures connected with imbedding theorems" (In Russian), in: *Teoremy vlozheniya i ikh prilozheniya*. Nauka, Moscow, 1970.

[49] Maz'ya, V. G., and Khavin, V. P. "Nonlinear theories of potential" (In Russian), *Uspekhi matem. nauk* **27** (1972), 66–138.

[50] Maz'ya, V. G. "Einbettungssätze für Sobolewsche Räume". Teil I, in: Teubner-Texte zur Mathematik. Teubner, Leipzig, 1979.

[51] Maz'ya, V. G. "Einbettungssätze für Sobolewsche Räume". Teil II, in: Teubner-Texte zur Mathematik. Teubner, Leipzig, 1979.

[52] Maz'ya, V. G. "On the extension of functions from the Sobolev spaces" (In Russian), *Zapiski nauchnyh seminarov LOMI* **1** (1981), 1–12.

[53] Miklyukov, V. M. "On removable singularities of quasiconformal mappings in a space" (In Russian), *Doklady AN SSSR* **188** (1969), 525–527.

[54] Mostow, G. P. *Quasi-conformal Mappings in n-space and the Rigidity of Hyperbolic Space Forms.* Publ. Math. de l'Institute des Hautes Etudes Scientifique, **34**, 1969.

[55] Nickolsky, S. M. "Properties of some classes of functions of many variables on differentiable manifolds" (In Russian), *Matem. sbornik* **33** (1953), 261–326.

[56] Nickolsky, S. M. "Inequalities for integral functions of finite degree and their application in the theory of differentiable functions of many variables" (In Russian), *Trudy MIAN SSSR* **38** (1951), 244–278.

[57] Nickolsky, S. M. *Approximation of Functions of Many Variables and Imbedding Theorems* (In Russian). Nauka, Moscow, 1977.

[58] Nevanlinna, R. *Eindeutige Analytische Funktionen*, 2. Aufl. Springer, Berlin-Göttingen-Heidelberg, 1953.

[59] Reshetnyak, Yu. G. "Some geometric properties of functions and mappings with generalized derivatives" (In Russian), *Siberian Math. J.* **7** (1966), 886–919.

[60] Reshetnyak, Yu. G. "Spatial mappings with bounded distortion" (In Russian), *Siberian Math. J.* **8** (1967), 629–658.

[61] Reshetnyak, Yu. G. "Generalized derivatives and differentiability almost everywhere" (In Russian), *Matem. sbornik* **75** (1968), 323–334.

[62] Reshetnyak, Yu. G. "Estimates for some differential operators with finite-dimensional kernel" (In Russian), *Siberian Math. J.* **11** (1970), 414–428.

[63] Reshetnyak, Yu. G. "On the estimate of stability in the Liouville theorem about conformal mappings of many-dimensional spaces" (In Russian), *Siberian Math. J.* **11** (1970), 1121–1139.

[64] Reshetnyak, Yu. G. "Some integral representations of differentiable functions" (in Russian), *Siberian Math. J.* **12** (1971), 420–432.

[65] Reshetnyak, Yu. G. "On the notion of capacity in the theory of functions with generalized derivatives" (In Russian), *Siberian Math. J.* **10** (1969), 1109–1138.

[66] Reshetnyak, Yu. G. "Stability in the Liouville theorem about conformal mappings for domains with nonsmooth boundaries" (In Russian), *Siberian Math. J.* **17** (1976), 361–369.

[67] Reshetnyak, Yu. G. "Integral representations of differentiable functions in domains with nonsmooth boundaries" (In Russian), *Siberian Math. J.* **21** (1980), 108–116.

[68] Tribel, H. *Interpolation Theory , Differential Operators.* VEB Deutscher Verlag der Wissenschaften, Berlin, 1978.

[69] Stepanov, V. V. "Sur les condition de l'existence de la differentielle totale," *Math. sbornik* **30** (1924), 487–489.

[70] Saks, S. *Theory of the Integral.* 1939.

[71] Stein, E. *Singular Integrals and Differentiability Properties of Functions.* Princeton University Press, Princeton, N.J., 1970.

[72] Sobolev, S. L. "On some estimates concerning families of functions having derivatives integrable with a square" (in Russian), *Doklady AN SSSR* **1** (1936), 267–270.

[73] Sobolev, S. L. "On a boundary-value problem for polyharmonic equations"(In Russian), *Matem. sbornik* **2** (1937), 465–499.

[74] Sobolev, S. L. "On one theorem of functional analysis" (In Russian), *Matem. sbornik* **3** (1938), 471–497.

[75] Sobolev, S. L. "On one theorem of functional analysis" (In Russian), *Doklady AN SSSR* **20** (1938), 5–10.

[76] Sobolev, S. L. *Some applications of functional analysis in mathematical physics* (In Russian). Leningrad, 1950; Novosibirsk, 1962.

[77] Solntsev, Yu. K. "On the estimate of the mixed derivative in $L_p(G)$" (In Russian), *Trudy Matematicheskogo Instituta im. V. .A. Steklova* **64** (1961), 221–238.

[78] Trotsenko, D. A. "Properties of domains with nonsmooth boundaries" (In Russian), *Siberian Math. J.* **22** (1981), 221–224.

[79] Ural'tseva, N. N. "Degenerating quasilinear elliptic systems" (In Russian), *Zapiski nauchnyh seminarov LOMI* **7** (1968), 184.

[80] Schwartz, L. *Analyse Mathématique.* Hermann, Paris, 1967.

[81] Ahlfors, L., and Beurling, A. "Conformal invariants and function theoretic null-sets", *Acta Math.* **83** (1950), 100–129.

[82] Aronszajn, N., and Smith, K. T. "Theory of Bessel potentials," Part I, *Ann. Inst. Fourier* **11** (1961), 385–475.

[83] Aronszajn, N., Fuad Mulla, and Szeptucki, P. " On spaces of potentials connected with L^p classes," *Ann. Inst. Fourier* **13** (1963), 211–306.

[84] Calderon, A. P. "On the differentiability of absolutely continuous functions," *Riv. mat. Univ. Parma* **2** (1951), 203–213.

[85] Calderon, A. P., and Zygmund, A. "Local properties of solution of elliptic partial differential equations," *Studia Math.* **20** (1961), 171–225.

[86] Calderon, A. P., and Zygmund, A. "On the differentiability of functions which are bounded variation in Tonelli's sense," *Rev. Unión mat. argent.* **20** (1962), 102–121.

[87] Calderon, A. P. "Lebesgue space of differentiable functions," in: *Proc. Conf. on Partial Differential Equations.* Univ. of California, Calif., 1966.

[88] Calderon, A. P., and Lewis, J. E. "On the differentiability of functions of several real variables," *Illinois J. Math.* **20** (1976), 532–542.

[89] Deny, J., and Lions, T. L. "Les espaces du type de Beppo Levi," *Ann. Inst. Fourier* **5** (1955), 305–370.

[90] Fleming, W. H. "Functions with generalized gradient and generalized surfaces," *Ann. mat. pura ed appl.* **44** (1957), 93–104.

[91] Fuglede, B. "Extremal length and functional completion," *Acta Math.* **90** (1957), 171–219.

[92] Gehring, F. W. "Lipschitz mappings and the p-capacity of ring in n-space," in: *Advances in the Theory of Riemann Surfaces.* Proc. Conf., N.Y., New York Stony Brook, 1969; *Ann. Math. Studies* **66**, Princeton; New York Univ. Press, N.Y., 1971, 175–193.

[93] Gehring, F. W. "Rings and quasiconformal mapping in space," *Trans. Amer. Math. Soc.* **103** (1962), 353–393.

[94] Giorgi, E. De. "Definizione ed espresione analitica del perimetro di un insieme," *Atti Accad. naz. Lincei Rend. Cl. sci. fis. mat. e natur.* **14** (1953), 390–392.

[95] Hasse, J. "A p-extremal length and p-capacity equality," *Arkiv. math.* **13** (1975), 131–144.

[96] Hedberg, L. I. "Removable singularities and condenser capacities," *Arkiv. math.* **12** (1974), 181–201.

[97] John, F. "Rotation and strain," *Commun. Pure and Appl. Math.* **4** (1961), 391–414.

[98] John, F. "Quasi-isometric mappings," *Commun. Pure and Appl. Math.* **21** (1968), 77–110.

[99] Jones, P. W. "Quasiconformal mappings and extendability of functions in Sobolev spaces." Preprint. Univ. of Chicago, Chicago, 1980, 1–30.

[100] Krickeberg, K. "Distributionen, Funktionen beschränkter Variation und Lebesguescher Inhalt nichtparametrischen Flächen," *Ann. Mat. pura ed appl.* **44** (1957), 105–133.

[101] Lewis, L. "Quasiconformal mapping and Royden algebras in space," *Trans. Amer. Math. Soc.* **158** (1971), 481–496.

[102] Martio, O., and Sarvas, J. "Injectivity theorems in plane and space," *Ann. Acad. Sci. Fenn., Ser. A1* **4** (1978/1979), 383–401.

[103] Meyers, M. G. " A theory of capacities for potentials of function in Lebesgue classes," *Mat. scand.* **26** (1970), 255–292.

[104] Meyers, M. G. Continuity of Bessel potential," *Isr. J. Math.* **11** (1972), 271–282.

[105] Nakai, M. "Algebraic criterion on quasiconformal equivalence of riemannian surfaces," *Nagoya Math. J.* **16** (1960), 157–184.
[106] Nirenberg, L. *Topics in Nonlinear Functional Analysis.* New York University, N.Y., 1974.
[107] Pauc, G. Y. "Functions with generalized gradients in the theory of cell functions," *Ann. mat. pura ed appl.* **44** (1957), 135–152.
[108] Rado, T., and Reichelderfer, P. V. *Continuous Transformations in Analysis.* Springer-Verlag, Berlin, Heidelberg, New York, 1955.
[109] Rickman, S. "Characterization of quasiconformal arcs," *Ann. Scand. Sci. Fenn., Ser. A1* (1966), 395.
[110] Smith, K. T. "Inequalities for formally positive integrodifferential forms," *Bull. Amer. Math. Soc.* **67** (1961), 368–370.
[111] Väisälä, J. "Two new characterizations for quasiconformality," *Ann. Acad. Sci. Fenn.*, Ser A1, **362** (1965), 1–12.
[112] Väisälä, J. "On null-sets for extremal length," *Ann. Acad. Sci. Fenn., Ser. A322* (1969), 1–12.
[113] Väisälä, J. "Removable sets for quasiconformal mapping," *J. Math. and Mech.* **19** (1969), 49–51.
[114] Vodop'janov, S. K. "Geometric properties of domains satisfying the extension condition for spaces of differentiable functions" (In Russian), in: *Nekrotorye prilozheniya funktsional'nogo analiza k zadacham matematicheskoj fiziki.* Novosibirsk, 1984, 65–95.
[115] Vodop'janov, S. K. "Geometric properties of domains and estimates from below of the norm of extension operator" (In Russian), in: *Issledovaniya po geometrii i matematicheskomu analizu.* Nauka, Novosibirsk, 1986, 117–142.
[116] Gol'dshtein, V. M. "Necessary extension conditions for functions of the Sobolev classes" (In Russian), in: *Teorija kubaturnyh formul i priloshen-iya funktsional'nogo analiza k zadacham matematicheskoj fiziki.* Nauka, Novosibirsk, 1981, 34–48.
[117] Gol'dshtein, V. M. "Capacity and extension of functions with generalized derivatives" (In Russian), *Siberian Math. J.* **23** (1982), 49–59.
[118] Gol'dshtein, V. M. *Imbedding Theorems, Extensions, and Capacity* (In Russian). Novosibirsk State University, Novosibirsk, 1982.
[119] Maz'ya, V. G. "On the extension of functions from the Sobolev spaces" (In Russian), *Zapiski nauchnyh seminarov LOMI* **1** (1981), 231–236.
[120] Shwartzman, P. A. "Extension theorem for one class of spaces definable by local approximations" (In Russian), in: *Issledovaniya po teorii funktsiy mnogih veshchestvennyh peremennyh.* Yaroslavl', **2** (1978), 215–242.
[121] Shwartzman, P. A. "Local approximations of functions and extension theorems" (In Russian), *Deponirovano v VINITI* **2025-83** (1983), Dep. 30 p.

INDEX